Charles Darwin's Life with Birds

Other books by Clifford B. Frith

Cape York Peninsula: A Natural History. Reed Books, Sydney, Australia. 1995, with Dawn W. Frith.

The Birds of Paradise: Paradisaeidae. Oxford University Press, Oxford, England. 1998, with Bruce M. Beehler.

The Bowerbirds: Ptilonorhynchidae. Oxford University Press, Oxford, England. 2004, with Dawn W. Frith.

The Ornithologist's Dictionary. Lynx Edicions, Barcelona, Spain. 2007, with Johannes Erritzoe, Kaj Kampp, and Kevin Winker.

Bowerbirds: Nature, Art & History. Frith & Frith, Malanda, Queensland, Australia. 2008, with Dawn W. Frith.

Birds of Paradise: Nature, Art & History. Frith & Frith, Malanda, Queensland, Australia. 2010, with Dawn W. Frith.

The Woodhen: A Flightless Island Bird Defying Extinction. CSIRO Publishing, Melbourne, Australia. 2013.

Charles Darwin's Life with Birds

His Complete Ornithology

Clifford B. Frith

OXFORD
UNIVERSITY PRESS

Oxford University Press is a department of the University of Oxford. It furthers the University's objective of excellence in research, scholarship, and education by publishing worldwide.Oxford is a registered trade mark of Oxford University Press in the UK and certain other countries.

Published in the United States of America by Oxford University Press
198 Madison Avenue, New York, NY 10016, United States of America.

CIP data is on file at the Library of Congress
ISBN 978-0-19-024023-3

1 3 5 7 9 8 6 4 2
Printed by Sheridan Books, Inc., United States of America

The author dedicates this book to
Dawn W. Frith

The Dawn, and love, of my life.

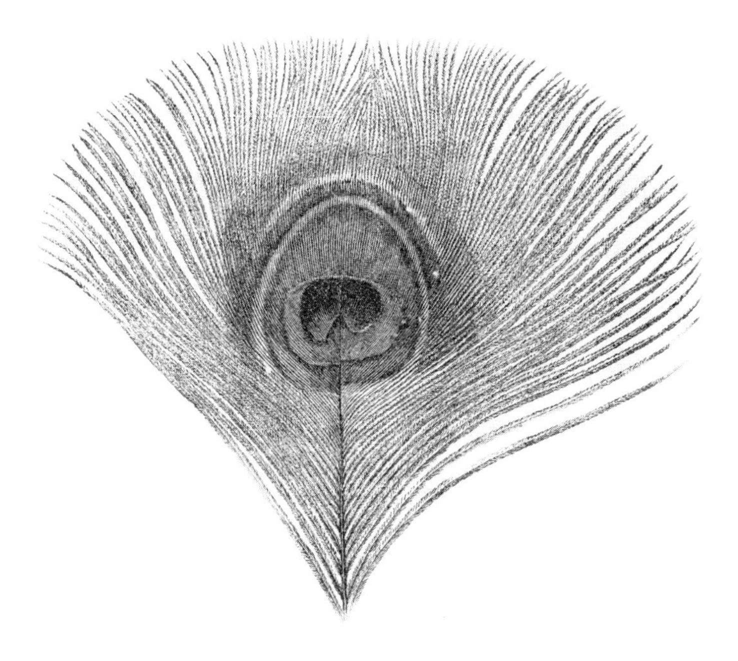

CONTENTS

FOREWORD

I was very fond of collecting eggs, but I never took more than a single egg out of a bird's nest, except on one single occasion, when I took all, not for their value, but from a sort of bravado. [Charles Darwin [CD] in F. Darwin 1950: 15]

A great deal of literature has been published, and much content of it frequently repeated, about Charles Darwin, his ancestors, immediate family and his own children, social background, religion, politics, morals, correspondence, working and thinking life and resultant publications, death, and funeral. Much is also written of his influence on his own species and how other species are perceived as a result of his remarkable studies of and theories about nature. This book is not a study of Darwin the evolutionist, and his theoretical work is not discussed other than so far as it pertains to birds. I deal almost exclusively with those aspects of his background, voyaging, working life, and publications that directly relate to him as a fledgling, aspiring, field working, thinking, and reflecting student of birds: Darwin the ornithologist. There are of course many other aspects of that remarkable and great man's inspirational life and work. These include travel, antislavery, geology, volcanos, coral reefs, barnacles, insects and particularly beetles, earthworms, numerous other invertebrates, fishes, mammals, climbing and insectivorous plants, orchids, emotions in vertebrates including humans, origin of species, natural and sexual selection, voluminous correspondence and much more—that are all beyond the scope of this work.

Without wishing to appear pedantic, I am bound to note here that the words "ornithology" and "ornithologist" have come to be widely used most loosely and inaccurately. They are often now applied to bird watching, or birding, or to bird watchers, birders, "twitchers," and so forth, which in many cases have little or nothing at all to do with ornithology. The origin of "ornithology" is the Greek *ornithologos*, and the late 17th-century Latin "*ornithologia*," meaning "bird science." In English language usage, "ornithologist" specifically means a person that studies birds scientifically, and it is in this context that I write of Charles Darwin herein. Having written this paragraph thus far, I was delighted to then discover that the very same distinction had been made in an admirable 2014 book about ornithology since Darwin, edited by Tim Birkhead and others. Darwin undoubtedly graduated from a bird egg-collecting boy, to a teenage bird hunter and watcher, to a critical bird observer, and then on to become an ornithologist of substantial scientific significance.

Darwin's youthful interest in birds is widely reported to have initially involved little more than the collecting of eggs, with this activity followed by the shooting of game birds. This is a quite unjustified oversimplification, as Darwin would certainly have learnt much about the habits of his quarry and of additional bird species in the pursuit of them, both as an egg collector and as a hunter of birds with a gun. In addition, he would have learnt much of wild birds while walking and fishing, which he spent much time doing. He also specifically recorded that he enjoyed closely watching birds and observing and noting their behaviour during this early period of his life, as is discussed in chapter 1.

For various historical reasons, Darwin is today ornithologically most closely, and indeed often almost exclusively, associated in the popular literature with the little songbirds that were initially called Galapagos finches. They were to subsequently become widely known as Darwin's finches, of the Galapagos Archipelago in the eastern Pacific Ocean. The great impact that this group of drab small birds has had on evolutionary science is widely thought to have been, at least in part, contemporaneous with Darwin's five weeks among them on their remote island homes. This is quite erroneous, however, as he was at that time too naive a bird student or evolutionist to appreciate what the various finches were potentially demonstrating by their restricted island distributions, plumages, and diverse beak shapes between their island populations; see chapter 2.

While at St. Helena in the South Atlantic Ocean in July 1836, towards the end of his voyage, Darwin did, however, write in his notebook words suggesting that he had come to appreciate that the various forms of mockingbird living on different islands of the Galapagos Archipelago had a novel and highly significant evolutionary story to tell. This appears to be the first time that Darwin committed to writing anything suggesting the notion that species could be other than immutable, or anything other than an expression of the hand of creation by God. But in this he was writing of the Galapagos mockingbirds and not of the finches.

Although Darwin was no ornithologist prior to boarding the HMS *Beagle*, and was but a budding one during the subsequent near five years of voyaging, he certainly bloomed into one of contemporary international standing upon settling into his sedentary life in England as natural history scientist, philosopher, and author; see chapters 3 and 4. His contributions as a collector of birds and as an ornithologist during the voyage of the *Beagle* were reviewed and scholarly assessed in 2004 by my esteemed colleague Frank D. Steinheimer. However, no review, assessment, and discussion of Darwin's entire published ornithology has previously appeared. In this book I not only do this, but I attempt to do so comprehensively with respect to every bird species and ornithological topic or consideration published by Charles Darwin; see chapters 5 to 7 and appendix 1.

Finally, I overview and discuss the role and significance of birds, ornithology, and other ornithologists in Darwin's life and work. In this text, I also assess Darwin the ornithologist; his ornithological knowledge, observations, interpretations, experimental and theoretical considerations; and their significance in the context of his biological work and specifically to ornithology then and subsequently; see chapter 8.

The constituent chapters of this book deal with Darwin's ornithological observations and writing in the chronological order in which they were made or published during his lifetime. This is the most logical and convenient way to deal with them, as it places his experiences with and thinking about birds in historical context, thereby showing his ornithological development. It does mean, however, that some bird species or groups and some other ornithological topics appear in more than one chapter, although the aspects of them covered therein predominantly differ: such as field biology as presented in his *Journal of Researches* in chapter 2 versus descriptions and taxonomic discussions in his *The Zoology of the Voyage of the Beagle* in chapter 3. Although I have sought to avoid repetition of Darwin's own text, this is not entirely avoidable, given the content of his books in chronological order and context. I refrain from littering the text with numerous and distracting "(see chapter . . .)" cross referencing, as the Index enables the reader to quickly locate subjects appearing in several places within the chapters of this volume. I use superscript numbers within the body of text to indicate sources of information, which are listed in the usual author and year of publication format at the end of each chapter and enable the reader to easily find the full citations in the References.

The impressive number and diversity of wild bird species; the domestic forms of several others; and the biological, anatomical, and theoretical considerations of them that Darwin covered in his published works represent, by any measure, a most substantial contribution to ornithology. His major books alone contain reference to, and consideration of, over 600 bird species as well as interesting and pertinent discussion of over 100 ornithological topics. Additional species and topics are to be found within his published zoological results of the *Beagle* voyage and in a few scientific papers published in scholarly journals. Some species are mentioned on numerous pages of various publications. Thus, for example, domesticated pigeons and their single ancestor species are discussed in great depth on well over 100 dispersed pages of Darwin's three major pertinent books, and on 93 pages in *The Variation of Animals and Plants under Domestication* alone. These are synthesized into coherent summarized accounts in chapters 2 to 6, as with all other bird species written about at any length by Darwin. Additional species that he wrote about only briefly are comprehensively included in appendix 1, together with all other bird species. The avian order and family to which any bird species mentioned by Darwin belongs to according to today's classification can be determined from the content of appendix 1.

In this book, I comprehensively detail Charles Darwin's published ornithology, with abbreviated reference to the location of each mention of every species within his various published works; see appendix 1. By Darwin's published ornithology, I mean those works authored by Charles Darwin in his lifetime. This therefore excludes most other publications that are about him, because their inclusion would require a vast and complex, multivolume work. Where possible, I have attempted to use nontechnical English terms throughout. However, Darwin discussed avian taxonomic issues that I must include to be comprehensive; and I have also updated the nomenclature of these for the benefit of the user. Those brief parts of the text dealing with the complexities of bird names, orders, families, genera, and species and their systematic positions within the class Aves may be passed over by those uninterested in them.

Darwin wrote a great deal about artificial selection bringing about numerous distinct breeds of the Rock Dove, by way of providing a fundamental foundation for his subsequent presentation of his theory of natural selection. It is therefore inevitable that its wild and domesticated forms feature extensively in this book. As I cannot omit any such material from a work dealing comprehensively with Darwin's published ornithology, I beg the readers' indulgence, and anticipate that some such texts might also be passed over by those less interested in them.

Numerous lists of the birds of the world exist in print, and more recently online, each expressing the various and varied taxonomic and nomenclatural opinions of their author(s). I have chosen to use that of the International Ornithologists' Union's World Bird List, version 5.2 (Gill and Donsker 2015, http://www.worldbirdnames.org or http://dx.doi.org/10.14344/IOC.ML.5.2) for the taxonomy, systematic order, and scientific and vernacular bird names in updating those applied by Charles Darwin. This I do because it is a most recent, widely available, much used list that reflects the work and views of a highly respected international body of eminently qualified ornithologists.

Given that Darwin collected hundreds of bird specimens from places previously little, if ever, visited by biologists, it is not surprising that a number of novel ones were named after him by those ornithologists describing them. These, and additional bird species named in honour of Darwin, are listed as appendix 2.

It is important to note that Darwin, or his shipboard assistant Syms Covington on his behalf, collected and preserved birds throughout the voyage of the *Beagle* opportunistically, and at times systematically. Although I may mention the odd incidence of this, I make no attempt to include all of them, as specimens were sometimes unidentified or unremarked on by Darwin at the time of collection. To include details of all of his bird collecting activity would burden the text with laborious detail. Rather, I provide a comprehensive listing of his bird specimens as appendix 3. What cannot be overstressed here is that although Darwin collected some birds as specimens, he was continuously seeing and observing living birds of numerous additional species throughout the voyage, both at sea and on land. Given his undeniable interest in birds from a young age, his curious and probing mind with respect to all living things, the accumulative experience and insight gained during the voyage stood him in good stead for his increasingly sophisticated ornithological thinking throughout his life.

Appendix 4 presents a brief discussion of the recent controversial proposition that English ornithologist and bird artist John Gould composed his paintings of British birds, and of bowerbirds in his work about Australian birds, as anti-Darwinian propaganda. The Glossary provides brief interpretations of words that might be unfamiliar to some readers. The names of people mentioned in the text, the dates of their birth and death when known, and the page(s) they are to be found cited within this volume, appear within the Index to Chapters.

I often abbreviate the (typically long) full titles of Darwin's various books, but which work is meant by these remains unambiguous. I use "Charles Darwin" but also abbreviate it to "CD" to save space. I also simply use "Darwin" or "Charles" as appropriate, purely to improve readability as the context of the text permits, with no lack of respect or any familiarity intended.

In writing of the bird species that Darwin details in his publications in the following chapters, it must be borne in mind that the text alludes to what he wrote about them—unless made clear that this is not the case. This avoids my having to constantly state that Darwin said so-and-so about a particular bird and so forth. It must be kept in mind that Darwin was writing in the light of available knowledge or local advice and that he was not always right in doing so, but was predominantly correct. Where I make a correction, addition, or observation within a quote, I do so within [square parenthesis].

As Darwin used the spelling Galapagos, and as this is also applied to common bird names today, I use it throughout rather than the strictly correct Galápagos.

Charles Darwin's major publications containing ornithological matter are cited within the text by the following abbreviations in bold typeface following (but in normal typeface in the body of the book); see also appendix 1 and References.

Darwin 1837 = Darwin, C. Notes on *Rhea americana* and *Rhea darwinii. Proceedings of the Zoological Society of London* Part V, No. 51 (read 14 March, 1837): 35–36.

Darwin 1838 = Darwin, C. *Life*, 1838: in *Charles Darwin Evolutionary Writings* by Secord (2010: 351–354); see References.

Gould and Darwin 1838–1841 = Gould, J. and Darwin, C. eds. *The Zoology of the Voyage of H. M. S. Beagle, Under the Command of Captain Fitzroy, R. N., During the Years 1832 to 1836. Part III: Birds*. Smith, Elder & Co., London, pp. 1–164. (As Darwin wrote much, if not the majority, of the text for this work, I cite it as by Gould and Darwin and not as by Gould alone.)

Darwin, C. 1839a = In Fitzroy, *Voyages of the Adventure and Beagle Between the Year 1826 and 1836 . . . Volume 3. Journal and Remarks 1832–1836*. Henry Colburn, London.

Darwin 1839b = Darwin, C. *Journal of Researches into the Geology and Natural History of the Various Countries Visited by H. M. S. Beagle Under the Command of Captain FitzRoy, R. N. from 1832 to 1836*. 1839, first edition. Henry Colburn, London. (See References for details of other editions of this title.)

Darwin 1839c = Darwin, C. R. *Questions About the Breeding of Animals*. Stewart & Murray, London.

Darwin, C. 1845 = Darwin, C. *Journal of Researches into the Natural History and Geology of the Countries Visited During the Voyage of H. M. S. Beagle Round the World Under the Command of Captain FitzRoy, R. N.* Second edition. John Murray, London.

Darwin 1849 = Darwin, C. R. Section VI: Geology. In Herschel, J. F. W. ed., *A Manual of Scientific Enquiry: Prepared for the Use of Her Majesty's Navy and Adapted for Travellers in General*. London: John Murray, pp. 156–195.

Darwin 1859 = *On the Origin of Species by Means of Natural Selection, or the Preservation of Favoured Races in the Struggle for Life*. John Murray, London. Facsimile edition with an introduction by Ernst Mayr. Harvard University Press, Cambridge, Mass., 1964.

Darwin 1860 = *Journal of Researches into the Natural History and Geology of the Countries Visited During the Voyage of H. M. S. Beagle Round the*

World Under the Command of Captain FitzRoy, R. N. A new edition. [Tenth Thousand.] John Murray, London.

Darwin 1868a = Darwin, C. R. *The Variation of Animals and Plants Under Domestication*. Volume 1. John Murray, London.

Darwin 1868b = Darwin, C. R. *The Variation of Animals and Plants Under Domestication*. Volume 2. John Murray, London.

Darwin 1871 = *The Descent of Man and Selection in Relation to Sex*. John Murray, London.

Darwin 1881b = Darwin, C. R. The Parasitic Habits of Molothrus. *Nature. A Weekly Illustrated Journal of Science,* **25**, 51–2.

Darwin 1889 = Darwin, C. R. *A Naturalist's Voyage: Journal of Researches into the Natural History and Geology of the Countries Visited During the Voyage of H. M. S. "Beagle" Round the World. Under the Command of Capt. FitzRroy, R. N.* 1889, new edition, John Murray, London. (Note: I use this printing of this edition, which differs very little in its ornithological content to that of the 1845 printing; see References.)

Darwin 1901 = Darwin, C. R. *The Descent of Man and Selection in Relation to Sex*. 1901, second edition. John Murray, London, second edition [first published 1871].

Darwin 1902 = Darwin, C. R. *The Origin of Species by Means of Natural Selection or the Preservation of Favoured Races in the Struggle for Life*, December 1902 reprint with "Additions and Corrections" and "An Historical Sketch," John Murray, London, sixth edition [first published 1859].

All other cited works appear in the References. In addition to the preceding, I also occasionally refer to letters written by or to Charles Darwin that contain further pertinent information. These letters were mostly found within *The Correspondence of Charles Darwin*. This vast work was not of course published by Darwin but was, and continues to be, compiled by a team of dedicated people under the editorship of Frederick Burkhardt and others during 1985 to 2014 so far, as an invaluable resource to scholars. This admirable work is cited as follows, but with the pertinent specific volume number and pagination also indicated:

Burkhardt, F. et al., eds. 1985–2014 = *The Correspondence of Charles Darwin*, 21 volumes, Cambridge University Press, Cambridge.

Clifford B. Frith
Malanda, Tropical North Queensland, Australia

ACKNOWLEDGEMENTS

Dawn Frith, my dearly beloved friend and colleague of the past four decades provided valuable discussion, advice, companionship, generosity, partnership with index compilation, and support throughout the research and writing of this work, which she read in draft and helpfully commented on.

I owe my good friend and esteemed colleague Johannes Erritzøe of the House of Bird Research, Taps, Denmark, my sincere gratitude for reading and constructively commenting on my draft texts, resulting in corrections and improvements, and for taxonomic discussion and advice on pertinent literature.

Clemency Fisher, Curator of Vertebrate Zoology, National Museums Liverpool World Museum, Liverpool, England critically reviewed some of my text concerning John Gould and provided additional information concerning it, for which I am most grateful.

My long-standing good friend and admired colleague Mary K. LeCroy of the American Museum of Natural History, New York, was, as ever, most kind and helpful in dealing with nomenclatural problems confronting me and in supplying obscure literature. Frank D. Steinheimer, and Franz Bairlein, Editor-in-Chief of the *Journal for Ornithology* of The German Ornithologists' Society, kindly consented to my including his meticulous work that is appendix 3 herein, including updates on three Darwin specimens kindly provided by him. Frank Steinheimer and John van Wyhe, both contributors of valuable literature on Charles Darwin and his works, kindly provided literature and helpful correspondence; John van Wyhe also most generously assisted my accessing pertinent illustrations.

My good friends Stanley and Kaisa Breeden provided enthusiastic interest, discussion, advice, literature, scanning facilities, and expertise; and Stanley kindly read and commented on a draft chapter. My neighbour, friend, and colleague Thane Pratt kindly offered advice on taxonomic literature; and Steffan Andersson kindly provided bibliographic details. Tim Birkhead of Sheffield University; Brian Gill of the Auckland Museum; Leo Joseph of the CSIRO, Canberra; and Richard and Barbara Mearns in England kindly provided helpful correspondence, taxonomic, and/ or bibliographic advice and literature. I thank Gregory Estes and K. Thalia Grant for stimulating conversations about Charles Darwin during my visit to the Galapagos Islands. Chris Tsilemanis of Booklark Books, Yungaburra, Australia, kindly assisted by seeking out and providing books.

Robert Prys-Jones and Joanne H. Cooper of the Natural History Museum ornithology department kindly provided literature and advice on Darwin's domestic pigeon specimens and, together with Emily Beech of Image Resources at that institution, access to images. My long-term friend, one-time British Museum of Natural History, Bird Department, colleague, friend, and fellow ornithological expeditionary, Daniel Freeman, kindly assisted with obtaining a portrait of Darwin. I am particularly pleased to be the first to include in any book a reproduction of the meticulously accurate painting of H. M. S. *Beagle* at anchor in Sydney harbour by Australian marine artist Frank Allen. I thank him for that privilege and for his generosity in providing the image.

The magnificent biographies *Darwin* by Adrian Desmond and James Moore of 1991 and the two volume *Charles Darwin* by Janet Browne of 1996 and 2002 were a constant rich source of information, inspiration, and pleasure, as were numerous other books about the man. A bird book that provided much novel and interesting information and joy was the 2013 *Birds and People* by Mark Cocker. The 2014 book, *Ten Thousand Birds: Ornithology Since Darwin*, by Tim Birkhead, Jo Wimpenny, and Bob Montgomerie, came to my notice in April 2014, when I had prepared complete first drafts of my chapters for this book. The timely appearance of this most interesting and enjoyable work was stimulating and helpful.

The dedicated, helpful staff of the State Library of New South Wales kindly provided access to the first 22 volumes of *The Correspondence of Charles Darwin*. I have enjoyed the support and interest of Jeremy Lewis with every aspect of preproduction, of Prabhu Chinnasamy during production, and of Erik Hane with contractual details, at Oxford University Press, New York.

CHAPTER 1

The Fledgling Bird Watcher

From reading White's Selbourne, I took much pleasure in watching the habits of birds, and even made notes on the subject. In my simplicity I remember wondering why every gentleman did not become an ornithologist. [CD in F. Darwin 1950: 18]

Charles Robert Darwin was born at Shrewsbury, Shropshire, in England, on the border of north Wales, on 12 February 1809, the same birthday as Abraham Lincoln. Both would come to share an abhorrence of slavery. Charles was the fifth of six children born to his mother Susanna Wedgwood and her medical doctor husband Robert Waring Darwin, a man of vast proportions and weight. Robert Darwin was no naturalist, but he enjoyed and appreciated nature as he travelled the English countryside, attending to his many admiring patients, and about his Shrewsbury country home estate of The Mount at other times.

A relative with the very same name of Robert Waring Darwin, son of Robert Darwin of Elston, published a book titled *Principia Botanica: Or, a Concise and Easy Introduction to the Sexual Botany of Linnaeus* in 1787. Charles Darwin's paternal grandfather, Doctor Erasmus Darwin, was an entirely different matter. Erasmus was a physician, womanizer, erotic poet, and much else; but he was also a botanist and zoologist. Of greatest significance to his grandson Charles was Erasmus's remarkable two-volume work *Zoonomia; Or the Laws of Organic Life,* published during 1794–1796, which included discussion of the evolution of life on earth. In 1803, Erasmus published his *The*

Temple of Nature,[1] which included the poem *Production of Life*, containing the following evolutionarily pertinent verse:

> *Organic life beneath the shoreless waves*
> *Was born and nurs'd in Ocean's pearly caves;*
> *First forms minute, unseen by spheric glass,*
> *Move on the mud, or pierce the watery mass;*
> *Then as successive generations bloom,*
> *New powers acquire and larger limbs assume.*

As a student at Edinburgh University, Scotland, Charles Darwin was to closely study his grandfather's book *Zoonomia*, which exposed him to ideas concerning natural laws, sexual reproduction, and the steady progressive change in life forms over geological time. Interestingly, the theoretical content of *Zoonomia* was coined "darwinising" by the poet, literary critic, and philosopher Samuel T. Coleridge. Charles admired this work of Erasmus then and was to consult it, now and again, during his future life as an evolutionary biologist. Thus he was not without some truly significant family background in natural history appreciation and philosophy, if not specifically in ornithology.

CHILDHOOD TO TEENAGE

In his autobiography, which he titled *Recollections of the Development of My Mind and Character*, Charles Darwin recalled that by the time he started attending day school, in the spring of 1817, he had, as an eight-year-old, a well-developed taste for natural history.[2] This interest was initially and primarily expressed through the identification of plants and the collecting of various natural objects during long walks—including shells, minerals, and birds' eggs—but at other times also seals, franks, coins, and similar inanimate cultural objects.[3] In writing his autobiography, Darwin stated "I was very fond of collecting eggs, but I never took more than a single egg out of a bird's nest, except on one single occasion, when I took all, not for their value, but from a sort of bravado."[4] This brief, and disarmingly frank and telling, sentence speaks volumes about the character of young Charles Darwin and his attitude to life and nature. Another revealing story, told by his son Sir Francis Darwin, relates how as an older man Charles killed a cross-beak, presumably the Red Crossbill *Loxia curvirostra*, a type of finch of the family Fringillidae, with a stone—something he was skilled at as a boy. As a result, "He was so unhappy at having uselessly killed the cross-beak that he did not mention it for years, and then explained that he should never have thrown at it if he had not felt sure that his old skill had gone from him."[5]

Charles Darwin's parents kept fancy tame pigeons at the family home; and his mother Susannah, who died when he was a little over eight years old, was particularly fond of caring for and enjoying them. Given his age and intense childhood interest in natural history, his parents' pigeons would presumably have had a significant impact on him as not only pleasing and interesting birds but also as being creatures worthy of the attentions and emotions of his much-loved adult family members. Later in life at his own family seat, situated in lovely rural English countryside rich in diverse wild bird life, Charles deeply enjoyed keeping and breeding pigeons himself—albeit for rather more serious and scientific reasons. Indeed, his pigeon keeping, breeding, and studies were to play a fundamental role in laying the foundations for the complex case for his theory of the origin of species by natural selection. His pigeon work demonstrated to the reading public the role of morphological variation within a species as the prerequisite for the action of artificial selection on domesticated animal species by people.

In writing his brief essay *Life* in 1838, Charles Darwin wrote of being eight and a half years old in 1817 and of "seeing game and other wild birds, which was a great delight to me.—I was born a naturalist." By the age of ten, his natural history collecting was mainly of minerals and insects in general but was soon to become specifically directed to the beetles, or Coleoptera, of the United Kingdom. He wrote, in remembering of being ten and a half years old, "The memory now flashes across me, of the pleasure I had in the evening or on [a] blowy day walking along the beach by myself, & seeing the gulls and cormorants wending their way home in a wild & irregular course."[6] In addition to this pleasure in bird watching, he also had a "strong taste for angling, and would sit any number of hours on the bank of a river or pond watching the float"; and he also had "a strong taste for long solitary walks."[7] He would doubtless have observed numerous birds of many species, and would probably have mentally noted their habits, during these frequent much-loved pastoral activities. The ornithological experience he gained during them was to be expressed now and again many years later by his comparing British bird species with unfamiliar ones that he observed in foreign lands visited by the HMS *Beagle*.

Darwin's childhood included various other ornithological influences. As his wealthy father's substantial library contained many, then fashionable to own, natural history books, it doubtless included volumes detailing British and exotic birds. So would have the libraries of other relatives and family friends, for a good number had by then been published—if largely nomenclatural catalogues of species or descriptive illustrated works.[8] It was indicative of his father's ability to invest in the finest of natural history publications that his sister Catherine was to write on 14 October 1832, to Charles, then aboard the *Beagle*, telling him that their father intended to commit the then impressive sum of 40 guineas to subscribe to the complete, massive, double elephant folio work on the *Birds of America* by John James Audubon.[9] Putting the great value of this work into today's financial terms is the fact that in December of 2010, a copy of this book sold in London for 7.3 million pounds Sterling (US$11.5 million).

In July of 1819, Darwin actively sought out and watched seabirds on the Welsh coast, where his elder sisters had taken him for three weeks and where he at times apparently "swore like a trooper."[10] Here he very probably also observed Rock Doves (*Columba livia*) in their wild habitat and at their nesting sites on coastal cliffs. This modest-looking pigeon was to loom large in his evolutionary thoughts, studies, and writings as the bird he came to so eloquently demonstrate to be the sole ancestral species to all domesticated pigeon "breeds."

Towards the end of his school life, Charles Darwin became "passionately fond" of shooting birds. He recalls being so excited at killing his first snipe (the Common Snipe *Gallinago gallinago*) that he wrote, "I had much difficulty in reloading my gun from the trembling of my hands."[11] At this time he also read Gilbert White's *Natural History and Antiquities of Selbourne*[12] and as a result gained pleasure from carefully watching and noting bird behaviour, as opposed to just seeing, hunting, and shooting them, "and even made notes on the subject." As he clearly recalled wondering to himself at that time "why every gentleman did not become an ornithologist," there can be no doubting his very real early interest in the subject as being worthy of serious attention and ambition.[13] The fact that ornithology was one of the first of the scientific disciplines to emerge within zoology in the early nineteenth century would probably have influenced Darwin's interest in the field.[14] As a result of reading the accounts of naturalist travellers and adventurers, he long thought of, and hoped to visit, tropical places and so see their natural attributes in person.

Darwin arrived at Edinburgh University in October 1825. His elder brother Erasmus had commenced his studies there in February of 1822. Erasmus was present at Edinburgh University with Charles only during his first year there; but after his departure, Charles came to know several young men who shared his interest in nature, and he much enjoyed their company and intellectual stimulation.

While in Edinburgh, on 18 January 1826, Charles wrote in his diary, "Saw a Hedge Sparrow [Dunnock *Prunella modularis*] late in the Evening creep into a hole in a tree; where do most birds roost in winter?" He shortly thereafter wrote "Heard a lark [Eurasian Skylark *Alauda arvensis*] singing at 20' past 7 o'clock & two bats at ½ 8 o'clock"; and on 1 May when back in Shrewsbury, "Susan heard a Cuckoo [Common Cuckoo *Cuculus canorus*]"; and the next day, "I believe I saw a swift [Common Swift *Apus apus*] late in the evening but am not sure." As the season progressed he also recorded Corn Crakes (*Crex crex*), Kitty Wrens (Winter Wren *Troglodytes hiemalis*), flycatchers (Spotted *Muscicapa striata* and/or European Pied *Ficedula hypoleuca*), very few sand martins (*Riparia riparia*), Lesser whitethroats (*Sylvia curruca* and/or Common whitethroats *S. communis*), stonechats (European Stonechat *Saxicola rubicola*), green linnets (European Greenfinch *Chloris chloris*), and a nuthatch (Eurasian Nuthatch *Sitta europaea*). On 22 May, he recorded shooting a Red-backed Shrike (*Lanius collurio*), and on 12 June "a bird with bright red breast & crown of head. Sonnet [Common Linnet *Linaria cannabina*] or Redpole [Common Redpoll *Acanthis flammea*]?" He shot a cormorant (Great Cormorant *Phalacrocorax carbo*) on 26 June. By autumn, he recorded shooting, on 1 September 7½ brace of partridge (Grey Partridge *Perdix perdix*); next day, 6½ brace; a couple of days later, 2½ brace; on 5 September 4 brace; and so forth. By late December, he recorded seeing a "Larger Titmouse [Great Tit *Parus major*], Grey wagtail [*Motacilla cinerea*]," and a "Water Ouzel [White-throated Dipper *Cinclus cinclus*]"; and on the 25th, he "Saw a hooded crow [*Corvus cornix*] feeding with some rooks [*Corvus frugilegus*] by the seashore near Leith."[15]

In February, Charles took lessons from a freed Negro slave named John Edmonstone who had retired as a servant to the Edinburgh medical lecturer Dr. Duncan. Edmonstone had travelled with the naturalist and explorer Charles Waterton and returned with him to England from Guiana in South America. Waterton had taught Edmonstone to become highly skilled at skinning and stuffing birds. Darwin enjoyed the taxidermy lessons from Edmonstone, which he privately commissioned for himself at one guinea for an hour every day for two months, as well as the company of a person he described as a "very pleasant and intelligent man." This social experience, as well as the anti-slave sentiments of his two grandfathers, could probably also have played a significant role in Darwin's subsequent strong personal objection to slavery.[16] In addition to learning to skin birds, which Charles clearly saw as highly germane to his interests and future in ornithology, he learnt about tropical South America and its rainforests from his conversations with Edmonstone. Apparently he was to subsequently put his bird skinning skills into practice during the voyage of the *Beagle*, at least until he was able to secure the paid help of a member of the ship's crew as assistant collector and preparer.

Charles was clearly thinking about ornithology at this time and was instructing himself by copying out a hundred genera of birds from the great book of 1760, *Ornithologie ou Méthode Contenant la Division des Oiseaux en Ordes, Sections, Genres, Espèces & Leurs Variétés*, by the French author Mathurin Jacques Brisson. As the Natural History Museum of Edinburgh University, created by Professor Robert Jameson, then included a substantial collection of bird specimens, it is quite possible that the student Charles Darwin took advantage of this collection to examine species unfamiliar to him.

In his personal diary for early 1826, Darwin asked of himself, among his occasional ornithological observations, "where do most birds roost in winter?" On 25 April, he wrote "no swallows, or rather the genus Hirundo, have appeared in or near Edinburgh," presumably because he knew very well that swallows (the Barn Swallow *Hirundo rustica*) would by then already be preparing to nest again at his family home of The Mount in Shrewsbury, to the slightly warmer south of the United Kingdom.[17]

Darwin wrote, about the summer of that same year, of walking 30 miles on most days on a tour of North Wales with two friends, during which he would undoubtedly have observed many and varied bird species.[18] Of only a few Charles Darwin manuscripts surviving from this early part of his life is one consisting of a brief list of northern bird species that he had copied from the Mathurin Brisson's, excellent for its time, 1760 six-volume book *Ornithologie*.[19] Another book that stimulated great interest in birds and their study, and thus the publication of other ornithological titles available to Darwin prior to his voyage, was the finely illustrated *Histoire Naturelle des Oiseaux* by Georges-Louis Leclerc de Buffon.[20] Buffon was a highly significant author of his time in that he did not merely describe and depict bird species but also included information of their distributions, ecology, and behaviour. Many subsequent books of the period dealing with birds were based on Buffon's much-admired and successful publication.

As interest in the birds of the world grew greatly during the eighteenth century, there was a reasonable number of books available to young Darwin on the birds of Britain—for example, by White, Bewick, and Montagu[21]—and also for elsewhere. Gilbert White's famous writing included descriptions and discussion of topics such as migration, general ecology, seasonality, and nesting biology. George Montagu's books paid particular attention to life histories and sexual dimorphism, and these works would presumably have had influences on Darwin's views of ornithology. There were other bird books written and published on the European continent as well as general zoological works including birds, many finely illustrated with the relatively new technology of lithography. Many were predominantly little more than descriptive works with black and white illustrations, with fewer being hand coloured, of bird species known at the time for particular countries or global areas rather than about particular groups of related birds. One such was the *Planches Enluminées* by Edmé-Louis D'Aubenton,[22] which when completed contained 1,008 hand coloured engravings, of which 973 illustrated 1,239 birds. Some other books available at the time dealt with the birds of North America, South America, Africa, Asia, and Australia. John Latham's works,

A General Synopsis of Birds and *A General History of Birds*,[23] were also highly significant at the time, earning Latham a great reputation and particularly so in Britain.

Thus the period of 1780–1830, up to the year before Darwin sailed from England on the *Beagle*, saw ornithological collections, studies, and publications (too numerous to detail here) dramatically increase and improve in quality. That said, it is considered that ornithology did not become an established discipline of the natural sciences until two decades into the nineteenth century: less than two decades before Darwin returned to England.[24] Darwin was therefore to write and publish his various works on birds at a significant time in the history of ornithology.

In his second year at Edinburgh, in 1826, Darwin spent time with William MacGillivray, curator of the museum there, who was kind to him and discussed much about natural history in conversation. As MacGillivray subsequently published a large book on the birds of Britain, which Charles considered to be excellent, it is more than likely that birds featured in at least some of their conversations.[25]

During the week following the 1 September opening day of the 1826 shooting season, Charles shot 55 partridges, three hares, and a rabbit. Indeed, so intense was his youthful enthusiasm for hunting that about this time, his father, deeply concerned about his then seemingly wayward teenage son's future, told him "You care for nothing but shooting, dogs, and rat-catching and you will be a disgrace to yourself and all your family."[26] Charles was cut to the quick by this remark from his much beloved, admired, and respected (if not also slightly feared at that moment) father—and it may well have had a considerable influence on his future attitude to life and work.

In mid-1828, Darwin spent time with two slightly older peers and their tutors in Wales, and this included him shooting various birds atop Cader Idris, at 3,000 feet or 980 m above sea level, specifically to stuff them for his collection and to give to, or trade with, fellow natural history collectors.[27] One of his travel companions wrote of Darwin telling him that he "used to sit in a natural 'chair' on the edge of the cliff, where he shot any bird on [the] wing below him, which he wished to secure, and the guide who was at the foot of the cliff had to pick it up & carry it home for preserving."[28] Apparently none of the British birds that CD collected and preserved as skin specimens have survived to today.[29] In addition to his frequent bird shooting, the teenage Charles Darwin spent much time walking, riding, and fishing in the English (and during visits there also the Scottish and Irish) countryside, during which he doubtless noted and learnt of the species of local birds and their ecology and habits. He was also a not infrequent visitor to zoological garden collections where he would seek out and observe living individuals of exotic bird species.

The young Charles Darwin can be mentally pictured sitting stock-still on a stool among the rank bank vegetation of a narrow English river, intent on his line and float like all good anglers, his fishing rod extending out over the water. He suddenly hears the high-pitched call of a Common Kingfisher (*Alcedo atthis*) flying downstream towards him and then sees it as a flash of brilliant blue as it perches on an exposed tree root opposite him—or even, conceivably, on his fishing rod, as such does occasionally happen. The gem of a bird, clearly reflected on the dark surface of the river, bobs its head and turns to give young Charles the benefit of its entire glorious plumage. This causes him to think to himself, "thank you God for creating such a beautiful creature for the pleasure of your human flock" with absolutely no conception of how utterly different his view of such experiences was to become; for example, see Alcedo (the genus of the Common Kingfisher) in appendix 1.

Writing of the history of ornithology, my one-time Natural History Museum colleague Michael Walters stated, "until the time of Darwin, there was no real need to enquire why certain animals were found in one place and not in others. The answer was clear: God had decreed that this should be so, therefore it was so."[30] And this is precisely how Darwin saw the natural world as he strode from England's soil across the gangplank and onto the, to him, depressingly small deck of the *Beagle*. At that time, Charles was a committed Christian to the extent of being fully convinced that upon his return to Britain, his only ambition was to find himself a peaceful parish in the English countryside where he could attend to his flock of worshipers and the local natural sciences if, perhaps, not necessarily in that order of priority. Nothing was then further from his mind than the giddying heights of scientific theories concerning geology, zoogeography, evolution, the origin of species, natural and sexual selection, and much more. They were, however, to come to him impressively quickly during the voyage—at least with respect to geology and, to a lesser extent, zoogeography.

The young student that travelled to Scotland was far from the familiar image of a frail, elderly, bearded Charles Darwin that typically comes to mind upon the mention of his name today. As a young man, prior to and during the entire voyage of the *Beagle* and for some years thereafter, he was a remarkably fit and vigorous, pleasant looking, fresh-faced individual that could out-walk, out-ride, out-shoot, and out-think most of his peers (colour figure 1.1). Moreover, he was far from immune to the charms of the fairer sex. He was quite besotted by and smitten with Fanny Owen, the flirtatious daughter of his father's friend and neighbour William Mostyn Owen of Woodhouse, and his feelings for her were no secret to his sisters. But it was to be his cousin Emma Wedgwood, of the famous pottery family, that he was to eventually be married to.

Having started at the university in Edinburgh studying medicine, Charles found it most unpleasant due to an abhorrence of blood, and thus especially of practical surgical procedure demonstrations. He therefore complied with his father's wishes, which were in part based on fears of his son becoming an idle sporting man, by moving to Cambridge University to study for the cloth with the Anglican Church in January 1828. He attended Christ's College, as did his second cousin William Darwin Fox, who arrived a year before him and who was particularly interested in birds. They became the closest of friends. Christ's College was ideal for Darwin, as it was "amenable to wealthy young men devoted to hunting and shooting" and was at the time more liberal and tolerant than most of the other colleges at Cambridge. That said, he did not study theology or divinity per se but rather was enrolled for a regular Bachelor of Arts Degree, after which he never did pursue study courses towards Holy Orders.[31] Although Charles's father was an extraordinarily wealthy man, he was certainly not prepared to see his son take to an idle way of life; and so after initially pressing him to read medicine, eventually to no avail, he insisted that he study for the church. Ironically, Charles's five-years-older brother Erasmus, known as Ras or Eras by family members, having read medicine at Edinburgh, dedicated his life from the age of 25, in part due to poor health, to high society as a much-loved, extravagant, prematurely retired social host at his elegant Queen Anne Street, London, residence. At that time in his life, Charles wrote of himself, "as I did not then in the least doubt the strict & literal truth of every word in the Bible, I soon persuaded myself that our Creed must be fully accepted."[32]

EARLY MANHOOD

During his first year at Edinburgh University, 17-year-old Charles attended a Plinian Society meeting on 27 March 1827, at which the habits of the Common Cuckoo (*Cuculus canorus*) were discussed and to which he contributed his own observations or opinions—quite possibly as his first public ornithological utterances. At about this time, he saw and heard the colourful, famous, and much-admired ornithologist and bird artist John James Audubon speak at a Wernerian Natural History Society meeting on the habits of North American birds. Audubon was born at Les Cayes, Saint-Domingue, now Haiti, of French parents but settled in America as an American. He exhibited a drawing of a "buzzard" and detailed his novel method of wiring dead birds into life-like postures to be able to draw them for his paintings.[33] As this Audubon lecture is specifically noted in his autobiography, penned in 1876 at 67 years old, it is safe to assume that it had a significant impact on Charles and his then growing ornithological interests. This is hardly surprising, for not only was Audubon revolutionizing the illustration and textual description of birds and their ecology and behaviour, but he was also a remarkably colourful, buckskin-wearing "backwoodsman" and entrepreneur attracting considerable international attention.

Darwin commenced his second year at Cambridge University on 24 February 1829, his second cousin and good friend William Fox having left with his degree in January. On 8 June Charles left Cambridge, went to London and then on to Shrewsbury to collect beetles and shoot birds until mid-June when he again toured North Wales, this time with entomologist Frederick William Hope. After about a week, however, he had to return to Shrewsbury due to severe dermatitis of his lips.

During 1829, Darwin attended a talk by the Reverend Leonard Jenyns on the "divine design of feathers" at a meeting of the Cambridge Philosophical Society. There is no record of what he thought about Jenyns's postulated "intelligent design" of feathers, but 42 years later he wrote at length, and in stark contradiction, about the evolution of feathers and their markings through natural and sexual selection in his book *The Descent of Man*. During this period, Darwin and his cousin William D. Fox wrote to one another about bird specimens they had bought or traded for themselves or for one another and about unusual British bird records as fellow bird enthusiasts.[34]

A great, and principally botanical, influence on young Darwin the budding naturalist at Cambridge was Professor John S. Henslow. Charles gained great pleasure and much knowledge and biological insight by joining pupil field excursions with Henslow, as they sought out and discussed relatively rare plants and animals in the British countryside. He became a favoured student companion to Henslow during the professor's biological field ramblings. It was while at Cambridge that his beetle collecting, and entomology in general, became most intense. One of his fellow avid beetle collectors was Albert Way of Trinity College. As Way was to subsequently become a well-known ornithologist,[35] he would with little doubt have talked about birds with Charles.

Interestingly, in assessing his own and Charles Darwin's contributions in 1908, a quarter of a century after Darwin's death, Alfred Russel Wallace noted that he considered their common interest in beetle collecting early in life was "most important." As the enormous number of beetle species exhibits "endless modifications of structure, shape, colour, and surface-markings that distinguish them from each other, and their innumerable adaptations to diverse environments," these insects impressed both men because of their intense interest in the variety of, and variation within, animal species. Wallace thought that their mutual fascination with "the outward forms of living things" was the only thing that ultimately led them "towards a solution of the problem of species."[36]

Darwin again left Christ's College, Cambridge, on 3 June 1830, going initially to London for several days prior to going on to Shewsbury. He yet again visited North Wales in August and there collected beetles, fished, and then hunted birds in Shropshire and Straffordshire. He returned to Cambridge on 7 October for the start of the academic term. In January 1831, Charles and some seven student friends that had formed the "Glutton, or Gourmet, Club" on New Year's Day 1830 to enjoy eating unusual and exotic fare, tried to eat an "old brown owl" probably a Tawny Owl (*Strix aluco*), but without any pleasure or success. That bad experience brought about an end to the club feasts. Other birds previously eaten by them included a hawk and a bittern.[37]

While developing an ever-stronger interest in natural history, and in particular geology and entomology, he simultaneously became disenchanted with the notion of becoming an ecclesiastic. But at least the latter vocation offered the promise of a peaceful country backwater where he might pursue his other interests while attending to the needs of a small flock of hopefully undemanding worshipers. Charles continued to believe that all animals and plants were created by God, to exclusively occupy the geographical distributions and habitats they were then found in, and had always been as people knew them to be in their appearance and habits.

Darwin was not a happy, contented, or settled university student. He had often dreamt of visiting the tropics to see their rich and exotic natural history, and this left him unsettled. During 1831, having read Alexander von Humboldt's massive *Personal Narrative*[38] about his travels in South America, a work he then greatly admired and continued to admire for the rest of his life, he became determined to visit Tenerife in the Canary Islands; Humboldt, having visited these islands en route to South America. Darwin encouraged several university colleagues to become interested in joining him there for the month of July, including his professor John Henslow; and having been given some funds to clear his debts by his father, he made plans for this expedition. He even began to learn to speak Spanish; and although he found the lessons "intensely stupid," he could not then know that they would prove useful to him in South America in the future.[39] But as time passed, several of his proposed fellow travellers fell by the wayside (the tutor Marmaduke Ramsay quite literally so because he died) due to other responsibilities, interests, and distractions: in Henslow's case, the birth of a child. It then proved too late to go anyway because he found that boats were scheduled only for June departures. Then Darwin's own life was suddenly thrown into frantic preparations for a far greater expedition to the tropics, about the entire globe, that totally eclipsed any brief jaunt to Tenerife (see following).

Much of Darwin's autumn pleasure each year, and notably through to 1830 and 1831, resulted from his bird hunting and shooting interests. This sporting activity meant a great deal to him, and he kept meticulous records of the precise species and numbers of birds that he "bagged." He was particularly fond of opening the season at his uncle Josiah Wedgwood's Maer Hall estate and also shooting at Woodhouse. It is very clear that he was a highly competent hunter and shooter of British game birds, an ability requiring a level of understanding of the habitat, ecology, and behaviour of the various species concerned. He kept an "exact record of every bird which I shot" throughout an entire season, making knots in a piece of string tied to a button hole of his jacket to record bird numbers killed in the field, a pedantic idiosyncrasy that he was at times ribbed about by his relatives and friends.[40] In September of 1830, he was bragging of killing ten brace of partridge on the very first day of the shooting season.[41] Harking back to his youthful boasts as a good shot in hunting British game birds, Darwin is reported to have told his children, in later life at Down House, how in South America he once "killed twenty-three snipe in twenty-four shots." And "in telling the story he was careful to add that he thought they were not quite so wild as English snipe."[42]

In addition to Common Pheasants (*Phasianus colchicus*) and partridges (Grey and Red-legged *Alectoris rufa* Partridges) he at times also shot Black Grouse (*Tetrao tetrix*), Eurasian Woodcock (*Scolopax rusticola*), and snipe. His knowledge of these and of other birds that he observed while hunting would have stood him in good stead in terms of locating and understanding their relatives and ecological counterparts beyond the shores of Britain.

Darwin again visited North Wales in August 1831, this time on a geological surveying tour under the guidance of Professor Adam Sedgwick, an experience that provided knowledge of much significance to his future studies in geology. There is surprisingly little else recorded that provides significant further insight into Darwin as a budding ornithologist prior to his receiving his life-changing offer of a berth on the HMS *Beagle's* global scientific voyage, intended to last some two years (*pace* five years; cf. Thomson[43]) from Captain Robert FitzRoy RN. News of this momentous event first came, as a bolt from the blue in the form of a letter from John Henslow, dated the 24th and received 29 August 1831. It forewarned him of a pending offer to join a British Naval vessel sailing to South America and home via the East Indies, a circumnavigation of the tropics. In giving this shocking advice, Henslow told Charles that he considered him the best qualified person likely to undertake the voyage. He was not at all well qualified as a sailor, however, having only ever been to sea once, on a return trip across the English Channel to Paris with his uncle Josiah who was bringing his daughters home from Geneva (one of which was to become Charles's wife more than a decade later).

The formal offer to join the *Beagle* voyage came in the same envelope as Henslow's letter, sent by George Peacock, another Cambridge professor; but this was unfortunately immediately followed by a temporary change of heart by FitzRoy that caused Charles intense, if brief, heartache. Upon then being interviewed by FitzRoy who, quite contrary to his expectations, found Darwin agreeable, he was once again offered the position as naturalist aboard the *Beagle*, due to sail near the end of September. Unfortunately, this was followed by Charles's father strongly objecting to the entire venture on a number of grounds. These included that it would be disreputable to a future clergyman (as was intended for CD by his father), would not further any useful goal in life for Charles, and it was a wild scheme. Other objections included that, as he was obviously not the first choice for the post, other potential candidates (there were four, including John Henslow and the Reverend Leonard Jenyns, who declined the invitation due to other commitments or ill health and parish duties respectively, and who recommended Charles Darwin in their stead) must have found objections to the voyage, that the vessel might thus be ill-equipped and not up to the task, that Charles would be poorly accommodated, and that he would prove unable to settle down to life in England after such an experience.

The fact that the ship concerned, the bark HMS *Beagle*, was of the Cherokee class of vessels that were commonly referred to within the navy as "coffins" or "coffin brigs," in the light of their unseaworthiness, would not have reassured anyone. Most considerately, however, Robert Darwin told his despondent son that "If you can find any man of common sense who advises you to go I will give my consent." It became clear that Robert was in all probability subtly alluding to Charles's much loved and highly respected uncle Jos Wedgwood; and most fortunately, for him and for the biological and geological sciences thereafter, his uncle responded brilliantly in the positive; and so his father removed his objections to Charles going. More than offering his, now unreserved, acceptance of his son's pending part in the voyaging venture, his father also indicated that he would be generous in his financial support of it. Perhaps the fundamental reason behind Charles's father's initial objection to his undertaking such a voyage was the fear of his son being seriously injured, if not lost to him entirely, by misadventure.

The euphoria experienced by Darwin at his father's change of heart was followed by repeated postponements of the sailing date of the *Beagle* throughout October and November of 1831. At one point during this period of delays and associated boredom, homesickness, loneliness, and misgivings about leaving his family and friends, and the risky venture he was committed to undertaking, Charles developed a rash about his mouth and experienced heart pain and palpitations that he was sure indicated heart disease. He was so determined to sail with the *Beagle*, however, that he kept his physical condition and his doubts to himself, improved, and eventually departed England—in the fullness of time to change human perceptions of the natural world, their own species, and themselves, forever.

While kicking his heels during a return to London (a quick trip took 24 hours for the 250 mile, or 400 km, horse-drawn coach journey), Darwin sought any and all experience, advice, knowledge, books, and articles that would be of use to him during his voyage. This included a visit to the Regent's

Park Zoo's West End Museum at Piccadilly, where some 4,000 bird skin specimens could be examined. If he did look through this collection, he would have seen specimens of numerous bird families, genera, and species new and exotic to him. He also met with the kind of experienced people he really needed to talk with. One such person was Benjamin Leadbeater (who had an Australian cockatoo species named after him, then Leadbeater's but now Major Mitchell's Cockatoo *Cacatua leadbeateri*) who showed him how to pack bird skins appropriately for successful shipping.[44] Again, specifically with respect to his future work in preserving bird skins, CD was fortunate enough to be instructed about the use of arsenical soap and preserving powder. He also first met William Yarrell here in London, a man he was to deal with later in life and who published significant work of his own on British birds.[45] At this time in London he may well also have seen the first parts of John Gould's folio publication *A Century of Birds from the Himalaya Mountains*, published during 1830–1833, which doubtless would have greatly impressed him. Charles did not then, however, meet the rapidly ascending British ornithologist John Gould in person, who had already published this lavish bird book at 27 years of age, before his departure aboard the *Beagle*. It was, however, to be Gould who examined and classified Darwin's bird collection upon his return from the voyage nearly five years later.

It is stated that Charles Darwin had an "extraordinary devotion to angling, which started at an early age, and apparently lasted until well into the *Beagle* years"; but this is then partly contradicted by the statement that "CD remained devoted to angling until shortly before the voyage of the Beagle."[46] In any event, he was apparently much more than a casual angler prior to the departure of the *Beagle* from England. An interesting observation that indicates something of young Charles's attitude to life is that he killed his bait worms in salt water to spare them any pain in being pierced by his hooks.[47] He was to become profoundly interested in the study of living earthworms later in his life, resulting in a major and fascinating book about them.[48]

Another example of CD's sensitivity to pain and suffering in animals was recited by John Herbert who was at Cambridge with him. While hunting with William Owen, Charles "picked up a bird not quite dead, but lingering from a shot it had received on the previous day; and that it had made and left such a painful impression on his mind, that he could not reconcile it to his conscience to continue to derive pleasure from a sport which inflicted such cruel suffering."[49] Charles did, however, continue to enjoy shooting birds before, and during, his voyage of discovery aboard the *Beagle*.

There can be little doubt that at that point in his life, Charles Darwin, sitting on the bank of an English river on a fine sunny day to be confronted with the feathered jewel that is a Common Kingfisher, as hypothesized previously, would have been completely comfortable with the conviction that it had been placed there by God for the pleasure and appreciation of himself and other civilized people of the British Empire. He could not, of course, have conceived of the possibility that he would see equally, and indeed far more, beautiful birds in places never inhabited by humans or populated only by people so primitive that their only interest in birds of any hue would have been the possibility of killing and eating them. His view of such things was to change profoundly as the horizon of his mind was broadened by significant experience and knowledge gained during the voyage of the *Beagle* and his philosophical life thereafter. Similarly, when Alfred Russel Wallace first saw an adult male King Bird of Paradise (*Cicinnurus regius*) in the flesh, while living among "rude uncultured savages" on the Aru Islands off western New Guinea, he "thought of the long ages of the past, during which the successive generations of this little creature had run their course—year by year being born, and living and dying amid these dark and gloomy woods, with no intelligent eye to gaze upon their loveliness; to all appearances such a wanton waste of beauty."[50]

The content of the quote heading this chapter and of the chapter text itself can leave no doubt that Charles Darwin paid close attention to bird life, went out of his way to have himself taught how to prepare freshly killed birds as scientific specimens, and looked to learning far more about them before the voyage of the *Beagle* became a prospect. Observations that he makes, in comparing attributes of bird species encountered during the voyage with birds familiar to him from his time preceding it in England, indicate much more than a passing acquaintance with bird life and indeed more the eye and mind of a potential field ornithologist. But two examples of a good number of instances indicative of such knowledge noted during his land travels from the *Beagle* involve the Grey-headed (*Halcyon leucocephala*) and Common (*Alcedo atthis*) Kingfishers in the Cape Verde Islands (see the section "Atlantic Island Birds," chapter 2) and his describing the call of a South American mockingbird by comparing it with that of the Sedge Warbler (*Acrocephalus schoenobaenus*[51]) of his young life in England. It is thus apparent that he had a personal ambition to become, among other things, an ornithologist—a person that studies birds scientifically. He was to become a particularly widely travelled and a richly experienced one at that.

BEAGLE BOUND

The *Beagle* did not finally depart Devonport, Plymouth, for foreign lands until 27 December 1831, with Charles and 72 other men, including two Fuegians, and one young Fuegian girl, aboard;[52] see colour figure 1.2. Charles was effectively then an almost 23-year-old medical student dropout, his having obtained a BA in Theology, Euclid, and the Classics, but with no degree in the sciences. That said, the view has been expressed that when Darwin boarded the vessel, which was to be his home base for the best part of the next five years, he was "an extraordinarily well-trained natural scientist"[53] and "probably the best-educated naturalist of his age in Britain at the time, particularly skilled in invertebrate zoology and with some knowledge of geological surveying and natural history collecting."[54] He was not to return to the shores of his beloved England until 2 October 1836. Long before this return, he was deeply disappointed, if not heartbroken, to learn from a letter by one of his sympathetic sisters that his much-admired neighbour Fanny Owen had married during the early part of his nautical absence.

INITIAL MIGRATION

After seemingly endless frustrating delays in the HMS *Beagle* being able to depart England's shores, due to additional fitting out of the ship and subsequent prolonged foul weather, Christmas Day of 1831 arrived. Given that holiday date, many of the ship's crew went ashore and got seriously drunk, only to return to the vessel significantly late and to then be out of order to downright insubordinate in addressing their superior officers. The first morning at sea proper thus involved the lashing of men as punishment for their crimes against the strict British naval disciplinary code. Four of them received an average of 33.5 lashes.[55] This experience was to the considerable shock, discomfort, and disapproval of young Charles Darwin. It was a most depressing start to his voyage, immediately following what he found to be tediously repetitive delays under the most trying of circumstances.

Having finally actually departed Plymouth, the cries of its familiar resident Herring (*Larus argentatus*) and Common Black-Headed (*L. ridibundus*) Gulls eventually lost to their hearing as land was left in their wake, the crew of the *Beagle* did not step foot on land again until reaching the Cape Verde Islands in mid-January 1832. They had intended to go ashore at Tenerife, having failed to do so at Madeira, and CD was terribly excited about this, as he would have been following in the pioneering footsteps of Alexander Humboldt who he greatly admired and would thus be able to see the wonders described in the great man's oft-read writings. Alas it was not to be because a small boat promptly put out of Santa Cruz to approach the arriving *Beagle* with instruction from the consul in residence that no ship's personnel were to be permitted ashore. This was for fear of them introducing cholera, which had recently broken out in several English cities. The ship's captain and crew were much disappointed, and Charles quite devastated, by this most unfortunate circumstance.

Darwin proved to be, and to remain, for the entire four-and-three-quarter-year voyage, at the complete mercy of the sea, and his constant bouts of severe seasickness proved most debilitating and distracting for him. This and the tiny proportions of his sleeping and work spaces aboard ship aside, however, he quickly knuckled down and adjusted to the life of a working naturalist aboard one of His Majesty's survey vessels in the finest tradition of the British Royal Navy. His exclusive advantage among the ships company as invited naturalist was the highly significant one of also being the captain's personal gentleman companion and, to some degree, confidant, and therefore able to enjoy the singular privilege of eating at the captain's table. In addition, he was able to use the skipper's cabin as a peaceful place to retreat to and relax and read in when it was not required by FitzRoy for his sole use. This was a great boon to what would otherwise have been a far less pleasant voyage because six-foot, or 1.8 m, tall Charles shared his tiny (11 x 10 foot or 3.3 x 3 m) working and sleeping poop cabin with John Lort Stokes and Philip Gidley King, in which he had to stoop and had only "just room to turn round & that is all."[56]

Other than providing "genteel" and intellectual company and sharing concerns and confidences with the captain, apparently free of any and all other social or service obligations to the crew and the rest of the ship's company and the Royal Navy, Darwin was at liberty to investigate natural history and to collect nature's productions as he saw fit. He was also assured the full support of the ship's officers and crew in every aspect of his work, just so long as it came second to the well-being and correct functioning of the ship as a tasked naval survey vessel. It was a plumb position for him because in not being paid but rather paying for his food and board, via his wealthy medical doctor father, he answered only to Captain FitzRoy on all significant day-to-day issues. Charles clearly greatly impressed FitzRoy because he was described by him in a letter to Beaufort at the Admiralty, his seasickness notwithstanding, as "a very sensible hard-working man, and a very pleasant mess-mate. I never saw a 'shore-going fellow' come into the ways of a ship so soon and so thoroughly as Darwin."[57]

The status of Charles Darwin aboard the *Beagle* has long and widely been described as unofficial naturalist; this implying that he was primarily, if not exclusively, an invited travel companion to Captain FitzRoy.[58] This has, however, recently been thrown into doubt by evidence strongly supporting the view that he was indeed the official naturalist aboard the *Beagle* from the Admiralty's point of view.[59] It is suggested that because Darwin declined to receive any salary for his voyage, as a naturalist could be paid or unpaid, and in fact paid for his food and drink aboard, he therefore gained the right to choose which British institution his specimens would be presented to—just so long as it was a public institution and collection.[60] He also insisted that he retain the right to quit the *Beagle* and its commissioned voyage "as soon &

wherever" he chose.[61] This last condition of Darwin's is acknowledged by John van Wyhe three times,[62] despite it being a seemingly surprising one for what does otherwise appear to be an officially appointed naturalist for a British Navy voyage. That this "escape clause" was clearly understood beyond Darwin, FitzRoy, and the Admiralty is indicated by the fact that Professor John Henslow encouraged him to persist with the voyage and not to leave it.[63]

It is well established that Captain FitzRoy, having read some of Darwin's diary aboard the *Beagle*, asked Charles if he would kindly consider making it available to him for incorporation into the official narrative of the voyage, to be published by the government. It would thus appear clear that Darwin's status as official naturalist aboard the *Beagle* did not, as with other official shipboard naturalists of the era,[64] officially oblige him to write an account of his results in the form that FitzRoy politely requested that he undertake. His participation in the *Narrative* publication was voluntary and could, perhaps, be seen as a personal favour to Captain FitzRoy and/or a form of polite appreciation to the Admiralty.

Given the rigidly strict hierarchy among officers and crew of British Navy vessels, not to mention the all-pervasive British social class distinctions of the time, and the fact that a ship's captain had to remain absolutely aloof of all other ship's company, a civilian companion such as Darwin represented for FitzRoy far more than mere companionship. It should be taken into account that this voyage of several years was in the face of no communication with people beyond the vessel other than when in port (i.e., no radio or digital communications as there are today). The immense pressure of responsibility under Admiralty orders, and the social isolation and resulting loneliness and stress for a naval captain at sea for extended periods of time, had resulted in the suicide of a number of skippers under active service. Such suicides included that of the previous captain of the *Beagle*, Pringle Stokes, at Port Famine, Strait of Magellan, on 2 August 1828.[65] Adding to this grim history was the fact that Captain FitzRoy's own uncle, Lord Castlereagh, had, after long dedicated British political service, cut his own throat on 12 August 1822. This was to prove the very same bloody method that FitzRoy was himself to employ for his own suicide, in London on 30 April 1865.[66] It has been suggested, as supposition, that these suicides were a significant influence on FitzRoy's desire to have a gentleman companion aboard the *Beagle* "who could keep him sane."[67]

As Captain FitzRoy was not only a keen amateur naturalist but also a naval captain eager to see the results of his voyage of navigation include the discovery of as much as possible within the natural sciences, he was most eager to assist Darwin in every practical way. Charles found himself on an incredibly good wicket but the innings was to last well over two years longer than the two that he had signed up for or could have ever imagined being involved in. This was a period of voyaging far in excess of what young Darwin actually wanted to spend away from his beloved family and familiar, tame, English green and pleasant land. He, like the rest of the world, could have no idea of the intense and immense impact his personal voyage was to have on his own faith, knowledge, thinking, health, subsequent life, and the natural sciences in general—including ornithology—and ever since.

Meaningful ornithological publications were not really generally available until about the middle of the seventeenth century. It was then that the work *Ornithologiae* appeared,[68] and that only to a limited audience. Little followed this until the late eighteenth century when *Ornithologie ou Méthode Contenant la Division des Oiseaux en Ordes, Sections, Genres, Espèces & Leurs Variétés*;[69] *Systema Naturae*;[70] and *Histoire Naturelle*[71] were published and bird study started to flourish.[72] Ornithology at that time involved little more than the naming, listing, describing, and illustrating of species with their collecting locality, and with some attempt to define their then grossly inadequately known distributions. Thus, the main value of such books as were available to Darwin prior to and during his voyage would have been the description and depiction of bird species together with some idea of where on earth they might be expected to be encountered. There were many well-illustrated works on British birds but also a good number on birds of the world in general, leading the American ornithologist and bird artist John James Audubon to write in London in 1835 that three works were then being published on the birds of Europe (one being by John Gould) and that "Works on the Birds of *all the World* are innumerable."[73]

Books in the *Beagle* library included numerous accounts of extensive voyages of exploration and discovery, in English and other languages, as well as volumes on general natural history and geological, entomological, meteorological, and other subjects. Titles of ornithological significance available to Darwin prior to or during part (as volumes were sent to reach the ship at various ports) or all of his voyage aboard the *Beagle* included Brisson,[74] Buffon,[75] Forster,[76] Labillardière,[77] Humboldt,[78] Molina,[79] Pallas,[80] Cuvier,[81] Wied-Neuwied,[82] Burchell,[83] Saint-Vincent et al.,[84] Quoy and Gaimard,[85] Spix and Martius,[86] Griffith et al.,[87] King,[88] Lesson,[89] Latham,[90] Spix,[91] and Blainville et al.[92] In the case of some larger multivolume works, only those published prior to the *Beagle* sailing, and any sent on to it during the voyage, could have been available aboard. This was the kind of reference material available to CD to assist him in the identification of birds he saw and shot, as well as the numerous voyage and land travel book accounts that included mention of birds observed.

As the eighteenth century came to an end, finely illustrated books dedicated to a particular group or groups of birds, termed monographs, started to appear. One notable such title was the splendid *Histoire naturelle et générale des Collibris, oiseaux-mouches, jacamars et promerops* (predominantly about the hummingbirds and jacamars),[93] which set a benchmark for fine publishing that was to be followed by similar monographic books. Although two catalogues that were known to detail the books held on the *Beagle* during Darwin's voyage were unfortunately lost, we do know that there were "upwards of 400 volumes" that primarily involved works of travel and natural history aboard.[94]

It has been stated that Darwin "hardly made any use of the literature on board the 'Beagle' in identifying or learning more about birds,"[95] but this cannot be known to be factual. After all, without consulting books aboard the *Beagle*, how would Darwin have made the identification of the birds that he did during the voyage? One eminent Darwin student stated that CD was "quite knowledgeable as a [avian] taxonomist, and he generally managed to classify his *Beagle* specimens under the appropriate family, genus, and sometimes even species using the published guides available to him on the voyage."[96] As Darwin reported that he often read aboard, and specifically recorded that he found the work of Molina[97] most helpful to his ornithological work, it is surely more likely than not that he did indeed consult books dealing with birds as a matter of course, as and when required during the voyage. In using the work by Molina on board, he found it a puzzle that no mention was made of the members of the tapaculos family Rhinocryptidae and critically wrote in his notebook, "Was he at a loss how to classify them? and did he think that silence was the more prudent course? It is one more instance of the frequency of omission by authors, on those very subjects where it would be least expected." He had additional criticisms to make of Molina's text used on the voyage.[98]

Friends and colleagues did try to keep Darwin informed of ornithological developments during his time aboard the *Beagle*, notwithstanding the extremely prolonged time it took for books and letters to be delivered to ports by other sailing ships that the *Beagle* was bound for. For example, his cousin Thomas Eyton wrote to him on 12 November 1933, telling him that John Gould was producing his splendid *Birds of Europe* and had published "an excellent monograph on the Toucans" since Charles had departed England.[99]

Initially Charles was able to use Henry Fuller, one of the ship's "boys," to assist him with trivial matters in sorting and preparing collected zoological material. But as the voyage progressed, he quickly came to realize that he would need some better and full-time assistance on board and ashore in performing his diverse work. He therefore asked Captain FitzRoy if a suitable member of the crew might be made available for him to employ as a personal servant. Seventeen-year-old Syms Covington, another of the ship's boys, became that servant. Upon Darwin's request for Covington's help, FitzRoy had apparently asked thirty pounds, but perhaps Darwin raised his annual fee over time. It is recorded, however, that Covington received thirty[100] or sixty[101] pounds Sterling a year from Charles for his services. It is perhaps surprising that in writing of this scenario, Richard Keynes chose to quote a work of fiction in stating that Darwin, in writing to his father from Maldonado to put the proposal of employing Covington to him, wrote "The man is willing to be my servant & *all* the expenses would be under sixty £ per annum, so that it being hopeless from time to time to write for permission I have come to the conclusion you would allow me this expense."[102]

Covington's subsequent help in collecting and preserving bird specimens was invaluable, so much so that he performed the great majority of this work after 1833. Moreover, he continued to work for Darwin in London after the voyage. Covington was even to send specimens of barnacles to Charles from Australia in 1839, where he had moved to for his remaining life, three years after the *Beagle* voyage had ended. Darwin and Covington were not, however, the only bird collectors aboard the *Beagle*, as Captain FitzRoy, the steward Harry Fuller, and the clerk Edward Hellyer also collected the odd specimen. Hellyer was to die attempting to retrieve a duck that he had shot from the Falkland Islands coast as it floated just offshore directly above a bed of giant kelp that entangled the poor man and drowned him. Syms Covington, whom Darwin did not find particularly likable, collected the odd bird now and then specifically for his own personal collection and disposal.

In chapter 2, I return us to Darwin's immediate future—to his initial ornithological observations of the voyage of the *Beagle*, in the Atlantic ocean and then onward to South America and beyond.

NOTES

1. E. Darwin 1803.
2. F. Darwin 1950: 13.
3. Desmond and Moore 1991: 13.
4. F. Darwin 1950: 15.
5. F. Darwin 1950: 74.
6. In Secord 2010: 353–4.
7. F. Darwin 1950: 15–6.
8. Farber 1982; Walters 2003.
9. Audubon 1827–1838.
10. Desmond and Moore 1991: 16.
11. F. Darwin 1950: 18.
12. White 1789.
13. F. Darwin 1950: 18.
14. Farber 1982.
15. Brent 1981: 40–1.
16. See Clements 2009: 11; Desmond and Moore 2009.
17. Darwin's Diary: 129, in Browne 1995: 65–6.
18. In Secord 2010: 374.
19. Browne 1995: 78.
20. Buffon 1770–1783.
21. White 1789; Bewick 1797–1804; Montagu 1802, 1813.

22. D'Aubenton 1771–1786.
23. Latham 1781–1785, 1787–1801, 1821–1828.
24. Farber 1982.
25. MacGillivray 1837–1852.
26. F. Darwin 1950: 17.
27. Desmond and Moore 1991: 29, 62, respectively.
28. Wyhe 2014: 45.
29. J. van Wyhe, personal communication, email, December 2015, J. van Wyhe–C. B. Frith.
30. Walters 2003: 9.
31. Wyhe 2014: 19, 26.
32. Secord 2010: 375.
33. F. Darwin 1950: 22.
34. E.g., Burkhardt et al. 1985–2014, vol. 1: 104–5, 117.
35. Huxley and Kettlewell 1974: 16.
36. Wallace in Berry 2002: 67–8.
37. Wyhe 2014: 82.
38. Humboldt 1805–1837.
39. Keynes 2003: 17.
40. F. Darwin 1950: 24.
41. Desmond and Moore 2009: 59.
42. F. Darwin 1950: 81.
43. Thomson 1995: 137, 143–4.
44. Desmond and Moore 1991: 109.
45. Yarrell 1837–1843.
46. Pauly 2004: xvii and 7, respectively.
47. In Desmond and Moore 1991: 13.
48. C. Darwin 1881a.
49. In Brent 1981: 76.
50. Wallace 1883: 444.
51. Darwin 1839b: 63.
52. Nichols 2003: 134.
53. Thomson 1995: 144.
54. Secord 2010: xi.
55. Nichols 2003: 144.
56. Burkhardt et al. 1985–2014, vol. 1: 176–7.
57. Cf. Keynes 1979, in Thomson 1995: 156.
58. E.g., Brent 1981: 107; S. Gould 1991: 30–31; Clements 2009: 32.
59. Thomson 1995; Thomson and Rachootin 1982; Wyhe 2013.
60. Wyhe 2013: 319, 324.
61. CD in Desmond and Moore 1991: 105.
62. Wyhe 2013: 4, 9.
63. Desmond and Moore 1991: 150.
64. John Wyhe 2013.
65. Nichols 2003: 3–4.
66. Gribbin and Gribbin 2003: 14, 283.
67. Clements 2009: 32.
68. Willughby 1676.
69. Brisson 1760.
70. Linnaeus 1758, 1766.

71. Buffon 1770–1783.
72. Walters 2003: 10.
73. In Smith 2006: 92.
74. Brisson 1760.
75. Buffon 1770–1783.
76. Forster 1778.
77. Labillardière 1799–1800.
78. Humboldt 1814–1829.
79. Molina 1809.
80. Pallas 1811.
81. Cuvier 1817, 1829, 1816–1845.
82. Wied-Neuwied 1820–1821.
83. Burchell 1822–1824.
84. Saint-Vincent et al. 1822–1831.
85. Guoy and Gaimard 1824.
86. Spix and Martius 1824.
87. Griffith et al. 1828.
88. King 1827.
89. Lesson 1828.
90. Latham 1821–1828.
91. Spix 1824–1825.
92. Blainville et al. 1834.
93. Audebert and Vieillot 1802.
94. Burkhardt et al. 1985–2014, vol. 1: 553–4; Grant and Estes 2009: 68.
95. Steinheimer 2004: 303.
96. Sulloway 1982b: 9–10.
97. Molina 1809.
98. Barlow 1963: 216, 256.
99. Burkhardt et al. 1985–2014, vol. 1: 350.
100. Desmond and Moore 1991: 138, 692.
101. Browne 1995: 228.
102. McDonald 1998, in Keynes 2003: 146.

CHAPTER 2

The Voyaging Bird Observer

The number, tameness, and disgusting habits of these birds, make them pre-eminently striking to any one accustomed only to the birds of Northern Europe. [C. Darwin 1839b: 63]

During the homeward part of their circumnavigation of the globe aboard the HMS *Beagle*, Captain Robert FitzRoy asked his companion Charles Darwin to contribute texts—based on his diary, field notes, and observations made during the voyage—to be incorporated into an account of the voyage by FitzRoy once they were settled back in England. This publication was to be the official results of the *Beagle* voyage for the British Admiralty. FitzRoy was formally obliged by his service to produce a work based on the ship's log, the results of the Admiralty-ordered surveys, and his own and his officers' notes and observations. As this commitment of Fitzroy's also involved his editing a volume based on the account of the first surveying voyage of the *Beagle* (sailing together with the HMS *Adventure*), under Captain Phillip Parker King, the naturalist's account by Darwin would enrich the whole project. Charles was not happy about his work being used in this way but felt obliged to accept it as his fate, given his debt to the captain and the navy for his part in the voyage. Most fortunately for him, however, once back in England for some time, FitzRoy changed his mind and suggested that Darwin's contribution constitute a separate, third, volume on the results of the two *Beagle* voyages.

While Darwin completed his volume for FitzRoy and the Royal Navy promptly in September of 1837, less than a year after his return, the captain, after an initial burst of writing, proved to drag his quill most frustratingly for Charles; and although his own manuscript went to the printers long before FitzRoy finished his, it languished there for over a year until the captain caught up.

Unfortunately, Darwin deeply offended FitzRoy with a grossly inadequate acknowledgement, as seen in the proof pages of his preface to his *Journal and Remarks*, of the generous hospitality and assistance that he had received from the captain and officers of the *Beagle* throughout the voyage. Although Darwin was able to put this to rights in both his *Journal* and his *Zoology of the Beagle*, this initial and careless oversight by Charles opened a wound that was never to fully heal.

Charles's volume eventually appeared as the unimpressively titled *Journal and Remarks, 1832–1836*[1] as the third volume of the *Narrative* of the expeditions.[2] Fortunately for Darwin, his *Journal and Remarks* was reissued in its own right, as a single independent volume under the title *Journal of Researches into the Geology and Natural History of the Various Countries Visited by H.M.S. Beagle* (hereafter *Journal of Researches*), several months later.[3] The serious schism, after such a long good relationship, between the captain and his shipboard naturalist—if at times fraught by tempestuous outbursts by FitzRoy—was doubtless not helped by the fact that Darwin's third volume of the official narrative set met with critical acclaim, with 1,337 copies sold over the first three years.[4] As the *Journal of Researches* it also enjoyed subsequent success, with new editions and reprints appearing, and the book remains in print to this day. FitzRoy's work, which included some rigid interpretations of biblical texts, received, for the most part, an at best lukewarm reception.

Darwin's *Journal and Remarks*, or *Journal of Researches*, initially released in June or July of 1839,[5] was his first completed publication with any significant ornithological content. Whereas this appeared some 32 months after his return to England he had written it in only six[6] or seven[7] months while continuing to oversee the publication of the multi volume work on *The Zoology of the Voyage of the Beagle* (of which two parts of the birds volume appeared before the *Journal of Researches* did). The ornithological content of Darwin's *Journal of Researches*, or what subsequently became *A Naturalist's Voyage*, commenced with observations of a kingfisher in the Atlantic Ocean.

ATLANTIC ISLAND BIRDS

It is significant that the first ornithological observation published by Charles Darwin, in his 1839a,b *Journal of Researches*, shows that he had become a close observer of British birds; that text involves basic comparative observations on the appearance and ecology of the Common Kingfisher with that of the Grey-headed Kingfisher (*Halcyon leucocephala*). He sighted the latter bird as an entirely new to species to him in life near Porto Praya on the island of St. Jago (now Praia on São Tiago) in the Cape Verde Islands, where the *Beagle* remained for three weeks. He recorded of it "The commonest bird is a kingfisher (*Dacelo jagoensis*), which tamely sits on the branches of the castor-oil plant, and thence darts on the grasshoppers and lizards. It is brightly coloured, but not so beautiful as the European species: in its flight, manners, and place of habitation, which is generally in the driest valleys, there is also a wide difference."[8] Thus on but the second page of his first published book, Darwin notes not only the tameness of an island-dwelling bird (a subject about which he was to write much in future years—see following and chapter 3) observed in January 1832, but also something of its habits, feeding ecology, and habitat in a comparative way. The comparative vein of his observations involve the widespread Common Kingfisher of Great Britain (where CD was obviously familiar with it), the Palaearctic, Asia, and to the Solomon Islands. The Common Kingfisher is not the most abundant or easily seen of British birds; and Darwin's familiarity with it doubtless reflects the many hours he spent walking along and fishing rivers, such as the one below The Mount. It is noteworthy that David and Mary Bannerman, a late and great married couple of British ornithological fame, wrote of Darwin's preceding quoted observations "We were to prove all these statements one hundred and thirty-four years later when we were in the same dry but now highly cultivated valley of San Domingos which was to impress Darwin as 'possessing a beauty totally unexpected from the prevalent gloomy character of the rest of the island'."[9]

On the very next page of his *Journal of Researches* book, Darwin writes of the Cape Verde Islands: "Near Fuentes we saw a large flock of guinea-fowl—probably fifty or sixty in number. They were extremely wary, and could not be approached. They avoided us, like partridges on a rainy day in September, running with their heads cocked up; and if pursued, they readily took to the wing."[10] These were in fact Helmeted Guineafowl (*Numida meleagris*), an African bird that had been introduced and became established there as a feral species. This is a highly sociable species of medium-large ground bird that typically associates in flocks. Darwin's words might suggest that he and his companions had been trying to approach these birds close enough to shoot them, perhaps for the *Beagle* mess table, but just possibly also for the preparation of a skin or two for his initial natural history collection. His observation that they avoided him "like partridges on a rainy day in September" is a delightful reflection on the great enthusiasm for, and knowledge of, game shooting he had gained in Great Britain prior to boarding the *Beagle*.

On 16 February, the *Beagle* hove-to near St. Paul's Rocks, immediately north of the Equator in the mid-Atlantic between the easternmost tip of South America and westernmost point of Africa. The only two bird species

Darwin observed on St. Paul's he described as a "booby," which is a kind of gannet, and a "noddy," a kind of tern. These would most likely have actually been the Brown Booby (*Sula leucogaster*) and the Brown (*Anous stolidus*) or Black (*A. minutes*) Noddy. He found them to be so tame that he "could have killed any number of them with my geological hammer" and anthropomorphically described them as being of "stupid disposition."[11] Later in life, fully aware of the evolutionary and behavioural significance of such "tameness" in birds on remote and predator-free islands, he would probably have preferred to apply a word such as "naive" rather than "stupid" for this unexpected bird behaviour.

In the Abrolhos Islands in March, Darwin recorded that species of "Gannets, Tropic birds & Frigates" that he did not further identify were "exceedingly abundant."[12] These birds could potentially have been any or all of the Masked, Red-footed, or Brown Boobies (*Sula dactylatra, S. sula*, and *S. leucogaster*); Red-billed or White-tailed Tropicbirds (*Phaethon aethereus* and *P. lepturus*); and Magnificent (*Fregata magnificus*) Great (*F. minor*) or Lesser (*F. ariel*) Frigatebirds, respectively.[13] All would have been a delightful new experience to him unless seen previously at sea.

SOUTH AMERICAN BIRDS

A few days after arriving at Rio de Janeiro, on 8 April 1832, Darwin first entered what he described as the South American "woods" on a "powerfully hot" day to find it motionless except for large butterflies fluttering about. He later entered what he described as a forest, "which in the grandeur of all its parts could not be exceeded."[14] He found, and was to always find, birds rather difficult to see in wet forest on this continent, and his expressed disappointment in this might be seen as reflecting his active interest in birds. It has been stated that in all his time in Brazil, CD collected but only one bird,[15] but he in fact collected specimens of some additional half a dozen species there (see appendix 3).

On his second day of travel out of Rio de Janeiro, near Mandetiba, north of Cape Frio, on 9 April, CD noted "beautiful fishing birds, such as egrets and cranes" near the coast.[16] It is difficult to be sure which egret species he in fact saw, but it could have been one or both of the Great White Egret (*Egretta alba*) and/or Snowy (*E. thula*) Egret. However, the only cranes in the Americas are the Sandhill (*Grus canadensis*) and Whooping (*G. americana*) Crane, which do not occur south of the areas of Houston, Texas, and Cuba, respectively. This was an understandable mistake given that time in ornithological history and that CD had never seen a living wild crane or stork, unless in captivity in England. Darwin's use of the name crane therefore reflects either a looser application of it by him than is used today or his ornithological inexperience in mistaking a larger heron or stork for a crane, not that anyone at that time could necessarily have confidently known that no crane species inhabits South America.

During a walk to "Gavia, or topsail mountain" Bahia, north of Rio de Janeiro, Darwin found that the movements and habits of hummingbirds reminded him of sphinx moths: "Whenever I saw these little creatures buzzing round a flower, with their wings vibrating so rapidly as to be scarcely visible, I was reminded of the sphinx moths, their movements and habits are indeed, in many respects, very similar."[17] Here, then, is another early comparative ecological and behavioural observation by CD, which actually involves an example of what would come to be known by subsequent evolutionary biologists as "convergent evolution." This type of "convergence" occurs where and when two unrelated species deal with a similar biological issue, in this case, the ecological one of feeding on flower nectar from flowers too frail to perch on by evolving a similar adaptation: the adaptation in this case being wings of a particular shape and ability, given the associated evolution of other anatomical adaptations to control them, to enable static hovering flight while feeding on flower nectar.

Darwin had remained at Rio, renting a cottage on the coast where he busied himself with natural history, before and while Captain FitzRoy took the *Beagle* back northward along the coast to Bahia to repeat some navigational calculations. In this Charles was probably extremely fortunate because just before going back to Bahia, eight crew members had hunted snipe on some local wetlands. There they contracted malaria, and as a consequence, the *Beagle* returned to Rio with three less crewmen who had died as a result of contracting the disease while hunting snipe at Rio. Had Charles joined them, he may too have contracted malaria.

Actually, Darwin recorded little of bird life during the first year of the voyage; but during the next two years, in and about South America, his ornithological recording not only increased but also included taxonomic, ecological, and behavioural observations and interpretations;[18] see appendix 1. It is remarkable to see that in his field notebooks, he would often quickly note no more than a bird's name, or a name with only a couple of additional words.[19] From these "bare bones" he could, however, obviously recall events well enough to flesh them out, as he did in both his *Journal of Researches* and *The Zoology of the Beagle*. It is clear, too, that he corrected some of his impressions gained in the field once able to communicate with more experienced bird people, such as John Gould, back in England. For example, in his "Rio Notebook," he recorded seeing "Toucans & bee eaters" on 15 April 1832.[20] As no bee-eaters occur in the Americas, CD must have mistakenly identified birds that are generally similar in appearance and behaviour to them, such as the jacamars of the not distantly related non-passerine or non-perching, bird order Piciformes (woodpeckers and relatives). In all probability, he would not have had any personal experience with living wild bee-eaters, the European species being very rare in Britain and in the Cape Verde Islands where he failed to note it. He would have therefore identified his Rio birds from the literature available to him at the time, or from the memory of book illustrations or specimens examined by him in libraries and collections back in England. With the benefit of expert advice, such as that of John Gould, he was able to avoid including his few avian misidentifications in his publications.

On the third day of horse riding out of Maldonado, where he spent 10 weeks, CD saw many "ostriches (*Struthio Rhea*)" and some "flocks contained as many as twenty or thirty birds [Greater Rheas (*Rhea americana*)]. These, when standing on any little eminence, and seen against the clear sky, presented a very noble appearance. I never met with such tame ostriches in any other part of the country: it was easy to gallop up within a short distance of them; but then, expanding their wings, they made all sail right before the wind [but see "Encounters with Rheas" following], and soon left the horse astern."[21] We can see here that CD had by this time become disposed to use the odd nautical turn of phrase. He noted that the gauchos use a simple form of the bolas for catching rheas by throwing them at their legs, from horseback.

Over the next two days of horse riding, Darwin also saw "great numbers of partridges (*Tinamus rufescens*). These birds do not go in coveys, nor do they conceal themselves like the English kind. It appears a very silly bird. A man on horseback by riding round and round in a circle, or rather in a spire, so as to approach closer each time, may knock on the head as many as he pleases. The more common method is to catch them with a running noose, or little lazo, made of the stem of an ostrich's feather, fastened to the end of a long stick. A boy on a quiet old horse will frequently thus catch thirty or forty in a day. The flesh of this bird, when cooked, is delicately white."[22] The bird concerned is now called the Red-winged Tinamou (*Rhynchotus rufescens*), and it is not at all related to the true partridges but is one of the tinamou family Tinamidae, many members of which are partridge-sized, whereas others are smaller or larger. Contrary to CD's perfectly understandable perception of the relationship between partridges and tinamous at the time, the two groups are actually members of entirely different avian orders. Thus his observation that tinamou behaviour was significantly different to that of partridges is not at all surprising in the light of today's appreciation of avian taxonomy. The latter indicates that tinamous are among the most primitive of birds in being most closely related to the ratites and are thus close relatives of the ostriches, rheas, cassowaries, Emu, and kiwis rather than being close to any other group of birds.[23]

While at Maldonado, Darwin paid particular attention to collecting birds and mammals and within a "morning's walk" of the place, he procured 80 bird species "of which many were exceedingly beautiful—I think even more so than those in Brazil."[24] He made observations on various songbirds, known to form the vast majority of the passerine birds of the order Passeriformes, including American blackbirds, cowbirds, mockingbirds, and tyrant flycatchers; and he also summarized his knowledge of some birds of prey. Notable among the latter were species of caracara and vulture, including the great Andean Condor (*Vultur gryphus*).

Writing of birds "extremely abundant on the undulating grassy plains around Maldonado," on the northern bank of the La Plata River, Darwin observed that "Le Troupiale commin" (the Shiny Cowbird [*Molothrus bonariensis*]) feeds in large flocks with other birds; perches on the backs of cows and horses; and while preening in the sun produces a hissing call. He noted that the species is reported to deposit it's eggs in other birds' nest, "like the cuckoo"; and that one authority remarked that except for *Molothrus pecoris* (now the Brown-Headed Cowbird [*Molothrus ater*] of North America) the cuckoos are the only birds that can truly be termed parasitical; perhaps true then but not today (see chapters 3, 4, and 8). Although closely allied to the Brown-Headed Cowbird, the species he observed on the plains of the La Plata only differed by being larger and in the colour of its plumage, but they were obviously of a distinct species in CD's eyes. Darwin was intrigued to find that both the passerine cowbirds and the non-passerine cuckoos, although different in almost every other habit, lay their eggs in the nests of other species, this being another example of convergent evolution.[25]

The thrush-sized Great Kiskadee (*Pitangus sulphuratus*) was, under the name *Saurophagus sulphureus* by Darwin, noted as typical of the American tyrant-flycatchers and prominent on the plains. He observed that it is structurally like the true shrikes of the Old World (except two species in North America) but in habits was like several bird species. He described its hovering and stooping foraging technique as being like that of some birds of prey, for example, the kestrels of the genus *Falco*. At other times he saw it foraging at the edge of water, like a kingfisher. Its habits and calls were described by CD, who interestingly noted that individuals of the species were kept in cages or "in courtyards with their wings cut" and became tame as amusing pets.[26] As CD's observations implied, it eats a particularly broad range of prey. This bird is the best known and most popular species in Brazil, in part because of its tame familiarity with people, which is in stark contrast to its intense aggression towards other birds and animals.[27]

At this point, Charles also first discussed in print two South American mainland birds, noting, if briefly, their remarkable but slightly different vocal abilities and habits; and John Gould subsequently assured him that despite their similarity in appearance, they constitute two distinct species of mockingbird as their different vocalizations suggested.[28] The mockingbirds of the

Galapagos Islands were, however, to subsequently become of far greater significance to the embryonic but then developing theory of the origin of species in Darwin's mind;[29] see "The Zoology of the Voyage of the Beagle" section of chapter 3.

The caracaras are seen as primitive, mostly carrion eating, falcons; and of the 11 species of them, Darwin appears to have observed the Mountain (*Phalcoboenus megalopterus*), White-Throated (*P. albogularis*), Striated (*P. australis*; colour figure 3.2), Southern Crested (*Polyborus plancus*), Yellow-Headed (*Milvago chimachima*), and Chimango (*M. chimingo*) Caracara. He noted that these birds fill the ecological niches of "carrion crows, magpies, and ravens; a tribe of birds which is totally wanting in South America." In this he was not strictly correct, however, as some 15 species of jays occur in the Neotropics, which are members of the "crow tribe" or the same family, the Corvidae, if not feeding like the more typical corvids as CD was possibly implying. Nevertheless, by his statement, Darwin was unconsciously intimating convergent evolution in the ecology of these two groups of species. He observed aggressive interactions between sympatric caracara species at sources of food and that these opportunist birds would eat the scabs off sores on horses' backs. They would follow hunting parties of people and dogs during the day and perch hopefully about a sleeping man to see if he failed to wake, and scavenge high-tide marks for beach-washed carrion—thus generally being vulture-like in habits. One or more of them will attack and kill injured animals. Several examples of the crafty and inquisitive nature of these birds are cited by CD as well as their distributions, habitats, and general habits, with nesting in a couple of species being detailed.

In marked contrast to a number of instances in which Darwin insightfully notes that unrelated bird species or groups show the same adaptive behaviour and associated morphology in completely different parts of the world, that is, demonstrating convergent evolution, he suggested that the walking and running of caracaras might be indicative of "an obscure relationship with the Gallinaceous order."[30] Although perceived as closer in CD's time, the caracaras and their immediate relatives, the falcons, are in fact very far removed from the gallinaceous birds within the non-passerine birds;[31] see also "The Caracaras" section of chapter 3.

In early August 1833, Darwin rode to large salt lakes some 15 miles, or 24 km, from El Carmen or Patagones Town, in the Rio Negro area. He observed that Chilean Flamingos (*Phoenicopterus chilensis*) were "singularly attached to salt lakes" and also that the lakes supported minute crustaceans in great abundance. He even observed, as a footnote, that as flamingos and such crustaceans occur in salt lakes on two continents, "we feel sure they are the necessary results of some common cause"; but he did not consider one being the food of the other but rather thought that flamingos search "probably for the worms which burrow in the mud."[32] In fact, burrowing worms are an unlikely typical food of the Chilean Flamingo, whereas crustaceans constitute their primary diet.[33]

Darwin decided to ride on northward to Bahia Blanca and then on to Buenos Aires. On 12 August, he noted that the plains he rode across were "inhabited by few birds or animals." But he did note that "the little owls of the Pampus (*Noctua cunicularia*) [Burrowing Owl (*Athene cunicularia*)] . . . are obliged to hollow out their own habitations" due to an absence of the burrowing mammal called "Bizcacha (*Calomys bizcacha*)." The latter, now called Vizcaca, is a rabbit-like mammal, which the owl takes advantage of where living together with it by occupying its burrows;[34] see appendix 1. At the end of September 1833, CD recorded the owl as exclusively inhabiting holes of the Bizcacha on the plains of Buenos Aires and detailed some of its life history, including that the owl eats mice and snakes. The Bizcacha has a habit of accumulating objects, including bones, stones, lumps of earth, dry animal dung, and so forth, about the entrance to its burrow. Darwin likened this behaviour to how male Australian Spotted Bowerbirds (*Chlamydera maculatus*) accumulate similar and additional items about the stick avenue bowers that they build to attract females;[35] (see colour figure A.2 of appendix 4). He would go on to discuss bowerbirds in much greater detail in his 1871 book *The Descent of Man and Selection in Relation to Sex*; see "Love-antics and dances" section of chapter 6. Interestingly, Darwin also mentions that an unidentified owl killed in the Chonos Archipelago of Chile had a stomach full of "good-sized crabs" of all things.[36] This remains a novel observation today because the two owls presently recorded on the Chonos Archipelago, the Rufous-Legged Owl (*Strix rufipes*) and Ferruginous, or Austral, Pygmy-Owl (*Glaucidium* [*nanum*] *brasilianum*), are not known to eat crabs;[37] see "New World Owls" section in chapter 3.

ENCOUNTERS WITH RHEAS

In chapter 5 of his *Journal of Researches*, Darwin gives "an account of the habits of some of the most interesting birds" he found to be common on the wild plains between Bahia Blanca and Buenos Aires. He gives details of the range, habits, food, vocalization, sexual dimorphism in size, and wary solitary nature of the Greater Rhea (*Rhea americana*) known to him, including their eating small fish and how local people hunt them. Although "so fleet in its pace, it falls a prey, without much difficulty, to the Indian or Gaucho armed with the bolas. When several horsemen appear in a semicircle, it becomes confounded, and does not know which way to escape. They generally prefer running against the wind; yet at the first start they expand their wings, and like a vessel make all sail." In describing their swimming ability, as well as that of the Emu (*Dromaius novaehollandiae*) of Australia, CD notes that little of their body appears above water; and their neck extends a little forward and their progress is slow. Twice he watched birds swim across the Santa Cruz River at a point "about four hundred yards [438 m] wide, and the stream rapid."[38]

The Gauchos that rode with Darwin showed him how to use the bolas, "two rounded stones, covered with leather, and united by a thin plaited thong, about eight feet long," thrown from horseback in hunting rheas to entangle them about their legs. In one attempt, he managed to get a ball and thong caught about a hind leg of his mount, which was fortunately experienced enough to immediately stop galloping. Whereas this greatly amused his Gaucho companions, Charles appears to have fancied himself as a bolas practitioner. He was sufficiently impressed by his experience of this hunting method to have had a long-haired Gaucho depicted using the bolas to catch a rhea from horseback decorate the front cover of the 1889 edition of his *A Naturalist's Voyage* in gold gilt; see colour figure 2.1.

On 16 September 1833, Darwin was told of a local storm that occurred on the previous evening, of which he had witnessed the distant lightning, involving hail stones "as large as small apples, and extremely hard." The hail was seen to have killed at least 20 deer and some 15 Greater Rheas (part of one of which CD ate), with others blinded in one eye, and numbers "of smaller birds, as ducks, hawks, and partridges [tinamous]" killed.[39] Such localized high mortality might occasionally occur due to wildfire in such a habitat but must surely be relatively rare as a result of falling ice.

At Bahia Blanca in September–October of that year, Darwin found the eggs of Greater Rheas "in extraordinary numbers . . . all over the country," which failed to hatch; and "Out of the four nests which I saw, three contained twenty-two eggs each, and the fourth twenty-seven." The Gauchos unanimously informed him "that the male bird alone hatches the eggs, and for some time afterwards accompanies the young"; and "that several females lay in one nest"; and that four or five hens will go to the same nest in the middle of the day, one after another. He also cites a published reference to the belief that in the Ostrich of Africa, two females lay in the one nest. He notes that it is likely that individual rheas lay large numbers of eggs in a season but that this must involve a long period of time; and he cites one bird in captivity laying 17 eggs, each at three-day intervals. Thus he observes that if individual females had to incubate their own eggs, the first-laid ones would probably be addled before the last was laid. If, however, each female "laid a few eggs at successive periods, in different nests, and several hens, as is stated to be the case, combined together, then the eggs in one collection would be nearly of the same age. If the number of eggs in one of these nests is, as I believe, not greater on an average than the number laid by one female in the season, then there must be as many nests as females, and each cock bird will have its fair share of the labour of incubation; and that [was] during a period when the females could not sit, on account of not having finished laying."[40] It is now known that the breeding biology of the Greater Rhea is in fact more complicated than this, with one study showing that in a given year, only a small proportion of males attempt to breed; and that some 30 percent of females, and only 4.5–6.1 percent of males breed successfully.[41]

Darwin was told of male rheas being fierce and potentially dangerous in defence of a nest, even leaping and kicking at mounted horses. Although what the Gauchos told him about basic Greater Rhea nesting biology during his voyage was subsequently found to be true, he of course had until the writing of his *Journal of Researches* text to consider its significance and to add such considerations to it. He also wrote in his account for August 1833, "I understand that the male emu, in the Zoological Garden, takes charge of the nest: this habit, therefore, is common to the family"; but this was presumably knowledge gained once back in England and then included in his *Journal of Researches*.[42] At that time he was able to observe that the "countless herds of horses, cattle, and sheep" had utterly altered the vegetation and ecology of the pampas and had "almost banished the guanaco, deer and ostrich."[43]

Following the preceding, Darwin gave an account of the discovery of the Lesser, or Darwin's, Rhea (*Rhea pennata*), which the Gauchos at Rio Negro had repeatedly told him about. They described both the bird and its eggs as being distinct from the larger Greater Rhea. The first of this smaller rhea species that he saw was shot by Conrad Martens, replacement artist aboard the *Beagle*, at Port Desire, Southern Patagonia, sometime during 23 December 1833 to 4 January 1834. Charles did not then appreciate that he was looking at a specimen new to science, as he initially thought it was a two-thirds grown Greater Rhea; and so the bird was promptly cooked and partly eaten before he remembered the smaller type of rhea of the Gauchos' conversations. Luckily the head, neck, legs, wings, many larger feathers, and a large part of the skin were preserved, and from these a "very nearly perfect specimen has been put together" (by John Gould), which was exhibited in the Zoological Society museum in London.

Darwin's notebook entry for 23 April 1834, suggests that this is when he first saw a living Lesser Rhea,[44] which would have been on an excursion up the Santa Cruz River. A Patagonian half Indian man in the Straits of Magellan informed CD that only the Lesser Rhea occupied the southern areas and that their nests typically contain no more than 15 eggs and that these are laid in the nest by more than one female.[45] In the latter detail about females he was correct, but in fact they may collectively lay up to 50 eggs in a single male's nest.[46] In writing to John Henslow in March 1834, from the East Falkland Islands, CD reported that he had acquired the pieces of the smaller rhea and observed that "The differences [from the Greater Rhea] are chiefly in color [*sic*] of feathers & scales on legs, being feathered below the knees; nidification & geographical distribution."[47]

In describing the new smaller rhea species for science, John Gould named it *Rhea darwinii* after Darwin, but as it had previously been described and alluded to as *Rhea pennata* (but not collected), this latter name had priority as the original one and is thus the official scientific name under the rules of zoological nomenclature.[48] All populations of the Lesser Rhea are now in serious decline due to hunting and habitat loss, with its northern subspecies in danger of extinction.[49]

The mutually exclusive distributions of the two species of rhea on the South American mainland taunted CD's mind as potentially indicating something profound about the speciation of vertebrates; but he could not at the time put his finger on what it was. This was well in advance of what was to strike him about island speciation, in mockingbirds, on the Galapagos Archipelago;[50] see the "Mockingbirds" section in chapter 3. For more about these ratites, see "The Rheas" section of chapter 3.

SEEDSNIPE AND OVENBIRDS

On first encountering seedsnipe, Darwin wrote to John Henslow at Cambridge, in self-depreciation, "There is a poor specimen of a bird, which to my unornithological eyes, appears to be a happy mixture of a lark pidgeon [*sic*] & snipe. . . . I suppose it will turn out to be some well-known bird although it has quite baffled me."[51]

The Least Seedsnipe (*Thinocorus rumicivorus*) was described by Darwin as "A very singular little bird. . . . It nearly equally partakes of the character of a quail and a snipe." He details its appearance, range, habitat, and habits. Having observed that in structure, the seed snipe "has a close affinity with quails" (presumably the true quails), he went on to state that the bird in flight, with "the long, pointed wings, so different from those in the gallinaceous order, the irregular manner of flight, and plaintive cry uttered at the moment of rising, recal [*sic*] the idea of a snipe." Indeed, the "sportsmen of the *Beagle* unanimously called it the shortbilled snipe."[52] In fact, the four species of seedsnipe, constituting the family Thinocoridae, are very far removed from the quails: Quails belong to the family Phasianidae, which makes them members of the basal or lower non-passerine order Galliformes, and thus closely related to the grouse, pheasants, guineafowls, and so forth. Seedsnipe, on the other hand, are only one to three families removed from the true snipe within the same non-passerine order, the Charadriiformes, and are thus closest to the waders and related birds;[53] see the "Seedsnipe" section of chapter 3 and "Charles Darwin as an Ornithologist" section of chapter 8.

A group of conspicuous ground-frequenting songbirds that Darwin could hardly have overlooked were then known to him as members of the genus *Furnarius*; this genus now consisting of six species of hornero, of the family Furnariidae. These fascinating birds were once known as ovenbirds, and their Latin name *Furnarius* means "baker" or "one who keeps an oven."[54] This alludes to the remarkable mud nests that they build and that then become baked by the sun into rock-hard terracotta-like structures impenetrable to most predators. Thus CD was able to note that locals knew the best-known species, the Rufous Hornero (*Furnarius rufus*), as the *Casara* or housemaker. Its mud and straw nests are built in the most exposed of sunlit situations, atop a post, rock, or cactus.

What Darwin thought to be another species of *Furnarius* was in fact the Common Miner (*Geositta cunicularia*), which is a member of the same family. He noted that the Spanish named it *Casarita*, and he described how it placed its nest at the bottom of a long cylindrical hole, said to extend horizontally into a bank to "nearly six feet [2 m] under ground."[55] For more about these birds, see the "Ovenbirds and Tapaculos" section of chapter 3.

STILTS, LAPWINGS, SKIMMERS, AND IBIS

The White-backed Stilt (*Himantopus melanurus*), described by Darwin as "a kind of plover" that appeared to be "mounted on stilts," was common in large flocks on the pampas between Bahia Blanca and Buenos Ayres. Stilts are not plovers; but in sharing the family Recurvirostridae with the avocets, that is placed next to the plover family Charadriidae, they are most closely related. Contrary to some authors, who had described the bird's wading gait as inelegant, Charles thought its movements "far from awkward." He found that their chorus of calls sounded like "the cry of a pack of small dogs in full chase" and would hear this on waking at night. Another bird that would disturb his slumber was indeed a plover—locally called the *teru-tero*, or Southern Lapwing (*Vanellus chilensis*), named after the sound of its loud call. In appearance and habits, this bird, like most of the lapwings, resembles the peewit as Darwin knew it in England (i.e., the Northern Lapwing [*Vanellus vanellus*]) but is armed with a sharp spur on the bend of each wing. Darwin had these birds constantly pursue him and his companions as they rode the grassy plains; and by this, I suspect that he would have meant mostly by one angry territorial pair after another because pairs of lapwings are notoriously noisy and aggressive in harassing potential predators within their territories. This behaviour is most annoying to hunters, and bird photographers, because the screaming birds notify all birds and other animal life in their area of approaching potential danger. This aggressive behaviour makes the species one of the most appreciated by Brazilian ranchers. Individual birds, denied flight by the removal of some of their feathers, are raised to become avian guards that loudly announce the approach of other people.[56]

Darwin reported Southern Lapwings to feign injury by performing a "distraction display" with an apparent broken wing to draw potential predators away from the immediate area of their nest, eggs, or young, their eggs being much esteemed as a delicacy by local people.[57]

During mid-October, near Punta Gorda on the Parana River, Darwin observed tern-like Black Skimmers (*Rynchops niger*) flying powerfully only just above the water while creating a furrow in it by holding the tip of their lower mandible into the surface. When the lower mandible hits a fish, the upper one instantly snaps closed to grasp and hold it in the bird's beak before being swallowed. Having seen flocks of these birds flying out to sea in the evenings, CD concluded that they fish at night "at which time many of the lower animals come most abundantly to the surface."[58] Given that their feeding technique is a tactile one, and thus not necessarily visual, Charles's conclusion was logical and correct: and skimmers do indeed also feed crepuscularly and nocturnally. A "M. Lesson" [*sic*] (in error for René-Primevère Lesson) wrote that these birds open shellfish buried in coastal sand banks,[59] but CD politely considered it "very improbable that this can be a general habit." He possibly thought, as this author does, that Lesson was confusing skimmers with oystercatchers, as the three South American oystercatcher species also have long bright red beaks not unlike that of the skimmer.[60] See also the "Skimmers" section of chapter 3.

Darwin observed only three other birds on the Parana River "whose habits are worth mentioning," which he does but briefly. These are the Green Kingfisher (*Chloroceryle americana*), Monk Parakeet (*Myiopsitta monachus*), and the Swallow-tailed Flycatcher (*Tyrannus forficatus*);[61] see appendix 1. The Monk Parakeet is the only stick nest-building parrot, as all 357 other species nest in holes.

At the end of December 1833, Darwin noted that the Black-Faced Ibis (*Theristicus melanopsis*) was not uncommon in most desert parts of Patagonia he visited, but he incorrectly wrote of it being "a species said to be found in central Africa." It is in fact endemic (i.e., peculiar, or unique) to South America, and CD might therefore have had the Olive Ibis (*Bostrychia olivacea*) of central Africa in mind because it has a similarly distinct and discrete black face. This error possibly suggests that he consulted books aboard the *Beagle* at the time that contained descriptions or illustrations of the Olive Ibis. He found that the food of the handsome Black-Faced Ibises included grasshoppers, cicadas, small lizards, and scorpions; that they seasonally flock and get about in pairs; and that they have a very loud cry like the "neighing of the guanaco";[62] see colour figure 2.3.

CONDORS AND OTHER VULTURES

On 27 April 1834, Darwin shot an Andean Condor (*Vultur gryphus*) on the Santa Cruz River, which he described and then went on to write of the range, habitats, habits, soaring, and growth of individuals of the species. He noted that they require high vertical cliffs to take flight and soar from, where pairs or groups sleep and breed; and he concluded, from what he could observe of them, that they breed once every two years, this being subsequently confirmed to be true. The country folk of Chile informed him they make no nest, and he personally saw none on ledges that young birds stood on; lay two white eggs on a bare rock shelf in November and December; and the young cannot fly for their first full year of life. In this he was misinformed, however, as we now know that a pair lays but a single egg clutch during each breeding attempt[63] and that the young are capable of flying at six months of age.[64] Once fledged and flying as well as adults do, the immature birds continue to roost with their parents on the same rock ledges and forage with them by day, but will often forage independently before their conspicuous ruff of longer neck feathers becomes white.

Darwin considered that the condors on the cliffs of Santa Cruz fed together on guanacos dying natural deaths or having been killed by pumas. He reported that, in addition to eating carrion, these condors attacked young goats and lambs, and so the shepherds trained their dogs to bark at the sight of them. Of course the shepherds killed condors when they could by baiting them and then riding them down on horseback or by marking their roosts during the daytime to then return at night to climb to and noose the birds resting there. Charles could not recall having ever seen a condor flap its wings in flight except to initially rise from the ground. He often saw them soaring in circles at great heights, and he recalled coming suddenly to the brow of a precipice to see 20 to 30 condors "start heavily from their resting-place, and wheel away in majestic circles."[65] This must have been a magnificent experience to have had in the wild, with so many of one of the world's most impressive large birds fleeing and flying together.

The condors, like typical or Old World vultures, which they are certainly not immediate relatives of (see chapter 3), quickly attend a corpse to feed on it and often will strip a skeleton of its meat *before* it starts to smell. Captain FitzRoy wrote of two guanacos being shot by Darwin and Stokes and covered by them with bushes so they could return to the boats to get help; but by the time they returned, four hours later, condors had "eaten every morsel of the flesh of one animal."[66]

Andean Condor populations are declining; and despite the bird being admired and considered semi-sacred, it is killed because of its alleged attacks on domestic stock and is abused for various festivals involving cruelty to, and the death of, birds. Its cultural significance to people is documented back to 2500 B.C.[67]

Darwin recalled that John James Audubon[68] maintained that the vultures of the Americas had little sense of smell, an opinion agreed with by some but not all ornithologists at the time,[69] so he carried out a simplistic experiment with a line of captive condors tied on the ground by their legs in a Lima market. Darwin is said to have wrapped some meat in white paper and presented it closely to the condors and found that they ignored it until able to touch the package with their bill, when they then immediately tore it open to eat the meat. Physician George Paget recalled, however, that Darwin subsequently recited to him over a meal at Christ's College that he "walked before them and past them [the condors] with a large piece of meat in his pocket. The birds took no notice whatever. He then threw within their reach the piece of meat wrapped up in paper. Still the birds took no notice. But when with the end of a stick, he uncovered a part of the meat they instantly rushed to seize it."[70] As a result of Audubon's assertion that the Turkey Vulture *Cathartes aura* could smell little, Richard Owen[71] dissected a specimen and found obvious olfactory nerves indicative of an acute sense of smell in the species, and Darwin noted this.

As a result of Owen reading his notes on the subject at the Zoological Society, it was reported that the smell of human bodies left too long in a West Indian home attracted Turkey Vultures that could not possibly have seen them. This led Darwin to write that it would appear carrion feeding hawks "posses [*sic*] both the sense of sight and smell in an eminent degree."[72] We now know that the basic field experiments that Audubon performed misled him into thinking that American vultures cannot smell because he used flesh that was far too putrid—for these birds avoid such strongly contaminated carrion if at all possible. Also of fundamental significance here is that although Audubon stated that he had experimented on the olfactory ability of the "turkey buzzard," he apparently actually did so on the Black Vulture (*Coragyps atratus*), which is far less sensitive to smell than is the Turkey Vulture.[73]

To test the ability of the Turkey Vulture to smell carrion, chicken carcasses were more recently laid out in tropical forest in Panama and then closely observed. It was found that the vultures were best at finding the chickens when dead for one full day but that they rejected them when they became rotten. The results of the study indicated that the vultures relied "almost entirely on the sense off [*sic*] smell to locate food."[74] The interesting, and surprising, finding is that although the three *Cathartes* "turkey" vultures do indeed have a keen sense of smell, as was indicated by Owen's work, the condors, King Vulture, and American Black Vulture do not.[75] Thus, FitzRoy's observation of condors and dead guanacos—which in any event may not have been covered well enough with bushes to prevent them being seen by the birds—the negative result of CD's experiment with Andean Condors, and Audubon's with the Black Vulture, are not surprising. See "New World Vultures" of chapter 3.

GEESE, DUCKS, WOODLAND BIRDS, AND KELP

The *Beagle* reached the Falkland Islands on 4 March 1833. Darwin described several goose species in the east of these islands, including the Upland and Kelp

Geese. He found the Upland Goose (*Chloephaga picta*) to be a common resident in pairs and small flocks but that it nested on small outlying islets where "foxes" could not reach them. The "foxes" were actually the Falkland Island Wolf or Dog, Antarctic Wolf, or Warrah (*Dusicyon australis*). He noted at that time that "As far as I am aware, there is no other instance in any part of the world, of so small a mass of broken land, distant from a continent, possessing so large a quadruped peculiar to itself." He also noted of the Warrah that "Within a very few years after these islands shall have become regularly settled, in all probability this fox will be classed with the dodo, as an animal which has perished from the face of the earth."[76] He was quite right, for sheep farming brought about a bounty offered for every Warrah killed. The last individual animal was shot in 1876. Given that this carnivore would have undoubtedly widely and intensively predated all birds accessible to it, its extinction will have had a profound influence (or, in some respects, perhaps a lack of it) on the ecology in general, and the avifaunas in particular, on those Falkland Islands it used to occur.

Darwin went on to give details of the endemic and remarkable Fuegian Steamer Duck (*Tachyeres pteneres*; see colour figure 2.4). He stated that these ducks, which attain up to 22 lb. or 10 kg in weight, were "very abundant" and were once called "race-horses" because of the way they paddled and splashed on the water while vigorously flapping their evolutionarily reduced wings. Darwin tells of the shellfish diet and foraging technique of this intriguing, heavily billed, duck. He then observed that three South American birds use their wings for bodily movement other than flight: "the penguin as fins, the steamer as paddles, and the ostrich as sails."[77] This is the kind of observation that might have had young Darwin's mind groping towards thoughts of the slow and steady evolution of adaptation as expressed in the limbs of various vertebrates. He subsequently, and rather callously, added that the head of the Fuegian Steamer Duck is so strong that "I have sometimes scarcely been able to fracture it with my geological hammer"![78]

Under his June 1834 *Journal* entry, Darwin discussed the birds of the "gloomy woods" of Tierra del Fuego:

> Occasionally the plaintive note of a white-tufted tyrant-flycatcher [White-crested Elania (*Elania albiceps*)] may be heard, concealed near the summit of the most lofty trees; and more rarely the loud strange cry of a black woodpecker [Magellanic Woodpecker (*Campephilus magellanicus*); see colour figure 2.5], with a fine scarlet crest on its head [this being worn by adult males only]. A little, dusky-coloured wren (Scytalopus fuscus) [Dusky Tapaculo] hops in a skulking manner among the entangled mass of the fallen and decaying trunks. But the creeper (Synallaxis tupinieri) [Thorn-tailed Rayadito (*Aphrastura spinicauda*)] is the commonest bird in the country. Throughout the beech forests, high up

> and low down, in the most gloomy, wet, and impenetrable ravines, it may be met with. This little bird no doubt appears more numerous than it really is, from its habit of following, with seeming curiosity, any person who enters these silent woods; continually uttering a harsh twitter, it flutters from tree to tree, within a few feet of the intruder's face. It is far from wishing for the modest concealment of the true creeper (*Certhia familiaris* [Eurasian Treecreeper]), nor does it, like that bird, run up and down the trunks of trees; but industriously, after the manner of a willow wren [Willow Warbler (*Phylloscopus trochilus*), of Eurasia], hops about, and searches for insects on every twig and branch. In the more open parts three or four species of finches, a thrush, a starling (or Icturus), two furnarii [referring to American blackbirds and ovenbirds], and several hawks and owls occur.[79]

This brief summary by CD includes several ecological observations, comparisons, and insights indicative of a serious ornithological observer and philosopher for his time.

In noting the ecological significance of kelp forest, of the giant rock-anchored seaweed, Darwin wrote of his June 1834 experiences: "Amidst the leaves of this plant numerous species of fish live, which nowhere else would find food or shelter; with their destruction the many cormorants, divers, and other fishing birds, the otters, seals, and porpoises, would soon perish also; and lastly, the Fuegian savage, the miserable lord of this miserable land, would redouble his cannibal feast, decrease in numbers, and perhaps cease to exist."[80] Thus, in discussing kelp, he succinctly expresses conservation sentiment, ecological relationships, and an example of human ecology and population.

On 7 September 1834, while horse riding inland from Santiago, Chile, Darwin telegraphically noted in his "Santiago notebook" seeing a "Goosander in river back white.—belly brown—breast black. Top of head black, beneath white bill, plain cry—runs quick. Very active in the rapids."[81] This description of the bird, although not too accurate and a little ambiguous, together with the described particular activity in rapids, suggests that it could have been an adult male of the Chilean form of the Torrent Duck (*Merganetta armata*). This is a species that was not collected by Darwin or the *Beagle* crew and one not mentioned in Nora Barlow's[82] review of "Darwin's Ornithological Notes" or in CD's published works. It is worthy of mention here, as no name appears to have been given for this bird sighting previously, which is endemic to the length of western South America. Darwin would have used the name Goosander because this is given to a bird of Britain and Europe that is similar in general appearance to a Torrent Duck and presumably was familiar to CD, if not in the wild then from the literature, and is now called the Common Merganser (*Mergus merganser*).

BIRDS OBSERVED IN CHILE

The next local avifauna that Darwin summarized briefly is that of Chile. He observed how Andean Condors will feed on a carcass that had been covered with "many large bushes" by a Puma (*Felis concolor*). This might suggest that the birds could smell the carcass, although their eyesight is most remarkably good (but see "Condors and Other Vultures" previously). Here he noted that adult female condors have bright red eyes and males have yellowish brown ones, and that a young female over a year old had a dark brown eye. Local people told him that captive birds could live for six weeks without food.[83] He noted that a "Sir F. Head [Francis B. Head] said that a Gaucho in the Pampas, upon merely seeing some condors wheeling in the air, cried 'A lion!'," this suggesting that the birds often fed on the prey of the cat.[84]

Perhaps the most conspicuous of all central Chilean birds were two species of the non-passerine tapaculo family, Rhinocryptidae. What Darwin referred to by the local name of "el Turco" is now the Moustached Turca (*Pterophtochos megapodius*) and is as large as a typical thrush but with much longer legs, shorter tail, stronger beak, and with a reddish brown plumage. It is a ground-dweller, living among thickets scattered across "dry and sterile hills." Given its long legs and habit of carrying its tail vertically erect as it scurries from bush to bush in great haste, Darwin found its appearance suggested "the bird is ashamed of itself, and is aware of its most ridiculous figure" and that on first seeing it, one is tempted to humorously exclaim "A vilely stuffed specimen has escaped from some museum, and has come to life again!" Individuals could not be forced to fly without the greatest trouble, and it only hopped rather than ran. Its calls were "as strange as its whole appearance"; and it was said to build its nest in a deep hole beneath the ground, which it does. Charles dissected several to find their muscular gizzards contained beetles, vegetable fibres, and pebbles.[85]

The second species was called the Tapacolo by CD, so called because the name means "cover your posterior" or, according to Darwin, "cover your arse" in the local vernacular—and is used as slang for gay men. He wrote of it "and well does the shameless little bird deserve its name; for it carries its tail more than erect, that is, inclined backwards towards its head." It is now called the White-throated Tapaculo (*Scelorchilus albicollis*) and is most closely related to the Moustached Turca. Darwin found it very common beneath "hedge-rows, and the bushes scattered over the barren hills, where scarcely another bird can exist. Hence the tapacolo is conspicuous in the ornithology of Chile";[86] see chapter 3.

Two hummingbirds were common, with a third observed "within the Cordillera, at an elevation of 10,000 feet [3,048 m]." The Green-backed

Firecrown Sephanoides sephanoides (see colour figure 2.6 for a fine example of a hummingbird) was present in various habitats and at Tierra del Fuego was described to Darwin by Captain Phillip Parker King as having been seen "flitting about in a snow storm." He found the species perhaps more abundant than any other bird in the humid forest of Chiloe Island. At that time, flowers were few, and CD thought the birds could not be surviving on only "honey," or nectar; and upon inspecting the stomach of one, he could "plainly distinguish, in a yellow fluid, morsels of the wings of diptera—probably tipulidae." He also found insect remains in several other hummingbirds he dissected elsewhere.

Also recorded by Darwin was that Green-backed Firecrown hummingbirds migrate to central Chile in autumn and in the spring begin to depart when the larger Giant Hummingbird (*Trochilus gigas*) replaces it there. He came to believe that the smaller hummingbirds did not breed in Chile, as "during the summer, their nests were common to the south of that country." He observed that the migration of hummingbirds "exactly corresponds to what takes place in this southern continent. In both cases they move towards the tropic during the colder parts of the year, and retreat north before the returning heat. Some, however, remain during the whole year in Tierra del Fuego; and in Northern California,—which in the northern hemisphere has the same relative position which Tierra del Fuego has in the southern,—some, according to Beechey [presumably Frederick William Beechey], likewise remain."[87]

Going on to discuss the Giant Hummingbird, Darwin's notes once again reflect his discerning eye, ecological insights, and comparative observations:

> In the neighbourhood of Valparaiso, during this year [1834], it had arrived in numbers a little before the vernal equinox. It comes from the parched deserts of the north, probably for the purpose of breeding in Chile. When on the wing, the appearance of this bird is singular. Like others of the genus, it moves from place to place with a rapidity which may be compared to that of Syrphus amongst diptera, and Sphinx among moths; but whilst hovering over a flower, it flaps its wings with a very slow and powerful movement, totally different from that vibratory one common to most of the species, which produces the humming noise. I never saw any other bird, where the force of its wings appeared (as in a butterfly) so powerful in proportion to the weight of its body. When hovering by a flower, its tail is constantly expanded and shut like a fan, the body being kept in a nearly vertical position. This action appears to steady and support the bird, between the slow movements of its wings. Although flying from flower to flower in search of food, its stomach generally contained abundant remains of insects, which I suspect are much more the object of its search than honey is. The note of this species, like that of nearly the whole family, is extremely shrill.[88]

The species is actually a primarily nectivorous one, and it obtains insects more by hawking than by hovering.[89] For more about these birds, see the "Hummingbirds" section of chapter 3.

In January 1835, en route from Cape Tres Montes to the Chonos Archipelago, Chile, Darwin reported seal herds being watched by Turkey Vultures: "This disgusting bird, with its bald scarlet head, formed to wallow in putridity, is very common on the west coast, and their attendance on the seals shows that they are dependant [*sic*] on their mortality."[90] Thus his observations of their carrion feeding habits brought about an appreciation of the adaptive nature of the bird's featherless head, also shown in the true vultures of Africa; and this similarity in the two geographically isolated groups thus is an example of convergent evolution. The "broken islets" of the Chonos Archipelago supported little of zoological interest to CD. His seeing a small mouse on several islands, however, got him wondering if hawks or owls ever deliver their prey alive, not necessarily deliberately, to their nest; because if they do, they might thus introduce the rodents to islands in this way—by their escaping from the hawk or owl nests.[91] This idea predated his deep and important considerations of, and experiments into, such potential and actual animal geographical dispersal via an agent species by many years (see the section "Birds as Dispersers of Plants and Animals" of chapter 4).

The kind of things that Darwin observed and recorded about South American birds—about their ecology, habits, nesting, and behaviour—suggest a real interest in living birds. A full-time dedicated bird watcher of today—with the benefit of the present state of ornithological knowledge and armed with modern field optics, video equipment, and one or more of the information-packed fine field guides to that continent—could doubtless contribute more observations of more species in several months than Darwin did while there during 42 months, on and off. Moreover, today we know precisely which genera and species occur there and, just as importantly, those that do not. However, what Darwin did manage to do with birds was by no means insignificant for the time, given what else he managed to achieve there in other fields of investigation.

GALAPAGOS ISLANDS ORNITHOLOGY

In mid-September 1835, Charles Darwin landed on the young volcanic Galapagos Islands, an archipelago that was never once part of the South American mainland but that emerged from the ocean only several million years ago. There can be little doubt that by this point in his circumnavigation, Darwin was sick and tired of the voyage, was desperately homesick, and would have been distracted by thoughts of homecoming, family, England, and his future life and work. The *Beagle* spent just over a month island-hopping within the archipelago, and Charles landed on only four of the islands. He was destined to be thereafter closely associated with those enchanted islands, following the publication of his *Journal of Researches*, *The Zoology of the Beagle*, and, to a lesser extent, *The Origin of Species by Means of Natural Selection, or the Preservation of Favoured Races in the Struggle for Life* (hereafter *The Origin of Species*). The most notable thing that he first observed about the natural history of the islands was "that the birds are strangers to man. So tame and unsuspecting were they, that they did not even understand what was meant by stones being thrown at them; and quite regardless of us, they approached so close that any number might have been killed with a stick."[92]

During nine days on James (now Santiago) Island, spent in company with Syms Covington, Benjamin Bynoe, and Harry Fuller in early October, Darwin found that "So damp was the ground, that there were large beds of a coarse carex in which great numbers of a very small water-rail lived and bred";[93] see colour figure 2.7. Today the Galapagos Crake or Rail (*Laterallus spilonotus*), the bird he saw in great numbers, is an officially "vulnerable," globally threatened, species with its declining world population estimated at between five and ten thousand at the turn of the millennium.[94] This reflects the negative result on birds—on but one of numerous islands about the world—of remote and isolated islands and island groups being eventually reached by people and their feral pests. Numerous species of rail in particular have suffered in this way, many of which had become flightless on their isolated predator-free island homes.[95] The Galapagos Crake can, however, fly, albeit not strongly.

In his account for October 1835, Darwin importantly wrote, *in hindsight* having returned to England, of the Galapagos Island avifauna:

> In my collections from these islands, Mr. Gould considers that there are twenty-six different species of land bird. With the exception of one, all probably are undescribed kinds, which inhabit this archipelago, and no other part of the world. . . . The only kind of gull which is found among these islands [Lava Gull (*Larus fuliginosus*)], is also new; when the wandering habits of this genus are considered, this is a very remarkable circumstance [but see following]. The species most closely allied to it, comes from the Strait of Magellan [Dolphin Gull (*Larus scoresbii*)]. . . . The general character of the plumage of these birds [within the general avifauna] is extremely plain and . . . possesses little beauty. Although the species are thus peculiar to the archipelago, yet nearly all in their general structure, habits, colour of feathers, and even tone of voice, are strictly American. The following brief list will give an idea of their kinds. 1st. A buzzard [Galapagos Hawk (*Buteo galapagoensis*)], having many of the characters of Polyborus or Caracara; and in its habits not to be distinguished from that peculiar South American genus; 2d. Two owls [Galapagos Barn Owl and Short-eared Owl]; 3d. Three species of tyrant-flycatchers [Vermillion Flycatcher, Galapagos Flycatcher and Eastern Kingbird]—a form strictly American. One of these appears identical with a common kind (Muscicapa coronata? Lath.)

[Black-crowned Monjita], which has a very wide range, from La Plata throughout Brazil to Mexico; 4th. A sylvicola [Yellow Warbler], an American form, and especially common in the northern division of the continent; 5th. Three species of mocking-birds [Charles, San Cristobal, and Galapagos Mockingbirds—but a fourth species, the Hood Mockingbird, is now also acknowledged], a genus common to both Americas; 6th. A finch, with a stiff tail and a long claw to its hinder toe [in fact a Bobolink (*Dolichonyx oryzivorus*), which Darwin initially thought to be a pipit (*Anthus* species)[96]], closely allied to a North American genus; 7th. A swallow [Southern Martin] belonging to the American division of that genus; 8th.A dove [Galapagos Dove], like, but distinct from, the Chilean species; 9th. A group of finches, of which Mr. Gould considers there are thirteen species; and these he has distributed into four new sub-genera. These birds are most singular of and in the archipelago. They all agree in many points; namely, in a peculiar structure of their bill, short tails, general form, and in their plumage. The females are gray or brown, but the old cocks jet-black. All the species, excepting two, feed in flocks on the ground, and have very similar habits. It is very remarkable that a nearly perfect gradation of structure in this one group can be traced in the form of the beak, from one exceeding in dimensions that of the largest gros-beak, to another differing but little from that of a warbler.[97] [See figure 2.8.]

Figure 2.8 The heads of four Galapagos finches: (1) Large Ground-finch (*Geospiza magnirostris*), (2) Medium Ground-finch (*Geospiza fortis*), (3) Small Tree-finch (*Camarhynchus parvulus*), and (4) Warbler Finch (*Certhidea olivacea*). From *A Naturalist's Voyage* (by Darwin of 1889: 379).

Parts of this text leave no doubt that the South American character of several of the endemic Galapagos bird species was appreciated by Darwin, which he was also aware of to some degree while actually in the Galapagos; see "The Zoology of the Beagle" section of chapter 3.

Darwin was incorrect in stating that only one species of gull, the endemic Lava Gull (colour figure 2.9), was to be found in the Galapagos: The Swallow-tailed Gull (*Larus furcatus*) not only breeds on almost all of the islands of the archipelago, but it too is endemic to the Galapagos. It is therefore surprising that he did not record this conspicuously different, blackish-headed, pale grey-backed, and pure white-bellied gull. Three other gull species have subsequently been recorded on the Galapagos islands; but as two of these are but winter visitors, and the third was only a single individual that having reached them stayed for two years there, it is no surprise that Darwin did not see them.[98]

Although I continue to refer to the Galapagos finches as finches, Galapagos finches, or, as they continue to be most widely known today, Darwin's finches as appropriate for context and convenience hereafter, they are today known to not be finches of any kind; see "Galapagos Finches and Relatives" of chapter 3. The "finches" were not, however, seen to be a closely related group of birds by Darwin *while he was in the Galapagos Islands*; and indeed, only six species were appreciated to be such by him while there and until John Gould demonstrated more to be so when CD got back in England; see chapter 4. Charles's subsequent observations were thus made with the benefit of advice from John Gould who examined the official *Beagle* collection of 31 finch specimens in London, together with a small number of additional skins collected by others aboard, in preparation of the *Birds* volume of *The Zoology of the Voyage of H. M. S. Beagle* with Darwin. In time, CD was to come to appreciate that having reached the far-flung Galapagos Islands where few other small land birds lived, the colonizing ancestral one or more finch species was, or were, able to subsequently reach various other islands; and the empty ecological niches available on them caused them to diversity in their island isolation. They diversified on the islands colonized by them by taking on the morphology, and particularly the bill size and shape, of other genera and species of small finch-like songbirds that might otherwise have lived there.

On pages 71 and 72 of his original field bird notes, written if not in situ on the Galapagos Islands then certainly shortly thereafter in 1835, Darwin made observations worthy of quoting in full:

> The constitution of the land is entirely Volcanic; and the climate being extremely arid, the islands are but thinly clothed with nearly leafless, stunted brushwood or trees. On the windward side however, & at an elevation between one & two thousand feet [305–610m], the clouds fertilize the soil; & it then produces a green & tolerably luxuriant vegetation. In such favourable spots, &

> under so genial a climate, I expected to have found swarms of various insects; to my surprise, these were scarce to a degree which I never remember to have observed in any other such country. Probably these green Oases, bordered by arid land, & placed in the midst of the sea, are effectually excluded from receiving any migratory colonists. However this may arise, the scarcity of prey causes a like scarcity of insectivorous birds & the green woods are scarcely tenanted by a single animal. The greater number of birds haunt, and are adapted for, the dry & wretched looking thickets of the coast land: here however a store of food is laid up. Annually, heavy torrents of rain at one particular season fall: grasses and other plants rapidly shoot up,—flower, & as rapidly disappear. The seeds however lie dormant, till the next year, buried in the cindery soil. Hence these Finches are in number of species & individuals far preponderant over any other family of birds.[99]

This brief account contains significant early insights into the zoogeography, climate, fauna, and flora of isolated, arid, islands—in this case, the approximately four- to five-million-year-old Galapagos Archipelago—and the ecological influences of, and interplay between, them and the resulting avifauna. This is clearly indicative of CD's increasingly sophisticated appreciation of ecological interactions in general, and of island biology and avian ecology in particular, during 1835.

On page 79 of his bird notes, Darwin intriguingly wrote of the frigatebird (the Magnificent [*Fregata magnificens*] and/or Great [*F. minor*]) that "The bird never touches the water with its wings, or even with its feet; indeed, I have never seen one swimming on the sea; one is led to believe that the deeply indented web between its toes is of no more use to it than are mammae ... in the male sex of certain animals; or the shrivelled wings beneath the wing-cases firmly soldered together of some coleopterous beetles."[100] Thus, this avian example of a character that has clearly become redundant to the animal's present lifestyle struck CD as indicative of observable dramatic morphological changes in a species. He was to subsequently raise this specific frigatebird observation, together with that of the same redundant nature of webbed feet in geese, in his *The Origin of Species*.

Interestingly, Darwin, or Syms Covington on his behalf, did not collect a single specimen of the Galapagos Penguin (*Spheniscus mendiculus*), which is endemic to that group of islands, and failed to mention it (or the previously reported endemic Galapagos Flightless Cormorant *Phalacrocorax harrisi*) in his notes or publications. It is the only penguin species that occurs north of the equator, if only just so; it was not to be described and thus made known to science until 1871.[101]

Darwin was to write to John Henslow from Sydney, Australia, on 20–29 January 1836, of his work in the Galapagos Islands that "I paid much attention to the Birds, which I suspect are very curious."[102] This clearly indicates how he saw the avifauna of the Galapagos Archipelago as particularly interesting and significant and not a subject of study that he took at all lightly—if not necessarily during his busy time among the islands, then certainly thereafter while still aboard the *Beagle*;[103] see chapter 8. Darwin's great reluctance to publish his ideas about the origin of species through natural selection in general, and in the Galapagos avifauna in particular, led Alfred Wallace to write in 1855, "Such phenomena [island speciation] as are exhibited by the Galapagos Islands, which contain little groups of plants and animals peculiar to themselves, but most nearly allied to those of South America, have not hitherto received any, even a conjectural explanation."[104] This seemingly harsh criticism of Darwin's delay in writing more about the significance of the Galapagos fauna and flora must have been hard for him to swallow.

THOUGHTS WITH THE GALAPAGOS IN THE *BEAGLE*'S WAKE

Under his October 1835 account of his *Journal of Researches*, Darwin reflected, again *in hindsight*, about his bird work in the Galapagos Archipelago by writing a critically important paragraph with respect to his understanding of, and insights into, speciation:

> It has been mentioned, that the [human] inhabitants can distinguish the tortoises, according to the islands whence they are brought. I was also informed that many of the islands possess trees and plants which do not occur on the others. . . . Unfortunately, I was not aware of these facts till my collection was nearly completed: it never occurred to me, that the productions of islands only a few miles apart, and placed under the same physical conditions, would be dissimilar. I therefore did not attempt to make a series of specimens from the separate islands. . . . In the case of the mocking-bird, I ascertained (and have brought home the specimens) that one species (*Orpheus trifasciatus*, Gould) is exclusively found on Charles Island; a second (*O. parvulus*) on Albemarle Island; and a third (*O. melanotis*) common to James and Chatham Islands. The last two species are closely allied, but the first would be considered by every naturalist as quite distinct. I examined many specimens in the different islands, and in each the respective kind was *alone* present [CD's emphasis]. These birds agree in general plumage, structure, and habits; so that the different species replace each other in the economy of the different islands. These species are not characterized by the markings on the plumage alone, but likewise by the size and form of the bill, and other differences. I have stated, that in the thirteen species of ground-finches, a nearly perfect gradation may be traced, from a beak extraordinarily thick, to one so fine, that it may be compared to that of a warbler. I very much suspect, that certain members of the series are confined to different islands; therefore, if the collection had been made on any *one* island

> [CD's emphasis], it would not have presented so perfect a gradation. It is clear, that if several islands have each their peculiar species of the same genera, when these are placed together, they will have a wide range of character. But there is not space in this work, to enter on this curious subject."[105]

The words that he actually wrote of these mockingbirds[106] in his telegraphic style, in his notebook in July 1836, were "In each Isd each kind is *exclusively* found: habits of all are indistinguishable. . . . When I see these islands in sight of each other, and possessed of but a scanty stock of animals, tenanted by these birds, but slightly differing in structure & filling the same place in Nature, I must suspect that they are only varieties. . . . If there is the slightest foundation for these remarks the zoology of Archipelagos will be well worth examining; for such facts undermine the stability of Species."[107] He also noted that the inter-island variation that he had observed in the Falkland Island fox-like Warrah presented the same kind of evidence; but he made no mention of the Galapagos finches in this context because he did not then appreciate that they constituted a single group of closely related species. While among the islands, he had no doubt that the obviously heavy-billed, and thus finch-like, species were indeed finches, whereas those with finer beaks he failed to see as being part of the same group.

Although in the subsequent editions of his *Journal of Researches* Darwin referred to 13 species of ground-finches within his account for October 1835 in the Galapagos, he did not in fact know that there were that many species during his time among the islands: nor did he then appreciate that the Galapagos finches were of a far greater significance than their small stature and drab appearance suggested to him. In view of widespread misunderstanding of the facts, it is worth reiterating and stressing them here. The importance that the morphological diversity of the Galapagos finches in their island isolation might have to play, although briefly alluded to under his October 1835 account of the *Beagle* among the Galapagos Islands, was *not* appreciated by Darwin at that time but in fact resulted from John Gould's determinations of the taxonomy and nomenclature of the finch specimens made after the *Beagle* had returned to England. It is evident from the conspicuous additions that Darwin made to his accounts of the various Galapagos finches in his revised editions of his *Journal of Researches*[108] to what appeared in the first edition[109] (see appendix 1) that only with the passing of the time between the writing of the two texts did he come to appreciate the significance of the island distributions and morphological variations between the various finches; see chapter 3. The one finch species far removed from the Galapagos, isolated on what was then Bow (now Cocos) Island of the Low Archipelago, off Costa Rica, and is the Cocos Finch (*Pinaroloxias inornata*), was discovered between editions of Darwin's *Journal of Researches*.

TAMENESS OF ISLAND BIRDS

Darwin wrote over three pages about the tameness of birds on the Galapagos, Falkland, and other remote islands. He noted how island-endemic species show no fear of man, whereas a migratory species to such islands—for example, the Black-necked Swan (*Cygnus melanocorypha*) on the Falklands—tellingly "brings with it the wisdom learnt in foreign countries." Having noted how easy it was for him and others to catch and kill these tame bird species by hand or with a stick, he also observed that they were apparently learning over time because the accounts of earlier visitors to the islands made it clear that they were previously even more tame and trusting. He concluded with the insightful sentence, "We may infer from these facts, what havoc the introduction of any new beast of prey must cause in a country, before the instincts of the aborigines become adapted to the stranger's craft or power."[110] He subsequently also noted that Turkey Vultures on the Falkland Islands were typically shy, in marked contrast to the endemic land birds, and therefore he wrote, "may we infer from this that they are migratory, like those of the northern hemisphere?"[111] He was correct in this because the subspecies of Turkey Vulture present on the Falkland Islands is not endemic there but also occurs along parts of the Pacific coast of South America. Moreover, individuals of another, second, mainland subspecies of the Turkey Vulture may also occasionally reach the Falklands.[112]

Darwin observed that whereas bird species that he lists are tame on the Falkland Islands, both predatory "hawks and foxes are present" there also. Thus he concluded that the "absence of all rapacious animals at the Galapagos, is not the cause of their tameness there. The geese at the Falklands, by the precaution they take in building [nests] on the islets, show that they are aware of their danger from the foxes; but they are not by this rendered wild towards man. This tameness of the birds, especially the waterfowl, is strongly contrasted with the habits of the same species in Tierra del Fuego, where for ages past they have been persecuted by the wild inhabitants."[113] He added that the only way to account for observations of wildness towards mankind in birds is a "particular instinct directed against *him* [CD's emphasis], and not dependant on any general degree of caution arising from other sources of danger; secondly, that it is not acquired by them in a short time, even when much persecuted; but that in the course of successive generations it becomes hereditary. . . . In regard to the wildness of birds towards man, there is no other way of accounting for it. Few young birds in England have been injured by man, yet all are afraid of him: many individuals, on the other hand, both at the Galapagos and the Falklands, have been injured, but yet have not learned that salutary dread."[114]

From his observations and considerations of birds he encountered on remote oceanic islands and at sea, Darwin wrote, "I believe, the waders are the first colonists of any island, after the innumerable web-footed species. I may add, that whenever I have noticed birds, which were not pelagic, very far out at sea, they always belonged to this order; and hence they would naturally become the earliest colonists of any distant point."[115] This suggestion would seem reasonable, on the face of it, but of course it can only be applied to islands that support habitats that would permit waders to feed, let alone reproduce, on them.

TAHITI, NEW ZEALAND, AUSTRALIA, AND HOME

Before reaching the Galapagos Islands, Darwin had accumulated, mostly collected by Syms Covington, 327 bird skin specimens, namely, Cape Verde Islands, 5; St. Paul Rocks, 2; Brazil, 8; Uruguay, 115; Argentina, 115; Tierra del Fuego, 49; and Falkland Islands 33.[116] In excess of approximately 65 birds were collected on the Galapagos Islands for CD's collection.

After the *Beagle* finally departed the Galapagos Archipelago, on 20 October 1835, with 30 giant tortoises aboard as fresh meat for the crew, the increasingly homesick Charles Darwin wrote very little about birds that he encountered. Thus we learn almost nothing of the birds he must surely have seen on Tahiti, which took 20 days to reach from the Galapagos and where 11 days were spent and no birds were collected. In New Zealand, he "saw very few birds" in the woods;[117] collected but a single bird during his period of time there, from 21 to 30 December 1835; and made no mention of the kiwis with which that country is so closely associated.

Charles had been deeply depressed to find no family mail awaiting him in Sydney. Of birds seen in eastern New South Wales and Tasmania, Australia, where 39 days were spent but no birds collected, he mentions only some half a dozen common and widespread species by loose generic common names[118] (see appendix 1) including parrots, white cockatoos, and crows. The cockatoos were undoubtedly Sulphur-crested Cockatoos (*Cacatua galerita*) and the crows Australian Ravens (*Corvus coronoides*); and it has been suggested that the parrots seen by Darwin on his trip to Bathurst were most likely Crimson Rosella (*Platycercus elegans*), Eastern Rosella (*Platycercus eximus*), and Australian King Parrot (*Alisterus scapularis*);[119] but they may in fact have been one or all of these plus several other potential parrot species. As a result of his return trip to Bathurst from Sydney, he noted that the Emu (*Dromaius novaehollandiae*) "is banished to a long distance, and the kangaroo is become scarce; to both the English greyhound has been highly destructive."[120] He also made mention of the Emu dance of the Aboriginal people. Whereas he made very few ornithological contributions in Australia, some of the officers aboard the *Beagle*, and most notably John Lort Stokes, during visitations there during 1837 to 1843, subsequently made far more significant ones.[121]

Having departed Australia on 14 March 1836, the next landfall was (Cocos) Keeling Island. Having, quite surprisingly, collected no birds at all in Australia, he did collect a single Buff-banded Rail (*Rallus phillipensis*) on Cocos Keeling Island. After this, most ports of call were to be at better-known islands and increasingly less time was to be spent ashore by the homebound and homesick ship's company.[122] Darwin probably partly reviewed and arranged his ornithological notes[123] preceding his landings at Mauritius, Cape Town, St. Helena, Ascension (where he found eagerly awaited mail from home), Cape Verde and the Azores Islands, and other islands between Bahia, Brazil, and England, during which no ornithology was detailed other than the briefest cursory remarks about a few bird groups or species.

On Mauritius, where the *Beagle* visited during 29 April to 9 May, Darwin managed to take a ride on the only elephant on that island at the time. This experience caused him, quite possibly with his years of seasickness in mind, to suggest "that the motion must be fatiguing for a long journey."[124] No birds were collected during 10 days at Mauritius, from where he wrote to his sister Caroline that no place "has now any attractions for us, without it is seen right astern, & the more distant & indistinct the better. We are all utterly homesick."[125] No birds were collected during a week at St. Helena Island.[126]

Darwin's lack of bird notes towards the end of his voyage may have in part been because of his increasing weariness of the voyaging, possibly seasickness, his obvious homesickness, the limited bird life encountered (predominantly seabirds, many of which he may have already seen), and his ever-increasing interest in matters geological and anthropological over natural history subjects. He may have also been preoccupied and busy with sorting and compiling his notes made during well over four years of travel. During this part of the homeward-bound voyage, CD gave much of his attention at landings to the scenery, topography, vegetation, geology, marine life and coral reef formations, and native and immigrant peoples and their social organization, government, and daily lives. On New Zealand, he was, however, made aware of the Moa, a giant flightless bird by then extinct there (actually several extinct species, of the family Dinornithidae, were involved). On the *Beagle* calling in again at the Azores, he was sufficiently pleased to see some of his old familiar British avian friends to note sighting the Common Starling (*Sturnus vulgaris*), Grey Wagtail (*Motacilla cinerea*), Chaffinch (*Fringilla coelebs*), and Common Blackbird (*Turdusmerula*) there.[127]

As the *Beagle* sailed north on its homeward approach to England, Charles and Syms Covington together organized their specimens, labels, catalogues, logbooks, and field notes, but CD alone compiled the bird list.[128] This latter fact might suggest that his ornithological results were of particular personal interest and value to Darwin.

THE VOYAGE COMPLETED

The *Beagle* circumnavigation lasted 1,741 days, or four years and nine months; but of this total, Darwin was actually on land for 594 days (or 34.1 percent of total days), at anchor for 566 days, and at sea for 581 days.[129] During his one and a half years spent actually afloat at sea, he would have seen numerous birds of a good number of genera and species but many of them not at all easily identified by sight, given the taxonomic appreciation of pelagic seabirds and other groups at the time and particularly given the poor quality of optics available to him on deck.

Whenever ashore, Darwin was deeply engaged in the study of every aspect of natural history, including palaeontology as well as his substantial interest in geology, and apparently rode for some 2,000 miles or 3,200 km on horseback.[130] Given this scenario, the number and quality of his ornithological observations and considerations are quite impressive and involve approximately 180 bird species or groups and topics—as are summarized in taxonomic order in appendix 1. Darwin collected during his voyage 468 complete bird skins, parts of the Lesser Rhea, nest and eggs of 16 birds, and 14 entire birds and parts of four preserved in spirit fluid; but a grand total of some 512 to 515 items were probably collected.[131] This is not a large number for such a long period of exploration by a naturalist on land and sea, but this has to be seen in the context of the far greater number of other biological and geological specimens collected and time taken up by Darwin's field observations and the documentation of them. He, somewhat reluctantly, presented 450 bird (and 80 mammal) specimens resulting from his voyage aboard the *Beagle* to the Museum of the Zoological Society in London on 4 January 1837.[132] He had returned with 1,700 pages of zoological and geological notes and an 800-page diary.[133] The Zoological Society Museum in London was broken up in 1855, and the British Museum of Natural History (BMNH) acquired some of Darwin's *Beagle* voyage bird specimens; but the majority went to John Gould's personal collection, and further material was purchased by other private collectors.

Today specimens from Darwin's *Beagle* voyage are housed in at least seven institutions additional to the Natural History Museum in London. In addition to this material, the latter institution also holds 59 skins of domestic pigeons, 25 of Persian birds from Tchcran, 11 of wild ducks, 6 of domestic ducks, 51 skeletons of pigeons, and 25 of chickens from Darwin's personal collection, presumably mostly accumulated at Down House.[134]

In writing about the birds he observed during the voyage of the *Beagle*, Darwin alludes to his experience of almost 40 species or groups of birds familiar to him in Britain in making morphological, ecological, or behavioural comparisons of them with foreign birds he encountered. These are grouse, partridges, ptarmigans, pheasants, quail, English Woodpecker, European Kingfisher, Common Cuckoo, Common Wood-Pigeon, Spotted

Redshank, Common Snipe, Common Sandpiper, Northern Lapwing, Mew Gull, Common Black-Headed Gull, Osprey, Black-Billed Magpie, Jackdaw, English Rook, Eurasian Blackbird, Fieldfare, Mistle Thrush, Spotted Flycatcher, European Robin, Whinchat, Common Starling, Eurasian Tree-Creeper, Winter Wren, tit-mice, Long-Tailed Tit, Barn Swallow, Eurasian River Warbler, Sedge Warbler, Willow Warbler, Blackcap, Chaffinch, Pine Grosbeak, and Bullfinch (see appendix 1 for details and scientific names). He was no ornithologically naive recorder of bird species seen along the way but one experienced in watching British species closely; interested in the birds' interactions with their environment, with others of their kind, and those of other species; their feeding ecology, adaptive morphology, nesting, and behaviour; and their geographical distributions.

With the passing of the almost five years of the voyage, Charles Darwin the ornithologist had grown exponentially in experience and understanding. He was not, however, to become a documented convert to evolution until several months after disembarking from the *Beagle*, in March 1837 (see "Footnote").

Darwin's youthful wondering as to why every gentleman did not become an ornithologist was certainly not wasted on himself. His return to England was to mark the beginning of a considerable amount of ornithological thinking, experimentation, and writing by him. He was to prove himself both a gentleman and an ornithologist of great distinction.

FOOTNOTE

Writing the definitive work on the HMS *Beagle*, Keith Thomson recorded that the ship was launched at Woolwich Naval Dockyard on the River Thames as a two-masted brig on 11 May 1820, only to then remain at anchor and without commission for five years. Upon commission in 1825, the *Beagle* had to return to the dockyard to be rigged as a three-masted bark. Thomson had to report "that no physical relic of the ship survives" and that what became of her after being sold as a hulk was unknown.[135] In early 2004, the hull remnants of the *Beagle* were discovered, by the use of ground-penetrating radar, beneath the shallow waters of the Essex marshes near Potton Island in England. She had been transferred to the British Customs and Excise service and had functioned as an antismuggling patrol vessel *WV-7* on that coast. She was refitted as a coastguard vessel in 1845 and was renamed *Southend* in 1851. By the 1860s, she was permanently anchored in the River Roach estuary as a floating surveillance facility until sold for scrap by auction for £525 Sterling in 1870, to then be stripped by the scrap merchants Murray and Trainer. What remains of her is only the keel and some of the lower hull planking, presently resting beneath the shallow waters of the marsh, and it is intended that these be excavated.[136,137]

NOTES

1. C. Darwin 1839a.
2. FitzRoy 1839.
3. C. Darwin 1839b.
4. Thompson 1995: 205.
5. Browne 1995: 413; but late May according to Desmond and Moore 1991: 284.
6. Huxley and Kettlewell 1974: 50.
7. Desmond and More 1991: 228.
8. Darwin 1839b: 2.
9. Bannerman and Bannerman 1968: 388.
10. Darwin 1839b: 3–4.
11. Darwin 1839b: 9.
12. Keynes 2000: 34.
13. Harrison 1985; Orta 1992.
14. Darwin 1839b: 21.
15. Haupt 2006: 59.
16. Darwin 1839b: 22.
17. Darwin 1839b: 37; Gould and Darwin 1838–1841: 112.
18. Barlow 1963: 204.
19. Chancellor and Wyhe 2009.
20. Chancellor and Wyhe 2009: 44.
21. Darwin 1839b: 48.
22. Darwin 1839b: 51.
23. Harshman et al. 2008; Gill and Donsker 2015.
24. Darwin 1839b: 55.
25. Darwin 1839b: 61.
26. Darwin 1839b: 62.
27. Cocker 2013: 351–2.
28. Darwin 1839b: 63.
29. Gould and Darwin 1838–1841: 60–4; Grant and Estes 2009.
30. Gould and Darwin 1838–1841: 18.
31. Sibley et al. 1988; Sibley and Ahlquist 1990; Sibley and Monroe 1990; Gill and Donsker 2015; Jarvis et al. 2014.
32. Darwin 1839b: 77–8.
33. del Hoyo 1992: 508–26.
34. Darwin 1839b: 81–2, 143–4; Gould and Darwin 1838–1841: 31–2.
35. Darwin 1889: 125; Frith and Frith 2004, 2008.
36. Darwin 1839b: 145.
37. König et al. 1999; Couve and Vidal 2003; Marks et al. 1999: 200–201, 217.
38. Darwin 1839b: 105–6.
39. Darwin 1839b: 134.
40. Darwin 1839b: 107.
41. Fernández and Reboreda 1998.
42. Darwin 1839b: 106–7.
43. Darwin 1839b: 139; Gould and Darwin 1838–1841: 120–23; see chapter 3.
44. Haupt 2006: 102.
45. Darwin 1839b: 108–9.
46. S. J. J. F. Davies 2002.
47. Burkhardt et al. 1985–2014, vol. 1: 370.

48. d'Orbigny 1834; Darwin 1839b: 108–9; see colour figure 2.2.
49. BirdLife International 2000: 630.
50. Gould and Darwin 1838–1841: 123–5.
51. In Haupt 2006: 47.
52. Darwin 1839b: 110–11; Gould and Darwin 1838–1841: 117–18.
53. del Hoyo and Collar 2014, Gill and Donsker 2015.
54. Remsen 2003: 259–61; Jobling 2010.
55. Darwin 1839b: 112; Gould and Darwin 1838–1841: 64–6.
56. Sick 1993: 225; Cocker 2013: 201.
57. Darwin 1839b: 133; Gould and Darwin 1838–1841: 127.
58. Darwin 1839b: 161–2.
59. Lesson 1828, vol. 2: 385.
60. Gould and Darwin 1838–1841: 143–4.
61. Darwin 1839b: 162–3.
62. Darwin 1839b: 194.
63. Ferguson-Lees and Christie 2001.
64. Brown and Amadon 1968: 192; Houston 1994.
65. Darwin 1839b: 219–22; Gould and Darwin 1838–1841: 3–6.
66. Keynes 2003: 226.
67. BirdLife International 2000: 635–6; Werness 2004; Cocker 2013: 157–9.
68. Audubon 1826.
69. Birkhead 2012.
70. Wyhe 2014: 107–8.
71. Owen 1837.
72. Darwin 1839b: 222 and footnote.
73. Birkhead 2012: 132.
74. Houston 1986.
75. Houston 1994.
76. Darwin 1839b: 250.
77. Darwin 1839b: 257–8.
78. Gould and Darwin 1838–1841: 136.
79. Darwin 1839b: 301.
80. Darwin 1839b: 305.
81. Chancellor and Wyhe 2009: 383.
82. Barlow 1963.
83. Keynes 2003: 248–9.
84. Darwin 1839b: 328.
85. Darwin 1839b: 329.
86. Darwin 1839b: 329–30; Gould and Darwin 1838–1841: 70–74.
87. Darwin 1839b: 330–31.
88. Darwin 1839b: 331–2; Gould and Darwin 1838–1841: 111–12.
89. Schuchmann 1999.
90. Darwin 1839b: 346; Gould and Darwin 1838–1841: 8–9.
91. Darwin 1839b: 351 footnote.
92. Darwin 1839b: 455.
93. Darwin 1839b: 459.
94. BirdLife International 2000; del Hoyo and Collar 2014.
95. Taylor and van Perlo 1998; Frith 2013.
96. Barlow 1963: 265.
97. Darwin 1839b: 461–2.

98. Heinzel and Hall 2000.
99. In Barlow 1963: 261.
100. Barlow 1963: 267–8; see also Gould and Darwin 1838–1841: 146.
101. Williams 1995.
102. Burkhardt et al. 1985–2014, vol. 1: 485.
103. *Pace* Bourne 1992: 31; Steinheimer 2004: 309.
104. Wallace 1855: 188.
105. Darwin 1839b: 474–5; Gould and Darwin 1838–1841: 60–64.
106. Contrast S. Gould 1980: 62, who attributed them to the Galapagos finches.
107. Chancellor and Wyhe 2009: 419.
108. Darwin 1889: 379–80.
109. Darwin 1839b: 461–2.
110. Darwin 1839b: 475–8.
111. Gould and Darwin 1838–1841: 8.
112. Woods 1988: 154.
113. Darwin 1839b: 477.
114. Darwin 1839b: 478.
115. Darwin 1839b: 543.
116. Steinheimer 2004: 303.
117. Darwin 1839b: 511.
118. Nicholas and Nicholas 1989.
119. Nicholas and Nicholas 2002: 51.
120. Darwin 1889: 441.
121. Whittell 1954: 100–107.
122. Thomson 1995: 195.
123. Cf. Keynes 2003: 370.
124. Darwin 1839b: 537.
125. In Brent 1981: 208.
126. Steinheimer 2004: 303.
127. Darwin 1839b: 595.
128. Browne 1995: 339.
129. Chancellor and Wyhe 2009: 570.
130. Aydon 2002: xxiv.
131. Steinheimer 2004: 300, 306, 317.
132. Datta 1997: 50, 84; Sulloway 1982b: 20.
133. Aydon 2002: 107.
134. Steinheimer 2004: 317; Cooper 2010: 25.
135. Thomson 1995: 8, 269.
136. McKie 2004.
137. Thompson 2005: 736.

CHAPTER 3

The Bird Collector Returns and Writes about Them

This singular genus [*Geospiza* of the Galapagos finches] *appears to be confined to the islands of the Galapagos Archipelago. It is very numerous, both in individuals and in species, so that it forms the most striking feature in their ornithology.* [Gould and Darwin 1838–1841: 99]

Charles Darwin returned to England, initially to Falmouth on 2 October 1836, after close to five years voyaging aboard HMS *Beagle*, with a large collection of preserved mammals, birds, fishes, other animals, plants, fossils, and geological specimens. These were to be added to those previously sent to England by him aboard other British ships from various foreign ports as he travelled through them aboard the *Beagle*. He faced the formidable task of gathering this large total amount of material together in one place and sorting it into a manageable and workable collection. Then it had to be comparatively assessed, researched, written up—by himself and others that he invited to participate—and then published over future years. This was, however, to prove to involve more years than he could ever have imagined at that time of his gleeful homecoming.

Having caught up with his, now adoring and admiring, family at The Mount for three weeks, which he reached by coach on 5 October, Charles rented a small house on Fitzwilliam Street, Cambridge, from 16 December 1936. This he did to set to work on his collections and notes, and in this he was assisted by Syms Covington, who he had continued to employ after they had both disembarked the *Beagle*. He was able to do this due to his father's generous financial support in providing an allowance and also in gifting him some stocks and shares. After intensive initial work there on his material, and finding that he had to communicate with many people in London concerning it—including John Gould who he met during the second week of March—he decided he would have to move to the capital city.

On 13 March 1937, Charles and Covington took up residence at 36 Great Marlborough Street, down the road from the London residence of his beloved older brother Erasmus. Life was then both exciting and hectic, for various colleagues had ensured that Darwin's name was on the lips of many geologists, palaeontologists, and biologists in and beyond London by publishing some of his discoveries and accounts from letters he had sent from overseas while still voyaging. Charles approached a number of authoritative individuals, each a specialist in a particular biological field of study, in the hope that they might undertake the working and writing up of his specimens—but it proved difficult to elicit interest in some of the animal groups he had collected. He was also simultaneously working on producing a manuscript, from his own *Beagle* diary and field notes, for the account of the natural history observed during the voyage that Captain FitzRoy requested as part of the official narrative of that circumnavigation.

The first thing that Darwin actually published after his return was a brief paper on his experiences with the flightless rheas of South America, having read his notes on them at the Zoological Society in London on 14 February 1837.[1] On 10 May he again made comments at the Zoological Society, but this time on the Galapagos finch genera *Geospiza, Camarhynchus, Cactornis*, and *Certhidea*. His *Journal of Researches* (to subsequently become the *Naturalist's Voyage Round the World*), completed at Great Marlborough Street, was published in London in 1839, and thus during the appearance of the various parts of *The Zoology of the Beagle*.[2]

THE ZOOLOGY OF THE VOYAGE OF THE *BEAGLE*

In early 1837, William MacLeay, a prominent fellow of the Zoological Society, had suggested to Darwin that he might consider producing a series of volumes presenting the results of his zoological collecting and observations. After all, this manner of publishing the comprehensive result of such a British naval voyage of discovery had by then become traditional. Because of this tradition, financial backing might be sought from the Admiralty to cover some of the production costs. Darwin gave this idea serious consideration and, not blind to the significant amount of time and effort it would involve, liked the idea of marshalling a limited team of authors under his own name as general editor of the published zoological results. He sought the support of some of the most eminent members of several major scientific institutions in London by way of their providing testimonials for his project proposal to the Admiralty. He also asked well-connected academic colleagues to smooth the way with the Chancellor of the Exchequer, who happened to be Thomas Spring Rice, the Member of Parliament for Cambridge University. The result was a success beyond CD's expectations—a grant of £1,000, which he thought might cover not only the cost of the production and printing of colour plates for his *The Zoology of the Beagle* series of books but also see him publish a book on his geological observations and conclusions. He was understandably thrilled and quickly signed a contract with a publishing house experienced in producing government publications that included quality colour plates. He agreed to produce both the completed texts and all required illustrations within two years—heady times indeed for a young man of 28 back in England only some 11 months after nearly five years spent circumnavigating the globe.

The fossil mammal specimens that Darwin collected were handed over to Richard Owen, then Hunterian Professor at the Royal College of Surgeons in London, for study and the initial publication of his novel findings. Darwin discussed his geological specimens, notes, thoughts, and ideas with Charles Lyell, author of the *Principles of Geology*—the work so much used and admired by Darwin during his long voyage. He had first met Lyell in person at Greenwich, where the *Beagle* was then berthed, over dinner together with Richard Owen. His reptile specimens were worked on by Thomas Bell, who reported that each Galapagos Archipelago island had its own endemic form of iguana lizard; his insects and to some extent also his mammal species were examined by Frederick George Waterhouse; and other zoological material was examined by William Martin of the Zoological Society in London.

Darwin was a perfectly competent observer and recorder of bird distributions, habitats, ecology, habits, behaviour, and more, as his publications clearly attest; but he had little experience of avian classification and nomenclature. He therefore proved extremely fortunate in obtaining, sometime prior to 16 January 1837, the assistance of the established ornithologist John Gould, then Superintendent of the ornithological collection at the Zoological Society. Gould assisted by delineating, and where required describing for the first time, the genera and species included in CD's collection;[3] not just the Galapagos finches and mockingbirds as has been suggested[4] but almost all of his *Beagle* bird collection. Gould was both an entrepreneurial ornithologist and highly competent avian taxonomist who eventually became internationally so well-known and admired for his quality publications, both scientific and artistic, that he was widely called simply "the bird man". He had added to his reputation by displaying a vast number of hummingbird skins mounted in life-like postures within glass cabinets for the great Crystal Palace Exhibition, where they were viewed by countless paying patrons (as well as by the royal family) before being moved on to the Zoological Society of London gardens as an exhibition there. Darwin had become a corresponding fellow of the Zoological Society immediately prior to sailing on the *Beagle*, and as such would possibly have dealt with Gould as a correspondent before subsequently meeting him in person. Charles became a full fellow of the Zoological Society in 1839.

By this point in time, John Gould had already published magnificent large and gloriously illustrated works on birds of the Himalayas and also of Europe and was in the throes of publishing the parts of a monograph on the trogons as well as descriptions of new birds from various parts of the world, including Australia. Enthusiastic, ambitious, and impressively productive, Gould exhibited most of Darwin's bird specimens at meetings of the Zoological Society during the first two months of 1837, meetings that Charles did not usually attend. This exhibiting of CD's bird collection apparently did not include every one of his *Beagle* specimens, as it is recorded that Charles had "skimmed off numerous birds in spirits and skins for Tom [Thomas Campbell Eyton, his long-term friend and colleague]."[5] Darwin recorded in his notebooks that he was particularly busy working on his birds for *The Zoology of the Voyage of the Beagle* during April and June of 1837, but these would not have been the only periods dedicated to that project.

The earliest preserved letter between Darwin and Gould is one to Gould, dated February 1838, CD having just seen the publisher for his *Zoology of the Beagle*. In this, Darwin asked that Gould prepare 50 colour plates for the volume on his bird specimens and observations and that he should have the plates hand coloured by his head colourist Mr. Gabriel Bayfield, who was unheralded for his fine work at the time.[6] Gould's drawings for this volume were transferred to stone for lithographic printing onto heavy paper by his wife Elizabeth, for subsequent hand colouring, and this was to prove the only volume exclusively illustrated by the Goulds as a couple.[7] Gould was, however, also busy planning a prolonged trip to Australia to produce a grand multivolume publication on the birds found there, and this impinged somewhat on his work on Darwin's bird collection from the *Beagle* voyage. Gould departed England on 16 May 1838, and did not return until 18 August 1840.[8]

In fact Gould was absent for a significant part of the writing of the volume on the *Beagle* bird specimens, having produced only the descriptions of the morphological characteristics of the various species. Darwin thus wrote much of the text himself, most of his contributions being written in the first person (for this reason I cite the work as being by Gould and Darwin 1838–1841 and not as written by Gould alone). Because of this Charles sought and gained the cooperation of George Robert Gray, "ornithologist assistant in the Zoological Department of the British Museum." Gray helped with Gould's existing text and with the taxonomic order in which the species accounts were to appear, as well as separating previously known species from entirely new ones and checking that the correct generic names to be applied to them all. With regard to the latter, Darwin retained some errors in the final publication inasmuch as several species appear under different genera at different points in the work (see appendix 1), perhaps because Gray did not proof read the final draft or did so but overlooked the errors. In addition to this help Thomas Eyton, a friend of the Darwin family who studied the osteology of birds, contributed descriptions and observations on the anatomy of twelve species, which appear as an appendix immediately before the index of species in the back of *The Zoology of the Voyage of H. M. S. Beagle*.

John Gould described and named 37 new bird species from Darwin's *Beagle* collection and one additional species was described by George Gray.[9] Gould had concluded, to CD's great surprise and amazement, that the 31 various drab small songbird specimens that he collected from various islands of the Galapagos Archipelago all represented a previously unknown, closely related single group of finches, which he attributed to the "family Coccothraustinae" [*sic*];[10] see following. The name Coccothraustinae, originally applied by Gould,[11] is not that of a family but is that of a subfamily—the spelling of subfamily names always ending with "nae," whereas family names end with "dae."

Gould assigned all of the finches to the single genus *Geospiza*, subdivided by three closely related subgenera. They were subsequently treated as members of the family Fringillidae, subfamily Emberizinae, tribe Thraupini, by the influential revolutionary revision of the world's birds based on pioneering DNA hybridization studies.[12]

Given the state of avian morphological taxonomy at the time and the largely uniform drab appearance of forms of this, then completely novel, group of small and similar birds, it is remarkable that Gould's original designations of the finches remains as close as they do to today's classification of them. Although the number of finch species remains similar, some definitions and descriptions of species have changed since Gould wrote his original descriptions. Moreover, his task was complicated by the fact that Darwin did not at the time of their collection record which island each individual finch was obtained from but only gave each specimen label a reference number and noted they were from the Galapagos Islands. Thus, in this light, the suggestion that Gould has "been given much probably undeserved credit"[13] for his pioneering task, performed under the pressure of his impending departure for Australia, appears unwarranted.

The analysis of his small drab Galapagos bird specimens as all being closely related finches was a shock to Darwin because the great diversity in beak size and shape found across this group of species led him to believe that he had collected only six finch species plus representatives of wrens, warblers, and orioles, when he in fact collected nine finch species.[14]

John Gould had proclaimed, on 28 February 1837, that the specimens of mockingbirds that CD collected in the Galapagos differed so greatly between the islands that they in fact represented not varieties of one species but three distinctly different species. (Today four mockingbird species are acknowledged as living within the archipelago, and are most closely related to the Bahama Mockingbird *Mimus gundlachii*.[15]) But this view of his mockingbirds would have been far less of a surprise to CD than was Gould's interpretation off his finches because he had strongly suspected something of the sort about them by having referred to the Galapagos populations as "varieties or distinct species" in his notes[16] and previously seen and collected mockingbirds on mainland South America.[17] That he did so was correctly noted by Frank Sulloway[18] who then contradicted this by suggesting that Darwin was "frankly stunned . . . by the realization that three separate species of mockingbird indeed inhabited the different islands of the Galapagos" when John Gould confirmed their status as three different species.[19] Indeed it was CD's realization that the mockingbirds differed from island to island, together with information given him that the giant tortoises also did so, that led him to make a now famous entry in his ornithological notes some nine months after leaving the Galapagos and four months before returning to England. In referring to his notes about these birds and reptiles, he wrote, during September or October of 1835, "If there is the slightest foundation for these remarks the zoology of Archipelagoes—will be well worth examining; for such facts undermine the stability of Species."[20]

John Gould eventually described 25 of the 26 land birds that Darwin collected from the Galapagos Islands as being species both new to science and endemic to the archipelago, and 3 of the 11 water birds collected there also as new and endemic species.[21] Knowledge subsequently gained of distributions of some of these birds beyond the Galapagos resulted in a slight reduction in the number of land bird species considered as endemic to the islands, but the proportion of the avifauna that is unique to them remains impressively high.[22]

Some confusion exists in the literature regarding when and how many species of Galapagos finches John Gould described and named in the *Proceedings of the Zoological Society of London* journal in 1837, as sole author, that is, prior to them being subsequently treated in *The Zoology of the Voyage of the Beagle* by him under Darwin's editorship during 1838–1841. For example, according to Raymond Paynter,[23] it was nine species (of which two are today combined into one); to Tree,[24] it was 11; to Sulloway,[25] it was 13; to Weiner,[26] it was 14; and to Aydon,[27] it was 12—although Darwin clearly stated it was 13 species; see the "Galapagos Islands Ornithology" section of chapter 2. The confusion is not surprising, for Gould himself wrote of his finding 14 new ground finch species confined to the Galapagos Islands.[28] The contradictory numbers in Darwin's time might have been due to the minutes of the Zoological Society meeting or the manuscript of its proceedings being changed—or because Gould possibly amended the latter. Indeed, the unpublished minutes do record that Gould named 12 finch species, as he failed to appreciate the Warbler Finch (*Certhidea olivacea*) was one such until shortly thereafter.[29]

The more recent confusion is, perhaps, less understandable. Although Gould did describe and name 13 species of Galapagos finches on 10 January 1837, he added an additional species on 10 May. By the time that the pertinent pages of the *Proceedings of the Zoological Society of London* were formally published (not before 3 October 1837),[30] Gould had, however, reverted the number of species to 13 (not all valid as species today). Errors were almost inevitable, given the novel nature of these most confusing similarly drab small species and Gould's limited samples of them to hand—some 60 or so for all of the species in total; see following. Also there could be errors because several of his perceived species have subsequently proved to have been based on what are now seen as subspecies or anomalous specimens.

The two men met in March 1837 when Gould reported that it appeared to him, based on Darwin's information, that each of 13 Galapagos finch species (some previously considered as mere varieties by Darwin) was peculiar, or endemic, to its own island. He could not be absolutely sure about this, however, because CD had not recorded from which island he had collected each specimen. As noted, this interpretation of the classification of his finch specimens paralleled CD's view of the differing forms of the mockingbirds on the various Galapagos Islands. This set Charles to thinking again about the possibility of differing varieties, or forms, of birds (and giant tortoises) morphologically diverging further over millennia in their island isolation to eventually acquire the status of distinct good species: species that were not, therefore, created in their present location and form by any god but that had evolved there.

To confirm his taxonomic conclusions, Gould, together with Darwin, needed to examine a larger sample of specimens of Galapagos finches recorded as collected from specified islands. Captain FitzRoy had collected 21 finches and several of his officers collected a few, notably Benjamin Bynoe and Edward Hellyer, as well as FitzRoy's shipboard servant Harry Fuller, who collected eight finches for himself. These men had been more meticulous in identifying which island each of their finch specimens had been obtained from. Darwin approached FitzRoy for permission to view his Galapagos finches, be it the 13 that FitzRoy had already donated to the British Museum in London or all 21 of them, along with his own material for direct comparison together at one time and place. FitzRoy readily complied with this request. In addition, Syms Covington had personally collected four finches; and, as he too had recorded their particular island homes, they were also examined comparatively with all of the other specimens.[31] The island-confirmed specimens enabled Gould and Darwin, by comparing the morphology of them with that of their own birds, to work out which island the majority of CD's skin specimens must have come from, but not without some errors.

As Darwin's understanding of the Galapagos land birds *while in the archipelago* is a fundamentally important point in the legend-laden history of Charles Darwin and the significance of his perception of the evolution of Galapagos finches and mockingbirds, I must clarify the issue. Darwin had hypothesized the South American origin of some of the Galapagos fauna while there but not specifically of the finches. In his field notebook titled on its front cover "Galapagos, Tahiti, Lima" (stolen from Down House ca. 1983), CD writes on page 30b of the Galapagos Islands, "I certainly recognise S. America in ornithology."

Another note of CD's about the Galapagos Islands mockingbirds, thought to be made during the voyage, reads "These birds are closely allied in appearance to the Thenca [Chilean Mockingbird (*Mimus thenca*)] of Chile . . . or Callandra [Calandria; Chalk-browed Mockingbird (*Mimus saturninus modulator*)] of la Plata."[32] Moreover, the fact that Darwin erroneously thought one of the Galapagos finches to be an American blackbird species, by noting it as questionably being of the genus *Icterus*, while still in the islands is unequivocal evidence of him comprehending the South American character of some Galapagos birds.[33]

Although the mockingbirds raised his suspicions about the immutability of species being incorrect, Darwin did not become a true convert to evolution during his voyage. He recorded this as fact in writing to the German naturalist Otto Zacharia in 1887 thus: "When I was on board the Beagle I believed in the permanence of species, but, as far as I can remember, vague doubts occasionally flitted across my mind."[34]

Once advised about them upon his return to England, CD was able to confirm that the land birds of the remote Galapagos Islands, including the finches, did indeed have their closest relatives in, and were therefore derived from, the South American mainland.[35] Sulloway observed that "What had particularly impressed Darwin about his Galapagos collections was the fact that *virtually all the land birds (25 of 26 species), although closely related to South American forms*, were endemic not only to the archipelago as a whole, but, in some instances, to specific islands within that group" (emphasis mine).[36] Sulloway subsequently wrote that "Darwin [while in the Galapagos], on the other hand, is said to have thought the finches were derived from a mainland species and had become modified by their new surroundings."[37] He then, correctly, excluded the Galapagos finches from this by writing "Darwin's finches played no role in this aspect [i.e., their 'American character'] of his evolutionary insight";[38] for indeed, no direct evidence for this specific insight by Darwin while among the islands exists.

Darwin's thoughts inevitably, then, led him to the notion that perhaps ancestors of Galapagos birds and reptiles of today had somehow reached the islands from the South American mainland during the geological past. There they became isolated on one or more islands within the Galapagos archipelago to then differentiate and eventually become distinct species, or speciate, over the millennia. This was then an utterly revolutionary idea—but one that CD's preceding observations and thinking must surely have prepared him for to some, if but slight, degree. In the revised edition of his *Journal or Researches*, he wrote of the Galapagos, "The natural history of those islands is eminently curious, and well deserve attention. Most of the organic productions are aboriginal creations, found nowhere else; there is even a difference between the inhabitants of the different islands; yet all show a marked relationship with those of America, though separated from that continent by an open space of ocean, between 500 and 600 miles [805 to 966 km] in width. The archipelago is a little world within itself, or rather a satellite attached to America, whence it has derived a few stray colonists, and has received the general character of its indigenous productions. Considering the small size of these islands, we feel the more astonished at the number of their aboriginal beings, and at their confined range."[39] The vertebrates of the Galapagos Islands had, then, provided Darwin with the backbone of an embryonic and profound theory that he was to spend the next two decades or so fleshing out, testing, and accumulating hard evidence for.[40]

Just as Charles was deeply struck by the apparent speciation of the various finches among the islands of the Galapagos Archipelago so he was by John Gould's finding that the smaller, or lesser, rhea was not merely a variety of the larger, or greater, rhea but was a distinctively different species: The two kinds apparently occupying different parts of the South American continent. Charles was undoubtedly delighted when John Gould celebrated his discovery of the new ratite by naming it *Rhea Darwinii* after him (*pace* the *Struthio darwinii* of Huxley and Kettlewell[41]); see colour figure 2.2. Not only was Charles present at the 14 March 1837 meeting of the Zoological Society at which Gould presented his original description of the new rhea, but he immediately followed it with notes of his own on both rhea species living in the wild.[42] He described the Greater Rhea as shy and yet easy prey to people approaching it "on horseback in a semicircle."

The Zoology of the Voyage of H. M. S. Beagle, Part III: Birds, was published in five, quarto, parts: part 1 in July 1838; 2, 3, and 4 in January, July, and November 1839, respectively; and part 5 in March 1841. Although actually written by John Gould and Charles Darwin, with a brief appendix by Tom C. Eyton, the title page of the completed volume attributes the entire work to Gould (not George R. Gray[43]), edited and supervised by Darwin. In his "Advertisement" to the work, CD states that the 50 colour plates "were taken from sketches made by Mr Gould himself, and executed on stone by Mrs Gould with that admirable success, which has attended all her works";[44] see, for example, colour figures 2.2 and 2.7.

Aside from the taxonomic order, nomenclature, species descriptions, and discussion on these topics, predominantly the result of considerations by Gould and Gray, which Darwin added his views to, Charles was not without his own taxonomic opinions. For example, Gould thought that a single specimen of the Dark-faced Ground-Tyrant *Muscisaxicola mentalis* (now *M. maclovianus*) from the Falkland Islands was a distinct species; but CD disagreed and was correct in doing so.[45] Darwin's major contributions to this work are his notes on the distributions, ecology, breeding biology, and behaviour of the living birds. As many of these observations appear, often all but verbatim, in his previous *Journal of Researches* and/or his subsequent books, I include in this chapter reference to only the more substantial species accounts within his *Zoology of the Beagle* not detailed elsewhere in this book. Darwin's less substantial individual ornithological contents are to be comprehensively found succinctly summarized in appendix 1. I do not include details from species accounts within the *Birds* volume of *The Zoology of the Beagle* that contain little or no more than morphological or anatomical descriptions of specimens and their collecting locations.

New World Vultures

As long ago as the early 1850s, it was appreciated that the Old World and the New World vultures exhibit distinctive differences in their anatomy, suggesting that they may have different origins. The subsequent long-established view of the New World vultures was that they and the Old World vultures all belonged to the single "birds of prey" order Falconiformes—separated within that order by other groups of raptors. Recent genetic studies have shown that although all vultures belong to the single raptor order, now the Accipitriformes, the Old and New World vultures form two distinct families that are not particularly close within it. More recently, however, the New World vulture family Cathartidae was treated as constituting the order Cathartiformes and three odd vultures, the Palm-nut (*Gypohieraax angolensis*), Bearded (*Gypaetus barbatus*), and Egyptian (*Neophron percnopterus*) Vultures, separated from the true Old World vultures by the Secretarybird (*Sagittarius serpentarius*), Osprey (*Pandion haliaetus*), and also 15 species of raptors in five genera within the family Accipitridae.[46] The similarity in morphology, ecology, and behaviour between the two geographically and genetically discrete groups of vultures offer a remarkable example of convergent evolution on the continents of Africa and South America.

The Black Vulture (*Coragyps atratus*; colour figure 3.1), which Darwin often referred to by the local name of *Gallinazo*, was thought by him to never occur south of the "neighbourhood of the Rio Negro, in latitude 41°"; and he never saw one in Patagonia, Tierra del Fuego, or Chile. It does not occur at Tierra del Fuego but is now known at least in some places elsewhere, if rarely. In Charles's experience, they preferred damp places and especially rivers, accounting for the fact that he found them abundant on the Rio Negro and Colorado Rivers but not on the intermediate plains. In citing several publications, CD correctly notes that the species is more widely distributed in the northern than the southern half of continental South America. He noted the species as a gregarious one, not only at sources of food but also as flocks soaring in circles at great height in what he interpreted might be for sport or perhaps for pair formation.[47] Whereas the latter, unlikely, function is perhaps conceivable, the anthropomorphic former is not.

Of the Turkey Vulture (*Cathartes aura*), Darwin discussed its range, noting that it occurs in the extremes of Tierra del Fuego and the coast of West Patagonia, where the Black Vulture does not, and that it is in some areas migratory. It also reached the Falkland Islands, where it was "tolerably common" but shy, in marked contrast to most other birds there. Darwin therefore suspected they were migratory there, learning to fear people elsewhere than on the Falklands. In fact, the status of the bird on the Falklands is not fully resolved; for although it is described as resident there, the population is of the same subspecies as that on southeastern South America, which might suggest gene flow between the island and mainland populations. He found it to be solitary or in pairs and readily identified "at a great distance from its lofty, soaring and most graceful flight." In places, he thought birds must live there exclusively on what "the sea throws up, and on dead seals: wherever these animals in herds were sleeping on the beach, there this vulture might be seen, patiently standing on some neighbouring rock."[48]

The Caracaras

The caracaras are some 11, mostly long-legged, bird of prey species in five genera and long considered as evolutionarily basal within the falcon family Falconidae.[49] As a group, the caracaras occupy South America north to the southernmost United States, with two species on the Falkland Islands. Some are remarkably tame, bold, inquisitive, clever, and amusing; and although CD did find them entertaining at times, he also found them disgusting.

Darwin pointed out that the taxonomy of the caracaras was confused, with several new genera being lately introduced, and George Gray therefore helped him with a brief discussion of the topic. Notes, in some cases limited to little more than assumptions about distributional ranges, were presented under six species names (see appendix 1). Much was seen of the various species of caracara by Darwin. Of the handsome and widespread Southern Crested Caracara (*Polyborus plancus*), commonly called the *Carrancha*, he noted it to "generally follow man, but is sometimes found even on the most desert plains of Patagonia." Its habitats and feeding habits were described, including the fact that it visited estancias and slaughtering houses in great numbers to eat carrion, which it will eat in company with Chimango Caracaras and Black Vultures with aggressive interactions.

While gathering at a food source, the Southern Crested Caracara was not found to be gregarious; elsewhere birds were alone or in pairs. They often took carrion from salt and fresh water shores, and CD thought that in Tierra del Fuego, such must represent their exclusive diet. Crested Caracaras were reported to be very crafty, stealing numerous eggs and picking scabs from sores on horses and mules, killing wounded animals and rarely attempting to take a fit one as was once witnessed by Mr. Bynoe of the *Beagle*'s crew when a bird attacked a "live partridge." Darwin stated that a person lying down on the open plain will awake to find one of these birds watching from surrounding hillocks "with an evil eye." With their crop full to bulging through their breast plumage, individuals were seen as an "inactive, tame, and cowardly bird." Its flight was heavy and slow "like that of an English carrion crow [*Corvus corone*], whose place it so well supplies in America," seldom involving soaring but sometimes gliding at great height.

The Southern Crested Caracara runs across the ground but not as swiftly as do its close relatives. In calling, it throws its head back, beak wide open, until the crown of the head almost touches its back. Darwin briefly described the nest and nest site and noted that, in addition to carrion, this caracara eats worms, shells, slugs, grasshoppers, frogs, and destroys young lambs by tearing their umbilical. Birds will also chase vultures, herons, and gulls, alone or in groups, until they regurgitate food. Charles concluded the bird to be one "of very versatile habits and considerable ingenuity."

Darwin was convinced that the Chimango Caracara (*Milvago chimango*) never kills prey but predominantly eats carrion, being the last species to leave a skeleton, while also eating anything else as an omnivore—including bread thrown out of houses. They injure newly planted potato crops by digging up the roots and follow the plough to eat earthworms and insect larvae. Their flight is heavy, and CD never saw one soar. Individuals proved tame, gregarious, commonly perched atop stone walls rather than trees, and uttered a gentle, shrill, scream.

Having discussed its somewhat complex synonymy, Darwin wrote more about his *Milvago leucurus* (now the Striated or Forster's Caracara [*Phalcoboenus australis*]) than other caracaras. He was informed that Striated Caracaras nest on rocky cliffs of small outlying islets and noted this as an odd behaviour in such a fearless bird—but without mention of "foxes" as being potential nest content predators.

In driving a land rover to a King Penguin (*Aptenodytes patagonicus*) colony from Stanley, capital of the Falkland Islands, I personally witnessed the remarkable tameness of the Striated Caracara. Seeing a big dark bird of prey on the ground ahead, standing behind a large grass tussock, I stopped the vehicle some 10 m from it and got out with my camera. I crouched down a little and slowly approached the clearly inquisitive bird. When close enough to take a picture, I tossed a small stone along the ground toward the bird in the hope that it would disturb it sufficiently to have it move out from behind the tussock, and it did so only to run in an amusing gait directly towards me to investigate the stone; see colour figure 3.2.

Darwin was misled into thinking the plumage of the Striated Caracara sexes "differs in a manner unusual in the family," presumably because he examined few birds and paid attention to what local Spaniards said to him, as he clearly describes the adult and the juvenile plumages and not the adult sexes of this relatively short-legged species. Darwin also found the bird "exceedingly numerous" and fearless on the Falklands—and although it is found on islands adjacent to it, it does not occur on the mainland of Tierra del Fuego as CD was correctly informed. Its food is carrion and "marine productions"; and in some locations, only the latter would have been available to it. This caracara readily attacks other birds; and several were seen to chase and kill a wounded cormorant, geese, and a heron, while others waited at a rabbit burrow entrance to seize the mammals as they came out. They thus cooperate as predators. One pounced on a sleeping dog, and others attempted to fly aboard vessels at anchor where they would eat hide lashed to ropes and steal any food left on deck. Individuals would pick up almost any, not too heavy, object from the ground, including a large hat, a pair of heavy balls used to catch cattle, and a compass. Whereas their flight is heavy and clumsy, they run rapidly over the ground with head held forward like "a pheasant." Darwin found it a noisy species, uttering harsh cries including one like that of the Rook (*Corvus frugilegus*, a kind of crow) of Britain and Europe—and as a result sealers called it by that name.

Hungry sealers ate Striated Caracaras (colour figure 3.2) and found their cooked flesh white, as a fowl's, and very good; whereas Darwin and some of the crew of the *Beagle* caught ashore for a time found the flesh of the Southern Crested Caracara far less pleasant to eat. Little else of significance was written about the other caracara species as defined in the *Birds* volume of *The Zoology of the Beagle* other than descriptions and, now outdated, taxonomic discussion of their morphology.[50]

The Galapagos Hawk

Darwin's newly discovered Galapagos Hawk struck him as very caracara-like in many respects, particularly in its ability to run quickly over the ground.

It is not a caracara, however, but a typical buzzard of the genus *Buteo* of the family Accipitridae. It is a ubiquitous bird on the Galapagos Islands where its tameness continues to impress and entertain visitors today.

John Gould described the Galapagos Hawk as a new genus and species (*Craxirex galapagoensis*); and Darwin stated that on the islands, it "supplies the place of the Polybori and Milvagines," or the caracaras, because he found its habits and tameness more like those of caracaras than of buzzards. It is, however, morphologically more of a buzzard than a caracara; and CD was made aware of this and so presented it as a member of the subfamily Buteoninae, although noting he could not decide if it should in fact be placed within the subfamily Polyboinae. But it is indeed a buzzard and is thus now known as *Buteo galapagoensis*; see colour figure 3.3.

Galapagos Hawks were "excessively numerous" in all habitats on the archipelago, to which Darwin correctly thought it was confined as an endemic species. Individuals would eagerly attend giant tortoise kills and butcherings and sit on the ground or perch in adjacent trees to await undisturbed access to the reptile's intestines and any meat left attached to the carapace. They killed and ate hatchling tortoises, iguanas, and lizards; ate all offal, dead fish, and marine productions washed ashore; and were said to kill young doves and chickens. In flight, it is not elegant or swift, but it runs over the ground quickly. It is a vocally noisy species, with many different calls. The craw is feathered, and CD thought it did not protrude the crop when full, as does the unfeathered craw of caracaras. Nests are built in trees, and eggs were just beginning to be laid in October. On James Island, CD saw 30 birds standing within a hundred yards of the crew's tents, and their plumages led him to believe that females do not acquire full plumage until late in life. The pale-breasted female plumage is actually the same as that of juveniles and immatures (adult males being sooty black), Darwin apparently being led to believe that adult females also became dark-breasted like adult males.

Following his Galapagos Hawk account, Darwin briefly described other buzzards under the names *Buteo erythronotus*, *B. varius*, and *B. ventralis* from the Falkland Islands and South American mainland and little more than this; the first two names both now referring to the Red-backed Hawk (*B. polysoma*), and *B. ventralis* being that for the Rufous-tailed Hawk. He then also briefly detailed the two falcons and the only harrier that he collected, on South America.[51] See appendix 1 for updated names.

New World Owls

Of the six owl species that Darwin listed as collected, only four are now recognized as valid species; see appendix 1. He had little to say about them beyond describing and detailing where he obtained them except for the Burrowing Owl, which he named *Athene cunicularia*. The latter is a small owl now placed in its own genus *Speotyto*, most closely related to *Athene*, which Gould and

Darwin attributed it to (and some authorities continue to do so). It occurs throughout much of the open country of the southwestern United States and South America and some Caribbean islands.

Darwin pointed out that the Burrowing Owl was often mentioned by previous travellers of the pampas because of its conspicuousness there as a common and entertaining inhabitant. During bright daylight and evening, they stood in pairs on a low rise near their burrow; but when disturbed, they entered it or gave a shrill harsh call as they made a short undulatory flight to land on the ground and then stare back at the intruder. Having found mice in the stomachs of the specimens he collected, CD could not credit how they could survive in their open habitat until he became aware of just how numerous small rodents were. He also saw one kill and carry off a snake and was informed that reptiles are commonly taken, as are insects elsewhere, and that the birds may nest in Marmot and Prairie Dog burrows in the United States;[52] see chapter 2.

Mockingbirds

Today the mockingbirds are classified as members of the songbird family Mimidae and consist of 14 species in the single genus *Mimus*; but only six of these had been named prior to the voyage of the *Beagle*, and on that voyage Darwin discovered three that John Gould subsequently described as new species.[53] The mockingbirds are named for conspicuously vocally mimicking the calls of other birds, although the Galapagos populations are not known for this ability. It was this group of birds on the Galapagos Islands, and not the finches there, that first brought Darwin's attention to a group of vertebrate animals that had diversified on islands, within an archipelago, originating from a common ancestor that colonized the islands from a distant mainland. He was only too well aware of where such an ancestor originated from, having previously observed mockingbirds on both sides of the Andes in South America; see "The Zoology of the Voyage of the *Beagle*" section previously.

In first describing what is now the Chalk-Browed Mockingbird (*Mimus saturninus*), John Gould unfortunately wrongly attributed it to the Straits of Magellan, where it is not found.[54] Darwin found it extremely common on the banks of the La Plata where it is called the *Calandria*, observing that the Patagonian Mockingbird (*Mimus patagonicus*) replaces it only a few degrees to the south, and that it was tame and bold in Banda Oriental. The birds frequented country houses to pick at meat hung to dry, a resource they defended from other birds by chasing them off, and actively hopped about thickets and hedgerows, often elevating and slightly expanding their tails.

Darwin found the habits of the Patagonian Mockingbird slightly different to those of the Chalk-browed Mockingbird—it being shyer, not moving its tail as much, and living on the plains and valleys thinly scattered with stunted thorny trees. While perched at the apex of a bush, it often "enlivens the dreariness of the surrounding deserts by its varying song," which CD likened to that of the Sedge Warbler (*Acrocephalus schoenobaenus*, familiar to him in England) but much more powerful. The calls of the mockingbirds as a whole, which he said sang only in spring, struck him as superior to almost every other South American bird he heard. Interestingly, he concluded from only their differing habits that the Chalk-browed and Patagonian mockingbirds were two distinct species. When confronted by specimens of both, he changed his mind, however, only to find that as soon as John Gould examined them he proclaimed them to be different species—this causing CD to observe "a conclusion in conformity with the trifling difference of habit and geographical range, of which he [Gould] was not at the time aware." Thus Charles was learning significant lessons about defining species among closely related birds, both from his field observations and his collaboration with Gould, the avian taxonomist.

The Chilean Mockingbird (*Mimus thenca*) appeared to CD to be confined to the pacific coast, west of the mountains, where it replaced the Chalk-browed and Patagonian species of the Atlantic side. It occurred, he thought, south to about Concepcion, at latitude 37° (but is now known to the Puerto Montt area), where habitats changed from thick forest to open ones. The *Thenca*, as local Indians called it, was common in central northern Chile and was found near Lima, Peru. In habits, it was like the Patagonian Mockingbird; but Darwin saw many with their heads stained by yellow flower pollen and concluded this resulted from birds taking small beetles concealed within the flowers. In this he appears to have been correct, as no mention is now made of nectar feeding by mockingbirds.[55] Contrary to a report of it nesting within a long passage, CD was, correctly, informed that it makes a simple nest that is externally of small prickly mimosa branches.

Only a very brief description is given of the Charles Mockingbird, named for the Galapagos island, but now called the Floreana Mockingbird (*Mimus trifasciatus*; see colour figure 3.4). The Floreana Mockingbird became extinct on Floreana Island before 1888, possibly due to predation by cats, and now lives on only two small adjacent islands where it numbers less than 300 individuals as an endangered species.[56] Gould defined and briefly described birds from Chatham and James Islands as another species, now named the San Cristobal Mockingbird (*Mimus melanotis*; see colour figure 3.5), which does not occur on James Island.

The bird Darwin listed as *Mimus parvulus* is now known as the Galapagos Mockingbird, which lives on Darwin (Culpepper), Wolf (Wenman), Pinta, Genovesa (Tower) and Isabela (Albemarie), Fernandina (Narborough), Daphne, Santa Cruz (Indefatigable), and adjacent smaller islands. Under this account, CD reviewed the fact that each of his morphologically different mockingbird species came from different islands, detailing which bird for which island(s), and that their small distances apart have the islands in sight of one another. The habits of the various species are similar, including the Chilean Mockingbird, and CD observed "they evidently replace each other in the natural economy of the different islands." He imagined that the tone of their voices differed slightly and found them lively, inquisitive, tame, and active birds that run fast; one landed on a cup he held and drank from it. They seemed to prefer dry coastal habitats but also occupied higher, damper, ones where they frequented houses and clearings and where they tore bits of drying strips of tortoise meat. He was, correctly, told that their nests are simple and open structures.[57]

Ovenbirds and Tapaculos

The 15 or so species of *cinclodes* are members of the very large New World songbird family Furnariidae, or ovenbirds, all now members of the genus *Cinclodes*. They are ground-frequenting birds of South America, save one that also lives on the Falkland Islands, that are predominantly buff to brown in general plumages with buff, brown, or white wing markings, and typically with a white superciliary eye-stripe and throat.

The next 23 pages of the *Birds* volume of the *Zoology of the Beagle* predominantly deal with other members of the Furnariidae, and the closely related small tapaculos family Rhinocryptidae, all presented under the thrush family Turdidae: the latter merely reflecting the poor taxonomic appreciation of little-known bird groups at that time. Given the state of the taxonomy of these numerous and predominantly drably plumaged birds, Gould and Darwin presented a treatment of the representatives of 24 species collected on the *Beagle* voyage, admirable for its time. As a good deal of the text in some of CD's species accounts is pure taxonomic discussion, I summarize here only those with sufficient original biological content, bearing in mind that next to nothing was known of them as living birds at the time that he observed them.

Darwin saw a good deal of four species of cinclodes and gave some observations of their then little-known habits. Of the widespread and geographically variable Buff-winged Cinclodes (*Cinclodes fuscus*), he found its general habits and voice similar to those of the Common Miner (*Geositta cunicularia*) but those of other cinclodes different to it. In flight, he found its two red wing bars were conspicuous, making it identifiable at a distance. It does not walk but hops and feeds exclusively on the ground; and CD found only beetles had been eaten by those individuals he collected, and of these beetles, many were fungi-feeders. Birds often frequented lake edges, where they foraged among detritus at the water's edge, open grassy plains where they often turned over dry mammal dung, and also open habitat of higher mountains of Tierra del Fuego. It was not common on the Patagonian coast and, contrary to what CD thought, it does occur in Chile. In not frequenting coastal beaches, this bird differs conspicuously from the other cinclodes species that CD observed. Dark-Bellied Cinclodes (*Cinclodes patagonicus*) were found "extremely common" on the sea shores of Tierra del Fuego but were also seen in stony arid valleys of the Cordillera at 8,000 feet, or 2,600 m, above sea level. They are presently recorded as occurring up to 2,500 m.[58]

The Blackish Cinclodes (*Cinclodes antarcticus*) inhabits the Falkland Islands where it is extremely tame—so much so that in 1763, the author Antoine-Joseph Pernety is quoted by Darwin as having killed 10 in half an hour "with a wand" and as noting it was so tame that "it would almost perch on his finger." In being "in no doubt that it is peculiar to this [island] group," CD was wrong, as it does in fact occur, as a distinct subspecies, on several islands off the extreme tip of the South American mainland. This was an understandable error, however, as he did not visit these islands. Darwin collected a Seaside Cinclodes (*Cinclodes nigrofumosus*) at Coquimbo on the Chile coast (colour figure 3.6). Other than the Buff-winged, Charles found that the various cinclodes almost exclusively lived on the sea beach, be it shingle or rock, where they foraged immediately above the surf among detritus. Pairs were occasionally seen similarly foraging on large river pebble beds far, once 10 miles or 16 km, inland.

Cinclodes species in Tierra del Fuego and the Falklands were hardly ever more than 20 yards or 18 m from the beach and were seen "walking on the buoyant leaves of the *Fucus giganteus* [Giant Kelp (*Macrocystis pyrifera*)] at some little distance from the shore"—but it was possibly an exaggeration for Darwin to write "In these respects, the birds of this genus entirely replace in habits many species of Tringa." He found small crabs, little shells, and one *Buccinum* up to a quarter of an inch, or 0.6 cm, long in the stomachs of Dark-Bellied Cinclodes and cites Friedrich Heinrich Freiherr von Kittlitz as finding some small seeds in them. They are typically tame and solitary but infrequently occur in pairs, call seldom, and both hop and run quickly. A nest of this species containing young was found by Darwin on 20 September near Valparaiso, placed in a small hole in the roof of a deep cavern close to the bank of a pebbly stream. He found another nest in the Chanos Archipelago three months later with a pure white egg in a small hole beneath an old tree close to the beach.[59]

The oddly named Huet-Huet (*Pteroptochos tarnii*) of south Chile and adjacent Argentina is a member the passerine family Rhinocryptidae. Darwin recorded it as known by native Indians as *guid-guid* on Chiloe Island, where it was abundant all over. English sailors most appropriately named it the barking bird for its dog-like call. Its call causes people to search in thickets, for what they think is a dog, but in vain—whereas at other times it might approach a still and quiet person and perch close by with tail erect. It feeds on the ground beneath thick vegetation and rarely flies, and then only for short distances, but hops vigorously with its tail erect. The nest, according to local informants, was built "amongst rotten sticks, close to the ground" but this is incorrect, as it in fact builds its nest at the end of a burrow.

The Huet-Huet's closest relative, the Moustached Turca (*Pteroptochos megapodius*) that is called *El Tutco* locally, replaces it to the north in dry habitats of stony hills with sparsely dispersed bushes of central Chile; and its world distribution is tiny. It hops from bush to bush with tail erect and looking "very strange and almost ludicrous" to give calls from concealment as strange as is its appearance. Beetles, vegetable fibres, and pebbles were found in their "extremely muscular gizzards"; and CD observed that this and "the fleshy covering to the nostrils, and the arched, rounded wing, and great scratching claws" made it easy to imagine some distant relationship to the "Gallinaceous order." But in this he was surely noting convergent evolution rather than suggesting any real relationship. He was, correctly, told it nests at the end of a deep burrow, which it excavates in the ground.

The Chucao Tapaculo (*Scelorchilus rubecula*) was noted to have much the same range as the Huet-Huet, but in fact it is more restricted. In Chiloe, Charles found it called the *Cheucau*, where it frequents gloomy places in damp forest. Like the Huet-Huet, it proved at times extremely secretive but at others would approach a still person closely, hopping with elevated tail. Superstitious local people feared it because of its variable strange calls. It was reported to nest build among sticks close to the ground; but in fact, it too nests at the end of a burrow.

The White-Throated Tapaculo *Schelorchilus albicollis* is found to be extremely common in central Chile where it occurs well north of the Huet-Huet and Chucao Tapaculo but does overlap the northern range of the Moustached Turca. In feeding and habits, CD found it like the Moustached Turca and "very crafty" in freezing at the base of a bush when disturbed, until after a while it will "try with much address to crawl away on the opposite side." Its calls vary with the seasons; and CD was correctly informed that it nest builds at the bottom of a deep burrow, like the Moustached Turca, Huet-Huet, and Chucao. The various species accounts of the next 16 pages of the *Zoology of the Beagle* are brief, including descriptions, notes on distributions, habits, and in a few cases nesting[60] and are detailed in appendix 1.

Galapagos Finches and Relatives

Of fundamental significance among these birds of the *Beagle* voyage with respect to the Darwinian literature are the Galapagos finches. Having carefully examined and considered the near 60 individual drab little birds collected on the Galapagos Islands by Darwin and others aboard the *Beagle*, John

Gould described 13 new species, including the Warbler Finch, as Gould could see that it was in fact a fine-billed finch. These are detailed in the *Zoology of the Beagle* volume on birds.[61]

The finches were presented as members of the "family Coccothraustinae" [*sic*] containing the genus *Geospiza* (8 species), and the subgenera *Camarhynchus* (2 species), *Cactornis* (2 species), and *Certhidea* (1 species) by Gould and Darwin.[62] Since then, a string of differing classifications of these finches has appeared.[63] As a result of recent sophisticated genetic studies, the Galapagos finches are now known to be members of the tanagers and allies family Thraupidae, and thus are not finches at all. To complicate the morphological and taxonomic problems of so uniform, but complex, a group is that they include a small number of species now known to hybridize among themselves. Today some species have, to add to the complications, been subdivided into subspecies so there now being 26 of them,[64] these representing additional subtle variations that Gould and Darwin were unaware of while trying to make sense of the group.

Most of the species accounts in the *Zoology of the Beagle* provide little more than a brief description and a listing of which island(s) each Galapagos finch species occurred on: but not always correctly. However, CD's introduction to them offers more; and as this group has become so fundamentally significant to the literature on Darwin, I cover the salient points here. At that time it was thought that the finches were confined to the Galapagos Archipelago, but as noted previously, an additional species was subsequently described (as *Cactornis inornatus*, which is now *Pinaroloxias inornata*) from Cocos Island, located between the Galapagos and the Costa Rican coast.[65] Although it was long taken that these finches must have reached the Galapagos from the west of the South American mainland, as CD thought had to have been the case, their closest living relatives are now known to be several genera of similar superficially finch-like species from the Caribbean Islands.[66] All species except the dully olive-backed Woodpecker Finch (*Camarhynchus pallidus*, not known when CD was in the Galapagos) and Warbler Finch (*Certheida olivacea*, which is now split into two species: the Green Warbler-Finch [*C. olivacea*] and Grey Warbler-Finch [*C. fusca*]) involve drab plumages that are all black or are brownish above and dully whitish below, which may or may not be sparsely to heavily striated with brown to black markings.

The members of the "genus" (i.e. the finches, as Gould and CD saw them) were "very numerous, both in individuals and in species, so that it forms the most striking feature in their [the Galapagos] ornithology"; and the characters of the various species "run closely into each other in a most remarkable manner";[67] see colour figures 3.7–3.11. That CD did not at the time appreciate he had collected not six but nine species of finch "should by no means be taken as a sign that he was ornithologically inexperienced or inadept [*sic*]," as was correctly observed by Frank Sulloway.[68] Whereas Darwin later implied, in hindsight, that he suspected that "certain members" of the finch species were each confined to "different islands," he admits to not appreciating this while he was in the Galapagos Archipelago.[69] As a result, he did not record which specimen was collected from which island, and this was to result in much subsequent confusion, research, and publications;[70] see the "Galapagos Islands Ornithology" section of chapter 2. In fact, CD collected finches far too briefly among too few islands to be able to appreciate that most species occur on more than one island.

Darwin's observations suggested to him that all-black finches "were invariably the males," although Benjamin Bynoe of the *Beagle* crew found, on dissecting many, that a few black birds were female; CD put this down to the phenomenon of some birds "in a state of high constitutional vigour, assuming the brighter plumage of the male." It is now well known, however, that some female birds in many groups attain male plumage characters in their old age (as some elderly women acquire a moustache and/or beard). Charles also determined that in some species, black individuals were far fewer than brown ones, and he assumed this was because it took three years to become black. The time required to attain adult plumage varies, however, between species; while one work considers Bynoe's black females erroneous.[71]

Darwin found the various species of finches "undistinguishable from each other in habits," which is unsurprising given his brief experience of them. He noted that they intermingled in large flocks, and did likewise with doves, and frequented "rocky and extremely arid parts of the land sparingly covered with almost naked bushes, near the coasts" where they foraged for seeds in the soil using beak and feet. They often ate bits of succulent leaves, and CD thought this was probably for their water content;[72] see colour figure 3.7).

Charles Darwin's work on the Galapagos finches back in England clearly showed them to represent an island archipelago-isolated group of great significance to the potential understanding of the evolution of species. As a result, they have stimulated a long series of major, intensive, and long-term studies by dedicated and eminent ornithologists, and students of CD's work, which have made them one of the best-studied and understood groups of birds on earth.[73] It was the title of the great David Lack's pioneering book, *Darwin's Finches*, that cemented the name of Darwin's finches for them in ornithology, the popular media, and the minds of the general public. It also played a large part in establishing the legend that has it that it was these birds that brought Darwin to his theory of the origin of species. However, it was the ornithologist Percy R. Lowe who apparently first applied the name Darwin's finches, in giving a 1935 lecture at the British Association to celebrate the centenary of Darwin's visit to the Galapagos Islands.[74]

It is shocking to record that the dramatic discovery and description of the Galapagos finches as new to science, and Darwin's subsequent celebrity and association with them, saw 460 finches collected from the archipelago in

1868; about 1,100 of them in 1891; 3,075 in 1897; an incredible 8,691 during 1905–1906, by the California Academy of Sciences;[75] plus numerous others. Although these numbers are certainly excessive, it was only with larger sample numbers of the various species that their taxonomy, much complicated by their similar and confusing various and variable plumages, was eventually resolved. This enabled ornithologists to show that the 13 Galapagos finch species that Gould and Darwin thought their specimens then indicated (*pace* McCalman[76] who states more than 20 species) in fact represented only nine valid species. See chapter 4 for more of Darwin's finches.

American Blackbirds

The next bird that Darwin had anything significant to write about was the Shiny Cowbird (which he called *Molothrus niger*, but is now known as *M. bonariensis*). It belongs to the family Icteridae of some 107 species, loosely referred to as the American blackbirds, which is placed only five families removed from the apex (i.e., birds perceived to be the most advanced) of the avian evolutionary tree.[77]

Charles found the Shiny Cowbird common in large flocks on the grassy plains of La Plata, often together with Brown-and-yellow Marshbirds (*Pseudoleistes virescens*) and other species. He saw few brown individuals relative to black ones and so thought the former were young birds and was somewhat scathing in observing that Félix Manuel de Azara[78] should have appreciated this rather than suggesting that they were the females. In fact, both men were correct, as the young of both sexes look like the brown females. The Brown-Headed Cowbird (*Molothrus ater*) of North America has an extremely similar female plumage, as CD noted in pointing out their very close relationship and similarities in calls and habits.[79] Other notes on these birds appear under the "Cuckoo Instincts" section in chapter 4.

Hummingbirds

Darwin was delighted and enthralled to see living hummingbirds in their native habitats. The hummingbird family, the Trochilidae, now includes some 343 species.[80] Darwin collected and made notes on only three species; but of the Glittering-Bellied Emerald (*Chlorostilbon lucidus*), he reports only that it was not abundant at Monte Video.

Of his *Trochilus forficatus*, now the Green Backed Firecrown (*Sephanoides sephanoides*), Darwin noted it as occurring over a range of 2,500 miles, or 4,023 km, on the west South American coast "from the hot dry country of

Lima to the forests of Tierra del Fuego." On damp Chiloe Island it was as abundant as any bird, and commonly frequented open marshy ground supporting bromeliad plants. At that time, Chiloe Island had no flowers, and the hummingbirds were feeding on crane flies and other small insects. In central Chile, the migratory population appears in autumn, initially in mid-April, and starts to depart in September. Nests were numerous on Chiloe and Chonos Islands during the summer migration of the species. When it migrates south, the Giant Hummingbird replaces it.

Hummingbird migration, CD observed, on both the east and west coast of North America, corresponds to that taking place in the southern half of the continent, with all birds moving toward the tropic for the cold part of the year and returning before the hotter conditions prevail there. But some spend the entire year in cold and wet Tierra del Fuego, as do others in the opposite climate of northern California. He found a nest with eggs on south Chiloe Island on 8 December and noted that the nest was made entirely of fine cryptogamic plants. Elsewhere he also found a nest of the Giant Hummingbird, externally of woven fibrous grass and lined with the felt "of the pappus of some composite flower";[81] for other observations on this species, see chapter 2.

The Galapagos Dove

What little Darwin wrote about the Galapagos Dove, which he found to be one of the most abundant birds on the Galapagos, is worth quoting, as it emphasizes the remarkable tameness of birds on that archipelago at the time. He wrote of this lovely dove:

> It frequents the dry rocky soil of the low country, and often feeds in the same flock with several species of *Geospiza* [seed-eating Galapagos finches]. It is exceedingly tame, and may be killed in numbers. Formerly it appears to have been much tamer than at present. [The buccaneer William Ambrosia] Cowley in 1684, says that the 'Turtle doves were so tame that they would often alight upon our hats and arms, so as that we could take them alive: they not fearing man, until such time as some of our company did fire at them, whereby they were rendered more shy.' [William] Dampier (in the same year) also says that a man in a morning's walk might kill six or seven dozen of these birds. At the present time, although certainly very tame, they do not alight on people's arms; nor do they suffer themselves to be killed in such numbers. It is surprising that the change has not been greater; for these islands during the last hundred and fifty years, have been frequented by buccaneers and whalers; and the sailors, wandering through the woods in search of tortoises, take delight in knocking down these little birds.[82]

The last paragraph presents a scenario that has been repeated countless times ever since, with history recording the all-pervasive fear than humans inevitably come to elicit in almost every wild vertebrate they encounter. But slow as Darwin noted it to be, he was, unknowingly at the time, observing natural selection at work—as the slightly more wary individual doves lived to produce offspring, whereas the more "tame" ones were eliminated. In photographing this species myself on several Galapagos Islands in 2005, I found that I could approach them within several metres but that they were certainly not about to come any closer or permit me to grab at them (colour figure 3.12). That is not to say they might not have remained tamer in other parts of the archipelago.

Seedsnipe

The Least Seedsnipe (*Tinochorus rumicivorus*) clearly intrigued and puzzled Charles Darwin, and I add here only material additional to that cited as written by him in his *Journal of Researches*; see chapter 2. The four seedsnipe species constitute the family Thinocoridae, which is today placed among the "wading birds" of the order Charadriiformes.

The Least Seedsnipe (the "short-billed snipe" of Keynes[83]), which Darwin saw as being like both a wader and a gallinaceous bird, was found "wherever there are sterile plains, or open dry pasture land in southern South America" in places where "scarcely another living creature can exist." He observed them in pairs or in flocks of five or six but sometimes to as many as 30 or 40 birds. When approached, they "lie close, and are then very difficult to be distinguished from the ground; so that they often rise quite unexpectedly. When feeding they walk rather slowly, with their legs wide apart. They dust themselves in roads and sand places. They frequent particular spots, and may be found there day after day. When a pair are together, if one is shot, the other seldom rises; for these birds, like partridges, only take wing in a flock." Informants told CD that the bird nests on the borders of lakes (but not always) and that five or six white eggs (in fact typically four, sometimes three) spotted with red are laid. Upon dissecting birds, only vegetable matter was found to have been eaten by them; and they were "exceedingly fat, and had a strong offensive game odour; but they are said to be very good eating, when cooked."[84]

The Rheas

Six pages of the *Zoology of the Beagle* birds text are dedicated to the Greater and Lesser Rheas, and I summarize only those aspects of Darwin's observations of these primitive giant flightless birds of the family Rheidae that are not to be found in chapter 2.

Local people were able to distinguish the sexes of the Greater Rhea at a distance, the male being larger and darker and with a larger head. What Darwin considered to be males emit a "singular, deep-toned, hissing noise" that he initially thought was that of "some wild beast," being difficult to tell how far away and in what direction it originated. In one day's hunting on horseback, 64 eggs were found, 44 being in two nests and the rest scattered about the ground. It puzzled CD that so many eggs should be wasted by being laid distant from any nests of males; and he wrote, "Does it not arise from some difficulty in several females associating together, and in finding a male ready to undertake the office of incubation? It is evident that there must at first be some degree of association, between at least two females; otherwise the eggs would remain scattered at distances far too great to allow of the male collecting them into one nest. Some authors believe that the scattered eggs are deposited for the young to feed on. This can hardly be the case in America, because the *huachos* [the local name for such extra-nest eggs], although often found addled and putrid, are generally whole."[85] It is also hard to imagine, in any event, that young birds could break open the eggs of their kind without sophisticated adaptive behaviour (e.g., by kicking one against another or using a stone in some way as do some birds of prey to open ostrich and Emu eggs).

In one study, the mean clutch size in 28 Argentinean nests was 18.1 eggs, and that in completed nests was 26.1 eggs;[86] whereas in another study of 70 completed Argentinean nests, the mean clutch size was 24.87 eggs.[87] We now know that the scattered extra-nest eggs, or *huachos*, are typically fertile; but no explanation for this odd behaviour by the females in laying them outside nests has been forthcoming.[88]

Gould and Darwin repeat Gould's[89] original description of the Lesser Rhea [*Rhea darwinii*; colour figure 2.2) and his comparative morphological observations with the Greater Rhea. Charles tells how he first heard of this similar but smaller, darker, and less common kind of rhea, which is more easily caught with the bolas, from various people who called it *Avestruz Petise* and stated that it preferred to frequent plains near the sea. They described its eggs as little smaller than those of the larger species but being "of a slightly different form" and with a tinge of pale blue to the shell. How Charles nearly missed obtaining the first, if incomplete, specimen of this formally unconfirmed new species is recited in chapter 2. Here, however, CD records that while the French travelling naturalist M. A. D'Orbigny (apparently an error for Alcide C. V. M. D. d'Orbigny) observed Lesser Rheas at the Rio Negro but could obtain no specimens of it, he named it *Rhea pennata* in his book.[90] It is also noted that as long ago as 1749 it had been observed "that Emus [for rheas] differ in size and habits in different tracts of land; for those that inhabit the plains of Buenos Ayres and Tucuman are larger, and have black, white, and grey feathers; those near to the Straits of Magellan are smaller, and more beautiful, for their white feathers are tipped with black at the extremity, and their black ones in like manner terminate in white."[91]

At Santa Cruz, birds were "excessively wary"; and Darwin thought that they could see approaching people before they themselves were seen—he did, however, observe them in pairs and in groups of four or five. Some of the *Beagle*'s officers thought that, unlike the Greater Rhea, the lesser species did not expand its wings to initiate fleeing from potential danger, and CD agreed with this. He concluded that the Greater Rhea "inhabits the eastern plains of S. America as far as a little south of the Rio Negro, in lat. 41°"; and the Lesser Rhea "takes its place in Southern Patagonia; the part about the Rio Negro being neutral territory [in this he was close to correct but he was then unaware that the Lesser Rhea has an isolated population far removed to the north]. Wallis [Samuel Wallis] saw ostriches at Bachelor's river (lat. 53° 54′), in the Strait of Magellan, which must be the extreme southern possible range of the Petise [i.e., the Lesser Rhea]";[92] see chapter 2. In 1913 Charles Chubb of the British Museum of Natural History described and named two subspecies of the Lesser Rhea. Today some ornithologists treat the northernmost one of these (*Rhea pennata tarapacensis*) as a full species, which is called the Puna Rhea (*Rhea tarapacensis*).[93]

Petrels

Eight species of pelagic petrels, of the family Procellariidae, were included in the birds volume of the *Zoology of the Beagle*, but most only briefly. These seabirds would have been of some interest to Darwin, as he would not have seen a living member of this group unless he had watched Northern Fulmars (*Fulmarus glacialis*) on the British coast as a young man, which are atypical of the group in being more gull-like.

Writing of the widespread Grey Petrel (*Procellaria cinerea*, identified for him by John Gould upon his return to England) on the southwest coast of South America, Darwin observed that it "generally frequents the inland sounds in very large flocks: I do not think I ever saw so many birds of any other sort together, as I once saw of these behind the island of Chiloe. Hundreds of thousands flew in an irregular line, for several hours in one direction. When part of the flock settled on the water, the surface was blackened, and a noise proceeded from them, as of human beings talking in the distance. At this time, the water was in parts coloured by clouds of small crustacea." This was surely a great maritime ornithological event for any person to have had the good fortune to experience.

The Common Diving-petrel (*Pelecanoides urinatrix*) was commonplace in deep quiet coastal inlets and inland seas of Tierra del Fuego and on the west coast of Patagonia north to the Chonos Archipelago. Charles only saw a single bird in the open sea, between Tierra del Fuego and the Falkland Islands. He found it "a complete auk" in its habits and in its morphology, although clearly a petrel; and seen at a distance swimming and diving, it would, he said, "almost certainly" be mistaken for a grebe. When approached, birds would typically dive to a distance and take flight upon resurfacing in one movement; and having then flown some way, then drop like a stone on the water "as if struck dead" to instantaneously dive again.[94]

The Southern Giant-Petrel (*Macronectes giganteus*), then known as *Nelly* by the English and *Quebranta-huesos* by the Spanish, Darwin found common in southern latitudes of South America, frequenting both inland sounds and open ocean far from the coast. It struck him as albatross-like in general appearance and in its flight. He discovered the beak of a large cuttle-fish inside one, and reported one killing a "diver," and others killing and eating young gulls. Similar but black-plumaged birds seen (and one collected) in the southern oceans were thought a distinct species, but Captain William Low assured CD that "these black varieties were the one-year-old birds of the common greyish black Nelly." In fact the typical, but variable, adult plumage is dirty white on the head, neck, and upper breast; and the rest is a mottled greyish-brown, with more whitish on the belly—although there is a darker adult morph in which its juveniles are sooty black. There is also a rare (< 10 percent of the population) adult white morph, typically being pure white flecked with black.[95]

Darwin was obviously much taken by the "extremely numerous" lovely black and white Cape Petrel (*Daption capense*; see colour figure 3.13). He wrote the following:

> They are tame and sociable, and follow vessels navigating these seas for many days together: when the ship is becalmed, or is moving slowly, they often alight on the surface of the water, and in doing this they expand their tails like a fan. I think they always take their food, when thus swimming. When offal is thrown overboard, they frequently dive to the depth of a foot or two. They are very apt to quarrel over their food, and they then utter many harsh but not loud cries. Their flight is not rapid, but extremely elegant; and as these prettily mottled birds skim the surface of the water in graceful curves, constantly following the vessel as she drives onward in her course, they afford a spectacle which is beheld by every one [*sic*] with interest. Although often spending the whole day on the wing, yet on a fine moonlight night, I have repeatedly seen these birds following the wake of the vessel, with their usual graceful evolutions. I am informed that the *Pintado* arrives in Georgia for the purpose of breeding, and leaves it, at the same time with the *P. glacialoïdes* [Southern Fulmar *Fulmarus glacialis*]. The sealers do not know any other island in the Antarctic ocean except Georgia, where these two birds (as well as the

Thalassidroma oceanica [Wilson's Storm-Petrel (*Oceanites oceanicus*) of the family Hydrobatidae]) resort to breed.[96]

In fact, the Cape Petrel is found throughout the entire southern oceans and also breeds on several other islands as well as on the Antarctic Peninsular and the continent itself.[97]

Skimmers

The three species of skimmer—one each in the New World, Africa, and Asia—form a group of superficially large tern-like birds. They were indeed long thought of as aberrant terns but are now known to be closer to gulls than to terns and are so placed in the gull family Laridae.[98]

Darwin saw the Black Skimmer (*Rynchops niger*) on the east and west South American coasts, frequenting fresh and saltwater habitats. He observed their remarkable foraging for fishes by "skimming" the water surface, generally in small flocks (see chapter 2). He noted that the extreme length of their primary feathers was "quite necessary in order to keep their wings dry" when feeding by "skimming"; whereas their long tail feathers are "much used in steering their irregular course." He found them common far inland, along the Rio Parana, where, he was informed, they remain year round, rest in flocks on grassy plains, and breed in marshes. Given its rare foraging technique, CD speculated that its pliable bill is a "delicate organ of touch" and had Richard Owen examine the head of one he had preserved in spirit (figure 3.14). Owen failed to find the kind of sensitive "nervous expansions which are so remarkable in the lamelli-rostral aquatic birds" [*sic*] but could not "deny altogether, a sensitive faculty in the beak of the Rhynchops";[99] see chapter 2. In fact, the lower mandible has surface diagonal ridging rich in nerve endings that are

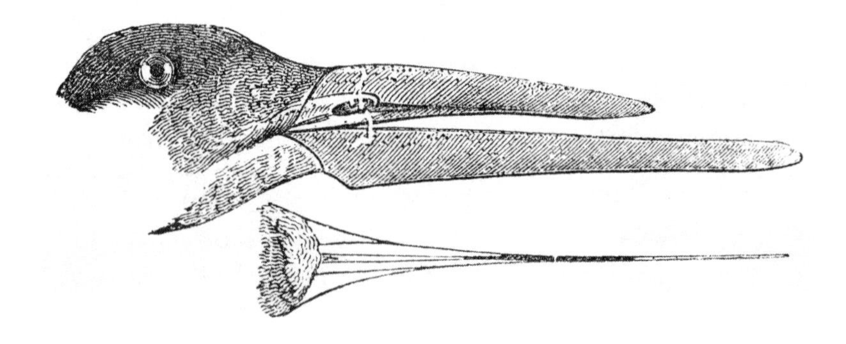

Figure 3.14 The head and beak of a Black Skimmer. From *A Naturalist's Voyage* by Darwin, 1889, p. 137.

highly sensitive to the touch, enabling the skimmer to instantly seize a fish touching the bill, and enabling birds to feed by night.[100]

SOME SUBSEQUENT DARWIN PUBLICATIONS

Darwin followed the publication of *The Zoology of the Voyage of H. M. S. Beagle* with a series of three books on his geological observations and interpretations from the voyage—on coral reefs, on geological observations on volcanic islands and parts of South America, and on his other geological observations on South America.[101] These were his primary years as a major contributor and authoritative figure in geology, and it was really after the dust had settled from his geological hammering and writing that he turned to his deeper biological considerations. In continuing his work, after the publication of his zoological results from the *Beagle* voyage, he consulted a great deal of literature; and typical of him as a meticulous bibliophile, he noted all those books that he wanted to read. In 1838, he started to list all books and periodicals consulted, including many containing ornithology.[102]

The publication *A Manual of Scientific Enquiry: Prepared for the Use of Her Majesty's Navy and Adapted for Travellers in General*, edited by John F. W. Herschel, was released by John Murray in 1849. Charles Darwin contributed Section VI to this volume titled Geology.[103] Within that text, he wrote under the heading "Distribution of organic beings" that "Any graminivorous [*sic*—CD might have meant to write granivorous here] bird, caught far out at sea, ought to have the contents of its intestines dried for the same object"—that object being to ascertain if it contained any seeds that might have been dispersed by the bird from one land mass to another in viable reproductive condition. He went on to write, "All facts or traditional statements by the inhabitants of any island or coral-reef, on the first arrival of any bird, reptile, insect, or remarkable plant, ought to be collected."[104]

Charles Darwin's world-shattering *On the Origin of Species* was first published in 1859 by John Murray of London, who was subsequently his publisher over many books and years and who also became his good friend. The ornithological contents of this immensely influential book are detailed in chapter 4.

NOTES

1. Darwin 1837.
2. Darwin 1839a,b; 1845; 1860.
3. Datta 1997: 88.
4. Cf. McCalman 2009: 298.

5. Desmond and Moore 2009: 215.
6. Jackson and Lambourne 1990; Jackson 1999.
7. Lambourne 1987: 46.
8. Sauer 1982; Tree 1991; Datta 1997.
9. In Gould and Darwin 1838–1841: 34.
10. Gould 1837b; Gould and Darwin 1838–1841: 146.
11. *Pace* Geospizinae of Keynes 2003: 324, 370.
12. Sibley and Monroe 1990.
13. Walters 2003: 130.
14. E.g., see Bowman 1963: 108, figure 1; Grant and Estes 2009: 145.
15. Sibley and Monroe 1990: 553; Gill and Donsker 2015; Grant and Estes 2009: 121–2.
16. Cf. Sulloway 1982b: 12.
17. Contrast Quammen 1996: 228.
18. Sulloway 1982b: 11–12.
19. Sulloway 1982b: 22; Grant and Estes 2009: 118–22.
20. Barlow 1963: 262.
21. Gould 1837b,c; Gould and Darwin 1838–1841; Tree 1991: 56.
22. Castro and Phillips 1996; Heinzel and Hall 2000.
23. Paynter 1970: 160–8.
24. Tree 1991: 55.
25. Sulloway 1982a.
26. Weiner 1994: 28
27. Aydon 2002: 120.
28. Gould 1837b.
29. Sulloway 1982c: 360.
30. Cf. Steinheimer et al. 2006: 182.
31. Sulloway 1982a.
32. Cf. Barlow 1963: 204, 262–3.
33. Darwin 1839b: 62–3.
34. In S. Gould 1985: 350.
35. C. Darwin 1839b: 474; 2002: 384; 1889: 380; 1902: 552; Sulloway 1979: 26.
36. Sulloway 1979: 26.
37. Sulloway 1982b: 7.
38. Sulloway 1982b: 22.
39. Darwin 1889: 377.
40. Desmond & Moore 1991; Browne 1995, 2002.
41. Huxley and Kettlewell 1974: 32.
42. Gould 1837a; Darwin 1837.
43. *Pace* Browne 1995: 408.
44. Gould and Darwin 1838–1841: ix.
45. CD in Gould and Darwin 1838–1841: 83.
46. del Hoyo and Collar 2014: 516–23.
47. Gould and Darwin 1838–1841: 7–8.
48. Gould and Darwin 1838–1841: 8–9.
49. Gill and Donsker 2015.
50. Darwin 1839b: 68; Gould and Darwin 1838–1841: 9–21.
51. Gould and Darwin 1838–1841: 26–31.
52. Gould and Darwin 1838–1841: 31–2.
53. Gould 1837c.

54. Gould 1836.
55. Cody 2005: 464–8.
56. Cody 2005: 488.
57. Gould and Darwin 1838–1841: 60–4.
58. Remsen 2003.
59. Gould and Darwin 1838–1841: 64–9.
60. Gould and Darwin 1838–1841: 70–74.
61. Gould 1837b; Gould and Darwin 1838–1841; Sulloway 1982a.
62. Gould and Darwin 1838–1841: 98–106.
63. E.g., Sclater and Salvin 1870; Salvin 1876; Ridgway 1897; Rothschild and Hartert 1899, 1902; Snodgrass 1903; Snodgrass and Heller 1904; Sushkin 1925, 1929; Swarth 1931; Lowe 1936; Hellmayr 1938; Lack 1945, 1947, 1969; Bowman 1961, 1963; Paynter 1970; Grant 1986; Sibley and Ahlquist 1990; Petren et al. 1999, 2005; Sibley and Monroe 1990; Zink 2002; Donohue 2011; Jaramillo 2011.
64. Gill and Donsker 2015.
65. Gould 1843.
66. Jaramillo 2011.
67. Gould and Darwin 1838–1841: 99.
68. Sulloway 1982b.
69. Darwin 1839b: 475.
70. E.g., Barlow 1963; Sulloway 1982a; Keynes 2000; Grant and Estes 2009: 145.
71. Lack 1945: 59; Grant 1986: 367.
72. Gould and Darwin 1838–1841: 98–106.
73. E.g., Lowe 1936; Lack 1947; Bowman 1961, 1963; Grant 1986; Grant and Grant 2010; Weiner 1994; Jaramillo 2011; and references therein.
74. Weiner 1994: 54.
75. Weiner 1994: 36.
76. McCalman 2009: 66.
77. Gill and Donsker 2015; Fjeldså 2013.
78. Azara 1801.
79. Gould and Darwin 1838–1841: 107–9.
80. Gill and Donsker 2015.
81. Gould and Darwin 1838–1841: 110–12.
82. Gould and Darwin 1838–1841: 115–16.
83. Keynes 2003: 222.
84. Gould and Darwin 1838–1841: 117–18.
85. Gould and Darwin 1838–1841: 120–23; see chapter 2.
86. Bruning 1974.
87. Fernández and Reboreda 1998.
88. Navarro and Martella 1998; S. J. J. F. Davies 2002.
89. Gould 1837a.
90. d'Orbigny 1834.
91. Dobrizhoffer 1749.
92. Gould and Darwin 1838–1841: 123–5.
93. Chubb 1913.
94. Gould and Darwin 1838–1841: 138–9.
95. Brooke 2004; Shirihai 2008.
96. Gould and Darwin 1838–1841: 137–41.
97. Harrison 1985: 237.
98. Gill and Donsker 2015.

99. Gould and Darwin 1838–1841: 143–4.
100. Podulka et al. 2004.
101. Darwin 1842, 1844, 1846.
102. See Birkhardt et al. 1988–2013, vol. 4: 434–573.
103. Darwin 1849: 156–95.
104. Darwin 1849: 183.

CHAPTER 4

The Sedentary Ornithologist Thinks of Origins

Great as the differences are between the breeds of the pigeon, I am fully convinced that the common opinion of naturalists is correct, namely, that all are descended from the rock-pigeon (Columba livia), *including under this term several geographical races or sub-species, which differ from each other in the most trifling respects.* [Darwin 1859: 23]

Charles Darwin apparently felt that he did not have enough to contend with in seeing his enormous and diverse collection of geological, botanical, and zoological specimens and notes on them, and his field observations on uncollected species organized, written up, and published—because he also found time to consider marriage. Much has been written about his objective contemplations and scribbling about as to if he should marry or not, given the pros and cons as he saw them in the context of his plans for his future as a "naturalist", and his subsequent proposal to his 30-year-old cousin Emma Wedgwood. He had thought that his older brother Erasmus had Emma in mind as a wife and had even suggested to his sister Catherine that he would return to England to find Erasmus married to, and sick of, her.[1] However, Charles and Emma came to be married quietly at St. Peter's church near Maer, Emma's family home in Staffordshire, on 29 January 1839, a couple of weeks before Charles also turned 30. Emma was no pauper, for her wealthy father Josiah gave her a bond of £5,000 Sterling. This allowed her an annual income of 400 pounds upon her marriage, which, in strictly legal terms of the day all became Charles's upon their marriage, was always treated by the couple as Emma's own money. Initially they lived together in a rented house in Upper Gower Street, near Bloomsbury in London, that they jokingly and fondly called Macaw Cottage because of its originally gaudy interiors and furnishings that were not at all to their taste.

Some parts of the volumes in the series of *The Zoology of the Beagle* dealing with the mammals and with the birds had appeared prior to Charles's marriage, but much remained to do on these and other animal groups for that publication. He settled down to also writing papers based on his geological and coral reef specimens, observations, and notes, among other subjects, while also enjoying the company of geological and biological colleagues in London. While in the capital, he also instigated wide-ranging enquiries into the breeding and hybridization of domesticated animals and cultivated plants with both colleagues in person and an ever-growing number of correspondents worldwide. His mind was presumably not only on the present and immediate future but was also looking well ahead to other potential writing projects.

Indicative of his intentions was that he had a questionnaire leaflet published titled *Questions about the Breeding of Animals*, which he distributed during April and May of 1839 from his London home. Specifically of birds (although many of his other questions could be applied to them), he asked, "If the cross offspring of any two races of birds or animals be interbred, will the progeny keep as constant, as that of any established breed; or will it tend to return in appearance to either parent?" Also, when breeds as different as pouter and fantail-pigeons are crossed, "are *their offspring* equally prolific, as those from between nearer varieties" (CD's emphasis). And "What are the effects of breeding in-and-in, very closely, on the males of either quadrupeds or birds? Does it weaken their passion, or virility? Does it injure the secondary male characters,—the masculine form and defensive weapons in quadrupeds and the plumage in birds? In the female does it lessen her fertility? Does it weaken her passion? By carefully picking out the individuals most different from each other, without regard to their beauty or utility, in every generation from the first, and crossing them, could the ill effects of inter-breeding be prevented or lessened?" Part of another, larger, question reads, "Have any cases fallen under your observation, of quadrupeds (as cats or pigs, &c.) or birds (fowls, pigeons, &c.) born in this country, from a foreign stock, which *inherited* habits or disposition, somewhat different from those of the same variety in this country?" (CD's emphasis). And finally, "Can you give the history of the production in any country of any new but now permanent variety, in quadrupeds or birds, which was not simply intermediate between two established kinds?"[2] These scientifically carefully worded queries attest to Darwin thinking deeply about issues that he was not to discuss the results of in his books until almost 30 years later; see chapter 5. They were topics given serious consideration or study by few other contemporaneous naturalists.

At this period, Darwin's mind was processing an incredible amount of biological material and ideas concerning it. It has been suggested that at about this time in his life, "birds were becoming his main interest, being the most prized and bred for beauty. Here if anywhere lay crucial clues to the origin of sexual selection and racial characters."[3] Although the last sentence was true in the context of his approach to the subject, I am not sure that there is any real hard evidence in support of the preceding sentence; for even though he sent his questionnaire out in 1839, it was not until the mid-1850s that he got down to work on domesticated birds.

The Darwins' first child, William, was born in December 1839. In 1840, the artist George Richmond painted a portrait of both Emma and Charles Darwin; see colour figure 1.1 for that of Charles. Their second child, Anne, was born in March of 1841. As Macaw Cottage was becoming too small for their growing brood, thoughts of moving to a larger house inevitably intruded into their busy and happy lives. Fortunately, his father's wealth and generosity meant that Charles could consider a property beyond his own means at the time, and so he and Emma started to look for one out of the city congestion, noise, and pollution.

At the start of January 1842, Charles sent his first completed book, on some of the geological results of his voyage, to his publisher. In May of that year, while on "holiday," he found time to outline a brief (35 tightly handwritten pages) sketch of his unfolding theory on species and how they come about without the influence of any god. This he initially wrote off the top of his head, without serious reference to his notebooks, and for the first time used therein the all-important term "natural selection" for his developing theory. He completed the sketch in June.

By the end of July 1842, the Darwins had experienced more than enough of dwelling in London, and its population of two million people, and so they more actively and eagerly sought a country retreat where Charles could work without social and other distractions, and Emma could raise her family in far less crowded, more quiet and clean, surroundings. As a result, they inspected a house at Down Village (but now Downe) in Kent with 18 acres, or 7.29 ha, of land that included a large, partly walled garden in a pleasant enough rural setting not too close to the small village. It was located 16 miles, or 26 km, from St. Paul's in central London, a two-hour journey at the time. The most convenient railway stations giving access to London were not as close to the house as they would have desired for their trips into the city, with Sydenham station being eight miles, or 13 km away. (In early 1856, Beckenham station was to open closer to Down House; and by 1865, Bromley station became available only six miles, or almost 10 km, away.)

Emma was not terribly keen on the property, known as Down House (see colour figure 4.1); but the couple discussed the place and concluded that, although not perfect for them, it would make a suitable home given that some changes and additions would have to be made to it in the future. The fact that Charles's father bought the house for them, at a cost of £2,200, would have made the decision all the easier. In broad general appearance, the front of this house was coincidentally not unlike that of Mount House in which Darwin was born, both being nearly square and of three floors and each floor with five windows across it. It became their property before the end of August, and they moved in during mid-September. Shortly thereafter, ornithologically aware Charles wrote of it: "larks abound here & their songs sound most agreeably on all sides; nightingales are common."[4] It was to be CD's family home and workplace for the remainder of his life, and both he and Emma came to love the place dearly. Of fundamental importance to the rest of their marriage was that their favourable financial circumstances meant that Charles could give himself to his family and his self-imposed work without concerns, beyond continuing to invest wisely. Emma was to give birth to eight more children there, in all six boys and four girls, although not all survived to adulthood: Mary Eleanor died only 24 days following her birth in 1842; Anne died aged 10 in 1851; and Charles Waring, her last child, was born in 1856 when Emma was 48 and died in 1858.

The year 1842 saw Darwin's book on coral reefs published, and the next year saw the series of books on *The Zoology of the Voyage of H. M. S. Beagle*, completed by the appearance of Part 5, number 2, on reptiles written by Thomas Bell and edited by Charles.[5] In 1844, his book on the geology of volcanos appeared; and in 1846, a book about the geology of South America was published.[6]

Although Darwin was no zoological or ornithological taxonomist, and in any event had lacked the time and in all probability the inclination to become one, he was giving serious consideration to the subject and was discussing it with colleagues. In writing to George Waterhouse on 26 July 1843 (in which he refers to a "Creator"), he observes, "The only doubt, which has ever occurred to me, lies in the rules being so exceedingly arbitary [*sic*] by which the value of groups are judged by. I mean, that perhaps many *orders* really exist, containing only one or two genera, but that from this very circumstance, they are not viewed as *orders* but only as families—Perhaps if the Goatsuckers [nightjars] & Woodpeckers, were varied into very many genera & very many species of each [now pretty much the case]—they would be looked on as orders equal to the hawks &c &c [now the case]—(Tell me what you think of this) This [*sic*] is a mere *illustration* of what I mean."[7] This text clearly demonstrates that although he was no authority on avian taxonomy, Darwin gave serious and knowledgeably critical consideration to the subject during the course of his early work.

In 1844, Darwin bent his mind to the external parasites of land birds on different continents; and in a letter of 7 November, he encouraged his correspondent Henry Denny to examine those found to be common to species in Great Britain and America for any geographically based morphological differences.[8]

Darwin was all too aware of his lack of experience as a biological taxonomist, and he was reminded of this by his close friend Joseph Hooker, botanist at the Kew Gardens in London, as being problematic for one so involved in species and their evolutionary origins. In October 1846, with the page proofs of his last geological contribution returned to the publisher, Charles cast about among his unexamined *Beagle* voyage specimens for a group of animal species he might study in detail as a self-imposed "apprenticeship" in basic morphological taxonomy, or systematics. He came across an atypical barnacle that he had preserved in spirit in the Chonos Archipelago, Chile, in November of 1834—a diminutive crustacean that lived within the shell of a mollusc. This was to be the beginning of eight years of dedicated, eye-straining barnacle studies: their morphology, taxonomy, and, often truly bizarre, sex lives.[9] The results of this magnificent, if extremely tedious and often intensely frustrating, meticulous desk-and-microscope-bound research resulted in four admirable volumes—two on the living and another two on the fossil barnacles.[10] With these intensive studies published, Charles, and also his eminent colleagues in the fields of biology, now believed himself to be well and truly qualified to discuss all taxonomic aspects of animal species.

Needless to say, Darwin the workaholic had been thinking about and making notes on numerous other biological subjects and issues during his "barnacle years," with the ultimate goal of a central theory to explain the origin of species through natural selection as the latter concept evolved in his mind. The results of his barnacle work he doubted were "worth the consumption of so much time."[11] He found time, however, to record the odd note about local natural history. One such ornithological note was jotted down at the Moor Park health spa in Surrey during a restful visit: "It has been stated that woodpeckers remove fragments [of wood that they displace in excavating a nest hole in a tree, presumably to avoid them attracting predators]. In 2 cases I can say this false for such fragments guided me to discovery of nest."[12]

Having widely circulated questions about breeding pigeons, and other birds and mammals, under domestication back in 1839, Darwin got very much hands on with that most useful subject in December of 1854 and during 1855. It was at the suggestion of his friend William Yarrell, a bird man that had no doubt that the various breeds of domestic pigeon derived from the Rock Dove as their only ancestral species, that Darwin took up pigeon keeping.[13] In his *The Origin of Species*, Darwin did, however, express doubts about a single common ancestor to all domestic pigeon breeds, as most pigeon fanciers believed that several ancestral species were involved.[14]

Whereas his mind and eyes were of course wide open to animals under domestication having nothing to do with the functioning of life in nature, Darwin could see their valuable potential in demonstrating selection, albeit artificial selection by animal keepers. Adding to his labours in experimenting with numerous plant seeds to test their viability after various periods of time in saltwater, he also started systematically preparing skeletons of the different breeds of domestic pigeon. Preparing pigeon, duck, and other skeletons was a messy and disgusting business that often had Darwin retching violently because of the appalling odours created by the process. At times, skeletons were even prepared in the Down House kitchen, to the great disgust and discomfort of his family. By the summer, he had come to find his domestic pigeon studies as enjoyable as they were instructive, and it was clear to him that they would be put to great service in articulating his ideas about selection in nature. He hated having to kill the birds for the preparation of their skeletons; and because of the awful odours involved, he eventually contracted people to produce the all-important bare bones elsewhere.

Although he may not have by this time been struggling so much with the contradiction between his religious beliefs and his ever-growing understanding of potential evolutionary processes, he did continue to suffer physically for his work, as he had from seasickness aboard the *Beagle*. In March of 1855, his first pigeon loft was erected in the garden;[15] and by November, he had some 13 or 14 pairs resident therein, which his daughter Henrietta, or Etty, came to significantly help him with in keeping them.[16] Contrary to his initial expectations of what he thought would be a boring experience, Charles deeply enjoyed his pigeon keeping. He wrote to Charles Lyell on 4 November 1855, "I will show you my pigeons! . . . which are the greatest treat, in my opinion, which can be offered to [a] human being."[17] It is reported that Charles had some 90 captive pigeons towards the end of the year,[18] but that number appears to have included other domesticated bird species.[19] Charles also set about seeking the help of people around the world with the supply of stuffed skins of as many breeds of domestic pigeon as possible for him to examine and compare[20] (see the "Lofty Pigeon Studies" section following).

On 9 December 1855, Darwin wrote to his cousin Tom Eyton, who wrote avian anatomical notes for his *Zoology of the Beagle* and provided CD with various specimens, telling him that he was looking at differences in the skeletons of "Pigeons, Poultry, Covey Birds & rabbits" but that he appreciated that he [Eyton] might publish on this before him.[21] He also sought advice from Eyton on preparing bird skeletons, as he was to publish on that esoteric subject.[22] On 30 August 1856, he sent a letter to William Tegetmeier and mentioned a paper that Tegetmeier was apparently preparing on the skulls of fowls, a fact that CD had only recently become aware of, and told him that Eyton was doing likewise; and he warned him against being forestalled by his own cousin.[23]

Charles contacted numerous other colleagues seeking information on a remarkably broad range of biological topics for his first and further editions of *The Origin of Species* and subsequent works. Among these sources of information was Edward Blyth, curator of the Museum of the Asiatic Society and based at Calcutta, India, who furnished CD with a great deal of invaluable ornithological facts and ideas; but also Abraham D. Bartlett, Rajah James Brooke, Sir Walter Elliot, Thomas Eyton, William D. Fox, Edward W. V. Harcourt (on Madeiran birds), William Jardine, Leonard Jenyns, Alfred Newton, Osbert Salvin, Philip Sclater, Robert Swinhoe, William Tegetmeier, Alfred R. Wallace, John Jenner Weir, and others.

An ornithological publication of much significance was Charles Lucien Bonaparte's (nephew of Napoleon) *Conspectus Generum Avium*, which started to appear in 1850.[24] This work included almost four times as many bird species as the previous widely used bird compendium books by Mathurin Jacques Brisson[25] and Georges-Louis Leclerc de Buffon.[26] Bonaparte died in 1857, before he could complete his vast work; but even in its incomplete form, it would have provided Darwin with a most helpful review of the known birds of the world for his subsequent works that included discussions on birds.

In mid-September of 1853, Charles lost a good friend and valuable ornithological colleague when Hugh Strickland, also a geologist and palaeontologist, was struck down by a train as he was studying the banks of a railway cutting. Strickland was an accomplished bird man who was an expert on the Dodo (*Raphus cucullatus*, an extinct giant flightless pigeon) and related extinct birds, avian osteology, and much more, and who would have been of further great help to Darwin in his ornithological studies had he lived longer.

The story of Darwin's years of dedicated hard work, family life, and sickness, at Down House leading up to the most shocking arrival of a handwritten manuscript by Alfred Russel Wallace in June 1858 from Ternate Island, in the wilds of the vast and distant Malay Archipelago, has been told countless times.[27] Wallace's paper, which he sent to CD to read and then have published in London if he felt it had appropriate merit, amazingly presented Darwin's own long-incubated theory of evolution by natural selection in words that Charles felt could have been his very own. This struck Darwin "like a thunderbolt from a cloudless sky!"[28] He was aghast, and beside himself with anguish and frustration, to discover that Wallace, who he had previously corresponded with and admired as a field naturalist, had come upon the very same revolutionary mechanism for the evolution of species. Wallace had quite remarkably done this while living in independent isolation and rugged primitive conditions in the wilds on the other side of the planet. As has also been told often, Darwin's close academic friends and colleagues Charles Lyell and Joseph Hooker found a reasonably diplomatic way to preserve and record his undoubted, but unannounced by publication, scientific priority by arranging for both his and Wallace's ideas on natural selection to be jointly read at a meeting of the Linnean Society in London on 1 July 1858. For this to happen, however, CD had to immediately knuckle down to write a brief outline of his theory under very great personal pressure, particularly as his young son and last child Charles Waring, born in 1856 and suffering from Down syndrome, became ill with scarlet fever and died on 28 June 1858.

Without the support of his good friends Lyell, Hooker, and Thomas Huxley, it is almost beyond doubt that Wallace's contribution on natural selection would have appeared before Darwin's, and the history of that most significant of contributions to the biological sciences would now tell a different story. As it was, the brief and terribly rushed paper that CD had sent to his friends for joint publication with that of Wallace was no satisfaction to him at all. In that paper, Charles cited the writing of the clergyman and political economist Thomas Robert Malthus, who argued that human population growth would outstrip food production unless somehow checked.[29] As a pertinent theoretical example of this, CD wrote that if, of eight pairs of birds at one place, only four pairs breed to rear only four young annually and those offspring reproduce at the same rate, then after only seven years (taken as a short bird's life), the result would be 2,048 birds. Because this is impossible growth to sustain, he wrote, birds must fail to raise some 50 percent of their young, or the average life span of them is not seven years. The great influence on CD of Malthus's writing about limitations on human populations, and coincidentally also on Alfred Russel Wallace, in formulating the notion of natural selection has been well documented.[30]

Charles also wrote of natural selection and the struggles of life generally being decided by "the law of battle, but in the case of birds, apparently, by the charms of their song, by their beauty or their power of courtship, as in the dancing rock-thrush of Guiana [Guianan Cock-of-the-rock (*Rupicola rupicola*), figure 4.2]." About natural selection on variation within species, he wrote, "An organic being, like the woodpecker or misseltoe, [*sic*] may thus come to be adapted to a score of contingences—natural selection accumulating those slight variations in all parts of its structure, which are in any way useful to it during any part of its life."[31] His barnacle studies had previously made him all too well aware of the intrinsic physical variation between individuals of any given species.

The jointly read Darwin–Wallace papers of 1858 provided the much-needed catalyst that prompted Charles to finally come to terms with writing the definitive account of his theory, supported by his long-accumulated wealth of evidence and his arguments for it. This grew and grew in the writing and eventually appeared as by far the most famous of all his books—*On the Origin of Species by Means of Natural Selection, or the Preservation of Favoured Races in the Struggle for Life*—in November of 1859. The word "evolution" does not, however, appear within the pages of this pivotal publication.

By the time Darwin came to write the introductory remarks to subsequent editions of his *The Origin of Species*, he had to record the fact that two

Figure 4.2 An adult male Guianan Cock-of-the-rock. From *The Descent of Man and Selection in Relation to Sex* by Darwin, 1901, p. 602.

Scottish authors had, independently, published relatively obscure or overlooked works that briefly suggested the notion of natural selection, but without discussing its significant implications.[32]

In working towards this monumental, world-shattering, volume, Darwin read and corresponded prodigiously and performed all manner of fundamental natural experiments at Down House. Today his experiments may strike many as simplistic and perhaps even somewhat rustic. Indeed, in writing to Joseph Hooker about some of them, he said "you would have a good right to sneer for they are so *absurd* even in *my* opinion that I dare not tell you [CD's emphasis]."[33] He recited to Hooker how partridges that he had shot after heavy rainfall, to see if they had plant seeds among the mud on their feet and legs, were carrying numerous seeds. Darwin, quite rightly, saw this as one of many means by which seeds could be geographically widely dispersed by animal agents. This is reminiscent of him asking himself way back in his 1839 book about the voyage of the *Beagle* if island-dwelling hawks or owls ever deliver live prey, caught elsewhere, to their nest; and in so doing, might thus introduce the occasional escapee species to islands?[34] (see chapter 2). This kind of idea about agent animal species dispersing other animal or plant species stimulated his numerous little "absurd" experiments at Down House.

Upon receiving the initial manuscript of *The Origin of Species*, John Murray sent it out to two referees to comment on. He read some of it himself and was not too impressed by what he thought to be an absurd theory. The reverend Whitwell Elwin, one of the referees and then editor of the *Quarterly Review*, was also unimpressed and was distressed by the text and suggested that Darwin confine the book to his pigeon work. The second referee, George Pollock, found the work admirable in addressing the numerous pros and cons for the theory objectively.[35]

In the context of knowledge of the natural sciences at the time, his home-spun experiments were remarkable and insightful, as the following account of the more substantial ornithological content of *The Origin* attests: the less substantial ornithological content is comprehensively incorporated into appendix 1. Because of the success of his previous publications and his ever-increasing public profile, all 1,250 copies of the first edition of *The Origin* were sold out on the very first day of publication, 24 November 1859, the retail price then being the high one of 14 shillings. This great success both thrilled and worried Darwin, however, because he was all-too well aware of the storm of outrage and criticism the content of his *The Origin of Species* would inevitably stimulate. During the first six months after its publication, CD read close to 200 letters and numerous reviews concerning the book; and yet this period can be seen as the calm before the storm for him and his great book of originality.

LOFTY PIGEON STUDIES

At the time that Charles Darwin took up the study of the variability in appearance and structure of the many domesticated pigeon breeds, keeping and breeding them and examining their comparative anatomy, the cultivation of these pigeons had been going on for centuries. There were then already numerous distinctive, and thus individually named, breeds commonly kept and "bred true" in Britain and Europe and elsewhere in the world. Almost all pigeon breeders, or fanciers, of that time were firmly of the view that the many highly distinctive breeds must have derived from a number of different ancestor species of wild pigeons. Most eminent naturalists, on the other hand, were strongly of the view that but a single species, the Rock Dove (*Columba livia*), was, as noted previously, the sole ancestor to all domesticated pigeon breeds despite their remarkable variation in size, proportions, shape, structure, and coloration, and Darwin's studies convinced himself of this too.

Of the 44 pages of his chapter 1, titled "Variation Under Domestication," of his *The Origin of Species* book, Darwin dedicates some 10 pages to domestic pigeons.[36] It has been suggested that he did not appear to have consciously remembered the pigeons that his mother kept and bred at his family home.[37] It is not inconceivable, however, that in initially marshalling his thoughts and evidence for his *The Origin of Species*, and subsequently for his *The Variation of Animals and Plants Under Domestication*, he at least subconsciously if not consciously recalled the pigeons that his mother had kept when he was a small child. In any event, he saw with great clarity that the various and extremely diverse breeds of the domestic pigeon, and those of other domesticated birds and mammals, offered a highly instructive and convenient means of demonstrating to his readership the importance of selection playing on the morphological variation inherent within individuals of a species—albeit artificial selection by humans over mere hundreds of years to several millennia. With this adequately presented, illustrated, explained, and digested, it would be no great stretch to then bring his readers to understand and appreciate the extremely slow and analogous process of natural selection on species over many millions of years in nature. The tactic of initially presenting and discussing variation and artificial selection under domestication was perhaps an obvious ploy, but it was also a brilliant one. He started writing his text about pigeons for his *The Origin of Species* on 14 June 1858.[38]

Darwin kept every "breed" of domestic pigeon he could "purchase or obtain" in addition to collecting preserved skins of them from numerous contacts within and beyond the British Isles. He bred the various breeds available to him and he also crossbred them to determine the morphology of the hybrid offspring. He noted that domestic pigeons had been kept for thousands of years and that many works had been published about them in various languages over "considerable antiquity," these providing an invaluable historical record of the pigeon "fancy." Darwin "associated with several eminent fanciers" and became a member of two of the then London Pigeon Clubs: the Southwark Columbarian and the Philoperista. The mental image of him in a smoke-filled public house full of gin- or beer-drinking "working class" pigeon keepers, who addressed him as "Squire," deep in conversation about their experiences with various pigeon breeds and their thoughts on their origins, is a delightful one. Of this kind of experience, Charles humorously wrote to a correspondent: "For instance I sat one evening in a gin-palace in the Borough amongst a set of pigeon-fanciers,—when it was hinted that Mr Bult had crossed his Powters [Pouters] with Runts to gain size; & if you had seen the solemn, the mysterious & awful shakes of the head which all the fanciers gave at this scandalous proceeding, you would have recognised how little crossing has had to do with improving breeds & how dangerous for endless generations the process was.—All this was brought home far more vividly than by pages of mere statements &tc."[39]

Charles found not a single fancier of pigeons, poultry, duck, or rabbit who "was not fully convinced that each main breed was descended from a distinct [wild] species." He was therefore only all too aware of the task ahead of him in convincing them, and the rest of the world, of the common descent of the various breeds of each domesticated animal from a single ancestor species. Most of his contemporary naturalists did indeed see all domestic pigeons as commonly derived from the Rock Dove, but CD succeeded in convincing the entire world of the truth of this, if not within the covers of *The Origin of Species* then certainly within those of his *The Variation of Animals and Plants Under Domestication*; see chapter 5.

Darwin was at pains to describe the remarkable differences between the breeds of domestic pigeon, not only in their plumage structures, colouration, and markings, but also in their gross physical structures (beaks, skulls, skeletons, bare and carunculated skin, eyelids, oil glands, and body proportions involving the wings, legs, feet, and crop) as well as peculiarities of breeds in flight and calls. One breed, the short-beaked tumbler, was so modified by artificial selection that a great number of them perish in the egg. This is because their beak is so reduced as to be inadequate for the hatchlings to break out of the shell unless the fanciers assist them in doing so. He also noted differences in the period required for the development of nestling down, and for fully mature plumage and sexual dimorphism, to be acquired between the sexes across the various breeds.

The observation was made by Darwin that the hypothetical result of showing specimens of at least 20 distinctive breeds of domestic pigeon to any ornithologist who was told they were of wild birds would lead that person to conclude that they were of distinct species involving several genera. He pointed out that if this were indeed the case, then at least seven or eight ancestral species would have to have been involved; but no such pigeon species with the appropriate characters existed and that to argue that all such ancestor species had, by truly remarkable coincidence, become extinct was beyond all straw clutching. I stress that his observation must be assessed in the context of knowledge and understanding of bird taxonomy at that time.

More telling evidence in support of a single ancestor species to all domestic pigeons included the fact that all of the morphologically, highly distinctive breeds remain interfertile, and also that the offspring of crosses between two breeds lacking any adult plumage traits of the ancestor Rock Dove nevertheless sometimes include individuals with the highly distinctive plumage characters of the Rock Dove. Finally, Darwin pointed out that, as domestic pigeons can be mated for life, breeds could be improved and bred true from a pair even while in a loft of mixed breeds. Moreover, as they can be bred in great numbers quickly with the fancier rejecting birds considered inferior, they have been, and remain, ideal subjects for such "artificial selection" practices.

In addressing the objection of many people to his theory of natural selection that a series of forms intermediate between two distinct species are not to be found in nature, Darwin used domestic pigeons as but one example among many to explain it. He pointed out that fantail and pouter pigeons are both descended from the Rock Dove and that if we had all intermediate varieties that ever existed between them, there would be an extremely close series between both and the Rock Dove but none directly intermediate between the fantail and pouter (e.g. individuals combining a partly expanded tail and an enlarged crop—the respective characteristic trait of the two breeds). The two breeds are now so modified that, lacking evidence of their origin, it would be impossible to determine from a comparison of their structure with that of the Rock Dove if they had descended from this or from some close relative, such as the Stock Dove (*Columba oenas*).

Domestic pigeons are again alluded to in Darwin demonstrating why a species once lost to extinction cannot recur in the identical form, even under the very same ecological conditions. He writes, "For instance, it is possible, if all our fantail pigeons were destroyed, that fanciers might make a new breed hardly distinguishable from the present breed; but if the parent rock-pigeon were likewise destroyed, and under nature we have every reason to believe that parent-forms are generally supplanted and exterminated by their improved offspring, it is incredible that a fantail, identical with the existing breed, could be raised from any other species of pigeon, or even from any other well-established race of the domestic pigeon, for the successive variations would almost certainly be in some degree different, and the newly-formed variety would probably inherit from its progenitor some characteristic differences."

In discussing the affinities of extinct species to each other and to living forms, Darwin wrote, "if the principle living and extinct races of the domestic pigeon were arranged in serial affinity, this arrangement would not closely accord with the order in time of their production, and even less with the order of their disappearance; for the parent rock-pigeon still lives; and many varieties between the rock-pigeon and the carrier have become extinct; and carriers which are extreme in the important character of length of beak originated earlier than short-beaked tumblers, which are at the opposite end of the series in this respect." Short sentences were not then the *de rigueur* that they are today.

One similar example is provided by the various extant and extinct rail species of the genus *Gallirallus* found on numerous western Pacific Ocean islands. The common ancestor of these rails was quite possibly the conspicuously successful and extant island-colonizing Buff-banded Rail

(*Gallirallus philippensis*) or a close relative of it. If the existing *Gallirallus* species were arranged as Darwin described previously for the breeds of the domestic pigeon, the arrangement would not necessarily accord with their relative temporal evolution as species—because the parent species still lives; others have become extinct; and, moreover, their differing bill lengths could not be taken into account, as these predominantly reflect ecological adaptation.[40]

Darwin wrote confidently, and justifiably, of having provided "conclusive evidence that the breeds of pigeon are descended from a single wild species." He measured various body proportions of the less than 12-hour-old hatchlings of seven fundamental breeds, the mature adults of which differ in proportionate body parts "in so extraordinary a manner" as to be ranked as distinct genera if found in the wild. The nestlings of these various breeds proved, however, hardly discernibly different in their proportions: with the single exception of the "short-faced tumbler," which showed a proportionate difference from the Rock Dove and other domestic breeds as in the adults. He explained this by observing that pigeon fanciers select for traits visible in mature adults and thus for ones that do not generally appear in early life, the short-faced tumbler being the exception to the rule. This rule applies to closely and many not-so-closely related wild species, their embryos and young being structurally extremely similar in relative proportions—but the pertinent morphological characters resulting from natural selection being apparent in older individuals.

Darwin concluded that he had discussed the origin of domestic pigeons "at some, yet quite insufficient, length"—something he was to put right within volume 1 of his *The Variation of Animals and Plants Under Domestication* published nine years later;[41] see chapter 5. Notwithstanding the "insufficient length" of Darwin's discussion of domestic pigeons in his *The Origin of Species*, Alfred Russel Wallace, his most dedicated of disciples, saw fit to recommend in a letter from Delli, Timor, in early February 1861 to his religious brother-in-law Thomas Sims, who was struggling to understand and accept the concept of natural selection, that he "read again Darwin's account of the Horse family & its comparison with Pigeons;—& if that does not convince or stagger you, then you are unconvertible."[42] Thus Wallace saw convincing evidence arising from Darwin's domestic pigeon studies of artificial selection as fundamentally significant support for the theory of the origin of species through natural selection.

COMPETITION AND SEX

The notion of the constant struggle for life by individuals and by species was the cornerstone of Darwin's theory of evolution by natural selection, or the survival of the fittest, and thus of his monumental book *The Origin of Species*. In addition to the immensely powerful, but typically painfully slow, process of natural selection through competition (at least in vertebrates) is sexual selection, which CD hypothesized as an extremely novel and controversial idea at the time.

In briefly introducing his chapter 3, on the "Struggle for Existence," of *The Origin of Species*, Darwin gives the "woodpeckers and the mistletoe" as examples of organisms exhibiting coadaptations; but he was not intending to indicate any ecological relationship between them. He later writes of the parasitic nature of the mistletoe on trees with which it does not really "struggle" but then makes this ecological observation: "But several seedling mistletoes, growing close together on the same branch, may more truly be said to struggle with each other. As the mistletoe is disseminated by birds, its existence depends on them, and it may metaphorically be said to struggle with other fruit-bearing plants, in tempting the birds to devour and thus disperse its seeds. In these several senses, which pass into each other, I use for convenience' sake the general term of Struggle for Existence."[43] This insight was expressed a very long time before most people appreciated any kind of "struggle" involving frugivorous birds and plant species and individuals of species all competing for food and seed dispersers, respectively—complex relationships splendidly demonstrated among tropical and British birds and plants by eminent English ornithologists David and Barbara Snow.[44]

In chapter 4 of his *The Origin of Species*, titled "Natural Selection; or the Survival of the Fittest," the last three words of which were to become forever intimately attached to his name, Darwin instigated his discussion of sexual selection. The latter theory was, however, presented in only three pages; and a full discussion of it with supporting evidence had to await the publication of his *The Descent of Man and Selection in Relation to Sex* in 1871. He observed that sexual selection in bird species is often more peaceful than in those of other vertebrate groups, involving singing between males to attract females, and that males of the Guianan Cock-of-the-rock (figure 4.2), birds of paradise, and of some other birds, congregate to display to females, who then choose the male most attractive to them. The plumage of males and females of a species compared to that of their young could partly be explained through sexual selection under such scenarios—as far more recent studies have demonstrated.[45]

Again using the invaluable analogy of artificial selection to hammer home his point, CD notes "if a man can in a short time give beauty and an elegant carriage to his bantams, according to his standard of beauty, I can see no good reason to doubt that female birds, by selecting, during thousands of generations, the most melodious or beautiful males, according to their standard of beauty might produce a marked effect."

In view of the preceding expressed sentiments, it was odd that Darwin then wrote that the "tuft of hair on the breast of the wild turkey-cock cannot be of any use, and it is doubtful whether it can be ornamental in the eyes of the female bird;—indeed, had the tuft appeared under domestication, it would have been called a monstrosity." It would appear that he erred in failing to heed his own repeated words, as previously, in emphasizing that beauty is in the eye of the beholding sex; for the Wild Turkey's breast tuft is indeed now seen as the result of sexual selection, although it can be argued that females select for it as an indication of male physical fitness, relative age, and so forth, rather than as an ornamental thing of beauty as such.[46] Of course we cannot definitively know what and how female turkeys might find an "attractive" trait in males of their kind.

The theory of sexual selection was greatly expanded on by Darwin in his *The Descent of Man*, but it continued to be long and hotly debated and seriously questioned even, and most vigorously, by Alfred Russel Wallace and other close friends and colleagues of its greatly respected author.

In chapter 5 of his *The Origin of Species*, Charles made his firm belief in the importance of competition between species as a significant influence on natural selection perfectly clear. He wrote, "We have reason to believe that species in a state of nature are closely limited in their ranges by the competition of other organic beings quite as much as, or more than, by adaptation to particular climates."[47] This view was, however, both ignored or disputed until quite recently[48] but is now widely, if not unanimously, held to be true of birds.[49]

FLIGHT CANCELLED

Flightless birds are briefly introduced by Darwin in his alluding to Professor Owen's truism that there is no greater anomaly in nature than a bird that cannot fly, and "yet there are several in this state." We now know that there are a good many more than several, for of the approximately 144 bird families, 24, or 17 percent, have or have had one or more flightless species among their numbers—the vast majority of them living on remote oceanic islands.[50]

Darwin observed that the "logger-headed duck of South America can only flap along the surface of the water, and has its wings in nearly the same condition as the domestic Aylesbury duck; it is a remarkable fact that the young birds, according to Mr. Cunningham, can fly, while the adults have lost this power."[51] This is a reference to the Fuegian Steamer Duck (*Tachyeres pteneres*), the name steamer being given to it and its close relatives because in rapid wing flapping over water, it brings to mind the steam-powered paddleboats of the era (colour figure 2.1). That its young can fly is, however, incorrect.[52]

As larger ground birds seldom fly except to escape danger, Darwin thought it probable that the near wingless state of birds inhabiting oceanic islands lacking predators was caused by disuse. Although he noted that the flightless "ostrich" inhabits continents where it is exposed to predators, it is able to defend itself by kicking "as efficiently as many quadrupeds." He suggested the Ostrich ancestor had habits like the bustard and that as "the size and weight of its body were increased during successive generations, its legs were used more, and its wings less, until they became incapable of flight."[53] Darwin was correct in thinking that the ostriches, and indeed other ratites, had lost their ability to fly rather than having never been able to do so. The result of studies of the genetics of the ratites shows that, contrary to a long-held view about their global ranges, their present distributions do not reflect geographical isolation brought about by the breakup of the supercontinent Gondwana (i.e., vicariance; e.g., cf. Quammen[54]) but by initial dispersal by flight and the subsequent loss of flying ability in (in some cases) predator-free isolation through, as CD thought, disuse.[55]

To the question "why has not the ostrich acquired the power of flight?," posed by an unidentified writer, Darwin responded "But a moment's reflection will show what an enormous supply of food would be necessary to give to this bird of the desert force to move its huge body through the air."[56] With a weight varying between 86 and 145 km, and averaging 111 km[57] —or 189.6 to 319.7 at an average of 244.7 lb.—the Ostrich is far heavier than the largest bird ever known to fly. The largest known bird to have flown is in fact the extinct seabird *Pelagornis sandersi*, discovered as a 25- to 28-million-year-old fossil in 1983 at Charleston, South Carolina. It had a wingspan of 20 to 24 ft., or 6.1 to 7.3 m, but an approximated weight of only 22 to 40 km, or about 48.2 to 88.4 lb.[58] Darwin's doubts about Ostriches taking to the air are perfectly understandable, given their huge size and weight, although their inability to ever do so in their present form is far more likely to have to do with weight and the laws of physics than with adequate food consumption sustaining flight.

Of the evolution of flightlessness, Darwin went on: "who would have ventured to surmise that birds might have existed which used their wings solely as flappers, like the logger-headed duck . . .; as fins in the water and as front legs on the land, like the penguin; as sails, like the ostrich; and functionally for no purpose, like the Apteryx [kiwi]? Yet the structure of each of these birds is good for it, under the conditions of life to which it is exposed, for each has to live by a struggle; but it is not necessarily the best possible under all possible conditions. It must not be inferred from these remarks that any of the grades of wing-structure here alluded to, which perhaps may all be the result of disuse, indicate the steps by which birds actually acquired their perfect power of flight; but they serve to show what diversified means of transition are at least possible."[59] And so here CD uses birds and the use or disuse of their wings to build his argument, with the judicial use of pedantic scientific qualification, to further advance his case for the evolution of morphological traits.

Henry H. Travers[60] stated that kiwis, "said by the Maoris to have been identical with a New Zealand species," once lived on the Chatham Islands, some 360 miles, or 579 km, to the east of central South Island, New Zealand, causing Darwin to puzzle over how the flightless bird reached there and to concede in a letter that it represented evidence of a land bridge having once existed and that permitted kiwis to reach the island on foot.[61] In fact, kiwis were never on the Chathams, no land bridge appears to ever have existed, and evidence now suggests the islands did not emerge from the sea until a mere 1 to 2 million years ago.[62]

ECOLOGICAL SPECIALIZATIONS

This is a topic that has been intensively and extensively studied in birds, and other animals, ever since Charles Darwin; and although it was not pioneered by him, the high profile of his publications, with their pertinent observations and ideas, resulted in further investigations with respect to the evolutionary potential in ecological specializations in species.

Of diversified habits in single animal species, Darwin recalls having watched a South American tyrant flycatcher foraging like a kestrel by hovering or by standing motionless at the water's edge to then dash into it for a fish like a kingfisher capturing prey. Also, he described a Great Tit (*Parus major*) in Britain climbing branches like a Eurasian Tree-creeper and killing smaller birds with blows to their head as does a shrike, or using the same kinds of blows to hammer at the seeds of a yew tree held on to a branch by foot and thus feeding like the Eurasian Nuthatch (*Sitta europaea*) does; and he similarly details several more species. In so doing, CD was hinting at the potential influence on eventual speciation among birds by what was to become known as "ecological isolation."[63] Indeed, he subsequently argued the hypothetical case that Great Tits so using their feet to hold nuts to hammer into the kernel would be selected for in individuals with larger feet and a beak modified for this technique, the feet also improving the birds' ability to clamber about tree branches, and thus the species to potentially become ecologically more like a nuthatch.[64]

OF DUCKS AND WHALES—THEIR BILL OF FARE

Having discussed baleen in the mouth of whales, and its possible evolutionary origin, Darwin asks if it might not have once been something more like the lamellated beak of a duck—as ducks, like whales, sift water to obtain their food by filtration.

Darwin argues:

> The beak of a shoveller-duck (*Spatula clypeata*) [now Northern Shoveler (*Anas clypeate*)] is a more beautiful and complex structure than the mouth of a whale.

> The upper mandible is furnished on each side (in the specimen examined by me) with a row or comb formed of 188 thin, elastic lamellae, obliquely bevelled so as to be pointed, and placed transversely to the longer axis of the mouth. They arise from the palate, and are attached by flexible membrane to the sides of the mandible. Those standing towards the middle are the longest, being about one-third of an inch in length, and they project 0.14 of an inch [3.6 cm] beneath the edge. At their bases there is a short subsidiary row of obliquely transverse lamellae. In these several respects they resemble the plates of baleen in the mouth of a whale. But towards the extremity of the beak they differ much, as they project inwards, instead of straight downwards. . . . The lower mandible of the shoveller-duck is furnished with lamellae of equal length with those above, but finer; and in being thus furnished it differs conspicuously from the lower jaw of a whale, which is destitute of baleen. On the other hand, the extremities of these lower lamellae are frayed into fine bristly points, so that they thus curiously resemble the plates of baleen. . . . From the highly developed structure of the shoveller's beak we may proceed, . . . without any great break, as far as fitness for sifting is concerned, through the beak of the Merganetta armata [Torrent Duck], and in some respects through that of the Aix sponsa [Wood Duck], to the beak of the common duck [Mallard (*Anas platyrhynchos*)]. In the latter species, the lamellae are much coarser than in the shoveller, and are firmly attached to the sides of the mandible; they are only about 50 in number on each side, and do not project at all beneath the margin. They are square-topped, and are edged with translucent hardish tissue, as if for crushing food. The edges of the lower mandible are crossed by numerous fine ridges, which project very little. Although the beak is thus inferior as a sifter to that of the shoveller, yet this bird, as every one knows, constantly uses it for this purpose. There are other species . . . in which the lamellae are considerably less developed than in the common duck; but I do not know whether they use their beaks for sifting the water.

Given that Darwin preserved in spirits the stomach content of a Greater Flamingo (*Phoenicopterus ruber*), collected in the Galapagos Islands (possibly by Harry Fuller[65]), it is surprising that he did not mention the remarkable baleen-like feeding adaptations in an extraordinary beak structure that is highly modified for filter feeding in this discussion of his; perhaps he failed to examine them.

Darwin goes on to observe that although the Egyptian Goose (*Alopochen aegyptiacus*) has a beak like the common duck, the lamellae are less numerous, inward pointing, and distinct from each other and that this goose uses its bill like a duck by throwing water out at the corners. The Common Goose (*Anser anser*) does not, however, sift water but uses its beak only to tear at or cut herbage. Thus, CD explains, there is a discernible gradient from the beak of the common goose, exclusively adapted to grazing, to that of the Egyptian Goose, adapted to both grazing and water sifting, to that of the Eurasian Shoveler adapted almost exclusively to sifting.[66] In so doing, he used the convergent evolution of a similar food-filtering mouth morphology, found in both whales and ducks, to illustrate the adaptive evolution of such traits.

CHICKENS THAT ARE NOT CHICKEN

Darwin appreciated just how instructive the variation in external appearance and structure between the various breeds, and between individuals of those breeds, of domesticated birds (i.e., artificial selection) could be in subsequently getting his theory of natural selection across to his readers. He also saw the advantage of demonstrating the way in which their biology and behaviour could also change under domestication.

In discussing changes in instinct in domesticated animals, Darwin observed that in some breeds of chicken the instinct to incubate their eggs is lost. Also, young chickens have lost "wholly by habit" their fear of dogs and cats, which "no doubt was originally instinctive in them; for I am informed . . . that the young chickens of the parent-stock, the Gallus bankiva [Red Junglefowl (*Gallus gallus*), Java subspecies], when reared in India under a hen, are at first excessively wild. So it is with young pheasants [Common Pheasant (*Phasianus colchicus*)] reared in England under a hen. It is not that chickens have lost all fear, but fear only of dogs and cats, for if the hen gives the danger-chuckle, they will run (more especially young turkeys) from under her, and conceal themselves in the surrounding grass or thickets; and this is evidently done for the instinctive purpose of allowing, as we see in wild ground-birds, their mother to fly away. But this instinct retained by our chickens has become useless [to the mother but surely not to her offspring] under domestication, for the mother-hen has almost lost by disuse the power of flight."[67]

CUCKOO INSTINCTS

Under the subheading "Special Instincts" in his chapter "Instincts," Darwin chose the cuckoo, and two sociable insects, to demonstrate how "instincts in a state of nature have become modified by selection." At the time, Darwin puzzled over the parasitic egg-laying habits of cuckoos, of the family Cuculidae, because the reproductive biology of the Common Cuckoo (*Cuculus canorus*), familiar to him in England as a spring and summer breeding visitor, was little understood, let alone that of other cuckoo species elsewhere.

It was to be 20 years after the 1902 publication of Darwin's sixth edition of *The Origin of Species* that Edgar Chance's admirable book, *The Cuckoo's Secret*,[68] would appear. This provided some firm facts about the subject and these were substantially added to 18 years later in Chance's controversial, in that it involved egg collecting, subsequent book, *The Truth about the Cuckoo*.[69] Since those pioneering studies, the improvement in our understanding of the biology of the Common and other cuckoo species has been profound.[70] At a fundamental level, we now know that only some 40 percent of the approximately 140 cuckoo species are actually parasitic, with most of them incubating their own eggs and raising their offspring in nests of their own construction.[71]

Some naturalists of Darwin's day thought that because the female Common Cuckoo lays eggs in other birds' nests at intervals of two or three days the laying of them in a nest of her own would require her to refrain from incubating the first-laid eggs, to avoid "eggs and young birds of different ages in the same nest." Thus, a long laying and hatching period could be problematic to such a migratory species. That said, CD observes that the "American cuckoo" (presumably the Black-billed *Coccyzus erythropthalmus* and/or Yellow-billed Cuckoo *C. americanus*) builds a nest and "has eggs and young successively hatched, all at the same time," but also that some individual females will lay eggs parasitically in the nest of another species—as has subsequently been confirmed. This led him to speculate that if the ancestral female Common Cuckoo might have only occasionally laid an egg in another bird's nest, like the American cuckoos do, they would have profited in being able to migrate earlier. Or if the resulting offspring were advantaged by being better provisioned by their foster parent (because the young cuckoo evicts the foster parents' young from the nest and thus lacks competitors for food) than their own mother could have done—"encumbered as she could hardly fail to be by having eggs and young of different ages at the same time"—then both the mother or her fostered-out young would be advantaged. That female cuckoo offspring so reared would inherit the habit of laying in other birds' nests now and then, and would thus be reproductively more successful, led CD to conclude, "By a continued process of this nature, I believe that the strange instinct of our cuckoo has been generated."

With these revelational words, Darwin lifted a cloud of incomprehension by clearly demonstrating the adaptive advantages of parasitism to the female cuckoo—in avoiding parental duties, and thus gaining the potential to lay more eggs, and thus being able to migrate south sooner while each of her young enjoys the undivided attention of two foster parents in feeding them.[72]

As noted by an eminent authority on parasitic birds, Charles Darwin packed more good ideas into four sentences on the Common Cuckoo than all previous commentators since Aristotle had done. Indeed that author, Nick

Davies, goes on to note that Darwin raises three fundamental questions about the Common Cuckoo (not all defined by Davies) and that these "form the basis for much" of his own fine book on the Common Cuckoo.[73]

The Common Cuckoo lays remarkably small eggs, and Darwin pointed out that this was an adaptation to its parasitic habits, to reduce the chances of the smaller host species rejecting the cuckoo's egg were it to be significantly larger than those of its hosts. He observed that the fact that the eggs of the two occasionally parasitic American cuckoos are more typical of birds of their hosts' size, which supports this conclusion. Of course a most remarkable adaptation to nest parasitism in the Common Cuckoo is that within three days of hatching, a young cuckoo evicts the foster parents' offspring by getting them onto its back and pushing them over the edge of the nest rim to fall to the ground. Thus the young cuckoo rids itself of any "sibling" competition for food delivered by its foster parents. Darwin saw no more difficulty in explaining the evolution of this instinctive behaviour than that of birds instinctively breaking out of their egg shells.

The various species of cowbirds of the New World were noted by Darwin to exhibit a range of nest parasitism even greater than that between the Common and American cuckoos. Some simply forcibly occupy the active nest of another species to egg lay in, evict the owners' eggs or young to then build their own nest atop the appropriated one, or lay one or more eggs in other birds' nests for them to hatch and foster the nestling(s). The Brown-headed Cowbird (*Molothrus ater*) has, however, perfected its parasitic habits, as has the Common Cuckoo, by laying but one egg in any host's nest. The female Brown-headed Cowbird will peck open or remove eggs of the host as she lays her own; and one hatchling cowbird ejected the host nestling, but this is apparently very rarely done. This cowbird species has been recorded to parasitize over 220 different host bird species, of which at least 144 have successfully fledged a young cowbird.[74] This observable variation in the "quality" of cowbird and cuckoo nest parasitism was seen by CD as grist for the mill in the evolution of such avian reproductive instincts and life histories.[75] The eggs of the Brown-headed Cowbird do not, however, mimic those of its hosts, which is only to be expected given that it parasitizes so many other bird species across diverse families and genera.

Ornithologists of the 1960s approached the problem of understanding brood parasitism by first assuming that it initially evolved as a result of birds responding to the loss of their own nest during their egg-laying period—by them then opportunistically taking over the use of the nest of another species. It came to be appreciated that it was not the cuckoo that evolved eggs that visually fooled their host species as Darwin thought, but rather that their egg mimicry was brought about by the host birds rejecting eggs that they could see were not their own and thus actively selecting for improved egg mimicry by the parasitic cuckoos through their relative hatching and nestling survival rates.[76]

The Common Cuckoo lays eggs at 48-hour intervals, while its hosts lay theirs at 24-hour intervals, and it was found that this gives the cuckoo embryo a 31-hour hatching advantage over that of the host species.[77] This earlier hatching, unappreciated by Darwin and his contemporaries, is advantageous to the parasitic hatchling in ensuring that it is better developed and equipped to overpower or eject the hosts' hatchling offspring. That is, by retaining their egg in their oviduct for an additional day, the females of at least some parasitic cuckoo species "internally incubate" it so that when it hatches, the hatchling is that much more physically advanced. It was concluded, however, that as opposed to being an adaptation to brood parasitism, such "internal incubation" is the result of a prolonged period between ovulation and egg laying. Because it results in early hatching, it may have predisposed some species to becoming brood parasites.[78] Thus, the evolution of this type of interspecific parasitism among birds has involved a kind of "arms race" in which both the parasitic and host species must continue to better adapt to the evolving situation to successfully parasitize, or avoid parasitism, respectively.

THE EVOLUTION OF BIRDS' NEST SOUP

In confronting some objections to his theory of natural selection as applied to instincts, Darwin briefly used the nests of various swift species, of the family Apodidae, as one of many examples enabling him to mount arguments in his own defence.

Some birds that build nests of mud add saliva to them as a binding agent. Darwin noted that one of the American swifts, the Chimney Swift (*Chaetura pelagica*) makes a nest of sticks agglutinated together with saliva. It is not difficult to imagine that circumstances, such as the need for a much smaller nest glued to a more vertical rock face by saliva, might result in positive selection for individuals that used more and more saliva and less and less other nest materials to better adhere to a smooth vertical surface. This scenario would progress until a species built nests almost exclusively of, what becomes to some degree hardened, saliva. Such nests, made exclusively of saliva, are precisely those built by three swiftlet species of one genus (*Aerodramus*) long used in the production of birds' nest soup. Their nests are said to now fetch something like their, very light, weight in gold on the regrettable, entirely unnecessary, and literally utterly tasteless market.[79] This illogical desire to eat the nests of swiftlets' saliva has caused massive recent declines of populations: for example, that of the Black-nest Swiftlet (*Aerodramus maximus*) in the famous Niah Cave in Sarawak, Borneo, was once estimated at 4,500,000 birds; but from the late 1980s and early 1990s, is at only 150,000 to 298,000 birds. The export value of edible swiftlet nests from Sarawak alone in 1995 was US$10,400,000. Indonesia is thought to export nests valued at as much as an incredible US$1,060,000,000 each year.[80]

THE EVOLUTION OF BIRD FLIGHT

In addressing a critic who could not see how the successive modifications of the forelimbs of a supposed or hypothesized bird prototype could have been of any advantage and thus eventually leading to wings capable of flight, Darwin invoked the penguin as an instructive example. He described the flippers of penguins as precisely intermediate between true arms and true wings, and yet they serve to sustain the lives of 18 species in 6 genera of the penguin family, Spheniscidae. Darwin stated that he does not suggest that penguin wings represent the "real transitional grades through which the wings of birds have passed," but he sees no difficulty in believing that "it might profit the modified descendants of the penguin, first to become enabled to flap along the surface of the sea like the logger-headed duck [see "Flight Cancelled" previously], and ultimately to rise from its surface and glide through the air."[81] In fact, the penguins evolved from ancestors that could fly, as did the ostrich, with the results of genetic studies suggesting that at about 70 million years ago, penguins shared a flying ancestor with the albatrosses.[82] The current taxonomy places the penguin order and family, however, after the Loon order Gaviiformes and before the order Procellariformes, consisting of the albatrosses, petrels, shearwaters, and storm and diving petrels.[83]

BIRDS AS DISPERSERS OF PLANTS AND ANIMALS

In his chapter "Geographical Distribution" under the heading "Means of Dispersal," in his *The Origin of Species*, Darwin presented some novel and telling observations that explain how plants and animals could well have been dispersed to far-flung locations, including remote oceanic islands, by air, sea, and via animal agents, including birds. He obtained some of these insights as a result of things seen and hypothesized during the voyage of the *Beagle*, and many were subsequently tested by basic experimentation at his Down House home over many years of his life.

Animal carcasses with seeds contained within their digestive tracts may float at sea for long periods of time to eventually wash up on some distant shore where the seeds may then germinate in some foreign soil to newly establish one or more plant species there. As early as 1855, however, Darwin had expressed the concept that was subsequently to become known as competitive exclusion under such a scenario with the words "But when the seed is sown in its new home then, as I believe, comes the ordeal; will the old occupants in the great struggle for life allow the new and solitary immigrant room and sustenance?"[84]

Dead pigeons with seeds in their crops were floated in tanks of artificial seawater by Darwin, who removed the seeds after 30 days and planted them in soil to find that nearly all of them successfully germinated. A floating dead bird can travel a very long way on ocean surface currents and winds during a month. Flying birds caught in gale force winds may travel at "35 miles [56 km] an hour" or more and could thus be displaced 500 miles, or 805 km, during the period that viable seeds remain in their digestive system. After such a wind-blown flight, an exhausted bird might then die and float for a long time; or, having reached land, be taken by a hawk and the seeds within it dispersed as the predator dismantled the corpse in eating it.

Individual racing, or homing, pigeons have recently been known to fly from Britain to Brazil, and another from Tasmania to the 1,500 km, or 932 mile, distant Subantarctic Macquarie Island.[85] Predation aside, fruit-eating pigeons could of course simply disperse any seeds (which, in eating fruits, they do not digest) by defecation or regurgitation. It is estimated that some 75 percent of the plant groups on Hawaii originally reached there by avian dispersal agents.[86]

Hawks and owls that swallow their prey whole disgorge pellets of indigestible food parts from roughly half to a full day later; and experiments at the London Zoological Gardens showed CD that these can include viable seeds that were within the prey species eaten, that is, small birds, mammals, or fishes. He fed dead sparrows with oats placed into their crops to both a large eagle and an owl and subsequently collected the pellets the birds regurgitated. Upon planting out the seed content of these regurgitated pellets, he found that some seeds survived to successfully germinate; one from the owl doing so after 21.5 hours in the bird's stomach—a period of time during which the bird could travel, he pointed out with ironic wording, "God knows how many miles." He was delighted with such findings, and his boyish enthusiasm for such positive results of his modest experiments comes through most clearly in his letters to friends and colleagues.

Soil not infrequently adheres to the beaks, feet, or legs of birds and may remain there for significant periods of time—and thus seeds held within such soil may be dispersed in the case of birds travelling distances in this condition. Darwin details cases of birds he examined with mud attached to them containing seeds that he removed and planted to find that they germinated successfully. Migratory birds as small as wagtails, wheatears, and whinchats were reported to CD to arrive on the British coast with cakes of mud attached to their feet. A leg of a Red-Legged Partridge (*Alectoris rufa*) with a 6.5 oz., or 184 g, ball of mud attached to it (this would seem a remarkable weight, but the "mass of clay" on this bird's leg measured 3 x 2.5 x 2 inches, or 7.5 x 6.5 x 5 cm[87]) was sent to Darwin. Three years later, this mud ball was broken, watered, and placed under a glass bell to result in 82 plants, 12 monocotyledons and 70 dicotyledons of at least three species, germinating. Thus there could be no doubting the potential for significant plant dispersion by birds, as numerous individuals of many species can be blown as far as across the Atlantic Ocean, from North America to Ireland and to England.[88] Some sticky seeds also frequently adhere to birds' plumage.

Beside plants, birds can also disperse animal life over considerable distances of the earth. In another novel basic experiment, Darwin suspended the removed feet of a duck in an aquarium where "many ova of fresh-water shells were hatching; and I found that numbers of the extremely minute and just-hatched shells crawled on the feet, and clung to them so firmly that when taken out of the water they could not be jarred off." These newly hatched molluscs survived on the duck's feet in damp air for between 12 and 20 hours, during which, CD observed, a duck or heron might fly 600 or 700 miles, or 966 to 1,127 km; and if blown to a more distant point or island, would surely alight on any available water and thus disperse the molluscs in a suitable microhabitat for their survival.

Birds disperse molluscs not only on their legs and feet, as Darwin observed. Two decades ago, a New Orleans academic collected migrating Upland Sandpipers (*Bartramia longicauda*) as they passed through the Gulf Coast to find that they invariably had small freshwater snails, apparently not likely of the local species, attached to their underwing feathers, with as many as 41 on a single bird—carried from Caribbean islands or more southern locations across the Gulf of Mexico.[89] Other wading birds, which frequent the muddy edges of coasts and ponds—as do other water birds that fly greater distances than most birds—doubtless disperse numerous plants and small invertebrate animals widely in this way.

Fish-eating herons likewise disperse those plant seeds that remain viable in the guts of fishes that they eat by swallowing them whole and then fly off with such "seeded" fishes inside them.[90] Many more examples of the remarkable dispersal of plants and animals, and studies and publications about them, have been recently and entertainingly reviewed by Alan De Queiroz.[91]

GALAPAGOS FINCHES AND MOCKINGBIRDS

Given that the volume of *The Zoology of the Beagle* dealing with the birds collected and observed by Darwin during the voyage was published during 1838 to 1841, in which the then 13 species of Galapagos finches were described, it is a surprising fact that he did not so much as mention them in his 1859 first edition of *The Origin of Species*. It is also worthy of note that the word "evolution" does not appear in that book, for which Darwin instead used "descent with modification," although the very last word in the book is "evolved": "There is grandeur in this view of life, with its several powers, having been originally breathed by the Creator into a few forms or into one; and that, whilst this planet has gone on cycling on according to the fixed laws of gravity, from so simple a beginning endless forms most beautiful and most wonderful have been, and are being evolved."[92] He retained this reference to the "Creator" within his final paragraph in following editions of *The Origin*, despite expressing regret over using the word "created" in his works where he actually meant by it "appeared"[93] (see "The Birds of the *Beagle* Voyage" of chapter 8).

In *The Origin of Species*, Darwin merely numerically includes the Galapagos finches in his discussion of total number of birds "created" on the archipelago.[94] This is thought to have been, at least in part, because John Gould's contemporary ornithological colleagues thought he was wrong in his naming of the many similar looking finches as distinct species. Some were particularly unconvinced by him including the fine-billed Warbler Finch among them. These concerns supposedly left Darwin reluctant to discuss them further in print. The fact that Darwin failed to record which island each of his finches were collected on also greatly restricted what he was able to observe and write about them with confidence.

John Gould determined that of 26 Galapagos land bird species, 25 were, in Darwin's words, "distinct species, supposed to have been created here; yet the close affinity of most of these birds to American species in every character, in their habits, gestures, and tones of voice, was manifest." Given his word "supposed" and context used, it is perhaps possible that by "created here" he (possibly cynically) meant by God, and his following words appear to demonstrate his doubt about this in his mind. (For more regarding his use of the word "created," see "The Birds of the *Beagle* Voyage" of chapter 8.) He goes on to underscore this growing serious doubt of his in god-given species by observing how the populations of mockingbirds differ in morphology on islands close to one another within the archipelago;[95] see "Mockingbirds" in chapter 3.

I must stress here that while the *Beagle* was in the Galapagos Archipelago, Darwin did not, contrary to much popular literature and erroneous perception, appreciate any significance of the finches there to any thinking about evolution and the origin of species. The Galapagos mockingbird populations, however, clearly did stir notions in his mind about their evolution among the islands into distinct, island-isolated, forms.

In the first edition of his *Journal of Researches*, Darwin[96] wrote of the Galapagos finch species, defined by John Gould, only that "It is very remarkable that a nearly perfect gradation of structure in this one group can be traced in the form of the beak, from one exceeding in dimensions that of the largest gros-beak, to another differing but little from that of a warbler." In his 1860 and subsequent editions of that work, however, an illustration of four Galapagos finch heads (figure 2.8) and the following notable text occurs: "Seeing this gradation and diversity of structure in one small, intimately related group of birds, one might really fancy that from an original paucity of birds in this archipelago, one species had been taken and modified for different ends."[97] The brief length of this newly added sentence belies its enormous historical, biological, and evolutionary significance. However, the surprising fact is that Darwin did not subsequently publish anything about the Galapagos finches in support of his theory of evolution by natural selection—until, that is, the posthumous publication of *Charles Darwin's Natural Selection*. In this, he wrote of his supposing that all bird species reaching the Galapagos islands "had to be modified, I may say improved by selection in order to fill as perfectly as possible their new places; some as Geospiza, probably the earliest colonists, having undergone far more change than other species; Geospiza now presenting a marvellous range of difference in their beaks, from that of a gros-beak to a wren; one sub-species of Geospiza mocking a starling, another a parrot in the form of their beaks."[98] It was only in the light of Gould's observations on his specimens and the hindsight so gained that CD was able to add his seemingly evolutionarily significant sentence—an important scenario overlooked by some authors.[99] Having added what he did in the subsequent editions of his *Journal of Researches*, the finches were subsequently studied by a number of ornithologists. The results of these studies enabled one of the finches' most eminent students to write of them that "Their special interest today is in providing the best example, in birds, of an adaptive radiation into different ecological niches that is sufficiently recent, geologically speaking, for intermediate and transitional steps to have survived."[100] In Darwin's day, however, he could not see the finches as offering strong evidence for his theory, simply because his limited knowledge of them in life led him to believe almost all species were similar in feeding habits and food: the single exception in his experience being the cactus finches (colour figure 4.3). He could not possibly have appreciated that their differing beak sizes and structures reflected competition, adaptation, and ecological isolation.

It was David Lack that first demonstrated it to be the case that no two bird species of very similar ecology can coexist in the same habitat and location, that is in sympatry, as was hypothesized by Georgy Gause.[101] Lack did this by presenting evidence of character displacement among the Galapagos finch species—for example, on islands where species of large, medium, and small ground finches (*Geospiza magnirostris, G. fortis*, and *G. fuliginosa*, respectively) coexisted, their bill shape and measurements were significantly different; whereas when one or two of these species are rare or absent, the bill of the only or dominant other species present has an altered bill shape of less restricted measurements that trend towards those of the other species.[102] Today the Galapagos finches are among the best long-term, intensively studied birds on earth, thanks in no small part to Peter and Rosemary Grant.[103]

BIRDS DISPERSED

Darwin continued with his considerations of "Geographical Distribution," under the heading "On the Inhabitants of Oceanic Islands," in his *The Origin of Species*. He briefly discussed endemism on remote islands and the lack of them on islands less remote because of various circumstances that do not necessarily involve their relative distances from mainlands.

Of the 26 land birds on the Galapagos Islands, "21 (or perhaps 23) are peculiar, whereas of the 11 marine birds only 2 are peculiar." Of course sea birds reach the islands far more frequently than do land birds. Bermuda is about as far from North America as the Galapagos is from South America, but it "has a very peculiar soil" and has no endemic land bird. This is because "very many North American birds occasionally or even frequently visit this island."[104]

Madeira, between the Azores and the coast of Morocco, northwest Africa, was made known to Darwin to be inhabited by 99 bird species, but of which only one was endemic there, and even that one being very closely related to a European species, while three or four others were confined to it and the Canaries. Thus, the avifaunas of both Bermuda and Madeira have been stocked from neighbouring continents, "which for long ages have there struggled together, and have become mutually co-adapted. Hence when settled in their new homes, each kind will have been kept by the others to its proper place and habits, and will consequently have been but little liable to modification. Any tendency to modification will also have been checked by intercrossing with the unmodified immigrants, often arriving from the mother-country." The influence of relatively regular gene flow from immigrants from mainlands therefore prevents speciation on such islands and the evolution of endemic bird species on them. This is in marked contrast to what has occurred on the Galapagos, and similarly geographically isolated oceanic, islands. Such remote isolation may also prevent "whole classes" of animals reaching them; "thus in the Galapagos Islands reptiles and in New Zealand gigantic wingless birds, take, or recently took, the place of mammals."[105] These were exciting and stimulating zoogeographical and ecological insights for the time.

EXTINCT BIRDS

It is hard to imagine that when he first published *The Origin of Species*, Charles Darwin was unaware of any fossil bird, although briefly and obscurely mentioning some "birds of the caves of Brazil" collected by Dr. William Lund.[106] It was not until 1861, two years after the first edition of that book, that the "feathered reptile" of Jurassic deposits in Bavaria was discovered. Darwin first heard of this fossil from Scottish geologist and palaeontologist Hugh Falconer in January 1863, which is now widely known as *Archaeopteryx*. In writing the sixth edition of *The Origin of Species*, Darwin was by then, however, aware only of *Archaeopteryx* as a fossil bird—this critically important extinct species presenting evidence of characters intermediate between those of reptiles and of birds. Since then, palaeontologists have described and named numerous fossil birds including many, notably from China and including passerine species, with such finely preserved feather detail that even sexual dimorphism in plumage of several species have been described.[107] Indicative of the immediately subsequent revolution in avian palaeontology was, however, that in March of 1880, when aged 71, Darwin was able to write to the American palaeontologist Othniel Charles Marsh to tell him how his work on fossil birds "has afforded the best support to the theory of evolution, which has appeared within the last 20 years."[108]

In 2007, protein was apparently extracted from the collagen of a *Tyrannosaurus rex* fossil in excess of 68 million years old. This protein was compared to that of chickens, and the results helped to confirm the by then widely held view that birds have descended from the dinosaurs.[109]

In his small personal pocket notebook identified by him as "D," Darwin telegraphically noted in August of 1938 that in a paper on the anatomy of the kiwi by Richard Owen, the "Osteology of birds to reptiles [is] shown in osteology of young Ostrich." In fact, Owen's pertinent words were, "The close resemblance of the bird to the reptile in its skeleton is well exemplified in the young Ostrich."[110] Such observations by Owen did not, however, get in the way of his rabid, religiously influenced, opposition to the theory of the origin of species by natural selection.

Charles Darwin wrote of extinction, "We may thus account for the distinctness of whole classes from each other—for instance, of birds from all other vertebrate animals—by the belief that many ancient forms of life have been utterly lost, through which the early progenitors of birds were formerly connected with the early progenitors of the other and at that time less differentiated vertebrate classes."[111] What intense pleasure and satisfaction he would enjoy could he know of our present much-advanced knowledge and appreciation of avian extinction, which is nevertheless surely but yet in its infancy in terms of the search for fossil bird families, genera, and species. Following the 1861 discovery of *Archaeopteryx*, so few other fossil bird species were found during the next 120 years that less than 10 were known in 1980; but then discoveries were made exponentially, with between 50 and 60 being known by about 2014.[112] A finely preserved 3 inches, or 8 cm, long fossil bird named *Pumiliornis tessellatus*, recently discovered in a shale pit at Messel, Hessen, Germany, not only has its feathers preserved but also flower pollen, indicative of nectar feeding, in its stomach.[113]

THE GROUNDED WOODPECKER

In Darwin's day, the number of known woodpecker species would have been but a part of the approximately 232 that are known today.[114] Nevertheless, the perception of their biology then was much as it is today—that all woodpeckers peck wood and must therefore climb on trees to do so and to feed as well as to nest in preexisting tree holes or ones that they excavate for themselves.

Darwin described the Campo Woodpecker, now the Campo Flicker (*Colaptes campestris*), as he observed it on the plains of La Plata "where hardly a tree grows" and noted its typical woodpecker tree-climbing physical adaptations (i.e., two-forward-two-backward pointing toe arrangement, stiff tail, and bill form) while stating that it "never climbs a tree!"[115] Eleven years later, popular ornithological author William H. Hudson stated in print[116] that the Campo Flicker does indeed climb trees and suggested that Darwin was not only wrong in this but "had purposely wrested the truth in order to prove" his theory. Darwin understandably took this accusation badly and expressed that he "should be loath to think that there are many naturalists who, without any evidence, would accuse a fellow worker of telling a deliberate falsehood to prove his theory."[117] In the 1872 sixth edition of his *The Origin of Species*, Darwin did, however, change his pertinent sentence to read "I can assert, not only from my own observations, but also from those of the accurate [Fèlix] Azara, in certain large districts it does not climb trees, and it makes its holes in banks!" He need not have done so to appease Hudson, for the standard modern literature on the Campo Flicker states that it forages "almost entirely," "virtually," or "almost exclusively" on the ground.[118] It nests in both earth banks and tree holes.

In his very last paragraph of *The Origin of Species*, Charles Darwin uses the vision of a hypothetically tangled, diversely vegetated, earth bank (perhaps one on his Down House Sandwalk [see "Footnote"]; see colour figure 8.4) inhabited by various insects, with worms in the damp soil, and "with birds singing on the bushes" to emphasize the ecological complexity of even microhabitats. As small and localized as this ecological example is, Darwin featured a bird at the top of the food chain.[119]

FOOTNOTE

Today Down House is a delightful place of pilgrimage for biologists, international visitors, tourists, and the general British public alike—now run by British Heritage as a wonderful monument to the life and prodigious and highly significant originality and productivity of Charles Robert Darwin; see colour figures 4.1, 8.3, and 8.4.

NOTES

1. Healey 2001: 136.
2. Darwin 1839c.
3. Desmond and Moore 2009: 140.
4. Browne 2002: 165.
5. Darwin 1842; Gould and Darwin 1838–1841.
6. Darwin 1844, 1846.
7. Birkhardt et al. 1988–2013, vol. 2: 376–7.
8. Birkhardt et al. 1988–2013, vol. 3: 75.
9. Stott 2003.
10. Darwin 1851a,b; 1854a,b.
11. F. Darwin 1950: 6.
12. In Browne 2002: 65.
13. Second 1981: 166.
14. Darwin 1859: 28; 1902: 32.
15. F. Darwin 1950: 145.
16. Keynes 2001: 245.
17. In Secord 1981: 166.
18. Desmond and Moore 1991: 445.
19. Cf. Browne 1995: 522.
20. Desmond and Moore 2009: 251–3; Cooper 2014: 15.
21. Birkhardt et al. 1988–2013, vol. 5: 522–3.
22. Eyton 1859.
23. Birkhardt et al. 1988–2013, vol. 6: 210–11.
24. Bonaparte 1850, 1857.
25. Brisson 1760.
26. Buffon 1770–1783.
27. E.g., Desmond and Moore 1991; Quammen 1996; Berry 2002; Browne 2002; Benton 2013 and references therein.
28. Wallace, in Marchant 1916: 91–2.
29. Malthus 1826.
30. E.g., Browne 1995: 385–90; Smith and Beccaloni 2008.
31. Darwin and Wallace 1858: 50, 52.
32. Darwin 1902: xx–xxiii; reviewed in S. J. Gould 1985: 336–46.
33. CD in Browne 1995: 519.
34. Darwin 1839b: 351.
35. Browne 2002: 75.
36. Darwin 1902: 23–33.
37. Healey 2001: 59, 230.
38. De Beer 1968: 149.
39. In Browne 1995: 523–4.
40. Taylor and van Perlo 1998; E. Mayr and Diamond 2001; Dutson 2011; Frith 2013.
41. Darwin 1902: 23–32, 135, 194–8, 202–3, 413–14, 456–7, 612–14.
42. Wallace in Wyhe and Rookmaaker 2013: 247.
43. Darwin 1902: 76, 78.
44. D. W. Snow 1976; B. Snow and D. Snow 1988.
45. E.g., D. Snow 1982; M. Andersson 1994; Frith and Beehler 1998; P. M. Bennett and Owens 2002; Frith and Frith 2004; Kirwan and Green 2011 and references therein.

46. Darwin 1902: 110.
47. Darwin 1902: 174.
48. E.g., Hesse et al. 1937; Andrewartha and Birch 1954.
49. E.g., Lack 1971; Diamond 1978; Birkhead et al. 2014.
50. Hume and Walters 2012; Frith 2013: 53.
51. Darwin 1902: 167.
52. Delacour 1954; Couve and Vidal 2003; Kear 2005.
53. Darwin 1902: 167–8.
54. Quammen 1996.
55. Phillips et al. 2010; De Queiroz 2014: 245–6; Low 2014: 146; Mitchell et al. 2014; Lee and Worthy 2014.
56. Darwin 1902: 281.
57. S. J. J. F. Davies 2002: 260.
58. Ksepka 2014.
59. Ksepka 2014.
60. Travers 1869: 178.
61. Birkhardt et al. 1985–2014, vol. 13: 278–81.
62. Aikman and Miskelly 2004.
63. Darwin 1902: 220–23; cf. Lack 1971.
64. Darwin 1902: 355.
65. In Sulloway 1982a: 88.
66. Darwin1902: 286–9, 310.
67. Darwin 1902: 328–9.
68. Chance 1922.
69. Chance 1940.
70. E.g., Berger 1964, 1968; Wyllie 1981; N. B. Davies 2000; Payne 2005; Erritzøe et al. 2012.
71. Erritzøe et al. 2012.
72. Darwin 1902: 330–36.
73. N. B. Davies 2000: 8–10; 2015.
74. Jaramillo and Burke 1999: 382–3.
75. Darwin 1902: 334–5.
76. Baker 1913.
77. Birkhead et al. 2011.
78. Birkhead et al. 2011; Birkhead et al. 2014: 305.
79. Darwin 1902: 355.
80. Chantler 1999: 414.
81. Darwin 1902: 442–3.
82. De Roy et al. 2013: 150.
83. Gill and Donsker 2015.
84. In Brent 1981: 402.
85. Low 2014: 189.
86. Carlquist 1974.
87. Burkhardt et al. 1985–2014, vol. 11: 251.
88. Darwin 1902: 509–14.
89. Beyer in De Queiroz 2014: 93.
90. Darwin 1902: 537–41.
91. De Queiroz 2014.
92. Darwin 1902: 669–70.
93. F. Darwin 1887, vol. 3: 18.

94. Darwin 1859: 398.
95. Darwin 1859: 398, 402.
96. Darwin 1839b: 462.
97. E.g., Darwin 1889: 380.
98. Stauffer 1975: 257.
99. Cf. Wyhe 2012.
100. Lack 1964: 178.
101. Gause 1934: 19–20.
102. Lack 1947: 81–90.
103. Grant 1986; Grant and Grant 2008; Weiner 1994, and references therein; and others: Donohue 2011, and references therein.
104. Darwin 1902: 543–4.
105. Darwin 1902: 543–5, 552.
106. Darwin 1859: 339.
107. Ackerman 1998; Birkhead et al. 2014: 1–41.
108. In Birkhead et al. 2014: 2.
109. McCarthy 2009: 184.
110. Owen 1838.
111. Darwin 1902: 592–3.
112. Birkhead et al. 2014: 20.
113. G. Mayr 1999; G. Mayr and Wilde 2014; A. R. Williams 2015: 22.
114. Sibley and Monroe 1990; Winkler et al. 1995; Gill and Donsker 2015.
115. Darwin 1859: 184.
116. Hudson 1870.
117. Darwin 1870: 706.
118. Short 1982: 388; Winkler et al. 1995: 325; and Winkler and Christie 2002: 515, respectively.
119. Darwin 1902: 669.

CHAPTER 5

The Variation of Birds under Domestication

I have been led to study domestic pigeons with particular care, because the evidence that all the domestic races have descended from one known source is far clearer than with any other anciently domesticated animal. [Darwin 1868a: 131]

The publisher of *The Origin of Species,* John Murray, had strongly urged Charles Darwin not to make that book too large, academic, and tedious a volume by including all of his substantial body of evidence and arguments in support of his central theory. Given the obvious great significance of the book, it was important to make it relatively brief and digestible so that it could reach as large an audience as possible. Murray suggested, with both good reason and commercial sense, that any such omitted material could well be presented in a subsequent book. Because Charles agreed to do this, he stated in his introduction to *The Origin of Species*, "I can here give only the general conclusions at which I have arrived, with a few facts in illustration, but which, I hope, in most cases will suffice. No one can feel more sensible than I do of the necessity of here-after publishing in detail all the facts, with references, on which my conclusions have been grounded; and I hope in a future work to do this."[1] His notion of "a few facts" was doubtless to prove highly amusing to a great many of his readers.

The "future work" that Darwin rather cavalierly mentioned proved, in the most broad sense, to come, in the fullness of time, to involve a book titled *On the Various Contrivances by Which British and Foreign Orchids are Fertilised by Insects, and on the Good Effects of Intercrossing* in 1862; two volumes on *The Variation of Animals and Plants Under Domestication* in 1868; *The Descent of Man and Selection in Relation to Sex* in 1871; *The Expression of the Emotions in Man and Animals* in 1872; *Insectivorous Plants* in 1875; *The Different Forms of Flowers on Plants of the Same Species* in 1877; *The Power of Movement in Plants* in 1880; and *The Formation of Vegetable Mould, Through the Action of Worms . . .* in 1881; all published by John Murray of London. It is a remarkable and well-documented fact that these works were produced by Darwin while suffering far from good health much of the time, which had begun to decline as early as late 1862, when he was 53 years old.[2]

With respect to his work on the variation of animals under domestication, Darwin was actively preparing to learn what he could in terms of domesticated birds. He contacted numerous correspondents seeking information about—and towards the end of 1855, specimens of—birds (and mammals) in domestication: predominantly ducks, fowls, and pigeons. This included people in South America and the Caribbean, Africa and the Middle East, India and Ceylon (Sri Lanka), China and Indonesia. Sir Walter Elliot alone was to send 29 skin and two skeleton specimens of domestic pigeons, involving nearly 20 breeds, to Down House from Madras in India, the two men meeting at a gathering of the British Association for the Advancement of Science in Glasgow, Scotland, in September of 1855.[3] Edward Blyth wrote to tell Darwin that he was sending some live domestic pigeons to him from India, via the head keeper at the Zoological Gardens in London.[4] In December 1855, Charles wrote to Alfred Russel Wallace in Sarawak asking for adult poultry of both sexes, with particular bones to be left in the specimens, and the provision of any native names applied to them; Wallace did respond, albeit almost a year later. Wallace was a busy explorer, living and working under the most trying of circumstances, and the postal service was indeed all but "snail mail" in those days. In August 1856, Wallace did, however, send to Darwin, via Singapore and then his agent Samuel Stevens in London, a domestic duck and a jungle cock specimen from Ampanan, Lombok (now in Indonesia).[5]

On 29 July 1856, Darwin attended the Anerley, near Sydenham, south London, poultry show. On 1 May 1857, he wrote to Wallace from Down House and stated "I have acted already in accordance with your advice of keeping domestic varieties & those appearing in a state of nature, distinct; but I have sometimes doubted of the wisdom of this, & therefore I am glad to be backed by your opinion.—I must confess, however, I rather doubt the truth of the now very prevalent doctrine of all our domestic animals having descended from several wild stocks; though I do not doubt that it is so in some cases."[6] On 10 August 1857, he made his way to the Crystal Palace poultry show.

In October, Alfred Wallace apparently wrote to Darwin suggesting that he reproduce domestic breeds of fowl and pigeon to be able to compare them with exotic breeds.[7] Darwin was to subsequently show which of several domesticated bird species with diverse breeds had single ancestors and which had more than one. In writing to Wallace on 22 December 1857, about his studies of the distribution of animals in the Malay Archipelago, Darwin expressed his view that "I am a firm believer that without speculation there is no good and original observation";[8] a view that remains valid, if unfashionable in some scientific circles, today. On 14 September 1858, the great aviculturalist William Bernhard Tegetmeier, who greatly assisted Darwin with all matters concerning the keeping and breeding of birds in domestication, visited him at Down House. This visit would have been highly significant to Darwin's plans for his studies of, and speculations about, domesticated birds.

At the end of 1859, Darwin's relationship with scholarly, religious, powerful, influential, and rather jealous and spiteful Richard Owen, who had become superintendent of the natural history department of the British Museum in 1856, had soured. In April 1860, Owen anonymously published a scathingly critical review of *The Origin of Species*, but it was obvious to his better-informed contemporaries that it was by him. This attack firmly cemented the two men as opponents, with respect to Darwin's belief in the origin of species and the influence of evolution by natural selection and Owen's belief in his god and creator, respectively, if not also personally. Following on this acrimony was the ever since world-famous meeting of the British Association for the Advancement of Science held at Oxford University in June 1860. At this meeting, Thomas Huxley, CD's "bulldog" of an advocate, debated Darwin's theory with Bishop Samuel Wilberforce who was known as Soapy Sam to his distracters for his mannerism of rubbing his hands together as if soaping them. Wilberforce, who it quickly became clear to Huxley and others present had obviously not bothered to read Darwin's book, had nevertheless been well primed by Richard Owen on pertinent anatomical arguments. However, the arrogant Bishop made the fundamental tactical error of making a personalized attack on Huxley, with reference to his family and descent from an ape. At the end of his eloquent and telling rebuttal of the factually erroneous content of Wilberforce's rabid speech, Huxley expressed the view that he would far rather have an ape for a grandfather than an intelligent man of power and influence who used his position and intellect to ridicule serious discussion of an important scientific theory and its respected, sincere, and objective author. This met with the riotous applause of the vast majority of the audience, which included many students.

Darwin had not attended this historic meeting in Oxford. Before he had long to enjoy the news and subsequent repercussions of Huxley's defence of his great theory, he became much involved in seeing translations of his *The Origin of Species* into several European languages. This was a task involving communications with eager potential translators and discussions and negotiations with his publisher. He also developed an interest in British carnivorous plants. He started to experiment with them until the illness of his daughter Henrietta and other pressing matters saw him shelve the subject early in 1861—but he was to return to these intriguing plants in a small way in mid-year, again briefly in 1864, and then in a substantial way in the early 1870s.

Also early in 1861, John Murray asked Darwin to prepare a third edition of *The Origin of Species*, which he did in short order, and a print run of 2,000 copies of that fully revised edition appeared in April. Charles was now a particularly busy naturalist and family man. By mid-1861, he had become enthralled by the study of orchids, their means of attracting insects and other pollinators, and the various and highly complex structures they have evolved to ensure the dispersal of their pollen. He also reverted to limited experimentation on carnivorous and other plants in his garden, greenhouse, and study. His orchid book, ten months in the writing and which CD saw as providing significant support for his origin of species theory, appeared in May of 1862.

In the middle of 1862, Alfred Russel Wallace visited Down House to meet Darwin in person for the first time; but shortly thereafter, CD's son Leonard fell ill, followed by Wallace doing so, which resulted in them not meeting again until at Charles's brother's London home the following winter. They got on well, and it would be likely that birds would have featured in their discussions, particularly as CD had no first-hand knowledge of the avifauna of the Malay Archipelago as then defined (i.e., Singapore to New Guinea), whereas Wallace had observed and collected numerous interesting birds across that archipelago. They might well have also discussed the wild and domesticated fowls and ducks of the islands, especially as Charles was becoming ever-increasingly involved with studying the variation of birds under domestication with particular reference to, among other birds, the domestic chicken and duck. Indeed, much of CD's years of 1862 to 1868 were spent reviewing, studying, and writing for what became his two-volume work on *The Variation of Animals and Plants Under Domestication*, which was to appear in 1868. In his introduction to this publication, he notes "the great delay in publishing this first work [i.e., after *The Origin of Species*] has been caused by continued ill-heath."[9]

John Murray, Darwin's publisher, was informed by a referee he had sent the text of *The Variation of Animals and Plants Under Domestication* to, but under a different title, that it was indigestible. Certainly it is particularly heavy reading in a good many parts, making it more difficult than some of his other works to deal with in this book. Charles completed his seven months of work on the proofs of the large work in mid-November 1867, by which time he harboured doubts about anyone reading it all. Writing to Joseph Hooker, he suggested that the best way to "read" it would be to "Skip the whole of Vol. 1., except the last chapter (and that only to be skimmed) [being about plant reproduction and thus of more interest to Hooker] and skip largely the 2nd volume; and then you will say it is a very good book." Once again, he emphasized its heavy going by writing to Fritz Müller: "The great part, as you will see, is not meant to be read."[10] John Murray intended to print a run limited to 750 copies; but then, with the title improved from his point of view, he changed his mind and doubled the number of copies to be printed. The book was published on 30 January 1868, some 13 years after Darwin started looking at domesticated pigeons. Murray was delighted to find his conservative estimate of demand for the book unfounded, for the printing of 1,500 copies sold within one week and a reprint was done only 11 days later.

VARIATION—VOLUME 1

Almost all of the substantial ornithological content of volume 1 of *The Variation of Animals and Plants* appears in three discrete chapters about birds and on the first 20 pages of a fourth chapter that also includes text on domesticated gold fish and insects. In his introduction to the book, Darwin noted that he could not possibly describe all of the races, or breeds, of all animals domesticated by humans; and that "in one case alone, namely in that of the domestic pigeon, I will describe fully all the chief races, their history, the amount and nature of their differences, and the probable steps by which they have been formed. I have selected this case, because, as we shall hereafter see, the materials are better than in any other; and one case fully described will in fact illustrate all others."[11] He then went on to state that a second and a third work were to appear at some future time, these proving to become in the fullness of time *The Descent of Man* and *The Expression of the Emotions in Man and Animals*, indicating an amazing appetite for long-term hard work in the face of frequent poor health.

An exception to the content of the first sentence of the last paragraph is that Darwin noted that in the song of the "mocking-thrush" and "carrion-hawk" on the Galapagos Islands, he "clearly perceived the neighbourhood of America." A second exception involves him once again alluding to the tameness of birds on the Galapagos Islands and that they have been slow to acquire and inherit a fear of humans.[12] Otherwise there are but the briefest mentions in passing of the odd domesticated bird elsewhere in the two-volume work, save a paragraph noting that in all domestic pigeon breeds and hybrids of them, some individuals will show the black wing bars indicative of their common ancestor; the Rock Dove (*Columba livia*).[13]

The Domesticated Pigeon

The two chapters on domestic pigeons present evidence and discussion of artificial selection that Darwin had intended for his "big book" before his publisher suggested he reduce it in size, to what became *The Origin of Species*. He did, however, retain some 10 pages on the subject in *The Origin*, and I do not repeat my text pertaining to it, found in chapter 4, here. Interestingly, Alfred Russel Wallace, having read Darwin's domestic pigeon account in *The Origin of Species* and found it particularly helpful in demonstrating variation within a species, wrote to Darwin from Sumatra on 30 November 1861, and suggested that the extended pigeon account he was preparing for his *Variation of Animals and Plants* would be "all the better for good illustrations."[14] This opinion may well have influenced Darwin, who included in that book six illustrations of various breeds and six showing variation of some bones found in various forms of domestic pigeons.

Darwin tells his readers that although the domestic pigeon is highly instructive because of its long history under domestication about the world, and the resultant great variation within its various distinctive breeds, "The details will often be tediously minute." I shall try to avoid the tedious details in the following pages.

Darwin kept living individuals of many pigeon breeds, including all English ones, and prepared skins and skeletons of each of them (see figures 5.1, 5.3, 5.4, 5.5,). He apparently promised William Tegetmeier that he would donate his skin and skeleton specimens of domesticated birds to the British Museum should they eventually amount to a useful collection.[15] Certainly at least some 60 of his pigeon skins and 53 skeletons, 25 chicken skeletons,

Figure 5.1 Some breeds of domesticated pigeon; from top left to bottom right: English Pouter, English Carrier, English Barb, English Fantail, African Owl, and Short-faced English Tumbler. From *The Variation of Animals and Plants Under Domestication*. Volume 1: pp. 137, 140, 145, 147, 149, and 152, respectively, by Darwin, 1868a.

six duck skins, and two canary skeletons were donated to that institution's collection by 1867–1868 and were detailed recently by Joanne Cooper.[16] In the late 1960s, I found his domestic pigeon skins in an apparently long-misplaced and overlooked large dusty cardboard box atop a 3 m tall bird specimen cabinet in the original Bird Room in South Kensington. The entire bird collection, including Darwin's domestic specimens, was moved to what was once Lord Walter Rothschild's Zoological Museum at Tring in Hertfordshire in the early 1970s and what is now the Bird Group of the Natural History Museum.[17] A "Catalogue of Down Specimens" compiled by Darwin is held by English Heritage at Down House.

Citing authorities that detailed 122 domestic pigeon "races" that "bred true," Darwin knew of several additional European ones and even more from India and elsewhere, concluding that there must have been "considerably above 150 kinds which breed true and have been separately named." A more recent guide to the domesticated pigeons of the world lists, and usefully illustrates in colour, 170 named breeds.[18] Today there are some 712 named breeds according to Wikipedia's "list of pigeon breeds"—but this number is inflated by the inclusion of the various forms to be found within the tumbler, pouter, homer, owl, roller, and so forth breeds.

The "magnificent collection" of skin specimens of the wild members of the pigeon family held in the British Museum, as Darwin described it, was examined by him. With the exception of a few atypical forms, he did not hesitate to conclude that some breeds of domestic pigeon differ as much from each other in external morphology as do the "most distinct natural genera" of wild, or natural, pigeon species. He observed, "We may look in vain through the 288 known [wild] species [now ca. 310] for a beak so small and conical as that of the short-faced tumbler; for one so broad and short as that of the barb; for one so long, straight, and narrow, with its enormous wattles, as that of the English carrier; for an expanded upraised tail like that of the fantail; or for an oesophagus [*sic*] like that of the pouter." How he could determine the latter character from skin specimens is unclear. Being the meticulous scientist that he was, he went on to qualify his statement: "I do not for a moment pretend that the domestic races differ from each other in their whole organization as much as the more distinct natural genera. I refer only to external characters, on which, however, it must be confessed that most genera of birds have been founded."[19] This was an impressive observation to put to an audience he was seeking to convince of the power and rapid visually obvious results of artificial selection. Of course the intention, in using the grasping of this principle, was to have his readers then more easily understand the potential results of natural selection.

Incredible for its time was a phylogenetic tree drawn by Darwin for the "Columba livia or Rock-Pigeon." This was a full-page graphic showing his view of the relationships to one another, including his indicating their relative distance apart by the length of lines drawn between them, of the various distinct breeds of domestic pigeon.[20] It was one of the first avian phylogenetic trees ever published, and the only specified bird phylogeny that CD ever had printed. It was to be followed in principle and function by countless such diagrammatic trees for some or all bird groups, or the members of a single group, by subsequent avian taxonomists. Darwin then listed and described all of the principle breeds of domestic pigeon known to him over some 24 pages, including some measurements of them, while always referring to the Rock Dove as his "standard" norm in measurements and appearance for comparisons and as being the wild bird species ancestral to all domestic breeds (colour figure 5.2). Following this, he gave five pages to discussing individual variability, noting the fundamental importance of this, as such variations can be "secured and accumulated by man's power of selection; and thus an existing breed might be greatly modified or a new one formed."

Noting that the number of wing and tail feathers in wild birds is typically constant, but can vary in species with many tail feathers, Darwin stated that in wild pigeon species, it is never less than 12 or more than 16. In the domestic fan-tail pigeon, the tail feathers vary in number, however, from 14 to 42; and an apparent consequence of this is that the oil or preen gland, located on the lower rump at the base of the tail, is lost in the breed. Individual birds of the same pigeon breed may show a good deal of variation in the beak and in other parts of the body and feather structure. Indeed, "there is hardly an exception to the rule, that the especial characters for which each breed is valued are eminently variable." Those characters defining the distinctive breeds are often most strongly evident in males; and some, such as facial wattles, increase in prominence with age, whereas in the ancestral Rock Dove there is no discernible sex or age difference.[21]

Pigeon fanciers cannot see, and therefore do not care about, any differences in the skeletons of their various breeds. Such differences between the Rock Dove, Stock Dove (*Columba oenas*), Common Wood-Pigeon (*C. palumbus*), and European Turtle-Dove (*Streptopenia turtur*), involving two genera, were found to be "extremely slight" by Darwin. He demonstrated and illustrated, however, and to the contrary, profound diversity in the structure and size of the skull, jaw, and other bones (figures 5.3–5.5) between mere breeds of domestic pigeon. He also described differences in some of their other bones—all being osteological differences brought about as by-products of artificial selection for exclusively external traits. He also described and discussed the various plumage structures and pigmentations, and the ages at which they are acquired, among the various breeds.[22] Eight years before this was published, he generously gave his pigeon skull measurements, which he had accumulated over considerable time and with much trouble, to Thomas Huxley for him to present in a lecture he was to give at the Royal Institution on 10 February 1860.[23]

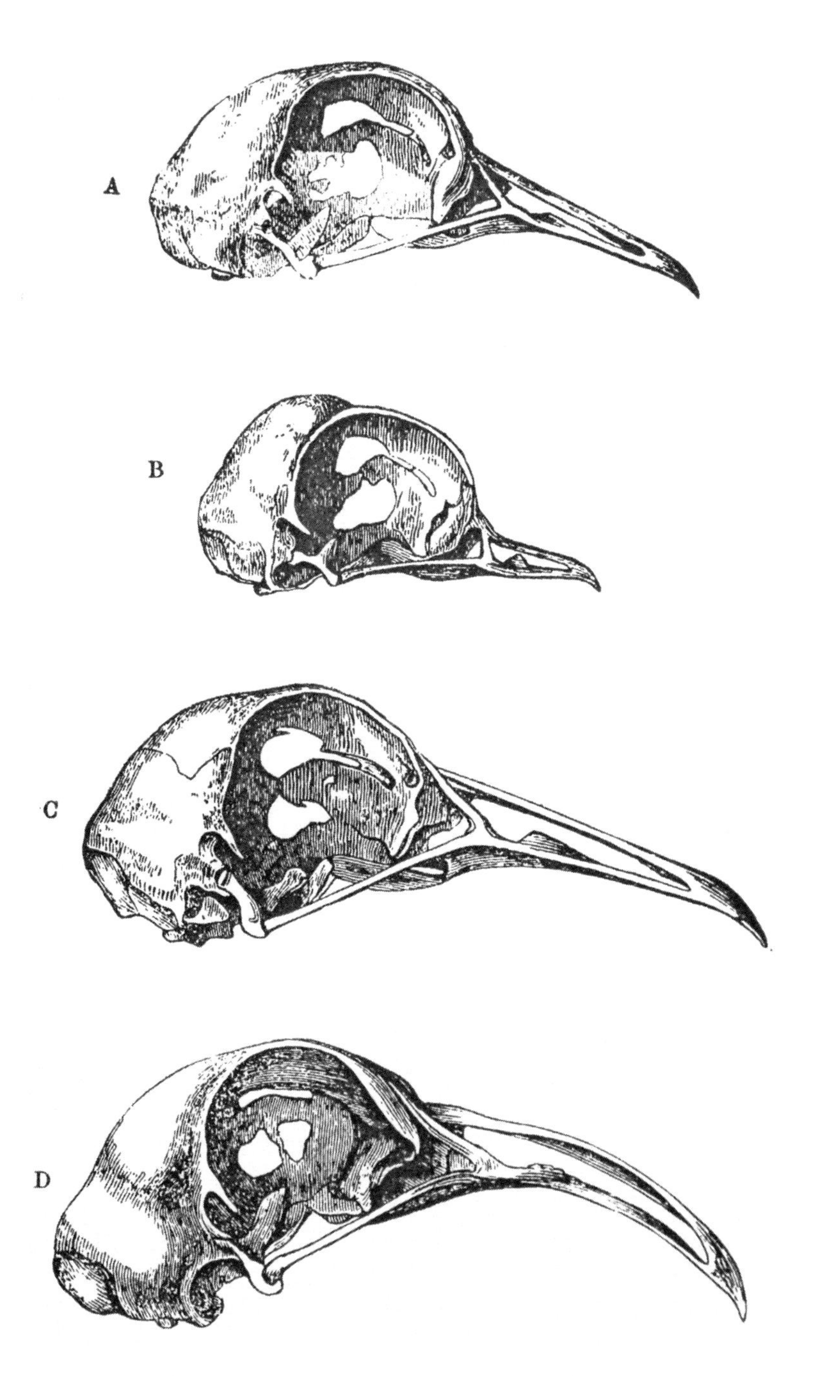

Figure 5.3 Skull of a Rock Dove and three domesticated pigeon breeds drawn from specimens prepared by Charles Darwin: (A) Wild Rock Dove, (B) Short-faced Tumbler, (C) English Carrier, and (D) Bagadotten Carrier; showing extreme structural differences. From *The Variation of Animals and Plants Under Domestication*. Volume 1: p. 163, by Darwin, 1868a.

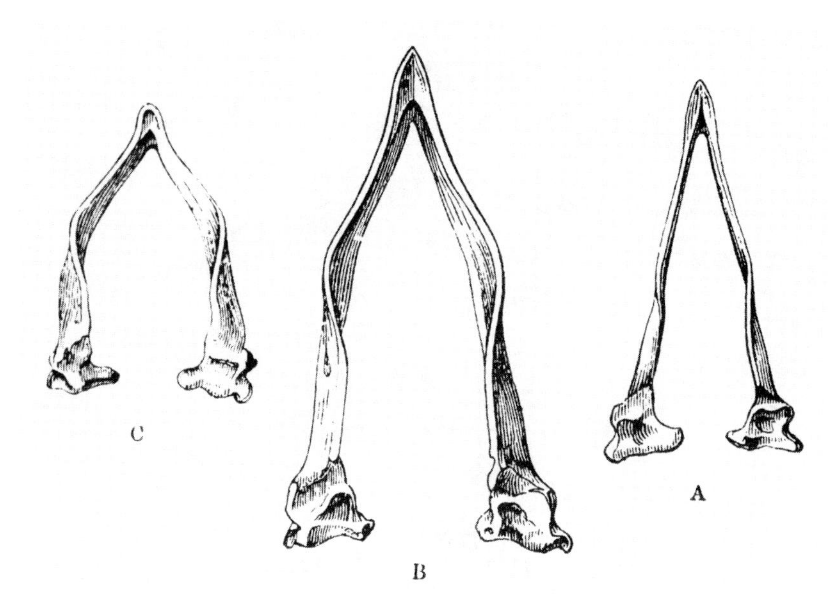

Figure 5.4 Lower jaw bones of a Rock Dove and two domesticated pigeon breeds drawn from specimens prepared by Charles Darwin: (A) Rock Dove, (B) Runt, and (C) Barb, showing extreme structural differences. From *The Variation of Animals and Plants Under Domestication.* Volume 1: p. 164, by Darwin, 1868a.

Observing that both the Dutch ornithologist Coenraad Temminck and John Gould had remarked that given a single ancestral species to the domestic pigeon, it must, then, have been one that nested and typically roosted on rocks rather than in trees. Darwin added that it must also have been a highly sociable species. In reviewing the only "five or six wild [pigeon] species" with these habits, Charles discounts all but the Rock Dove (including some populations then treated as distinct species but now, as CD had suggested, merely subspecies of the Rock Dove) as the single potential ancestor species that could also account for the known domestic breeds and their variations.

In consolidating his argument for the Rock Dove as the single ancestor to the eleven or so basic breeds of domestic pigeon, rather than several ancestral species, Darwin listed six reasons in support of it. First, if they had not arisen from variation in one species and its geographical races (subspecies), they must have descended from several extremely different species, as no amount of crossing between only some five or six wild species could produce forms as distinctive as are the breeds of domestic pigeons. Crossing wild species could not result in pouters or fantails because their traits do not occur in the other potential ancestor species. Moreover, they would have to be rock nesting and roosting social species that had become extinct, because no such species now exist.

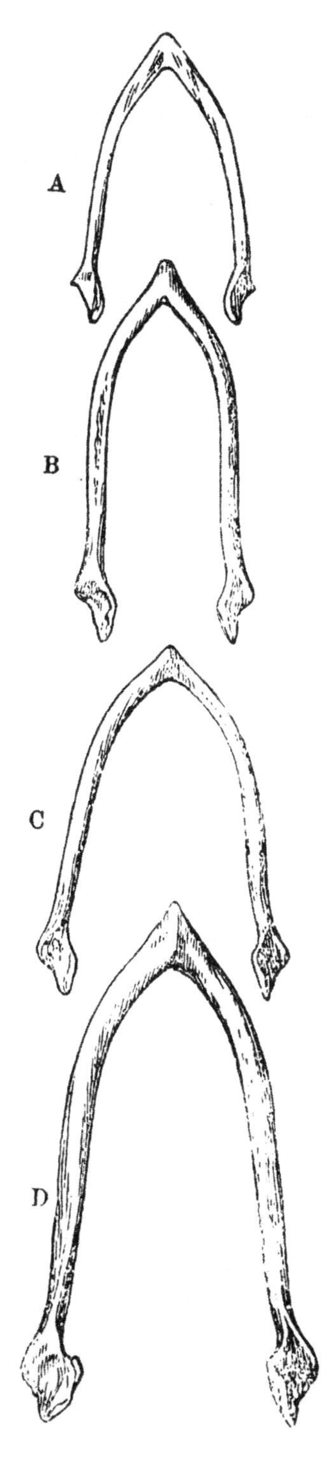

Figure 5.5 Furcula, or wishbones, of four domesticated pigeon breeds drawn from specimens prepared by Charles Darwin: (A) Short-faced Tumbler, (B) and (C) Fantails, and (D) Pouter, showing differences. From *The Variation of Animals and Plants Under Domestication*. Volume 1: p. 167, by Darwin, 1868a.

Second, the belief that several ancestor species were involved implies such species were sufficiently domesticated to breed in captivity, which at that and preceding time was most unlikely because although wild pigeon species could be tamed, they were extremely difficult to breed. The belief, however, dictates that almost a dozen wild species, now unknown, were fully domesticated and commonly bred as captives.

Third, the eleven or so basic breeds have never established themselves in the wild anywhere, despite their global domestication, because they have become so highly modified under artificial selection that they would not survive as feral populations.

Fourth, if people did choose several wild species as ancestors to the domestic pigeon, then they must have chosen a group of highly abnormal species, given the characters of the basic domestic pigeon breeds as compared with the more conservative characters of wild pigeon species known to Darwin. Moreover, as noted previously, those abnormal wild species would have to all have then become extinct.

Fifth, all of the domestic pigeon breeds will readily pair and breed with one another to produce fertile offspring, whereas hardly a case was known of hybrid offspring between two distinct wild pigeon species being fertile. (We do now know, however, of numerous hybrids between two wild pigeon species; but we do not know if the resulting offspring are fertile.[24])

Sixth, and finally, excluding certain important characteristic differences, the basic domestic breeds agree most closely both with one another and with the Rock Dove in all other respects. They are highly sociable, dislike perching and roosting in trees and will not nest in them, lay two eggs, have the same incubation period, are adaptable to a great range of climates, prefer similar foods, and eagerly take salt. All except two perform the same courtship postures and *coo* in the same unique manner; and all have the same peculiar metallic breast feathering, which is "far from general with pigeons." Each breed presents a similar range of colour variation and most show "the same singular correlation between the development of down in the young and the future colour of plumage." In all breeds, the proportional length of toes and primary feathers is nearly the same, characters prone to differ in several wild members of the pigeon family.

Thus, in summary, the breeds of domesticated pigeon differ profoundly in external characters that people have selected for as desired attributes, whereas they all retain their fundamental habits and biology. Having observed that all domestic breeds resemble the Rock Dove and each other, Darwin noted that the wild Rock Dove is a slaty-blue bird with two black wing bars, an upper rump varying from generally white in Europe to blue in India, the tail with a black subterminal band, and the outer webs of the outer tail feathers edged with white except at the tips. These combined characters are not found in any wild pigeon species other than the Rock Dove: A white upper rump and two black wing bars occur in only two wild pigeons—the alpine Snow Pigeon (*Columba leuconota*) and the Hill Pigeon (*C. rupestris*) of Asian topography higher than 1,500 m or 4,921 ft., above sea level.

Whenever a "blue" bird appears in any breed of domestic pigeon, it shows the two black wing bars, thus clearly emphasizing its distant Rock Dove origins.[25] Darwin concluded that with his arguments, which are far from comprehensively included here, fairly considered, it would take an overwhelming body of evidence to support the notion that several wild species were ancestors to the domestic pigeon breeds, as opposed to only the Rock Dove, and that absolutely no such evidence exists to support the notion. It was a most convincing case, notwithstanding the great variation within and across the breeds, and one that was quickly and widely accepted, even among those numerous members of pigeon fancier clubs across Great Britain and Europe.[26]

To indicate that the Rock Dove had been domesticated long enough for an adequate number of generations to enable artificial selection to bring about the existing various breeds, Darwin cited authorities that presented the earliest record of domesticated pigeons as the fifth Egyptian dynasty of ca. 3000 B.C. Another author cites pigeon as listed on a bill of fare in the previous dynasty, and it is suggested that this would probably have involved domesticated birds. Indeed, proof exists in the form of terra-cotta figurines that the Rock Dove was in domestication in Mesopotamia, or Iraq, at about 4500 to 4000 B.C.[27] The Romans apparently paid "immense" prices for pigeons of "pedigree and race." Darwin reports that in about 1600, the Akber Khan of India much valued pigeons and had some 20,000 birds carried about with his court; he goes on to cite other such cases of pigeons in captivity through history. Thus, pigeons had been systematically domesticated over much of the Old World for at least 5,000, perhaps as much as 10,000 years,[28] this period of time clearly permitting the occasional variation that might be preserved by artificial selection.[29]

In reviewing the history of the principle breeds of domestic pigeon, about which he found more was known than for any other domesticated animal, Darwin cited the works of a number of authorities that enabled him to conclude that they nearly all existed before the year 1600, and that given an average lifespan of about five or six years, some of them "retained their character perfectly for at least forty or fifty generations."[30]

The two pigeon chapters in volume 1 of Darwin's *The Variation of Animals and Plants* conclude with "Selection is followed methodically when the fancier tries to improve and modify a breed according to a prefixed standard of excellence; or he acts unmethodically and unconsciously, by merely trying to rear as good birds as he can, without any wish or intention to alter the breed. The progress of selection almost inevitably leads to the neglect and ultimate extinction of the earlier and less improved forms, as well as of many intermediate links in each long line of descent. Thus it has come to pass that most of our present races are so marvellously [*sic*] distinct from each other, and from the aboriginal rock-pigeon."[31] In this way Darwin used the humble domestic pigeon to show how artificial selection, be it systematic or unconscious, can bring about morphologically different forms so distinctive as to appear to have arisen from several species, as those people breeding them in CD's lifetime believed to have been the case. This clear demonstration of what artificial selection could bring about over a few centuries helped much in opening the eyes and minds of his readers to the potential of his, theoretical, natural selection. See the "Pliable Pigeons" section of chapter 8 for a discussion of published criticism of Darwin's pigeon-keeping expertise.

As much more about domestic pigeons appears in volume 2 of *The Variation of Animals and Plants Under Domestication*—but appears as only a page, a paragraph, or a sentence or two, or less, scattered at numerous points throughout the book under many different subject headings—I can but provide reference to their locations and context in appendix 1.

When Darwin was trying to explain variation and inheritance in pigeons and other domesticated birds, he was doing so entirely ignorant of genetic processes. Gregor Johann Mendel read the paper resulting from his meticulous long-term experiments on hybridization and inheritance in pea plants at a meeting in February of 1865. He published them in a journal in the following year,[32] but because they excited almost no attention, Darwin remained completely unaware of them as he published his two-volume work, *The Variation of Animals and Plants Under Domestication*. Many descriptions and observations that CD presents are now far better, if not fully, understood in the light of countless genetic studies.

The Domesticated Fowl

To add to his case on variation under domestication, Darwin reviewed the breeds of the domestic fowl, or chicken (*Gallus gallus*), their history, external and internal morphology, habits, and origins (figures 5.6–5.9). William Bernhard Tegetmeier, an established authority and the author of *The Poultry Book*,[33] greatly helped Darwin in doing this by providing him with information and discussion and supplying specimens. Tegetmeier exhibited in London some domestic fowl skins of new varieties, presumably to Britain, from Darwin's personal collection.[34]

Darwin noted that although most of the basic fowl breeds were then in England, many "sub-breeds" were probably unknown to him; and as a result, he could not claim his discussion to be complete. On the face of it, he wrote, the various main breeds seem to have "diverged by independent and different roads from a single type." He described 13 main breeds, finding the "game breed" the typical one in differing least from "the wild *Gallus bankiva*," now the Red Junglefowl (*Gallus gallus*), subspecies *bankiva*, of Sumatra, Java, and Bali.

As was the case with the domestic pigeon, the naturalists of the day believed the distinctive breeds of fowl derived from a single ancestor, whereas poultry fanciers thought that several ancestral species were involved and that most breeds are extremely ancient. Charles, however, found no evidence of selection (i.e., of discrete "true breeding" domestic breeds) prior to the Romans maintaining six or seven breeds at the start of the Christian era. Thus, artificial selection, through fowl keepers doing no more than occasionally destroying what they perceived as inferior birds and breeding those thought to be the better ones, would modify their stock. Charles cites the example of the Romans recording almost 2,000 years ago that they considered birds with five toes and a white ear lobe the best type.

In playing devil's advocate, Darwin first argued the case for multiple ancestors to the domestic fowl. He then reviewed all of the wild species of *Gallus* and argued the case for what was then *Gallus bankiva* as the single ancestor. He concluded: "From the extremely close resemblance in colour, general structure, and especially in voice, between *Gallas bankiva* and the Game fowl; from their fertility, as far as this has been ascertained, when crossed; from the possibility of the wild species being tamed, and from its varying in the wild state, we may confidently look at it as the parent of the most typical of all the domestic breeds, namely, the Game-fowl." He also found that, again as in domestic pigeons, odd individuals of even the most distinctive of domestic fowl breeds will show "reversion" to the appearance of their ancestor, the wild Red Junglefowl; and some individuals even do so as they age. Moreover, experimental crosses between various distinctive domestic breeds led to the resulting cock offspring showing reversion towards the adult plumage of the Red Junglefowl cock, although such reversion was not apparent in the hens.[35]

Figure 5.6 The head of a cock Spanish Fowl (left), Hamburgh Fowl (centre), and Polish Fowl (right) showing differences brought about by artificial selection. From *The Variation of Animals and Plants Under Domestication*. Volume 1: pp. 226, 228, and 229, respectively, by Darwin, 1868a.

Turning to the history of the fowl under domestication, Darwin reported the first credible reference to it as being illustrations on Babylonian cylinders of between the sixth and seventh centuries B.C. and on the Harpy Tomb in Lycia of approximately 600 B.C. Theognis and Aristophanes mention it between 400 and 500 B.C., and Charles felt confident that it reached Europe about the sixth century B.C. It was found in Britain by Julius Caesar, was domesticated in India at 800, possibly 1200, B.C., and reached China about 1400 B.C.[36] Since Darwin's time, we have discovered, however, that people had domesticated chickens in the Yellow River Valley of China as long ago as about 5400 B.C. Moreover, evidence arising from DNA studies suggests that the domestication of fowls had happened even earlier than this, in the region of Thailand or Vietnam; and that both the Red and the Grey Junglefowl (*Gallus sonneratii*) species were involved as their original ancestors.

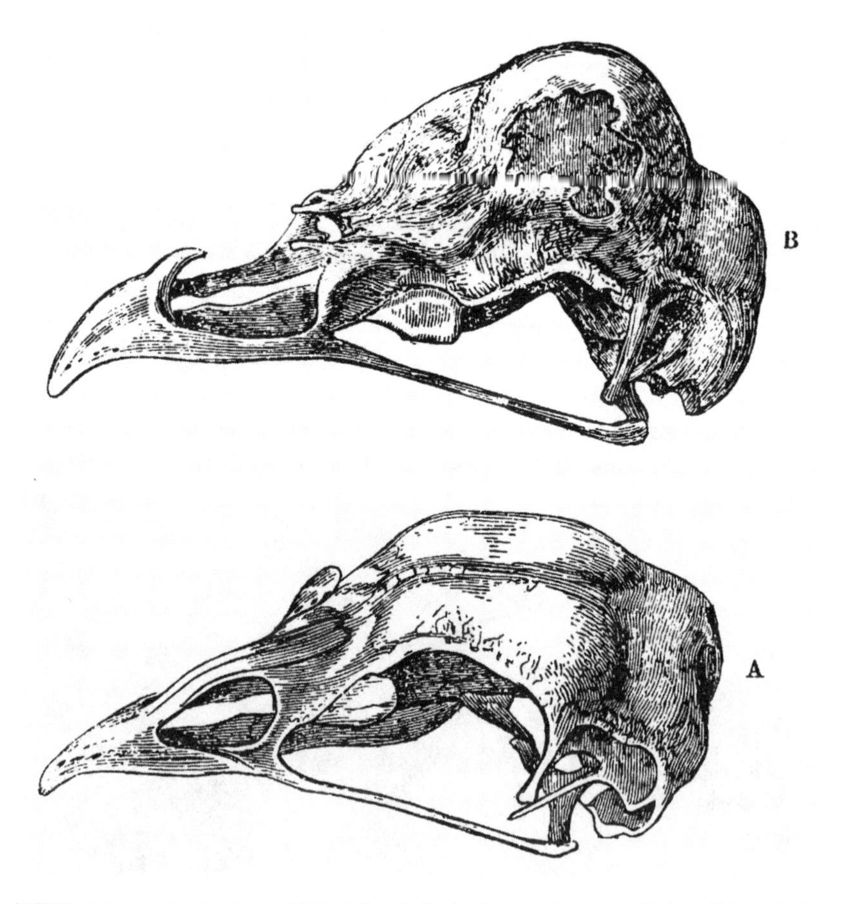

Figure 5.7 The skulls of a wild Red Junglefowl, Javan subspecies, below (A), and of a White-crested Polish Cock above (B), showing extreme structural differences. From *The Variation of Animals and Plants Under Domestication*. Volume 1: pp. 226–229 by Darwin, 1868a.

The pedantically thorough Charles Darwin goes on to discuss at length differences in external morphology between the various chicken breeds. He also discusses individual variability in eggs, chicks, and older birds; plumage acquisition ages; secondary sexual characters; external morphology not connected to the sexes, between breeds and individuals; skeletal differences; the effects of the disuse of body parts; and the correlation of growth in various characters. In making his osteological observations, based on his own collection of skeletons, he made all comparisons and discussion taking *Gallus gallus* as the "standard" or norm. He found the skull of the largest Cochin fowls twice as long, but not nearly twice as broad, as in Bantams, and illustrated a few of the more divergent skull forms (see figures 5.7–5.9). Darwin was struck by the "great variability of all the bones except those of the extremities" in domestic fowls, which he accounted for by their having been exposed to unnatural conditions, despite the fact that the fanciers do not intentionally select for skeletal change. In discussing the effects of disuse of body parts, he demonstrated this with weights and measurements of bones, most notably of wings and legs, given that domestic fowls use them far less than do their wild ancestors.[37]

Figure 5.8 Longitudinal sections of the skulls of a Polish Cock above (A) and of a Cochin Cock below (B), showing extreme structural differences. From *The Variation of Animals and Plants Under Domestication*. Volume 1: pp. 226–229 by Darwin, 1868a.

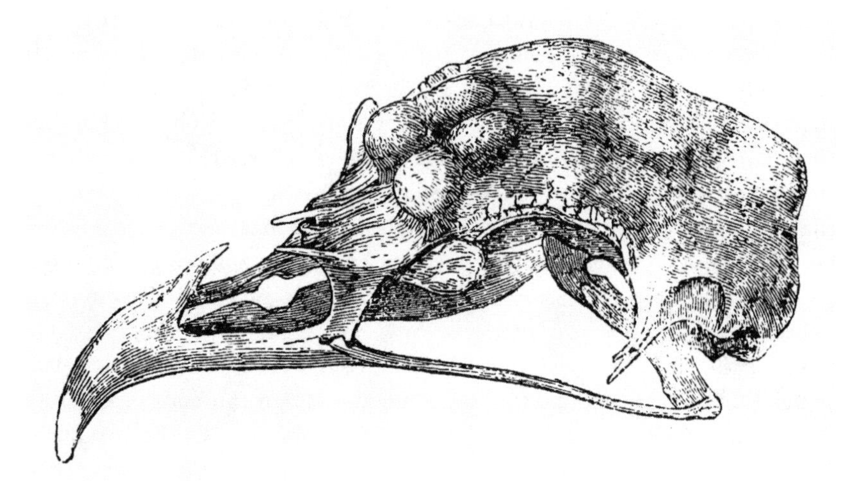

Figure 5.9 Skull of a Horned Fowl. From *The Variation of Animals and Plants Under Domestication*. Volume 1: p. 265 by Darwin, 1868a.

Among other impressive statistics that we can be amazed by today is that there are at least 12 to 30 billion domestic chickens on earth and that 24 million are killed and cooked every single day in the United States alone. In 2001, some 57 million tons of eggs were eaten by people each day, those of Britain alone eating 26 million, the vast majority involving those of domestic chickens.[38] In Australia, 560 million chickens are consumed each year:[39] chicken feed indeed! Some domestic chickens lay as many as 10 times more eggs than do their wild ancestors, with each egg twice as heavy.[40] The human race owes a very great deal to the evolution and subsequent domestication of this, all too often taken for granted, familiar bird (if only as a dead and featherless one). Unfortunately, countless millions of them, and other "table" bird species, are kept in appallingly inadequate conditions with respect to their physical, social, and mental needs and well-being.

As much more about domestic fowls also appears in volume 2 of *The Variation of Animals and Plants Under Domestication*, but appears as only a page, a paragraph, or a sentence or two or less, scattered at numerous points throughout the book under many different subject headings, I can only provide reference to their locations and context in appendix 1.

The Common Domesticated Duck

Having reviewed the pigeon and the fowl, six other domesticated bird species were dealt with by Darwin but far more briefly, the first being the common duck. As with the previous two species, the duck fanciers of Darwin's day believed the various breeds of duck to have derived from several ancestral species; whereas most naturalists acknowledged but a single ancestor, which is now known as the Mallard (*Anas platyrhynchos*).

After describing the four chief breeds of the domestic duck, Darwin stated that it was unknown to the Egyptians, who frequently depicted individuals or groups of wild duck species; to Jews of the Old Testament; and to Greeks of the Homeric era. The modern view is that ducks may have been domesticated, from the Mallard, for at least 3,500 years and originally most probably in China.[41]

Darwin refrained from applying all of the arguments he used against and for a single ancestor for the domesticated pigeon and fowl to the duck. He did, however, cite the observation of a Mr. E. Hewitt who kept pure-blooded wild Mallards from hatchlings and found that they changed and deteriorated in character over only two or three generations. After three generations, Hewitt's ducks "lost the elegant carriage of the wild species, and begun to acquire the gait of the common [domestic] duck. They increased in size in each generation, and their legs became less fine. The white collar round the neck of the [adult male] mallard became broader and less regular, and some of the larger primary wing-feathers became more or less white." Having made that statement, Charles had no doubt at all that the common domestic duck was derived from only the Mallard. He emphasized this view by observing the telling fact that of all of the "great duck family" (ca. 150 species today), only males of the Mallard have four central tail feathers curled upwards—and that these occur in all breeds of the common domestic duck.

As he did with the fowl, CD illustrated how dramatic and profound the result of domestication can be on the skull, a character not artificially selected for (see figure 5.10), and went on to compare other parts of the skeleton between wild and domestic ducks. He also cited the ornithologist Philip Sclater in observing that the structural changes in the skeleton of the all but flightless Tristan Moorhen (*Gallinula nesiotis*; see appendix 1) were fundamentally most similar to that in the common domestic duck, those in the latter also clearly resulting from less wing and more leg use.[42]

As more about domestic ducks also appears in volume 2 of *The Variation of Animals and Plants Under Domestication*, but only as brief accounts scattered through the book under diverse subject headings, I can only provide their locations and context in appendix 1.

The Domesticated Goose, Peacock, Turkey, Guineafowl, and Canary

Having reviewed the domestic pigeon, fowl, and duck, Darwin did the same for the other widely domesticated bird species; but as they had been

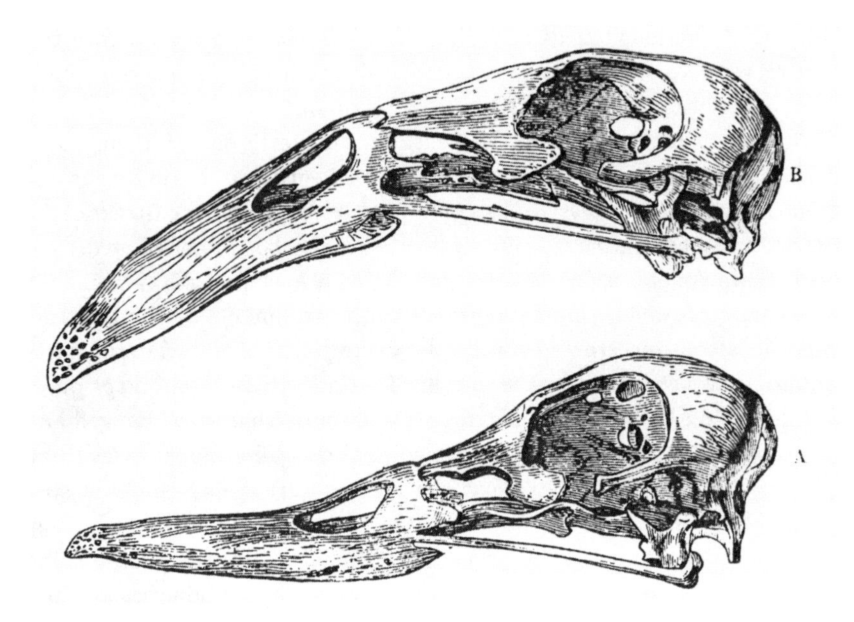

Figure 5.10 The skull of a wild Mallard duck, below (A), and of a Hook-billed Duck, above (B), showing extreme structural differences. From *The Variation of Animals and Plants Under Domestication*. Volume 1: p. 282 by Darwin, 1868a.

domesticated for far less time, or had varied little during the time that they had been kept under domestication, his text on each species was brief.

In reviewing the domestic goose, Darwin makes the observation that hardly any other anciently domesticated bird, as denoted by it being mentioned in Homer's verses and by being kept in Rome in 388 B.C. as a sacred creature, has varied so little in external appearance. Despite several very similar wild European goose species, there was, even then, general consensus among naturalists that the Greylag Goose (*Anser anser*), the young of which are easily tamed and were long domesticated by Laplanders, was the sole ancestor of the domestic goose. Its limited variation, compared with other long-domesticated birds, is described by CD; and he suggested that this is partly accounted for by less artificial selection having been applied, with the bird being kept more for food and feathers than for any diversity in appearance selected for by fanciers. That said, Darwin did note that the Romans preferred the taste of the liver of white-plumaged individuals.[43]

Modern authors cite the fact that Egyptian relief carvings dated at ca. 2649 to 2150 B.C. depict what appear to be both Graylag and Egyptian Geese being fed and reared on grain.[44] In not being a member of the genus *Branta*, the Egyptian Goose (*Alopochen aegyptiaca*) is not a true goose but is a species of its own monotypic genus, being most closely related to the South American geese of the genera *Neochen* and *Chloephaga*. As these latter geese are well removed from the true geese within the lower, and predominantly larger, waterfowl,[45] the Graylag Goose is considered the undoubted sole ancestor to the domestic goose.

Another bird that has hardly varied under domestication, except for the production of white and piebald "decorative" forms, is the Indian Peafowl (*Pavo cristatus*), which Darwin deals with in two pages. He expresses doubt as to the origin of birds in Britain and as to whether they descended from stock introduced to Europe by Alexander the Great or from birds imported subsequent to the Greek's considerable influence. Today it is known that awareness of peafowl under domestication dates back to at least 745 B.C. in Assyria,[46] with them reaching Europe by at least 500 B.C.[47]

Darwin stated that Indian Peafowl did not breed freely in the United Kingdom, where they were then seldom kept in large numbers. He reviewed the occasional appearance of "black-shouldered," or "japanned" individual birds, as sufficiently distinctive as to have been erroneously named as a separate species ("*Pavo nigripennis*") by no less an ornithologist than Philip L. Sclater. These birds differ from the typical Indian Peafowl in that the adult males have scapular, secondary, upper-wing covert, and thigh feathers that are blackish, and the females are paler. In some collections, such birds completely replace the normally plumaged ones. They can suddenly appear within breeding flocks of typically plumaged birds, and Darwin correctly concluded that these black-shouldered birds therefore represented no more than a variation under domestication, possibly induced by the British climate or some other cause "such as reversion to a primordial and extinct condition of the species."[48]

According to Darwin, the domesticated turkey (*Meleagris gallopavo*) was established by John Gould to have descended from the wild Mexican population. This population was at that time treated as a distinct species but is now considered to be only a subspecies of the North American Wild Turkey, as was intimated to be true by CD. It has white rather than tan tips to the tail feathers. Whereas others thought that a different ancestral population, or two, might be involved, Gould was correct in his determination. The bird was under domestication by American Indians before Europeans discovered that continent, at sometime between 150 B.C. and 400 A.D., in Mexico or what is now the southwestern United States.[49]

English domesticated turkeys of Darwin's time were smaller than wild American birds and had varied to no great extent, with but a few breeds named after English counties and a white variant, all of which bred true if maintained in genetic isolation. There had apparently once been a distinct "buff-yellow" breed in Holland with an "ample topknot" of elongate feathers, but it became extinct. A white form also had a crest of "feathers about four inches long, with bare quills, and a tuft of soft white down growing at the end," a character pointed out by Darwin to be found in the wild peafowl and several pheasant genera. Some turkeys kept "half wild" in England produced the odd variable individuals, which CD attributed to their novel climate and conditions. Ecological conditions in India had resulted in much smaller, flightless, and darker turkeys with "the long pendulous appendages over the beak enormously developed."[50]

Recent genetic research results suggest that the domesticated turkeys in Europe are probably older than those in America; the species being first taken from Mexico, as John Gould believed, to Spain early in the sixteenth century and then subsequently to England during the reign of Henry VIII, and on to colonies in the Atlantic.[51] Today, the 1873-established American Poultry Association acknowledges eight breeds of domestic turkey; five registered in 1874 (six years after Darwin's *The Variation of Animals and Plants* was published); and one each in 1909, 1951, and 1977.

Within only the one paragraph dedicated to the guineafowl, the common domestic form is correctly stated by Darwin to have originated from "*Numida ptilorhynca*" of Africa: now the Helmeted Guineafowl (*Numida meleagris*). Despite the climatic conditions of Europe, it had changed very little in appearance, other than by being paler or darker than the typical plumage, while varying more under climates closer to the African one in the "West Indies and on the Spanish main." Feral guineafowl on Jamaica and St. Domingo had become smaller with black legs, which Darwin found worthy of note, as feral animal populations are repeatedly said to revert in characters to their original type.[52] It is said that the guineafowl has been domesticated for some 4,000 years and that the first historical reference to them is in fifth dynasty Egyptian murals in the Pyramid of Wenis at Saqqara.[53] Apparently the fall of the Roman Empire brought about the loss of guineafowl in domestication in Europe, and they did not reappear there until the Late Middle Ages, when folk having sought gold and slaves in western Africa returned with guineafowl.[54]

The canary (Atlantic Canary *Serinus canaria*) is the last bird that Darwin discussed in his book on variation under domestication, giving but a single page of text to it. First and foremost, he stated it is but a recently domesticated bird, having been so only within the last 350 (and thus now ca. 500) years. It had by then been crossed with 9 or 10 other finch species, also of the same family Fringillidae, with some of the hybrids being "almost completely fertile" but without evidence of any of them resulting in the formation of distinct breeds. Many varieties of canary were produced within only 350 years; and prior to 1718, some 27 varieties were listed in France. In 1779, the London Canary Society published a lengthy listing of desired qualities in the appearance of canaries, enabling Darwin to stress that methodical artificial selection had been intensely applied for a good while, resulting in variations in shape, size, and stature, as well as in colour. That the birds produced three or four clutches a year helped significantly with the process of artificial selection. He notes that some breeds have "top-knots," that one had feathered feet, and that the character of black wing and tail feathers in some prize canaries was only retained by them until their first moult.[55] For more about Darwin and the domesticated canary, see the section "Red canary or red herring?" of chapter 8.

Additional odd pieces of information about these various domesticated birds also appear in volume 2 of *The Variation of Animals and Plants Under Domestication*, but as they are brief and scattered under different subject headings I can but indicate their locations and context in appendix 1.

VARIATION—VOLUME 2

The development and advances of the genetic sciences have corrected, updated, or clarified most of the pertinent, and predominantly domesticated, bird descriptions and observations by Darwin that are scattered through the various parts of *The Variation of Animals and Plants Under Domestication* by Darwin (for which see appendix 1).

In volume 1 of *The Variation*, the eight main species long propagated under domestication are dealt with by more substantial texts of a page and up to two entire chapters for pigeons, as detailed previously. In contrast, the numerous birds mentioned in volume two of the work are dealt with but very briefly, with these involving at least 158 of the 432 text pages constituting that volume. Many of these cited bird species involve little more than their name given in the context of some topic under consideration, an example being that 21 different bird species or groups are mentioned on the single page 154, under "Changed conditions of life." Domesticated pigeons, fowls, turkeys, and ducks in particular are mentioned but briefly in passing, within texts on various diverse topics, on far too many occasions to be able to deal with them here in any meaningful way, particularly as the treatment results in much repetition of the same points being made. Thus, I review following only those species or groups involving greater consideration and discussion by Darwin. All of the remaining taxa included in volume 2 are comprehensively listed and their context summarized in appendix 1.

One fascinating paragraph in volume 2 of *The Variation* book relates to human-induced changes in the colour pigmentation of bird plumages, included within a discussion of the external conditions of life. Darwin writes that it was well known (maybe then but, ironically, probably less so today) that hemp seed causes bullfinches and certain other birds to become black feathered. He goes on to cite Alfred Russel Wallace who recorded that Amazonian natives fed a common green parrot the fat of a fish that resulted in them becoming "beautifully variegated with red and yellow feathers."[56] The species involved was the Festive Amazon parrot (*Amazona festiva*), an all-green bird save for a red forehead and rump and with small blue areas about the head. Such manipulation of the feather colours of birds was long practiced, as the Spanish discovered upon arriving in South America that Inca bird handlers changed living birds' red and green feathering to golden yellow or salmon pink by rubbing the skin secretions from poison arrow frogs into the skin of the birds.[57]

Wallace also reported that South American Indians would pluck out feathers from part of the plumage of "many birds" and "inoculate the fresh wound with the milky secretion from the skin of a small toad" causing the new replacement feathers to be brilliant yellow. If these new and novel yellow-pigmented feathers were plucked out, they were replaced with more yellow ones. Natives of Gilolo in the Malay (now Indonesian) Archipelago similarly cause the plumage of the Chattering Lory (*Lorius garrulus*) to be altered and to "thus produce the *Lori rajah* or King Lory." However, when fed on "natural vegetable food, such as rice and plantains," these parrots maintain their normal plumage pigmentation.[58]

Another interesting Darwin paragraph relates the modification of skull form and structure in the crested Polish domesticated fowl breed. Their skulls "are perforated by numerous holes, so that a pin can [hopefully theoretically] be driven into the brain without touching any bone." As such skull perforation also occurs in tufted ducks and geese, Darwin found the correlation between it and the presence of a crest clear. He had also previously argued, in volume 1 of *The Variation of Animals and Plants*, that the crest in the Polish breed was probably at first small in the breed's development. By continued selection for it, the crest became larger "and then rested on a fleshy or fibrous mass; and finally, as it became still larger, the skull itself became more and more protuberant until it acquired its present extraordinary structure";[59] see figure 5.8.

In his final chapter, "Concluding Remarks," Darwin addressed at some length the "exposure to unnatural conditions of a large number of animals of the same species, allowed to cross freely, with no selection of any kind." He used a hypothetical population of 500 wild Rock Doves kept in a single aviary in their native land and fed as pigeons usually are but not allowed to increase in number. Given that pigeons in domestication propagate rapidly, he supposed that 1,000 or 1,500 would be killed each year "by mere chance" at the hands of their keeper to maintain the number at 500. After several generations, some young birds would vary, the variations would tend to be inherited, and many variations would "occur in correlation" as in wing/tail length, number of primaries, and number and breadth of ribs in correlation with body size and shape; the number of scales relative to foot size; length of tongue to beak length, nostril, eyelid, and lower jaw size and shape relative to facial wattle development; nakedness of young to future plumage colour; size of feet and beak; and so forth. In their captivity, they would use their wings and legs little, and associated parts of their skeleton would thus become reduced in size.

As many individuals in the hypothesized captive population would have had to be killed every year, the chances are "against any new variety surviving long enough to breed." Because any variations arising would be extremely diverse, the chances of two birds varying in the very same way pairing to reproduce would be most unlikely. However, a varying individual would occasionally transmit its novel character to its offspring, even if not paired with a similar variety. The offspring would in turn be exposed to the same conditions, bringing about the variation in its one parent, but they would also inherit from that parent a tendency to vary in the same way. Thus, if the conditions tended to produce a particular variation, then all birds might eventually become so modified. But a far commoner result, Darwin argued, would be that each bird would vary in its own way and, as they would all intercross, the result would be a population of but slightly differing individuals—but showing far more variation than did the ancestral Rock Dove while showing no tendency to form distinct breeds.

Given the domesticated pigeon evidence that he presented elsewhere (as well as that for domesticated mammals), Darwin suggests that if two separate pigeon populations be treated as described previously but with one kept in England and the other in the tropics, with each fed different foods, they would be differently modified, through variations, resulting from climate and food influences. Frustratingly, however, CD lacked adequate evidence of the effect of such changed conditions, and his examination of a large collection of domestic pigeons from India showed "remarkable similar" variation to that found in European birds.

Darwin extended the hypothetical scenario to two distinct breeds confined together in equal numbers. He suspected each would to some extent prefer to pair with an individual of its own breed, but that they would also intercross to produce "hybrids," and the greater vigour and fertility of these in the population would cause it to become an interblended one "sooner than would otherwise have occurred." As some breeds must be dominant over others, it would not follow that such an interblended population would be strictly intermediate in character. There would also be, as CD demonstrated elsewhere, a strong tendency to reversion to characters of the ancestral Rock Dove. Thus, he argued, this population would in time come to be little more heterogeneous in character than the first hypothetical captive population.

Getting to the crux of the matter, Darwin then changes the preceding scenarios, involving the indiscriminate killing of individuals to maintain a population of 500, to one of discriminate culling. An owner noting even slight variation in one bird that he wanted to then produce a breed from would "succeed in a surprisingly short time by carefully selecting and pairing the young. As any part which has once varied generally goes on varying in the same direction, it is easy, by continually preserving the most strongly marked individuals, to increase the amount of difference up to a high, predetermined standard of excellence. This is methodical selection." He goes on to observe that a pigeon fancier that did no more than constantly remove longer-billed birds from his loft because he liked shorter-billed ones more would in time significantly modify his stock without any thought of making a new breed. Several fanciers unconsciously favouring preferred characters in this way would probably select for subtle to grossly different characters and would thus ultimately bring about distinct breeds, as Darwin points out has actually occurred with pigeons, cattle, and sheep.[60]

In complete contrast to *The Origin of Species,* the two volumes of tediously detailed reporting on and discussion of *The Variation of Animals and Plants Under Domestication* did not rush off bookshop shelves as quickly, although a second edition did appear only a few months after the first. Only some 5,000 copies sold during Darwin's remaining lifetime. His subsequent book, *The Descent of Man and Selection in Relation to Sex*, which he started to write on 15 March 1868, proved once again to be eagerly purchased, read, and widely discussed and debated. In chapter 6, I review the ornithological content of *The Descent of Man*.

NOTES

1. Darwin 1902: 2.
2. Desmond and Moore 1991: 510; Browne 2002: 227.
3. Cooper 2010: 26; 2014: 15.
4. Cooper 2014: 16.
5. Wyhe and Rookmaaker 2013: 94.
6. Wyhe and Rookmaaker 2013: 129.
7. Wyhe and Rookmaaker 2013: 87.
8. *Pace* S. J. Gould 1985: 41.
9. Darwin 1868a: 2.
10. In Desmond and Moore 1991: 549–50.
11. Darwin 1868a: 1.
12. Darwin 1868a: 9, 20–21, respectively.
13. Darwin 1868a: 29, 37, 55, 61.
14. Wyhe and Rookmaaker 2013: 268.
15. In Haupt 2006: 229.
16. Cooper 2010: 24–5; 2014.
17. Stearn 1998.
18. McNeillie 1976.
19. Darwin 1868a: 131–3.
20. Darwin 1868a: 136.
21. Darwin 1868a: 134–62.
22. Darwin 1868a: 162–79.
23. Burkhardt et al. 1985–2014, vol. 8: 404.
24. McCarthy 2006.

25. Darwin 1868a: 180–95.
26. Darwin 1868a: 201–3; Mayr 1963: 134.
27. Simms 1979: 35; Cocker 2013: 233.
28. Johnston 1992.
29. Darwin 1868a: 204–7.
30. Darwin 1868a: 207–11.
31. Darwin 1868a: 224.
32. Mendel 1866.
33. Tegetmeier 1867.
34. Tegetmeier 1857.
35. Darwin 1868a: 225–45.
36. Darwin 1868a: 246.
37. Darwin 1868a: 248–75.
38. Jones 2008: 150; Cocker 2013: 65.
39. Low 2014: 313.
40. Jones 2008: 150.
41. Cocker 2013: 86.
42. Darwin 1868a: 276–87.
43. Darwin 1868a: 287–90.
44. Houlihan 1986: 55; Cocker 2013: 88.
45. Kear 2005: x–xi.
46. Armstrong 1975: 140.
47. Crocker 2013: 76.
48. Darwin 1868a: 290–92.
49. Schorger 1966; L. E. Williams 1981.
50. Darwin 1868a: 292–4.
51. Cocker 2013: 44.
52. Darwin 1868a: 294.
53. Hastings Belshaw 1985: 3.
54. Cocker 2013: 37.
55. Darwin 1868a: 295.
56. Darwin 1868b: 280.
57. Hanson 2011: 208.
58. Wallace, in Darwin 1868b: 280.
59. Darwin 1868b: 333.
60. Darwin 1868b: 420–24; Bodio 2009.

CHAPTER 6

The Ornithologist Thinks about Avian Sex

Secondary sexual characters are more diversified and conspicuous in birds, though not perhaps entailing more important changes of structure, than in any other class of animals. I shall, therefore, treat the subject at considerable length. [Darwin 1901: 549]

With the dauntingly massive task of writing and then proof reading the 900 pages of his *The Variation of Animals and Plants Under Domestication* behind him, Charles Darwin's health improved, and he was inclined to enjoy more family and social life than he had felt able to deal with for quite a while. The period around 1867 was one during which he was to meet in person a good number of eminent contemporary scientists and authors both with and without his family members present. This social contact with his biological peers was timely, as he was contemplating facts, thoughts, and theories of great significance to what were to become his future book projects. His need to discuss these with friends and colleagues face to face, in addition to writing to them and to others, was of critical importance to him. Thus, Henry Walter Bates appears to have joined him at his home on 15 March 1867. He then met with George Robert Gray on 23 March, and Edward Blyth appears to have visited him at home on 24 March 1868. These three men would have doubtless discussed ornithological matters with Darwin, Bates being familiar with South American birds, Gray with global avian taxonomy and nomenclature, and Blyth with the avifauna of India.

On 12 September 1868, Alfred and Annie Wallace, Edward Blyth, and John Jenner Weir visited Darwin at his home and, given their expertise, they would with little doubt have discussed birds with Darwin's work in progress in mind. Similarly, on 22–24 January 1870, Alfred Newton, Albert Günther, and Robert Swinhoe visited Down House presumably for a largely, if not exclusively, ornithological conference as Darwin worked on his *Descent of Man and Selection in Relation to Sex* draft chapters. Günther dropped in on him again on 15 August of that year, together with Rudolf Albert von Kolliker. The particular interests of all of these men would suggest that Darwin invited them to Down House with a view to discussing birds, presumably with his sexual selection theory primarily in mind. For the recluse that he typically was, this made his home a veritable hive of social and ornithological activity.

Darwin had, to date, carefully avoided confronting the explosively emotive and politico-religiously, highly controversial issue of the evolutionary origins of human beings and their relationship to the animal kingdom in general and to primates in particular. But he now started to give time to thinking about this most fraught of subjects with a, if reluctant, view to possibly publishing on them. He revived his thinking about the expression of emotions in animals and man, which he had previously accumulated notes on from his observations and thoughts about them (including of his own children). He also reviewed his concept of sexual selection, so briefly mentioned in but three pages of his *The Origin of Species*, which he found particularly exciting as it appeared to solve some major issues with his fundamental theory of natural selection. Sexual selection would, he thought, account for marked differences, in most cases between their adult males, in closely related bird and some other animal species: many as a result of one sex selecting for favoured characters in the other. A good example of this he saw as peahens choosing to mate with the better-plumaged and more vigorously displaying individual peacocks (see following).

Many scientific peers and friends, most conspicuously including Alfred Russel Wallace, disagreed with Darwin about this seemingly far-fetched idea of sexual selection; for how, they thought and argued, could one sex of a lowly animal possibly be so terribly discerning over almost imperceptibly slight differences in the quality of appearance between individuals of the opposite sex. Indeed, Wallace wrote to Darwin on 19 March 1868, "How can we imagine that an inch in the tail of the peacock, or ¼-inch in that of the Bird of Paradise, would be noticed and preferred by the female?"[1] Clearly, Charles had some work to do in consolidating his theory of sexual selection and in convincing even some of his closest colleagues and friends of its importance, let alone the scientific community and the public at large. This he did to the satisfaction of himself and that of many, but by no means all, within the pages of his book *The Descent of Man and Selection in Relation to Sex*.

Alfred Russel Wallace presented a copy of his new book *The Malay Archipelago* to Darwin in early 1869, which CD found to be magnificent in every respect. Among the wealth of novel zoology that Wallace described and discussed therein was much about bird groups and species previously unfamiliar to Darwin. A fall from his faithful and gentle old riding horse during the spring of 1869 did not help Charles's frame of mind and body, however, as he approached the task of revising his *The Origin of Species* for its fifth edition. This was to bring the number of copies printed of that title to the nicely rounded, and doubtless deeply satisfying, figure of ten thousand.

THE DESCENT OF MAN ASCENDS

In February 1871, John Murray of London published *The Descent of Man* in two 450-page volumes. Murray had objected to but a single sentence in the submitted manuscript and did so to the extent that he asked for it to be removed from the text. In this action, he did no more than reflect the moral perceptions of his day, as the offending sentence apparently implied the appalling observation that females might feel sexual desire.

Whitwell Elwin, who, as the publisher's referee, had suggested that Darwin restrict his *The Origin of Species* to a text about domestic pigeons, found the manuscript of *The Descent of Man* almost incomprehensible and little more than drivel.[2] Darwin had spent two years actually writing the manuscript of this work, but he had previously spent almost the same period of time in writing drafts on the topics therein for inclusion in his *The Origin of Species*, which, as is related at the beginning of chapter 5 herein, he proved unable to include within it for want of space. In fact, *The Descent of Man* includes evidence, ideas, and discussions that Darwin had been putting together in his mind and his notebooks for decades. Each of his 21 chapters deals with a discrete subject, with the three notable exceptions of two chapters dedicated to "Secondary sexual characters of insects," two to "Secondary sexual characters of man" [humans], and four chapters to "Secondary sexual characters of birds."

Birds are easy to see almost everywhere on earth. People can relate to their lives relatively easily, and Darwin's multiple chapters on them reflect the fact that the vast majority of birds are, like humans, warm-blooded, diurnal, and primarily visual creatures that behave in observable, colourful, and complex ways. Thus, many people had already recorded numerous facts about wild bird species as well as a handful of them very familiar as being long domesticated and selectively bred. The view has been expressed that the "clinching proof of sexual selection" came in these four chapters about birds by Charles Darwin.[3] That some of the bird species that he deliberately chose to dwell on at length were widespread and familiar ones doubtless contributed to his success in getting his profoundly novel ideas across.

Whereas some ornithological observations made by Darwin in his *The Descent of Man* resulted from his personal experiences, he also gleaned numerous facts and ideas from copious personal correspondence and what must have been an exhaustive review of the then available literature. The vast majority of text about birds in Darwin's *Descent* appears in the four chapters dedicated to them, these amounting to 213 pages in total. In keeping with this book to this point, space limitations do not permit me to summarize, or even allude to, all of the bird species, bird groups, and other ornithological topics contained in *The Descent of Man* within the constraints of this chapter and chapter 7. I must therefore limit myself to include here only those avian species, groups, and topics more fully treated by Darwin. All other pertinent material is, however, comprehensively detailed and briefly placed into context in appendix 1. In this book of mine, about what Darwin published about birds, I cannot possibly indulge in an exhaustive review and discussion of what is today the vast and diverse practical and theoretical study of avian sexual selection but must confine myself to a few pertinent post-Darwin remarks.

By late March 1871, an impressive 4,500 copies of *The Descent of Man* had come off the presses and to its author's delight had earned him close to £1,500. He had become an eagerly sought—if most controversial, admired, or detested—author.

SEXUAL SELECTION

In opening his first chapter on birds, Darwin made the observation that is quoted at the head of this chapter, followed by "On the whole, birds appear to be the most aesthetic of all animals, excepting of course man, and they have nearly the same taste for the beautiful as we have."[4] Although he does not attribute this aesthetic appreciation as being an exclusively visual one, he has been recently interpreted as "depicting" it as "driven by the eyes of the female rather than by either her head or her reproductive organs."[5] This is, however, quite misleading, as Darwin writes of female birds preferring or being "*unconsciously excited* by the more beautiful males"[6] (emphasis mine). Clearly, the eye cannot, of itself, be excited, whereas the mind and hormones can; and it is thus perfectly evident that this is what CD was alluding to (i.e., the head and reproductive organs, and not the eye, of the female being stimulated).

Birds of a Feather Select Together

Long before the 1871 publication of his *The Descent of Man*, Darwin worried and puzzled over what he saw at the time as evidence representing significant hurdles to his evolving ideas about sexual selection. On 3 April 1860, he wrote to his American colleague Asa Gray, "The sight of a feather in a peacock's tail, whenever I gaze at it, makes me sick!"[7] He penned this because at that time, he had not developed what were to become his firm thoughts about the kinds of selection pressure that could bring about such extravagant and beautiful plumage in adult male birds, be it by natural or sexual selection.

A highly pertinent part of Darwin's text about birds in general and Indian Peafowl in particular is worth quoting here: "Are we not justified in believing that the female exerts a choice, and that she receives the addresses of the male who pleases her most? It is not probable that she consciously deliberates; but she is most excited or attracted by the most beautiful, or melodious, or gallant males. Nor need it be supposed that the female studies each stripe or spot of colour; that the peahen, for instance, admires each detail in the gorgeous train of the peacock—she is probably struck only by the general effect. Nevertheless, after hearing how carefully the male Argus pheasant [Great Argus (*Argusianus argus*)] displays his elegant primary wing feathers, and erects his ocellated plumes in the right position for their full effect; or again, how the male goldfinch (*Carduelis carduelis*) [European Goldfinch] alternately displays his gold-bespangled wings, we ought not to feel too sure that the female does not attend to each detail of beauty."[8] This might appear a slightly contradictory text to some in that it, scientifically correctly, initially avoids stating that females might consciously deliberate on the quality of male attributes but then indicates that small details of male plumage could well influence female choice of mates; but in this, Darwin was suggesting the female response is instinctive rather than conscious.

Darwin had suggested to the great bird keeper William Tegetmeier four years previously, in 1867, the experimental trimming of tail feathers of already successfully mated domestic cockerels to see if females then came to find them less attractive than before their artificial loss of plumage; but it was not put into practice. In a 27 February 1868, letter to John Jenner Weir, Darwin long foresaw a highly informative experimental test of female choice by writing that it would "be a fine trial to cut off the eyes of the tail-feathers of male-peacocks";[9] see following.

Darwin came to think that females were far more demanding in their choice of mates than were males. He proposed his theory of sexual selection to involve females preferentially selecting from competing males by discerning differences between them in their plumage and courtship but was unable to convincingly suggest how females benefited from this process. Alfred Russel Wallace was particularly critical of this failing and was unconvinced by the theory. The notion of sexual selection remained in obscurity for a century after it first appeared in print.

With all of the preceding, and much more, filling his head, Darwin still found significant time to continue observing and experimenting on plants in his garden, greenhouse, and study. His large garden and adjacent areas of woodland and pasture also provided him with daily sightings of a diversity of wild British bird species. This was particularly so as he took his regular exercise and contemplative walks around his partly wooded circular Sandwalk, beyond the long vegetable garden behind Down House, thus keeping his ornithological experiences and musings active; see colour figures 4.1 and 8.4.

It took a considerable time for zoologists to attempt to test Darwin's hypothesis that peahens can indeed discern subtle differences between individual peacock plumage elaboration. When they eventually did so, they applied his suggested feather cutting technique and the results appeared unequivocal, with both the simplicity of the elegant studies and their logical and brief titles of resulting publications being admirable. First, it was established that the number of ocelli, or eyespots, on the males' feather train correlated with increased male age up to, but not beyond, four years.[10] Furthermore, both the number and the symmetry in distribution of the eyespots on an individual male's train also proved to be closely correlated.[11]

Studies of 10 courting feral peafowl in England at their lek (a site traditionally used by competitive aggregations of one sex, typically males, to attract and court the other sex) demonstrated that those adult males with more eyespots were more successful in mating females but that those males that had their eyespots experimentally reduced in number suffered an associated loss in mating success.[12] Further research found that peacocks with poor mating success suffered higher predation than did their more successful peers, suggesting inferiority in genetic fitness in other ways;[13] and females that mated with males wearing better trains laid larger numbers of eggs.[14] Topping off this impressive train of peafowl studies was the finding that offspring of the most successful (i.e., best ocelli-plumaged) males grew and survived better than did those of lesser-endowed males.[15] Thus, female selection for male adornment was apparently also convincingly correlated with relative male (and female) genetic fitness. In this way, it was demonstrated how more discerning females can benefit significantly from their critical process of sexual selection (but see the end of this chapter).

More recently, however, studies of the same free-ranging peafowl population that Marion Petrie studied, but seven years later, by Mariko Takahashi and colleagues[16] led to their concluding that hens do not select cocks exclusively for their display plumage characteristics. Moreover, male plumages were found to show little variation across populations and did not correlate to the physical condition of individuals. These authors saw the peacock train and its eyespots as obsolete characters with respect to mate selection by peahens. Several subsequent authors expressed serious doubt about these findings in a published rebuttal and suggested that peahen choice might vary with ecological conditions.[17]

Another study repeated the reduction of eyespots on peacock train experiments that were instigated by Marion Petrie. The results this time indicated that peacocks that had their eyespots reduced did show a drop in mating success but that untouched males naturally had 165 to 170 eyespots; and, on average, those with more eyespots were no more successful in obtaining copulations than were males with less. It was therefore thought by the investigators that females do not exclusively select mates with the largest number of eyespots (i.e., female selection could not be explained by natural variation in eyespot numbers between males) and that additional plumage traits may also be attractive to females. Males lacking significant numbers of eyespots may, however, be the most unfit among males.[18]

Yet another investigation suggested that peacock display plumage has resulted from natural selection, with its display function being to intimidate rivals (in which case sexual selection might surely be involved, as adult males are all rivals) and predators.[19] Clearly the issue of the extent of the influence of natural and/or sexual selection involved in the displays and associated appearance of adult males of even such a common and well-known bird as the peafowl is not yet fully resolved.

Emphasizing the complexity of the peacock display, and our limited appreciation of it, is the fact that a study published in 2015 demonstrates for the first time that courtship-displaying Indian Peafowl produce infrasonic sounds with their train of upper tail covert feathers.[20] These deep rumbling sounds are inaudible to people, but they presumably must be of significance to females in their choice of mates.

Previously, in 1926, Hilda Cinat-Tomson of Latvia published the results of her innovative studies involving the manipulation of the number of the conspicuous black throat spots on male Budgerigars (*Melopsittacus undulatus*). This provided the first unequivocal proof of female choice in birds because male Budgerigars with the most spots proved most attractive to females. Males made more successful by the experimental increase of their throat spots that subsequently had them artificially reduced in number then became less attractive to females. Males that were less attractive to females before being spot-manipulated in any way became more attractive when the number of their throat spots was artificially increased.[21] Although this elegant and significant study attracted little attention, doubtless in part because of the relatively obscure journal in which it appeared, subsequent studies on the manipulation of tail length in male Barn Swallows (*Hirundo rustica*)[22] and African finches[23] did attract wide attention and confirmed Cinat-Tomson's findings; see the section "Female Choice and a Tale of Tails" following.

Darwin was then of the view that females selected males for their beauty and gallantry, as quoted previously, and in this he was more at odds with Alfred Russel Wallace than with most aspects of selection theory. Wallace believed that females must select not for beauty in males but for attributes in males of more direct value to themselves in producing offspring with good genes for survival and reproduction. In fact, both men were correct to some extent because although females do select for traits in males that reflect good genes, these are indicated in the beautiful plumage of adult males of some, predominantly polygynous, species as well as by other characteristics; see the section "Secondary Sexual Characters" following. A recent finding is that in several galliform birds—including the Wild Turkey, Sage Grouse, Common Pheasant, and Red Junglefowl but excluding the peafowl at least—females do not rely on plumage traits in choosing their mate but rather on unfeathered, and typically brightly pigmented, body structures.[24] The principle and function remains, however, fundamentally the same.

Sexual Selection and Sex Ratios

Darwin was one of the first biologists to appreciate the potential importance of intraspecific sex ratios to sexual selection, suggesting that a balanced sex ratio is adaptive. He reviewed the small amount of information about relative proportions of the sexes produced by birds but was severely hamstrung in doing so by the limited knowledge and understanding of the subject at that time. Other than the observation of the sexes of a large sample (1,001) of hatchling domestic fowls, in which the proportion of males to females was near equal, most of the other pertinent observations CD could quote involved subjective ones on the proportion of adult sexes in wild species populations. As the latter were subject to the difficulties of not being able to sex individual birds until they reached a certain age, as well as numerous environmental and social conditions affecting survival, such figures are of little worth in the context of a meaningful discussion of primary sex ratios.[25] Moreover, a number of his informants that recorded they saw more adult males than females of given species in the wild, most notably of hummingbirds, apparently took no account of the more cryptic appearance and behaviour of female-plumaged birds, and, to the contrary effect, that immature males wore female-like plumage for a year or more. In fact, although significant variation in the sex ratio of birds in general at hatching from near enough parity is unusual, it does occur in a small proportion of species.[26] Darwin considered sexual selection likely to be most effective in a bird species with an unbalanced sex ratio, typically fewer males than females, as in polygynous species.[27] Of monogamous species, typically with a nearly equal sex ratio, he thought that genetically fitter and earlier-breeding females may select the most fit males as mates. The adult sex ratio in socially monogamous (see chapter 7) species is, however, quite often not parity but in fact involves more males than females.[28]

Secondary Sexual Characters

Male birds of various species "charm" females by a variety of vocal or instrumental sounds and are ornamented by a great diversity of structurally modified and colourful feathering, combs, wattles, inflated areas of bare skin, and other such traits, including in at least one species, the Musk Duck (*Biziura lobata*) of Australia, the production of a peculiar pungent nuptial odour. In presenting these attributes to females, the males will incorporate into their courtship intricate dancing movements, posturing, and feather manipulation be it while standing, perched, swimming, or in flight. Darwin observed that the adult sexes of one species can be so different in appearance as to have long been thought of as representing two distinct species until convincing evidence to the contrary became available to ornithologists. He also noted that in some species, the adult sexes have beaks that are so unlike as to be ecologically significant, in that they are each presumably adapted to taking differing proportions of some food types.[29]

Law of battle

Under this subheading, Darwin noted how the males of almost all birds are "extremely pugnacious." As a result, the adult males of some species are exploited by people about the world for their entertainment by watching them and gambling on the outcome of pitching two males together in, sometimes mortal, artificially stimulated combat. Although this activity is well known to involve the use of male domestic fowls, or cockerels, birds as disparate as the Watercock (*Gallicrex cinerea*, a rail species) and bulbuls (small songbirds of the family Pycnonotidae) are also kept and used in this way.

The Ruff (*Philomachus pugnax*) is a wading bird that breeds in northwestern Europe and northern Asia and migrates to western and southern Europe, Africa, southwestern Asia, and beyond to spend the northern winters there. It is an extraordinary wader in that it reproduces polygynously, which is to say that the promiscuous adult males seek to fertilize as many females, or reeves as they are known, as they can during each breeding season. The females nest and raise their offspring alone and unaided by their father. The highly aggressive males are conspicuously larger than the females, and they gather each display and mating season to form competitive aggregations on the lek. In spring, the lekking adult males wear a great erectile ruff of elongated feathers and head tufts, the colour and markings of which vary conspicuously between individuals (figure 6.1). Females visit these leks to be courted by the vigorously competing males and from among which they select the male they find most impressive to be mated by, that male most often being located at the geographical centre of the lek.

Figure 6.1 Four adult male Ruffs display on their lek with a plain-plumaged female beyond them. From *The Descent of Man and Selection in Relation to Sex*: p. 553, by Darwin, 1901.

Darwin recorded that the male Ruffs fight like game-cocks, "seizing each other with their beaks and striking with their wings. The great ruff of feathers round the neck is then erected, and according to a Colonel Montagu 'sweeps the ground as a shield to defend the more tender parts;' and this is the only instance known to me in the case of birds of any structure serving as a shield. The ruff of feathers, however, from its varied and rich colours probably serves in chief part as an ornament." Charles was correct about the ruff for, although it may conceivably provide some slight softening of blows from rival males, there appears to be no suggestion of this being a function of the ruff feathering.[30] Today we know that the individuals with paler ruff plumage colours tend to be nonterritorial, younger, and peripheral, or "satellite," males that are smaller than the territorial and more dominant, darker-ruffed, centrally located males on each lek. It is also established that females do find leks with more individual males in attendance, including satellite males, more attractive than leks with fewer males and satellites.[31] The obvious benefit of this is that larger leks permit females to select from a greater number, or pool, of competing males that they can observe interacting and so use in their comparative choice of mate.[32]

A complicating and most interesting additional factor recently discovered among competing, lekking, male Ruffs is that some individuals, which are slightly larger than females, take advantage of wearing female plumage throughout their lives. This remarkable retention of female plumage is apparently a strategy that enables such males to obtain copulations within a lek of adult plumaged and more dominant males while wearing their "false colours." That is to say that by mimicking females in their external appearance, they avoid excessive aggression from rival adult males. These "faeder" males, as they are called, were found to have testes 2.5 times larger than those of normal males during the peak breeding time of late April and are delightfully termed "sneakers."[33]

It is the various polygynous bird species, such as the Ruff, that Darwinian sexual selection most strongly applies to. In these species, females attending the leks of multiple males can there observe aggressive interactions and displays between males, which typically differ from those directed at females, in addition to their courtship displays, before making a choice of mate. In mounting his case for sexual selection by female birds, and reviewing a range of species exhibiting greater or lesser sexual dimorphism in plumage characters, Darwin failed to make mention of the good number of species groups in which both sexes wear the same colourful plumage ornamentation during each breeding season: for example, grebes, herons, pelicans, and divers or loons. In these species, the unisex nuptial plumage features in mutual displays that reinforce the pair bond and stimulate and synchronize mating and avoid hybridization. Darwin may possibly have omitted discussion of such groups as being irrelevant to his case under consideration.

Sexual dimorphism in size

Darwin reviewed fighting among males of a number of other bird species. That the males are larger than females in many he put down to the result of generations of selection between fighting rival males. Contrary to this, the females in some species are larger than the males, this state being termed "reversed sexual dimorphism," such as is found in most birds of prey. Charles debunked the suggestion that this is because the larger females perform most of the work of raising offspring. He observed that the larger size of females functions just as it does in those species where the male is the larger sex: this being to compete with other females in acquiring mates. In this he was, always remembering the extremely limited knowledge of the subject at the time, quite wrong.

In reviewing the, by then, substantial knowledge of raptor sexual dimorphism and biology, British ornithologist and raptor specialist Ian Newton concluded that the relative degree of sexual dimorphism in size found within the group relates closely to diet—it being greatest in species that hunt large, fast, agile prey. It is also associated with differences between the prey species taken by each sex of raptor species. In raptors in which the females are the significantly larger sex, they physically dominate their mates, but this is probably an incidental result and not the reason for the reversed sexual dimorphism.[34] In addition, Newton found the reversed sexual dimorphism size difference in Eurasian Sparrowhawks (*Accipiter nisus*) correlated with the different roles of the sexes in breeding biology and sex-specific selection pressures. This suggested that it partly results from conflict between the need for hunting agility, males having more than females, and the opposing need for large body reserves, which females have more of than males as an aid to successful reproduction. Thus, "The compromise in body size is therefore drawn at different points in each sex."[35] The subject of reversed sexual dimorphism among raptors remains, however, highly complex and has stimulated numerous hypotheses to account for it, with no less an authority than Ian Newton himself describing it as "intractable."[36]

Weaponry in birds

In opening the topic of secondary sexual characters, Darwin noted that, if rarely, the males of some bird species wear weapons for fighting with each other—the leg spurs of the domestic and fighting cockerel undoubtedly being the most widely and commonly known example to most people. Some of the commonest forms of fighting weaponry in birds (other than the beak) consist of one or more leg spurs, most typically found in the gallinaceous groups, and used intraspecifically. Darwin briefly described some of these aggressive instances within a few species, as well as exceptional instances of leg spur use against individuals of other species.

Another form of spur that may be used in fighting is worn by some birds on the anterior bend of the folded wing, notably in some species of waterfowl, rail, jacana, and lapwing; see colour figure 6.2. In some of these species, the spurs are of similar size in both sexes all year round and would therefore appear to be purely for defensive purposes, and thus the result of natural selection. In other species, however, they become larger and sometimes also more colourful in one sex during breeding seasons, suggesting to Darwin that they also or exclusively represent a sexual character resulting from sexual selection.[37] Although some recent studies of spurred galliform species suggested that females select those males with longer spurs, it is, however, now considered likely that in almost every spurred species, the spurs typically function only for fighting among males.[38]

Vocal and instrumental music

Moving on to considering this subject, Darwin started with the statement that the voice of birds "serves to express various emotions, such as distress, fear, anger, triumph, or mere happiness. It is apparently sometimes used to excite terror, as in the case of the hissing noise made by some nestlings-birds [*sic*]." He also noted that some social birds "call to each other for aid; and as they flit from tree to tree, the flock is kept together by chirp answering chirp. During the nocturnal migrations of geese and other water-fowl, sonorous clangs from the van may be heard in the darkness overhead, answered by clangs in the rear." In this he is describing what are now widely known as "contact calls," which are used by flocking birds, but also by the members of a monogamous pair as they forage, to keep associating individuals in touch, particularly when out of sight of one another. Other vocalizations are given by birds as alarm call; but the "true song" of most species are often confined to the breeding season "and serve as a charm, or merely as a call-note, to the other sex." At the time, however, naturalists disagreed over the function of singing in birds.

That male birds typically perch in a conspicuous place to sing suggested to many people of the time that this was to vocally advertise their presence to females; whereas others, including Gilbert White of Selbourne in 1825,[39] thought it was to compete with rivals. We now know that such singing prior to pair formation may have both functions, as Darwin suggested; and that after pairing, the main, if not exclusive (for we now know that paired males of many species will opportunistically mate with additional females), function is territorial. However, territoriality in birds was not widely known at that time, and was not convincingly established until by Henry Howard in his 1907–1914 and 1920 publications, and so CD makes no mention of it with respect to avian vocalizations. He came close to appreciating it, however, by noting that bird catchers could trap large numbers of birds at the one spot by hiding a live singing captive and a stuffed bird placed on a conspicuous perch close to it (i.e., singing males attract rival males). Charles also noted that some naturalists believed song could not function to "charm" females because the males of some species, such as the European Robin (*Erithacus rubecula*), sing in autumn. He accounted for this by anthropomorphically and contradictorily writing "nothing is more common than for animals to take pleasure in practising whatever instinct they follow at other times for some real good" or that they should sing outside the courtship season "for their own amusement"—when, again, he was close to potentially becoming aware of territorial singing in birds in general and in European Robins specifically.[40]

In recording the fact that the "Order of Insessores" or songbirds (then almost exclusively the majority of passerines) have far more complex vocal organs than most other birds and sing well, Darwin noted that some of them, such as the crow family, do not. He observed that it is remarkable that "only small birds properly sing" and yet detailed Albert's Lyrebird (*Menura novaehollandiae*, a giant terrestrial songbird) in mentioning its wonderful singing and its amazing ability to mimic other birds' calls. Charles also cited John Gould[41] as stating that male Albert's Lyrebirds form "*corroborying places*, where it is believed both sexes assemble,"[42] which might imply lekking, to sing and display their tails. Gould actually specifically wrote of both lyrebird species being solitary in habits;[43] and of the Albert's Lyrebird, he wrote "Each bird forms for itself three or four '*corroborying places*'."[44] In the Australian Aboriginal context, a corrobory involves plural dancers, and this could have misled Darwin's interpretation of Gould's wording. Male lyrebirds typically sing from a traditional specific forest floor location and courtship display to visiting females alone.[45]

In addition to the vocal and instrumental sounds and some internal vocal organs of various birds, Darwin described examples of some of the grouse in which the males have additional vocal modifications. These are areas of bare colourful skin on the neck that are inflated, balloon-like, during courtship calling to produce more far-carrying "curious hollow" and other sounds. He illustrated the Greater Prairie-Chicken (*Tympanuchus cupido*) of North America doing this; and it, understandably, reminded him of air sacs inflated to the sides of a calling frog's mouth (see figure 6.3). This is another example of convergent evolution, of a similar structure and its function, in two completely unrelated animal groups or classes.

Figure 6.3 An adult male Greater Prairie-chicken in courtship display, with three female-plumaged birds present. From *The Descent of Man*: p. 569, by Darwin, 1901.

The throat pouch and appendage of the male Great Bustard (*Otis tarda*) of the Palaearctic and the Umbrella Bird (*Cephalopterus ornatus*) of South America were identified as another type of physical aid to enhancing courtship vocalizations by Darwin (colour figure 6.4). Grossly modified tracheas, involving some convoluted ones, found in a diversity of larger water bird species including swans, ducks, and cranes, in which it is more developed in males than in females, were also seen as adapted for the same purpose. After Darwin wrote on the subject, it was discovered that some monogamous manucode birds of paradise, notably the Trumpet Manucode (*Manucodia keraudrenii*), also have a convoluted trachea that is unique among song birds and that is larger in adult males than in females.[46] Although CD wondered whether the calls of these males during the breeding season serve as "a charm or merely as a call to the female," or express "love, jealousy and rage," he, once again and understandably for his time, failed to mention territoriality as a function.[47] On the other hand, it could, in the case of one or more species, permit the sexes of a monogamous pair to maintain contact over large foraging areas. Darwin's tentative allusion to sexual selection by female birds on the song of males[48] led to the assumption that it affects male mating success and that male song also functions in competition between rival males—and evidence accumulated since then supports this view.[49]

The appreciation of instrumental "music" or sound in birds was in its infancy when Darwin reviewed it in a little over half a dozen pages. He wrote of male peafowl and birds of paradise rattling feather quills together, turkeys scraping their wings on the ground, and some grouse species producing a buzzing by friction of the wings or a drumming of them by striking the wings over the back. Darwin also noted that groups of African weavers produce a "whirring sound like a child's rattle" by each male quivering their wings while singing in a gliding courtship flight display in turn. Male nightjars make a booming courtship noise with their wings, whereas woodpeckers use tree branches to hammer out a far-reaching percussion to impress females or warn rival males.[50] A male Eurasian Hoopoe (*Upupa epops*) uses indrawn breath to produce a vocal and instrumental sound. Darwin was informed by Robert Swinhoe that it calls as it taps its beak tip against a hard substrate;[51] but as this is not mentioned in modern literature, whereas that they incline the head downwards to increase the volume of their calls is, the beak tapping would appear to be a misinterpretation of observed calling birds.[52]

Various feathers of many other birds are modified in an impressive diversity of ways specifically to produce remarkable mechanical courtship sounds in intensive flight or in perch-based displays. Charles reviewed those known to him and illustrated some of them, including those of the small but highly colourful South American manakins (see figure 6.5). In applying his theory of sexual selection to these peculiar adaptations, Charles wrote convincingly: "we know that some birds during their courtship flutter, shake, or rattle their unmodified feathers together; and if the females were led to select the best performers, the males which possessed the strongest or thickest, or most attenuated feathers, situated on any part of the body, would be the most successful; and thus by slow degrees the feathers might be modified to almost any extent. The females, of course, would not notice each slight successive alteration in shape, but only the sounds thus produced."[53] It was a considerable time until ornithologists once again paid close attention to the oddly modified feathers of some of the adult male manakins and how they actually produce the sharp loud sounds that function in their courtship displays to impress females. The application of video recording cameras at the display perches of some of those species studied have enabled ornithologists, by being able to greatly slow down almost unbelievably rapid real time display movements in replay, to see how these diminutive birds physically incorporate mechanical sounds into their courtship.[54]

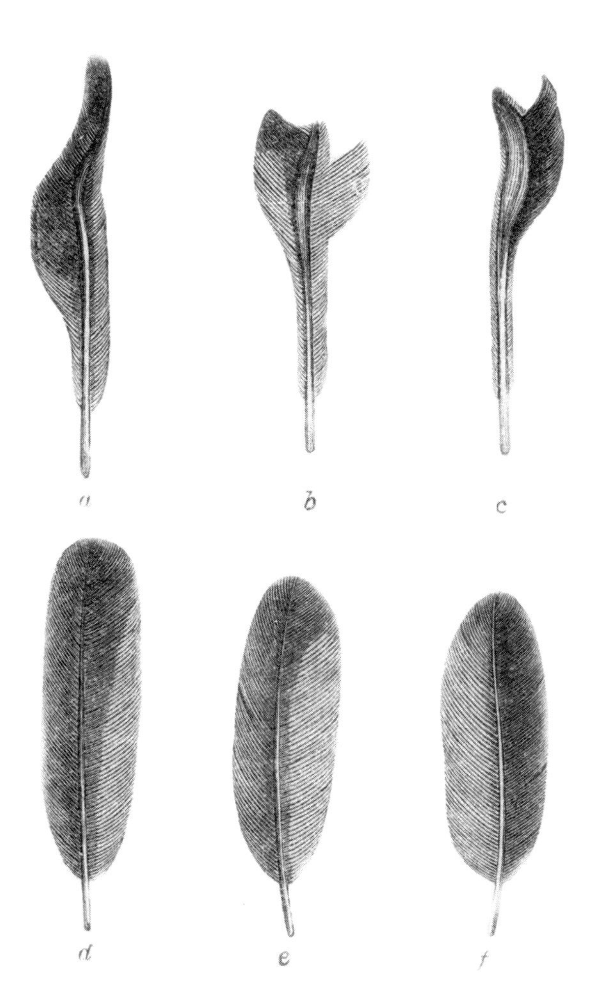

Figure 6.5 Secondary wing feathers of the Club-winged Manakin (*Machaeropterus deliciosus*), the three upper ones (a,b,c) being of the adult male and the three lower ones (d,e,f) being corresponding feathers of the adult female. Those of the adult male are modified for the production of mechanical sound during courtship display. From *The Descent of Man*: p. 579, by Darwin, 1901.

Love-antics and dances

Under this heading, within *The Descent of Man*, the communal and solitary male displays of a few grouse, heron, and New World vulture species are reviewed, including the brief flight displays of some warblers and bustards. The bowerbirds of Australia are, however, described as the most curious in this respect because of the sophisticated bower structures the males build, decorate, and court females at (colour figures 6.6, A1, A2). Because of John Gould's quote of what a Mr. F. Strange told him about captive Satin Bowerbirds (*Ptilonorhynchus violaceus*), Darwin wrote of both sexes building bowers but the male as being the main builder.[55] This was no doubt because the female-plumaged bird held by Strange in captivity was in fact a young male, or if it was indeed a female because he saw it do no more than peck at the bower walls rather than actually build any part of the structure, for female bowerbirds do not construct bowers.[56] In writing "These curious structures, formed solely as halls of assemblage, where both sexes amuse themselves and pay their court, must cost the birds much labour," CD was, other than the anthropomorphic "amuse themselves," more insightful than he could have realized.[57] We now know that the level of "cost" involved in the building, decorating, and attending bowers is "taken into account" by females with them selecting owners of superior bowers built by what is presumably the genetically fittest of available males to fertilize them;[58] see appendix 4.

Decoration

Under this subheading, Darwin offers a review of the extreme kinds of feather structure, pigmentation, ornamentation, and other types of structures found in adult males of various birds. These traits are brought about as a result, Darwin tells his readers, of sexual selection by females (see figures 6.7–6.17). He noted that some of these secondary sexual characters may be worn for only each single courting and breeding season or for life, depending on the species. With regard to the numerous examples of unique unfeathered structures of modified, often brightly pigmented, skin, CD offers the bellbirds of South America as a fine example. He points out that they present a classic example of "the common rule that within the same group the males differ much more from each other than do the females";[59] colour figure 6.7. There are many other examples of this, and perhaps none more dramatic and ostentatious than within the birds of paradise;[60] figure 6.8. Finally, Darwin reviews the number of annual moults that various bird species or groups undertake, be it single or double, with respect to the various types of plumage relative to age and climatic and sexual seasonality, and he gives a good number of specific examples.[61]

Figure 6.8 An adult male Lesser Bird of Paradise (*Paradisaea minor*) in courtship display. From *The Descent of Man*: p. 588, by Darwin, 1901.

Display of plumage by male birds

Turning to this topic, Darwin observes that although the display of their ornaments by males function to "excite, attract, or fascinate females," the males of some species will display in the absence of females, be they communal or solitary, and some will display to individuals of other bird species, or even to creatures of other animal classes, including people. I have seen relaxed male birds of paradise in the wild instantly adopt a full courtship display posture and direct it at any passing bird, mammal, or even larger flying insect! Wallace's description of 12 or more plumed male birds of paradise (*Paradisaea* species) displaying in a tree-top lek to a point of absorption so great that most could be shot by bow and arrow by Papuan hunters is cited by CD, as is the fact that captives spend much time in maintaining and cleaning their fine plumage. Constant feather maintenance by such adult male birds is fundamental to their successful courtship, as females are so critical in making their choice of a mate that is in fine and unblemished plumage; see figure 6.9.

A South American species, and according the Darwin "one of the most beautiful birds in the world," in which highly ornate and brilliantly orange feathered adult males also display to their drably coloured females on a traditional lek site is the Guianan Cock-of-the-rock.[62] These males, when fully occupied in their displays, can be easily killed by Indians, just as can the lekking birds of paradise by Papuans. Darwin went on to detail the courtship display of adult male pheasants, noting how they will laterally expand their

Figure 6.9 An adult male (below) and female (above) Tufted Coquette hummingbirds (*Lophornis ornatus*) at the nest, illustrated by Darwin to show male plumage decoration. From *The Descent of Man*: p. 590, by Darwin, 1901.

plumage and lean their body toward the watching female to create the best visual effect from the female's point of view; see figure 6.10. He also briefly described the fantastic display of the truly huge Great Argus pheasant and how its posturing exposes the otherwise typically concealed vast secondary feathers with their quite amazing "ball and socket" decorative markings; see figures 6.11–6.19. Given the complexity of the description and discussion of the evolutionary origins of these incredible Great Argus feather markings, CD had to deal with them later in his work; see the section "Selection on Variability, Graduation, and an Eye for a Bird" following. That these unique and elaborate plumage decorations of the adult male Great Argus are usually only ever exposed to view when they are presented to females, as males posture during courtship, led CD to observe that they serve exclusively in sexual

Figure 6.10 Two adult male (above and left) and a female (below right) Grey Peacock-Pheasant, illustrated by Darwin to show male decoration and display. From *The Descent of Man*: p. 604, by Darwin, 1901.

Figure 6.11 An adult male Great Argus pheasant, illustrated by Darwin to show one of its courtship display postures. From *The Descent of Man*: p. 606, by Darwin, 1901.

seduction. Thus, they offer the strongest evidence for them being the result of sexual selection by countless generations of discerning females.

Darwin reviewed some of the most spectacularly plumaged of bird species and their displays, almost all of which involve promiscuous males and a polygynous mating system. He went on to describe a selection of courtship displays by males of socially monogamous species in which the males lack lavishly ornate feathering and are only slightly more colourful than their females. He made it clear, however, that sexual selection by females is responsible for their only slightly more handsome partners. But the objective biologists of today could not accept the anthropological view that male birds "take delight in displaying their beauty" as did CD and "all naturalists" familiar with the habits of birds of his era. Moreover, we now know what

Darwin could not have known in his time, which is that in numerous socially monogamous passerine birds, males do not display their colours to females until after they have paired with a mate for a breeding season. Thus, in such cases, male display serves to strengthen the pair bond and to stimulate and synchronize mating rather than to initially attract and seduce a mate. But it almost certainly comes into play in males courting females for extra-pair copulations in the many socially monogamous species in which males typically "play away," or cheat on their nesting partner; see chapter 7.

In stating "The various ornaments possessed by the males are certainly of the highest importance to them, for in some cases they have been acquired at the expense of greatly impeded powers of flight or of running," Darwin was unknowingly close to grasping the fundamentals of the novel, intriguing, and controversial "handicap principle" that was proposed about a century later by the thoughtfully innovative Amotz and Avishag Zahavi.[63] Fundamentally, this idea argues that females select for mature males that are observably handicapped by elaborated body parts, or by constructed bowers in the case of bowerbirds, because their very continued survival in the face of such handicaps indicates to females their physical and genetic fitness. Indeed, the preceding quoted insightful sentence, written as it was by Darwin in the context of sexual selection by females, could have of itself stimulated such a subsequent, fascinatingly elegant, theory.[64]

Length of courtship

By the time Darwin wrote *The Descent of Man*, much evidence already indicated that in the vast majority of sexually dichromatic bird species it is the males that wear the brighter plumage and give better song or produce mechanical sounds. Such traits are predominantly typically apparent only during the courtship and breeding season of each year, when males are highly competitive for access to females. Under "Length of Courtship," CD reviews the breeding seasonality of a number of polygynous and monogamous bird species. In mentioning the Superb Lyrebird (*Menura novaehollandiae*) of Australia, he quotes a Mr. T. W. Wood who reports that "a traveller" claimed to have seen 150 adult males "ranged in order of battle, and fighting with indescribable fury."[65] I would say that it was "indescribable" because it never happened, as nothing remotely like aggregations of more than two, perhaps three, adult males are known to occur in this now particularly well-studied lyrebird species.[66]

Bowerbirds

Male bowerbirds are truly extraordinary and remarkable because in building bowers and decorating them, they are producing secondary sexual characters that are external to their bodies and thus provide indicators to females that are truly *symbolic* of their relative fitness.[67] Moreover, these symbolic characters can be viewed and assessed by females in the absence of the bower-owning male. Darwin was most interested in this, at the time little known and understood, bowerbird behaviour. In a letter of 16 February 1868, he wrote to Abraham Bartlett of the London Zoological Gardens asking if he would "have all the coloured worsted [cloth] removed from cage & the bower, & then put, all in a row, at same distance from bower, the enclosed coloured worsted, mark whether the Bird *at first* makes any selection. Each packet contains equal quantity; the packets had better be separated & each thread put separate, but close together; perhaps it would be fairest, if the several colours were put alternately,—one thread of bright scarlet, one thread of brown, &c &c. There are six colours.—Will you have the kindness to tell me whether the birds prefer one colour to another?"[68] Charles was clearly suggesting a well thought through, eloquently simple, controlled "natural" experiment, if in captivity, that was way ahead of its time. Sadly, no record survives of any response from Bartlett. In recent decades, a very great deal of intensive research, including numerous simplistic to technologically highly sophisticated field experiments, has been carried out on bower decoration by male bowerbirds.[69] See also the "Bowerbirds" section of appendix 4.

Since the appearance of Darwin's *The Descent of Man*, a veritable mountain of literature has appeared on sexual selection, its functions, mechanisms, and the incredibly diverse, often extraordinary, traits in appearance and display behaviours associated with it. A considerable proportion of this literature is about birds, and it is far too great in volume to attempt even a brief summary of here.

MAGPIE GATHERINGS, BIRD BRAINS, AND SPOUSE REPLACEMENTS

Citing the Reverend W. Darwin Fox, Charles Darwin records conspicuous, noisy, and animated aggregations of British Black-billed Magpies (*Pica pica*) that celebrated the "great magpie marriage," mostly early in spring. Other than Fox's allusion to some kind of courtship or pairing activity, Darwin offered no explanation for this phenomenon. It is now well understood to be a form of "territorial probing" that is predominantly instigated in late winter and early spring by the more dominant of nonbreeding magpies seeking to test the strength of territorial defence at the borders of adjacent established breeding pairs' territories. A pair initially doing this is very quickly joined by additional nonbreeding birds also seeking to establish a territory; the average number of individuals at 225 such temporary "gatherings" being nine.

Another kind of magpie "gathering" involves pre-roost assemblies of birds.[70] Darwin was unaware that the magpie gatherings consisted of nonbreeding birds lacking territory. It is ironic, therefore, that his immediately subsequent text, under his heading "Unpaired Birds," uses the Black-billed Magpie as one of a number of examples of bird species that have nonbreeding birds in the population that, in the event of one of an established breeding pair dying or being killed, will immediately be replaced by another. This occurs even if the surviving partner is attending an active nest containing eggs or young.[71]

MENTAL QUALITIES OF BIRDS, AND THEIR TASTE FOR THE BEAUTIFUL

Half a dozen subsequent pages of anecdotal and somewhat unsatisfactory observation on this topic in *The Descent of Man* involve the response of a bird to the loss of its mate; birds feeding other injured individuals; apparent curiosity; captive birds distinguishing people, cats, and dogs as individuals known and unknown to them; instinctive aggressive reactions to individuals of other species wearing plumage colours of their own kind; and birds "decorating" nests and bowers.[72]

FEMALE CHOICE AND A TALE OF TAILS

Writing about "Preferences for Particular Males by the Females," Darwin reviewed cases of pairs of wild and captive birds of two species mating to produce hybrid offspring. He then related some clear examples of females making a choice from several suitors in wild bird species and noted that domestic hens prefer to mate with the "most vigorous, defiant, and mettlesome male." Domestic pigeons, as he observed previously, prefer a mate of their own breed, but are much less fussed about male plumage colour, whereas female domestic turkeys prefer wild males when available to them.

With regard to the significance of male secondary sexual characters, Darwin cited an informer in South Africa who observed that female Long-Tailed Widowbirds (*Euplectes progne*) disown males robbed of their long nuptial tail feathers. In the 1980s and 1990s, Malte and Staffan Andersson experimentally studied wild Long-Tailed and Jackson's Widowbirds (*Euplectes jacksoni*) in Africa. By artificially manipulating the relative length of the tail feathers of adult males, they demonstrated the truth of what CD had reported—that successful long-tailed males lost their attractiveness to mate-seeking females if their tails were shortened. On the other hand, less successful males became more successful when their tails were artificially lengthened.[73] Thus longer tails on males meant much to females in their selection of a mate from the pool of available males of variable tail length.

Darwin was informed that a male Silver Pheasant (*Lophura nycthemera*) successful over all his rivals in mating females was immediately superseded by a rival upon his plumage getting spoiled. Yet another, much experienced, informer pointed out to Darwin that all of the numerous albino individuals of a number of species he observed failed to pair and mate, and CD put this down to their rejection by normally plumaged conspecifics. These and his previous considerations led Charles to conclude that the pairing of birds is not left to chance, "but that those males, which are best able by their various charms to please or excite the female, are under ordinary circumstances accepted. If this be admitted, there is not much difficulty in understanding how male birds have gradually acquired their ornamental characters." He then goes on to use his helpful artificial selection analogy yet again by adding, "All animals present individual differences, and as man can modify his domesticated birds by selecting the individuals which appear to him the most beautiful, so the habitual or even occasional preference by the female of the more attractive males would almost certainly lead to their modifications; and such modifications might in the course of time be augmented to almost any extent, compatible with the existence of the species."[74]

Darwin's concept of active female choice in selecting a mate in animal species was not received well by conservative western human societies at a time when women had few rights and whose opinions in significant matters were hardly listened to let alone taken seriously, although it was familiar to and accepted by John James Audubon[75] and by contemporary aviculturalists (who could witness mate selection among their captives).

SELECTION ON VARIABILITY, GRADUATION, AND AN EYE FOR A BIRD

The next topic that Darwin discussed was the important one of "Variability of Birds, and especially of their Secondary Sexual Characters." He first stated that slight differences in appearance between individuals of many wild species are not questioned, it having been frequently demonstrated, whereas marked differences are rare. Nevertheless, he provided examples in a good number of species and included discussion of some sexual and age dimorphic characters with respect to their origins being in morphological variation.[76]

There were inevitably numerous aspects of nature that countless people saw as being far too intricately complex, sophisticated, beautiful, and wonderful to have been brought about by a theoretical and then perceived to be random process of natural selection, or by sexual selection. Darwin was only too well aware of such serious potential threats to his case, but he did not shrink from them. This profound inability in people to accept that highly complex organisms, or parts of them, could be the result of selection in nature was famously expressed by William Paley.[77] Paley observed the hypothetical scenario that in stumbling across a stone on a heath, he might assume it to have lain there forever and that it would be difficult to prove the absurdity of this notion. However, had he stumbled on a watch, he would certainly not assume it had been there forever. Paley sees that the remarkably complex design of the watch "must have had a maker . . . who comprehended its construction, and designed its use." He observes that although nobody could argue with this conclusion, this is in effect what atheists do in attributing all unimaginably complex natural productions to the processes of nature. In supporting his assumption of a "designer" being behind the production of natural complexities, Paley cites the eye of higher vertebrates, and especially that of humans, which he compares to instruments of sophisticated human design to conclude that God and not nature is the only "designer."

Paley believed that an eye had to be fully developed to be of use. However, Darwin indicated how a far from perfectly developed eye can be of benefit to a lowly species, and have the potential to evolve into a better eye; and many scientists have since proved this to be true.[78] It is this kind of thinking of Paley that is behind the contemporary concept of "intelligent design," used by religious zealots to support the theory of the existence of a god and creator, as is splendidly addressed, reviewed, and assessed by Richard Dawkins in his book *The Blind Watchmaker*.[79]

One such problematic, complex and beautiful, natural trait that seriously worried Darwin, and had him thinking long and hard, was perhaps the most remarkable of all patterns of pigmentation found on birds—as well as on some mammals, reptiles, fishes, amphibians, and insects. This had to be addressed fully, at least in his discussion of sexual selection in birds, and this resulted in his text under the title "Formation and variability of the ocelli or eye-like spots on the plumage of birds." He defined an ocellus as consisting of "a spot within a ring of another colour, like the pupil within the iris, but the central spot is often surrounded by additional concentric zones"—giving the ocelli on the widely familiar tail coverts of adult male Indian Peafowl as an example; see figure 6.12. Having made comparisons with allied bird species showing pertinently similar markings, Darwin observed that "circular spots are often generated by the breaking up and contraction of stripes" and that "a dark spot is often formed by the colouring matter being drawn towards a central point from a surrounding zone, which later is thus rendered lighter; and, on the other hand, that a white spot is often formed by the colour being driven away from a central point, so that it accumulates in a surrounding darker zone. In either case an ocellus is the result." Charles also noted that given the ocelli of variable quality on an abundance of fowl forms, and on numerous insects, the formation of them cannot be an evolutionarily complex process.

Of his subheading "Gradation of secondary sexual characters," Darwin stated that "Cases of gradation are important, as showing us that highly complex ornaments may be acquired by small successive steps." He goes on to use the adult male Indian Peafowl and Great Argus pheasant to demonstrate how ocelli evolve on birds' plumages, considering those on the Indian Peafowl cock to be "certainly one of the most beautiful objects in the world." Importantly,

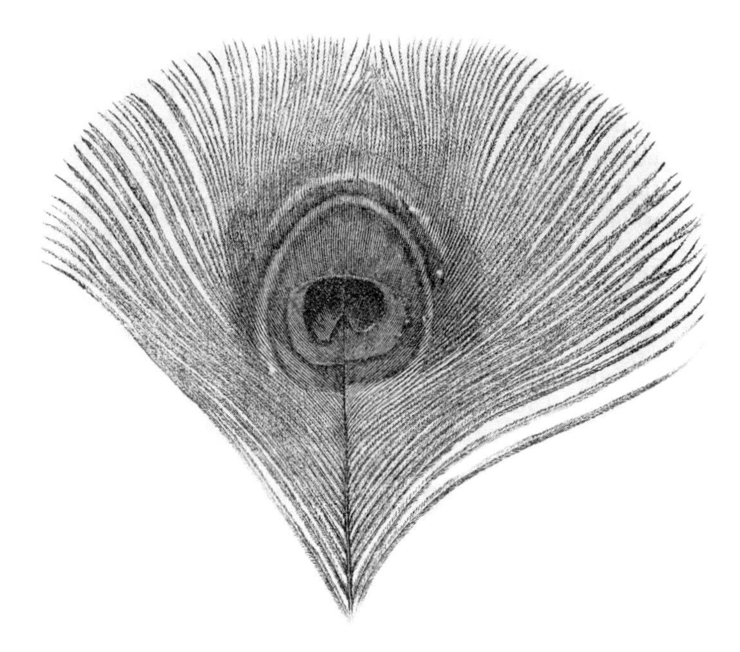

Figure 6.12 A tail covert feather tip of an Indian Peafowl cock, illustrated by Darwin to show that the "transparent zone is represented by the outermost white zone, confined to the upper end of the disc." From *The Descent of Man*: p. 657, by Darwin, 1901.

Darwin draws attention to the fact that in the male Indian Peafowl, the "lower margin or base of the dark-blue centre of the ocellus is deeply indented on the line of the shaft. The surrounding zones likewise shew [*sic*] traces, as may be seen in the drawing, of indentations, or rather breaks"; see figure 6.12.

If gradual evolution was involved, Darwin argued, then species must have once lived that exhibited "every successive step between the wonderfully elongated tail-coverts of the peacock and the short tail-coverts of all ordinary birds; and again between the magnificent ocelli of the former, and the simpler ocelli or mere coloured spots on other birds; and so with all other characters of the peacock." He therefore reviewed the plumage characters of the various species (now eight) of the genus *Polyplectron*, these being so closely related to the peafowl and argus pheasants as to be known as "peacock-pheasants." As a group, they not surprisingly show some similar tail and tail-covert plumage structures and patterns, crests, and display posturing to those of the peafowl and argus pheasants. Darwin found a graduation in male tail-covert length across the various species of peacock-pheasants. Looking closely at the circular or oval ocelli decorating their tail feathers, he was initially distressed to find that ocelli on the first species he examined (Grey Peacock-Pheasant [*Polyplectron bicalcaratum*]) differed fundamentally from those of the Indian Peafowl in that there were two on each feather; see figure 6.13. Charles then found that in other peacock-pheasant species, the two ocelli on a single feather were, however, much closer to each other. In the Malayan Peacock-Pheasant (*Polyplectron malacense*), they actually touch each other and that on the tail-coverts of that species they merge to create a deep indention on the line of the shaft of what was now a single ocellus (if showing its origin in two ocelli) just as in the Indian Peafowl cocks' ocelli; see figure 6.14.

The true tail feathers of the peafowl species are concealed beneath their grossly elongated tail-covert feathers, including when in display, and there are no ocelli on them. As their close relatives that lack elongated tail-coverts do have ocellated tail feathers, Darwin thought that these might have been lost in the peafowl. He therefore examined the ocelli on the tails of the various peacock-pheasants and found that, as he suspected, relative to the length of tail-coverts in the various species, some of the uppermost ocelli on the tail feathers are reduced or absent. Charles showed that this correlation between tail-covert structural modification and reduced tail ocelli was closer to that of the peafowl cock situation in the Malayan Peacock-Pheasant; see figures 6.12 and 6.14. Thus he demonstrated how sexual selection by females could bring about the ornamentation of the male peafowl from a peacock-pheasant-like ancestor, and the latter from even drabber pheasant-like species.

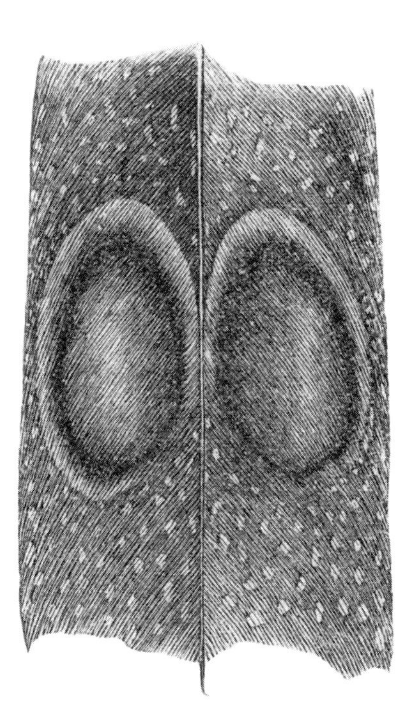

Figure 6.13 Part of a tail covert feather of an adult male Grey Peacock-Pheasant, illustrated by Darwin to show two ocelli. From *The Descent of Man*: p. 659, by Darwin, 1901.

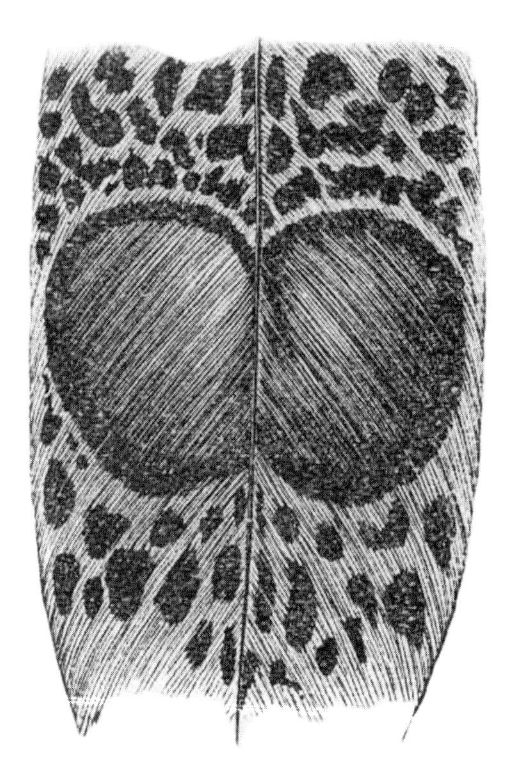

Figure 6.14 Part of a tail covert feather of an adult male Malayan Peacock-Pheasant, illustrated by Darwin to show two ocelli "partly confluent." From *The Descent of Man*: p. 659, by Darwin, 1901.

Not satisfied with this convincing male peafowl/peacock-pheasant demonstration, Darwin turned his attention to the mind-blowing ocelli on the wing feathers of the adult male Great Argus. These ocelli occur on the up to some 60 cm, or 2-ft., long secondary wing feathers and are made visible by the bird only during the remarkable courtship display; see figure 6.11. They are not beautiful for their bright saturated or iridescent colours, as are the ocelli of the male peafowl and peacock-pheasants, but rather for the remarkable shading of far more subdued pigmentation. This shading results in three dimensional-looking "balls lying loose within sockets" as Charles described them; see figures 6.15, 6.18, and colour figure 6.19.

In addition to reproducing his lovely illustrations, CD demonstrated over almost a dozen pages of dense and difficult text how the singularly impressive three-dimensional-looking ocelli of the male Great Argus were gradually derived from dark lines and spots and adjacent paler areas of more usual or ordinary kinds of feather marking; see figure 6.16. Given the nature of these wonderful, fully formed ocelli and adjacent markings, CD's explanation of their evolution, through graduation, was more technical and complex than in the case of the male Indian Peafowl's ocelli. This

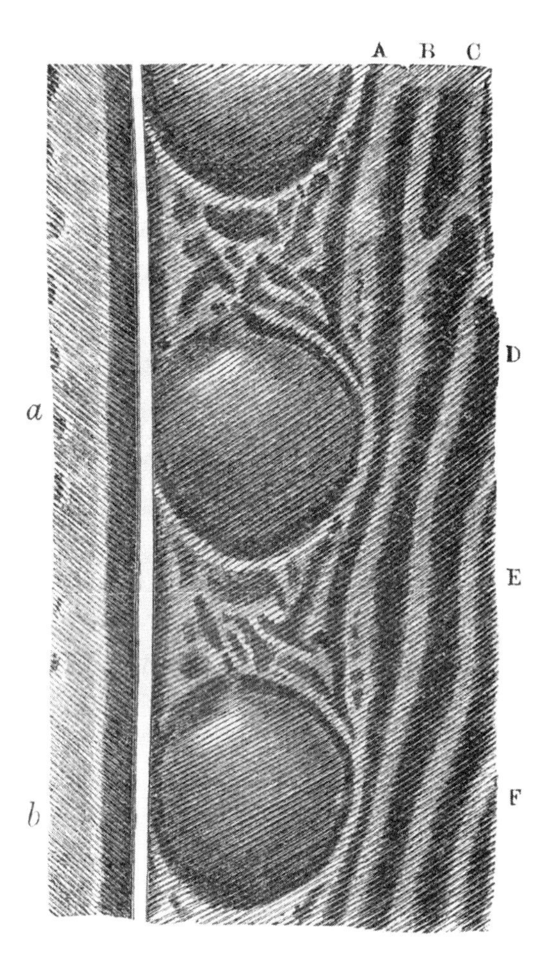

Figure 6.15 Part of a secondary wing feather of an adult male Great Argus pheasant, illustrated by Darwin to show "two perfect ocelli, a and b. A, B, C, D, &c., are dark stripes running obliquely down, each to an ocellus." From *The Descent of Man*: p. 663, by Darwin, 1901.

complexity was not helped by one of his supporting text figures having its critically instructive text labels, at the end of reference lines or arrows, omitted by the printer. Significant as part of his explanation was that the black "socket" ring surrounding each "ball" is usually broken or interrupted above the pale highlight of each ball, this being a remnant of the dark line from which the socket mark originated. He also demonstrated that although the highlights on the apparent "balls" on the ocelli do not do so when the wings are at rest or are not in the fully erect display position (see figures 6.11, 6.18, and colour figure 6.19), they do, incredibly, appear on the upper part of the "balls" during full display posturing. Thus they give the impression of being naturally lit from above, and thus to give the normal three-dimensional effect to the eye of the audience, only during the full courtship display posturing. Darwin concluded, "Thus almost every minute detail in the shape and colouring of the ball-and-socket ocelli can be shewn to follow from gradual changes in the elliptic ornaments; and the development of the latter can be traced by equally small steps from the union of two almost simple spots, the lower one [figure 6.16] having some dull fulvous shading on its upper side."[80] It has been suggested that these elaborately evolved Great Argus ocelli "may serve to mesmerize the female into immobility."[81]

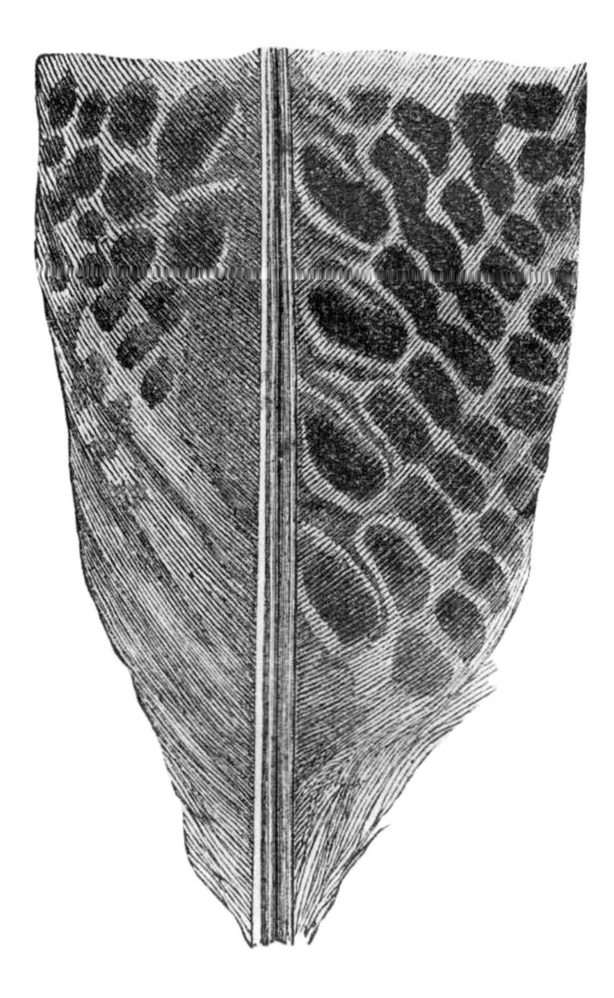

Figure 6.16 Basal part of the secondary wing feather, nearest to the body, of an adult male Great Argus pheasant, illustrated by Darwin to show "the first trace if [*sic*] an ocellus." From *The Descent of Man*: p. 664, by Darwin, 1901.

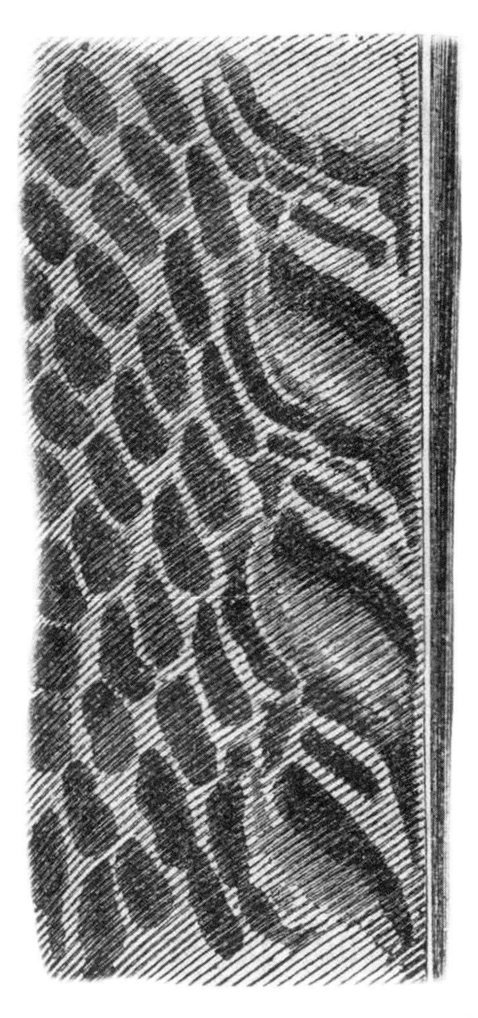

Figure 6.17 Portion of one of the secondary wing feathers near to the body of an adult male Great Argus pheasant, illustrated by Darwin to show "the so-called elliptic ornaments." From *The Descent of Man*: p. 665, by Darwin, 1901.

To end his first two chapters on birds in *The Descent of Man*, Charles Darwin summarized that from what he discussed about the principle of graduation, the laws of variation, changes clearly apparent in many domesticated birds, and the plumage of immature birds, the probable steps by which males have acquired their ornate plumage can be indicated with some confidence. In other cases, however, the evolutionary processes involved eluded his full understanding. As an example, he observed that in the Purple-bibbed Whitetip hummingbird (*Urosticte benjamini*), the adult male has the four central tail feathers tipped white, a trait not then known in any other hummingbird species, whereas the females have the three outer tail feathers tipped white. The

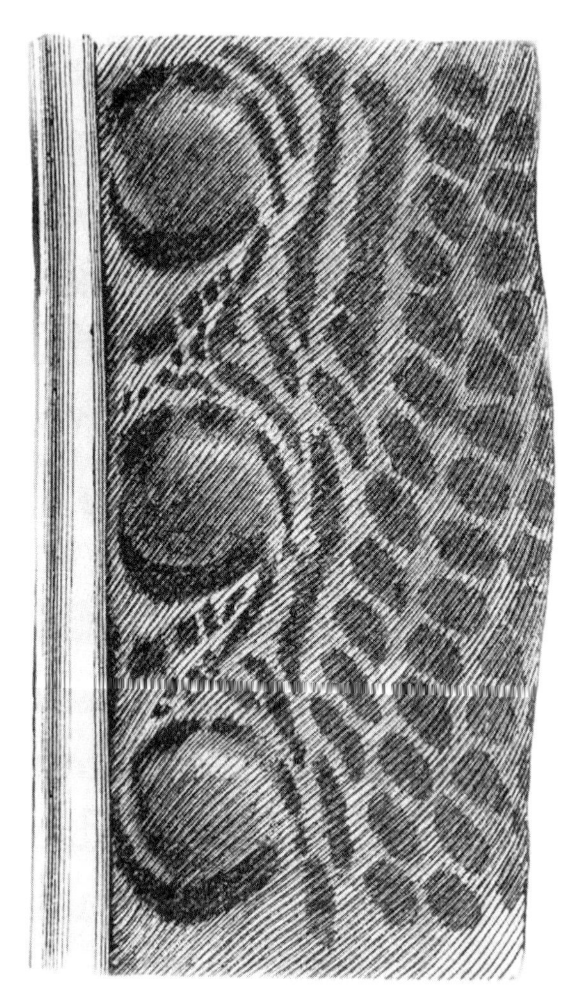

Figure 6.18 "An ocellus [on part of a secondary wing feather of an adult male Great Argus pheasant, illustrated by Darwin to show it] in intermediate condition between the elliptic ornament and the perfect ball-and-socket ocellus." From *The Descent of Man*: p. 668, by Darwin, 1871.

Duke of Argyll raised this apparent anomaly and stated that he could not see how natural selection could bring it about. Darwin agreed with him but then also suggested, and demonstrated, how sexual selection could indeed produce it in the light of existing variation in tail markings and their presentation during courtship displays within the hummingbird family as a whole. He concluded with this: "we can perceive that the males which during former times were decked in the most elegant and novel manner would have gained an advantage, not in the ordinary struggle for life, but in rivalry with other males, and would have left a larger number of offspring to inherit their newly-acquired beauty." This is a sentence of significance, and not without a beauty of its own.[82]

In the final "General Summary" to his entire *The Descent of Man*, Darwin wrote, clearly with an eye to critics including Alfred Russel Wallace, "Although we have some positive evidence that birds appreciate bright and beautiful objects, as with the bower-birds of Australia, and although they certainly appreciate the power of song, yet I fully admit that it is astonishing that the females of many birds and some mammals should be endowed with sufficient taste to appreciate ornaments, which we have reason to attribute to sexual selection; and this is even more astonishing in the case of reptiles, fish, and insects. But we really know little about the minds of lower animals. It cannot be supposed, for instance, that male birds of paradise or peacocks should take such pains in erecting, spreading, and vibrating their beautiful plumes before the females for no purpose."[83]

A most recent study of the genetics of six pertinent species of the Galloanserae, consisting of the waterfowl and landfowl (including the game-birds, pheasants, peafowl etc.), with respect to sexual selection surprisingly concluded that the ornately plumaged males of polygynous species do not pass on the best of possible genes to females. Such males have been found to have a genome that evolves more rapidly than in males of life pair-bonded monogamous species. The quickly evolving genome shows, presumably as a result, slightly negative mutations and thus resulting offspring of such promiscuous males will be less genetically fit. It was concluded that "sexual selection is a powerful force in shaping broad patterns of genome evolution."[84] However, how the negative genetic fitness situation uncovered by this study can be maintained under the influence of natural selection clearly requires study.

NOTES

1. Burkhardt et al. 1985–2014, vol. 16, part 1.
2. Cf. Brent 1981: 477.
3. Desmond and Moore 2009: 374.
4. Darwin 1901: 549–50.
5. J. Smith 2006: 116.
6. Darwin 1901: 757.
7. Burkhardt et al. 1985–2014, vol. 8.
8. Darwin 1901: 642–3.
9. Burkhardt et al. 1985–2014, vol. 16: 197–8; *pace* Birkhead et al. 2014: 327, 331.
10. Manning 1989; Petrie 1993.
11. Manning and Hartley 1991.
12. Petrie et al. 1991; Petrie and Halliday 1994.
13. Petrie 1992.
14. Petrie and Williams 1993.
15. Petrie 1994.
16. Takahashi et al. 2008.
17. Loyau et al. 2008.

18. Dakin and Montgomerie 2011.
19. Jordania 2011.
20. Freeman and Hare 2015.
21. Cinat-Tomson 1926.
22. Møller 1991, 1992.
23. M. Andersson 1982, 1994; S. Andersson 1992; S. and M. Andersson 1994.
24. In Ligon 1999: 104, 146–53.
25. Darwin 1901: 381–3.
26. Clutton-Brock 1986; Donald 2007.
27. Darwin 1901: 334.
28. Hill et al. 1994; Ligon 1999.
29. Darwin 1901: 549–51; Frith 1997.
30. Van Rhijn 1991.
31. Ligon 1999: 376.
32. Darwin 1901: 552–4.
33. Jukema and Piersma 2006; Erritzøe et al. 2007.
34. I. Newton 1979: 26–7.
35. I. Newton 1986: 323.
36. Furguson-Lees and Christie 2001: 35–9.
37. Darwin 1901: 554–62.
38. Davison 1985; Ligon 1999: 91, and references therein.
39. In Darwin 1901: 564.
40. Cf. Lack 1970.
41. Gould 1865: 308–10.
42. Darwin 1901: 619.
43. Gould 1865: 300, 310.
44. Gould 1865: 308.
45. L. H. Smith 1988; Lill 2004.
46. Frith 1994; Frith and Beehler 1998: 25.
47. Darwin 1901: 562–73.
48. Darwin 1901: 564.
49. Ligon 1999: 64.
50. Short 1982.
51. Darwin 1901: 575.
52. Cramp 1985: 795–6; Krištin 2001: 401–2.
53. Darwin 1901: 573–80.
54. Kirwan and Green 2011, and references therein.
55. Gould 1865, vol. 1: 444; Darwin 1901: 582–3.
56. Frith and Frith 2004, 2008, 2009b.
57. Darwin 1901: 580–83.
58. Frith and Frith 2004, and references therein.
59. See Snow 1982; Kirwan and Green 2012; Snow 2004.
60. Frith and Beehler 1998; Frith and Frith 2009a, 2010; Laman and Scholes 2012.
61. Darwin 1901: 583–600.
62. Darwin 1901: 601; see figure 4. 2.
63. Amotz Zahavi 1977; Amotz and Avishag Zahavi 1997, and references therein.
64. Darwin 1901: 600–15.
65. Darwin 1901: 619.
66. L. H. Smith 1988; Lill 2004.
67. Frith and Frith 2004.

68. Burkhardt et al. 1985–2014, vol. 18: 411.
69. Frith and Frith 2004, 2008, 2009b.
70. Birkhead 1991: 56–9.
71. Darwin 1901: 619–25.
72. Darwin 1901: 625–31.
73. M. Andersson 1982, 1994; S. Andersson 1992; S. Andersson and Andersson 1994.
74. Darwin 1901: 631–43.
75. Audubon and Macgillivray 1831–1839.
76. Darwin 1901: 643–52.
77. Paley 1828.
78. Ings 2007; Nilsson and Pelger 1994, and references therein.
79. Dawkins 1986.
80. Darwin 1901: 652–73.
81. Johnsgard 1994: 89.
82. Darwin 1901: 673–5.
83. Darwin 1901: 941–2.
84. Harrison et al. 2015.

CHAPTER 7

More Thoughts on Sex, and Emotions, in Birds

But the most pugnacious and the best armed males [of birds] rarely or never depend for success solely on their power to drive away or kill their rivals, but have special means for charming the female. [Darwin 1901: 756]

Chapter 15 of *The Descent of Man*, the third chapter dedicated to birds in that insightful work, starts with Charles Darwin stating he had to "consider why the females of many birds have not acquired the same ornaments as the male; and why, on the other hand, both sexes of many other birds are equally, or almost equally, ornamented?" He reminded his readers that in *The Origin of Species* he suggested that should females wear the structurally elaborated, ornate, and conspicuously coloured male plumage, it would be dangerous to them when at the nest, while incubating and feeding their offspring. This is because ornate and colourful females would potentially visually attract the attention of predators, and so the pressure of natural selection checked the "transmission," or inheritance, of such male traits to female offspring. Various facts that Darwin acquired, and his consideration of them, brought him to conclude that in sexually dimorphic species, successive variations have been limited to the sex in which they first appeared. In the meantime, Alfred Russel Wallace suggested that in almost all cases, successive variations were initially transmitted equally to both bird sexes, but that females did not acquire conspicuous male traits because of natural selection pressure on them during incubation for cryptic protective plumage,[1] just as Charles had initially thought and "in some few cases," continued to do so.

Darwin's theory of sexual selection experienced somewhat of an initial deafening silence, although a few close colleagues expressed their views about it to him. Wallace's published opposition to major aspects of this theory resulted in it spending a very considerable period of time out of favour with the vast majority of the growing number of evolutionary biologists.[2] Indeed, it remained largely in the zoological wilderness until as recently as 1930 when a book by Ronald Fisher resurrected the theory and presented his own novel concept of "runaway sexual selection," which did not become as influential as it should have until a second edition appeared almost 30 years later.[3] Darwin's sexual selection then became satisfactorily resolved.[4] From then on the intensity of research on animal mating systems, sexual selection, and associated morphology grew exponentially and, most significantly, raised and started to address the question of reproductive conflict between the sexes within species.

Darwin followed his broad introduction by what he described as "a tedious discussion on a difficult point, namely, whether the transmission of a character, which is at first inherited by both sexes can be subsequently limited in its transmission to one sex alone by means of natural selection." It was indeed a tedious discussion, and all the more so because of the lack of any understanding of genetics at the time. Most of Darwin's contemporary readers would have doubtless struggled much with the pertinent text. He pointedly reminded them of the fact that characters limited to one sex are always latent in the other sex, and that they may therefore often express themselves as visually apparent in older females, that is, as some traits typical of adult males.

PLUMAGE SEXUAL AND AGE DIMORPHISM

After an esoteric discussion of hypothetical examples of what might and might not be achieved by artificial selection on domesticated pigeons and fowls, Darwin felt that, even with unlimited time, it would perhaps be impossible to change one form of transmission into the other. He could therefore not accept that it had happened in nature. By successive variations limited to each sex of a species in transmission, males could, however, easily acquire conspicuous characteristics different from females of their species, with the females remaining drab through natural selection pressure for them to be visually cryptic. As males require bright nuptial plumage to compete with rivals for access to and the ability to visually impress females, it will be selected for whether it is transmitted only to males or to both sexes. Thus, females sometimes acquire colourful adult male plumage or other traits to limited, or even to near full, extent, as is the case in many species. But this can only happen when it is not detrimental to females at nests or at other times.[5]

Much of the "tedious" text alluded to by Darwin relates to his arguments against the opinion of Alfred Wallace—that when both bird sexes are conspicuously coloured, the nest or its situation, or site, conceals the incubating bird from predators. However, when the male is brightly coloured and the female dull, the nest is open and the sitting cryptic female is thus exposed but difficult to see, with, as always, some exceptions to the rule.[6] Darwin's, contrary, view was that colourful females "have acquired the instinct of building domed nests oftener than dull-coloured birds" and are thus concealed from view when incubating and brooding, enabling them to be brightly plumaged. He presented many examples of species providing evidence for and against both scenarios. Among these he erroneously stated that female pittas, of the passerine family Pittidae, are brightly coloured; and yet they build open nests, and would thus be conspicuous to predators when in the nest. In fact, all pitta species build substantial domed nests that completely conceal the eggs and the incubating bird from view.[7]

Most examples that CD presented he saw as providing evidence contrary to Wallace's view, but he admitted to the issue being most complex. He did observe that although domed nests conceal the eggs and incubating parent from view, they are larger and bulkier than are open, cupped, nests and as such may be significantly more obvious to predators, particularly tree-climbing carnivores. In addition to nest type and site, factors such as the various breeding systems involved, the age and season at which male plumage traits are acquired by one or both sexes, and the timing and number of moults have to also be taken into account. He made a point of noting that in many groups of closely related birds, and particularly in species of the same genus and thus with similar nests, sexual dimorphism varies between species from extreme, typically with adult males far brighter than females, to negligible.

Contrary to the views imposed on Darwin and Wallace by the then limitations of ornithological knowledge is that we now know that the adult sexes in many polygynous species, in which only one sex performs nesting duties, are largely similar in plumage. Also now known is that many more species than they were aware of with distinctly different adult sex plumages care for their offspring at the nest as a pair.[8]

One significant aspect of the issue of plumage colouration of the adult sexes that Darwin could not take into account, as it was unknown in his time, is that birds perceive colours quite differently than do humans, avian vision being far more sensitive to ultraviolet light.[9] The plumages of most bird species reflect ultraviolet light that is visible to birds but invisible to people.[10] Thus, predatory birds may see something very different than people do when viewing a habitat with what we see as colourfully or cryptically plumaged birds within it. A strong correlation has also been found between the ultraviolet reflective plumage of birds and courtship display presentation of such feathering, suggesting that ultraviolet signals are associated with courtship display posturing.[11]

This chapter 15 of *The Descent of Man* was continued by Darwin expressing serious misgivings about Wallace's views that weapons, ornaments, and bright colours apparent in male birds are the result of "conversion, by natural selection, of the equal transmission of characters to both sexes, into transmission to the male sex alone." He also doubted Wallace's view that the plumage colourations of females are due "to the preservation, for the sake of protection, of variations which were from the first limited in their transmission to the female sex." That said, CD then stated that he would defer further discussion of the issue until reviewing differences in plumage between the young and old in bird species in his next chapter.

Darwin concluded the chapter by briefly discussing moult seasonality, noting that as (adult) egrets, herons, and many other birds acquire "elegant plumes, long pendant feathers, crests, &c." only during each summer, they must be nuptial in function despite being worn by both sexes. He observed that although the females are thus made more conspicuous, this is of no great consequence, as these larger, heavy-billed, birds can defend themselves from most predators.[12]

Chapter 16 of Darwin's *The Descent of Man*, his last on birds, considers "the transmission of characters, as limited by age, in reference to sexual selection." In species in which the young differ in colour from the adults and that immature colour does not appear to have any special function (perhaps a questionable interpretation today), it may be attributed "like various embryological structures" to the retention of an ancestral character. That said, CD qualified this as only being confidently applied when the young of several species closely resemble each other and have the appearance of the adults of other species of the same taxonomic group because such latter adults prove "that such a state of things was formerly possible."

Darwin cited the familiar example that young lions and pumas are marked with faint stripes and/or rows of spots on their pelage, and that, as many related species are similarly to more strongly so marked in all ages, these markings in young lions and pumas indicate a striped ancestor to those cat species. The young retain vestiges of their ancestors' stripes, as do the kittens of black adult domestic cats. Thus, mature individuals of some species have altered colours, whereas their young have remained less altered from their progenitor through CD's, then, principle of inheritance at corresponding ages. He goes on to give some avian examples, as exhibited by both plumage and structure; for example, the beaks of young cross-bills (*Loxia curvirostra*) are initially conical, like those of related and ancestral finches, with their mandibles only becoming elongated to cross each other with increased age.

There are numerous birds in which the young closely resemble their parents, be they bright or dully feathered, and these provide no clue about their ancestral plumage. When both young and old are plumaged in a similar manner across a group of species, CD reasonably thought it probable that their progenitors were generally similarly plumaged.

Although admitting to the "extreme complexity" of the subject, and noting that it remained much needed attention by "some competent ornithologist," Darwin used the literature to attempt a review of the "classes of cases, under which the differences and resemblances between the plumage of the young and the old, in both sexes or in one sex alone, may be grouped." Prior to listing these "classes of cases," he observed that they may graduate into each other and that when young are said to resemble their parents, it is not implied that they are identical, as their feathers are typically less vivid, softer, and often different in shape. Each of these numbered categories were treated by Darwin in as many as 14 pages to as few as a single page, within which are given examples of them and brief discussion of some graduate ones among them and exceptions to them. I can include here only a few of the more lengthy and interesting observations that CD made, as follows:

1. *Adult male more beautiful or conspicuous than adult female; young of both sexes in first plumage like adult female (e.g., common fowl and peacock) or more so than like adult male.*

Of this class, CD observes that numerous examples across the birds of the world could be given, in all orders of both polygynous species, in which adult males are typically far more colourful than females, and monogamous ones in which males are most typically only slightly brighter than females but in some are much more so. One anomalous case Darwin noted is that of the Black-eared Fairy (*Heliothryx aurita*) hummingbird in which adult males differ greatly from adult females in having a splendid gorget and ear tufts. Females have a much longer tail than males. The young of both sexes are very much like adult females, including in having a long tail. This means the tail in males gets shorter toward maturity, and CD pointed out that this is most unusual in birds. It is now known to occur in some other bird species, however, including several birds of paradise.[13]

An informative case concerns three heron species, each a representative of the genus *Ardeola* on different continents, which are strikingly different in their nuptial plumage but are barely distinguishable, if at all, in their winter dress. The young of all three species closely resemble adults in winter plumage. In another two species of the genus, both adult sexes and young wear almost the same winter plumage as the previous three species all year round. As this winter appearance is worn by several species at different ages and seasons, Darwin logically concluded that it is probably indicative of that of their common ancestor's appearance. Differences between the females of these geographically isolated members of the same genus are slight compared with those between their adult males in nuptial plumage, because females are morphologically less modified than males.

Darwin noted that a few bird species present "a singular and inexplicable exception." He observed that female Greater (*Paradisaea apoda*) and Lesser (*P. minor*) Birds of Paradise differ more from each other than do their respective adult males. This was pointed out to him by Alfred Wallace in October 1868.[14] This specific case is not quite as clear-cut as Darwin suggested because adult males of the former species have a dark brown back and wings and yellow flank plumes, whereas those of the latter have a yellow back, much yellow in their wings, and conspicuous white ends to their yellow flank plumes.[15]

Given that the Lesser Bird of Paradise meets a different species of its genus with dark-breasted females at each extremity of its geographical range on New Guinea (Greater Bird to the southwest and Raggiana [*P. raggiana*] to the east), it's distinctive, white-bellied, female plumage could be seen to have evolved as an "isolating mechanism," that is, to avoid potential hybridization. This interpretation of such scenarios, of geographically adjacent morphological differences between congeneric species, is now seen as less convincing than was previously thought. It is now considered more likely to be the result of differing sexual selection by females for male traits within each species population.[16] That possibility aside, CD suggested such exceptions are analogous to those occurring independently of artificial selection in some sub-breeds of domestic fowl, between which females are different, whereas the males are hardly distinguishable.

Having accounted for differences between the adult males of closely related species by sexual selection, Darwin addressed those between their females. In considering the females of species within a genus, he was almost certain that their differing appearances were chiefly the result of greater or lesser transference to them of male characters acquired through sexual selection, and he discussed several example species. He appreciated that in some bird species, this happened very long ago and that the males subsequently underwent great changes without these being transferred to the females. For example, females and young of the Black Grouse (*Tetrao tetrix*) closely resemble both sexes and the young of the Willow Ptarmigan (*Lagopus lagopus*), suggesting to CD that the former descended from an ancestor in which both sexes were similar to the Willow Ptarmigan. He stressed and demonstrated that brilliant colours are rarely transferred to females, and young, and thought this was limited by general conditions of life, including the need for them to be cryptically pigmented while nesting (and caring for fledged young). Thus, females would acquire adult male plumage characters only to the degree that the conditions of life, through natural selection, would permit, given that they would retain some social or other function of benefit greater than that of an entirely cryptic dull plumage.

2. *Adult female more conspicuous than adult male; young of both sexes resemble adult male.*

This class applies to the relatively rare phenomenon in which adult females of species are brighter than the males and in which the males perform incubation: Their young look like the adult male. In these cases, the contrast between bright females and dull males is not as great as it is in the preceding class 1 species. As the three species of painted snipe[17] were seen by Darwin to offer a fine example, he detailed them. He pointed out that the females are both larger and more colourful than their respective males, and that in the Australian Painted-snipe (*Rostratula australis*), the trachea modification for sexual vocalization is simple in males but complex in females.[18] But we now know that this is not so in the trachea of the other painted-snipe species, thus emphasizing the species-specific nature of sexual selection. In fact, whereas adult females of the Australian and Greater Painted-snipe (*Rostratula benghalensis*) are more colourful than their respective adult males, this type of "reversed sexual dimorphism" is not true of the South American Painted Snipe (*Nycticryphes semicollaris*[19]).

Similar reversed sexual dimorphism, with larger and brighter females than males, occurs in the three species of phalarope (family Scolopacidae, genus *Phalaropus*) in which the duller males incubate the clutch. Alfred Wallace argued that here, again, is evidence supporting his idea that the incubating sex is rendered cryptically plumaged by natural selection. Darwin, on the other hand, believed that in these species, the females actively sought out and competed for access to more sedentary and passive males, who would choose to mate with bigger and more colourful females; and so sexual selection by males was responsible for the bigger and brighter females. The appearance of the males and young would therefore reflect a little modified ancestral plumage, once common to both sexes, rather than one modified to be protective at the nest. Certainly it is now assumed that males of polyandrous species take much care in evaluating potential mates, given that they alone must care for eggs and young, although a single explanation for the evolution of this reproductive strategy remains elusive.[20] Polyandry is known in approximately 30, or roughly 0.3 percent, of bird species on earth. In addition to the painted snipes and phalaropes, these include most ratites, some tinamous, jacanas, plovers, sandpipers, and the Plains-wanderer (*Pedionomus torquatus*) of Australia.[21] That the females are typically the larger and more colourful adult sex indicates sexual selection at work, be it female–female, male–female, or the influence of both.

Although the adult sexes of the Southern Cassowary, a giant flightless ratite, are similar in uniform black feather colouration the female is the larger sex, has the bare head and neck skin brighter than in males, and becomes pugnacious during breeding seasons. The smaller male incubates the clutch, fasting while doing so, and raises the offspring alone; see colour figure 7.1. Similarly, the female of the closely related Emu is the larger and more vocal sex, and the males incubate and raise the young. The male Ostrich is, however, an exception to the rule among the bigger ratites in that he is the larger and more colourful sex and yet undertakes the incubation of the clutch while both sexes care for the offspring.[22]

3. *Adult male resembles adult female; young of both sexes have own peculiar first plumage.*

These consist of species in which the adult sexes are similar to identical, and the young have a distinctive first plumage as, for example, in the European Robin (*Erithacus rubecula*) and "Scarlet Ibis" (*Eudocimus albus*). That the adults of this ibis become much duller scarlet in captivity led Darwin to take this as evidence that the bright colour was a sexual character. But it subsequently became known that this colour loss is the result of a lack of appropriate natural dietary components that pigment the plumage of the ibis. The scarlet ibis actually, and remarkably, came to be seen to represent a scarlet form of what is now called the American White Ibis but they have recently been again treated as two distinct species; the Scarlet (*E. ruber*) and White (*E. albus*) Ibis.[23] This understandable misinterpretation aside, CD concluded that in this class, the plumage exhibited by the young reflects the plumage type of the ancestor of the respective species.

In some species in which the adult sexes are similar, this monomorphism may function to make sex determination by other pairs of the species difficult and thus be useful in intersexual aggressive interactions such as territorial disputes.[24]

4. *Adult male resembles adult female; young of both sexes in first plumage resemble them (e.g., Common Kingfisher, many parrots, crows, hedge-warblers).*

The similarity between young and old is "never complete, and graduates away into dissimilarity." For example, the young of some kingfishers are not only less brightly coloured than their respective adults, but many of their ventral feathers are terminally edged with brown, "a vestige probably of a former state of the plumage." Frequently within one group, even a genus, the young of some species closely resemble their parents whereas those of other species differ from them much. Thus, both adult sexes and young of the Eurasian Jay (*Garrulus glandarius*) are closely similar; but the young of the Grey Jay (*Perisoreus canadensis*) of North America differ so much from their parents that they were formerly thought to be of a different species from their adults.

Further emphasizing the difficulty and unsatisfactory nature of the review attempted by Darwin is that he writes at this point "under the present and next two classes of cases the facts are so complex, and the conclusions so doubtful, that any one who feels no especial interest in the subject had better pass them over." He concluded this section by noting that the plumage of certain birds increases in beauty for many years after the first acquisition of mature feathering, citing as examples the male peafowl, some birds of paradise, and herons. Charles doubted that this typically resulted from the selection of successive beneficial variations (but thought it probable in the birds of paradise) or merely of continuous growth, his unstated implication being that it reflected sexual selection by females.

A recent study of species in which both sexes are brightly plumaged concluded that "The association of bright female plumage and female aggression in diverse avian groups suggests the female aggression may be the general explanation for the evolution of bright female plumage."[25] It may generally be the case, however, that aggressive interactions between females "are more intense among monogamous species with extensive paternal contributions[26] than among polygynous species with little or no paternal care, irrespective of female plumage colour."[27]

5. *Adults of both sexes have distinct winter and summer plumage, whether or not males differ from females. Young resemble adults of both sexes in their winter dress or, much more rarely, in their summer dress, or they resemble only females or have an intermediate appearance; or may differ greatly from both winter and summer adults.*

The cases in this class are, not surprisingly, complex, depending as they do on inheritance by sex, age, and season, with some individual birds passing through at least five distinct plumages. In species in which males differ from females during summer alone, or, more rarely, during both seasons, the young generally resemble females: for example, the American Goldfinch (*Carduelis tristis*) and Australian fairywrens of the genus *Malurus*. In species in which the sexes are alike year-round, the young may resemble the adults in their winter dress (e.g., Cattle Egret [*Bubulcus ibis*]) or, much more rarely, their summer dress (e.g., Razorbill [*Alca torda*]); may be intermediate (e.g., many waders); or may differ much from the adults, as in some herons and egrets. Darwin saw that all of these cases apparently depend on characters acquired by adult males having been variously limited in transmission according to age, season, and sex.

6. *Rarely, young in first plumage differ according to sex; males more like adult males and females more like adult females.*

Although seemingly intuitive, such cases are found in various groups, but not commonly so, Darwin detailing some in which the sex of nestlings or fledglings is apparent in their plumage. As examples, he mentioned the Eurasian Blackcap (*Sylvia atricapilla*); Common Blackbird (*Turdus merula*); and other thrushes, mockingbirds, and the Red-billed Streamertail hummingbird (*Trochilus polytmus*). This is in fact now known not to be true of the Common Blackbird in which the juvenile sexes cannot be identified with certainty due to considerable variability, although males do tend to be darker with stronger rufous streaking.[28]

In most of the species within five of the preceding six numbered classes, the adults of one sex or of both sexes are brightly coloured, at least in the breeding season, whereas the young are less colourful to dull. Darwin knew of no case in which the young of dull-coloured species wear bright colours, or of the young of bright-coloured species being more colourful than their parents. The exception in the fourth class is where the young of most species are like their colourful adults, and as these "form old groups, we may infer that their early progenitors were likewise bright."[29] This exception aside, the vast majority of bird species have apparently acquired more colourful plumage than their ancestors wore, with the duller plumage of the latter being indicated to some degree by the plumage of their young.[30] Darwin was convinced that colourful males of sexually dimorphic species were the result of sexual selection through female preferential choice of mates.[31] On the other hand, Wallace saw colourful male and dull female plumage in such species as functioning for species recognition/isolation and for protective cryptic plumage in nest-attending females resulting from natural, not sexual, selection. But his explanation of brightly coloured male plumage was unconvincing, and his notion that male adornment was the product of "superabundant health and vigour" was even more so.[32]

Recent research has not found convincing support for Wallace's view that differences between the adult sexes in plumage and other ornamentation function as species-specific recognition signals, or for species isolation. On the other hand, such research results do support his view that marked sexual dimorphism has not come about by, in most species, males evolving brighter colouration but rather by females evolving more cryptic plumage to protect them and their offspring while nesting. Thus sexual dimorphism in plumage colouration is now correlated with differences in parental care contributions by the sexes. A novel additional factor potentially involved here is that of extra-pair paternity, or cheating on their pair-bonded mate by breeding females; see the following.

The few, brief, additional references to birds in the last five chapters of *The Descent of Man* are to be found within appendix 1. In addition to the problems that Darwin alluded to previously was the highly significant one that in the vast majority of species known at the time, almost nothing was documented of their breeding systems and ecology, which fundamentally influence sexual dimorphism in birds.[33] The situation is very different today, to some degree because of interest and stimulation created by the ornithological content of Darwin's *The Descent of Man* published close to one and a half centuries ago.

A great deal of research has now been undertaken concerning the life histories, ecology, and reproductive behaviour of numerous bird species worldwide, enabling students to consider the dimorphism between the sexes, or the lack of it, in the light of knowledge gained. Moreover, the development of sophisticated studies of the genetics of bird species and individuals has resulted in remarkable and surprising insights into the mating systems of bird groups and species unimaginable to Charles Darwin, Russel Wallace, their contemporaries, and even the next generation. A result of this is that sexual selection for male traits by females has been time and time again demonstrated to be as powerful a pressure on males as Darwin thought it to be, in both polygynous and monogamous bird species.[34] The subject remains, however, a highly complex one.

The recent ability of biologists to genetically "fingerprint" individual birds has resulted in the need to redefine the simplistic application that Darwin gave to the words "monogamy" and "monogamous."[35] In his day, and for long after, these words implied that in bird species so assigned, the sexes formed a firm and devoted pair-bond for at least one breeding season, if not many successive ones, to raise their own offspring together. Many recent studies employing genetic fingerprinting have shown, however, that offspring in the nests of such species are very commonly not all fathered by the attending, pair-bonded, male "parent."[36] Thus, females commonly cheat on the male assisting them to defend their nesting territory and raise their offspring. This shocking revelation has resulted in the need for ornithologists to think and write about two distinctly different types of monogamy: genetic and social. A recent review of this remarkable phenomenon reported that pertinent genetic studies of a cross-section sample of the nestlings of "socially monogamous" female birds, previously perceived to be genetically monogamous, demonstrated that more than 85 percent of them were in fact sexually polygamous.[37] That is to say that at least one of their nestlings was not fathered by the male, nest-attending, territory defending, partner parent.

The revolutionary discovery of promiscuity by "pair-bonded" females (and thus, of course, by males too) reproducing within what, as a result, had to become known as socially monogamous pairs has significant repercussions for the Darwinian view of sexual selection in such species. This is because the females also exercise preferential choice as to which extra pair male or males they will mate with in addition to their cuckold nesting partner. This may conceivably express itself in greater sexual selection pressure by females for structural, colourful, or behavioural male traits than in genetically monogamous species. For example, the Australian fairywrens, of the songbird family Maluridae and genus *Malurus*, are species that nest as a cooperative social group involving at least one fully coloured adult male and several additional birds in dull female or female-like plumage. The extra-pair cooperative individuals typically have some kin relationship to the adult breeding pair. Nesting, egg-laying, female fairywrens have been known for some time now to be sexually polygynous, with up to 95 percent of broods in nests including at least one nestling not fathered by the attending bonded male partner or other members of the cooperatively nesting group.[38] The gloriously plumaged adult males of these fairywren species often add to their colourful courtship displays by seeking out and holding in their beak a brightly coloured flower petal to show to females. In some species, the flower petal colour contrasts with that of the male's plumage; whereas in others, it compliments it.[39] It has been suggested that the extra-pair copulations may be behind the evolution of this remarkably novel petal-using courtship behaviour.[40] Given that extra-pair copulations by nesting females of socially monogamous birds continue to be found to be typically practiced in a rapidly growing number of species, its potential significance with regard to sexual selection by females and the sexual dimorphism of each species has to be considered. What Darwin described as the difficult subject of sexual selection and dimorphism in birds remains difficult and is now far more complicated than he could ever have imagined possible. Indeed, the extent of extra-pair-bond paternity in birds is recently found to be correlated with the degree of sexual dimorphism in plumage colouration brightness within species.[41]

A fundamentally important historical and theoretical extension to the understanding of the power of sexual selection since Darwin is its potential to influence the evolutionary process of speciation. Basically, this is affected by the females of a population selecting for male morphological traits that eventually diverge sufficiently from those of the males of adjacent populations to bring about genetic isolation, and thus speciation.

THE EXPRESSION OF THE EMOTIONS IN MAN AND ANIMALS

Charles Darwin's book *The Expression of the Emotions in Man and Animals* first appeared in 1872, the year following publication of his previous work. As can be imagined from its title, it contains few, mostly brief, mentions of birds. What it does include can be comprehensively found in appendix 1, and I elaborate here only on any more significant bird accounts therein.

Of the way in which hatchling domestic chickens pick small food particles from the ground, Darwin cites a simplistic experiment involving the making of a noise with a fingernail on a board to imitate the sound of a feeding mother hen. The resultant response by the hatchlings suggested that the sense of hearing stimulated them to first peck at food. Unfortunately, CD did not state if the hatchlings could see the finger scratching the board, which might have visually stimulated them to peck rather than, or together with, the sound produced.[42] In fact, domestic fowl chicks instinctively peck at small items on the ground and thus find food for themselves, but they will also be stimulated to do so by the foraging actions of conspecifics.

Darwin then gave an instance of what he surprisingly calls "an habitual and purposeless movement." He observed that when a Common Shelduck (*Tadorna tadorna*) came across a worm cast on sands exposed by a receding tide, it patted the ground with its feet "dancing, as it were, over the hole." This action makes the worm come to the surface, as CD observed, and is thus decidedly purposeful. He reported that when tame shelduck came to their keeper seeking food, "they patted the ground in an impatient and rapid manner"; and he insightfully stated that this "therefore may almost be an expression of hunger" (or at least of anticipated immediately imminent food being forthcoming). An informant told him that flamingos and the Kagu (*Rhinochetus jubatus*, of New Caledonia) perform the same action when eager to eat.[43] It is probable that many birds that feed on mud-dwelling worms and the like will behave in the same, purposeful, way.

Kingfishers always vigorously beat fish they have caught on a stout perch to stun or kill them before swallowing them. When fed in an aviary, they do this to food items even when, in the case of mincemeat, it results in the loss of the food as it flies apart. This is of course an example of birds performing what is under natural, wild, conditions a beneficial behaviour resulting from natural selection—because swallowing living spiny fish or other so armed prey could result in injury or death.[44]

Darwin acknowledged birds typically adpress, or sleek, their plumage in fear, making them appear smaller; but in anger, they erect, or fluff, their feathering, making them appear larger. He illustrated this nicely in the case of a Mute Swan (*Cygnus olor*) driving an intruder away.[45]

Of the remarkable "Secretary-hawk," or Secretarybird (*Sagittarius serpentarius*), as this bird of prey is known today, Darwin noted that it has "its whole frame modified for the sake of killing snakes with impunity." In this it is primarily the enormous elongation of the legs and their heavy scalation, and that of the feet, and their claws, that are modified to this purpose and specialized diet.[46] This is indeed a remarkable African raptor that has become highly adapted to ground foraging and snake killing. It has legs that are more stork-like than hawk-like, their length keeping the bird's body well out of reach of striking serpents. Its handsome pale head is decorated with some half a dozen elongated, narrow, spatulate, black feathers that project backward from behind each ear region. These are reminiscent of early pen

Figure 7.2 An adult Mute Swan illustrated by Darwin to show it "driving away an intruder." From *The Expression of Emotions in Man and Animals*: opposite p. 100, by Darwin, 1904.

quills, often worn behind the ear of people when not in use, and bring about the bird's otherwise inexplicable name (see colour figure 7.3).

THE EFFECTS OF CROSS AND SELF FERTILISATION IN THE VEGETABLE KINGDOM

In 1876, Darwin's *The Effects of Cross and Self Fertilisation in the Vegetable Kingdom* was published, again by John Murray of London. In this book appears a paragraph in which CD cited long- and short-beaked hummingbirds feeding on the nectar of plants and, indeed, even having beaks adapted to some specified plant genera. He also recorded personally seeing *Mimus*, or mockingbirds, with pollen on their heads and cited an observation of "Nectarinidae," or sunbirds, also of the songbirds, fertilizing *Strelitzia* plants at the Cape of Good Hope. He did not doubt that many Australian flowers are fertilized by "honey-sucking birds"; and he reported that Alfred Russel Wallace often saw the beaks and faces of brush-tongued lories, which are nectar-eating parrots, with pollen on them in the Moluccas. Darwin cites an observation given to him in New Zealand that many *Anthornis melanura*, or New Zealand Bellbirds, had their heads covered with pollen of a native Fuchsia.[47] Although novel to Darwin at that time, this bellbird is actually a type of honeyeater, of the predominantly Australian family Meliphagidae, and as such it takes nectar from a wide range of plant species, native and introduced.

Some of the content of *The Descent of Man*, and of the subsequent *The Expression of the Emotions in Man and Animals*, has resulted in Charles Darwin being considered the father of the study of animal behaviour.[48] In particular, Oskar and Magdalene Heinroth, Henry Eliot Howard, Julian S. Huxley, Frederick B. Kirkman, Konrad Lorenz, Edmund Selous, and Niko Tinbergen would presumably have been inspired by Darwin's observations and interpretations of bird behaviour to become eminent early pioneering students in that field of study. The 1872 *The Expression of the Emotions in Man and Animals* was the last book by Darwin that contained any, if but little, ornithology. His last publication on birds was, however, a brief 1881 scientific paper on the parasitic habits of some cowbirds and cuckoos.

NOTES

1. E.g., Wallace 1868, 1889.
2. See Benton 2013: 107–29.
3. Fisher 1930, 1958.
4. Fisher 1930; Amotz Zahavi 1977; Hamilton and Zuk 1982; Amotz Zahavi and Avishag Zahavi 1997, and references therein.
5. Darwin 1901: 676–84.
6. Wallace 1889: 274–87; Darwin 1901: 689–90; Benton 2013: 110–24.
7. Lambert and Woodcock 1996; Erritzøe and Erritzøe 1998.
8. Bennett and Owens 2002: 130, and references therein.
9. E.g., Cuthill 2006.
10. E.g., Bennet et al.1996; Bennet and Owens 2002, and references therein.
11. Bennett and Owens 2002: 155.
12. Darwin 1901: 676–706.
13. Frith and Frith 1997; Frith and Beehler 1998.
14. In Burkhardt et al. 1985–2014, vol. 16: 785, but cited by CD as Wallace's 1869 *The Malay Archipelago*, vol. 2: 394.
15. Frith and Beehler 1998.
16. E.g., Bennett and Owens 2002: 138–50.
17. Lane and Rogers 2000; Christidis and Boles 2008; del Hoyo and Collar 2014.
18. Darwin 1901: 727.
19. Marchant and Higgins 1993; Kirwan 1996.
20. Cf. Ligon 1999: 224, 432–3.
21. Erritzøe et al. 2007: 203–4.
22. S. J. J. F. Davies 2002.
23. Hancock et al. 1992; del Hoyo and Collar 2014: 396.
24. Ligon 1999: 87.

25. Irwin 1994.
26. E.g., Arcese 1989.
27. Ligon 1999: 88.
28. Cramp 1988; Collar 2005; J. Erritzøe personal communication, email January 2015, J. Erritzøe–C. B. Frith.
29. Darwin 1901: 745.
30. Darwin 1901: 707–62.
31. Darwin 1871.
32. Wallace 1889: 295.
33. Bennett and Owens 2002: 130.
34. M. Andersson 1994.
35. E.g., Barash and Lipton 2001.
36. Black 1996, and references therein.
37. Bennet and Owens 2002: 79, and references therein.
38. Brooker et al. 1990; Mulder et al. 1994; Rowley and Russell 1997; Cockburn et al. 2013.
39. Rowley and Russell 1997: 75.
40. Rowley 1991: 79.
41. Møller and Birkhead 1994; Bennett and Owens 2002: 147.
42. Darwin 1904: 50.
43. Darwin 1904: 50.
44. Darwin 1904: 50.
45. Darwin 1904: 98–100; figure 7.2.
46. Darwin 1904: 111.
47. Darwin 1876: 371.
48. Birkhead et al. 2014: 248.

CHAPTER 8

Charles Darwin as a Fully Fledged Ornithologist

With such moderate abilities as I possess, it is truly surprising that I should have influenced to a considerable extent the belief of scientific men on some important points.
[CD in F. Darwin 1950: 71]

What Charles Darwin did, said, and wrote as a boy and a young man, up to his sailing from England's green and pleasant land aboard the HMS *Beagle*, leaves no doubt that he saw the study of birds as an interesting part of the natural sciences. He wanted to be a part of birds and their study, as an ornithologist. His intensive and extensive outdoor walking and sporting activities in rural England and Wales as a youth provided him with ample opportunity to observe and think about birds. Darwin clearly had an appreciation of them in the field that was as deep and keen as most naturalists of his age and day. As a late teenager, he asked himself probing questions about the habits of British birds, and he employed someone to teach him how to skin birds for scientific purposes in his own time and at his own expense. He contributed to a public discussion about the habits of Common Cuckoos as a mere 17-year-old. Several of his major books include substantial descriptive, biological, behavioural, and theoretical ornithological content. As a 67-year-old, he recalled his attending a lecture by the ornithologist and great bird artist John James Audubon, when he was about 18 years of age, as being worthy of mention in his brief autobiography.[1] At 72 years old, Darwin published, in the year before his death, a scientific paper on the biology of parasitic American cowbirds.[2]

THE BIRDS OF THE *BEAGLE* VOYAGE

As is readily apparent from the content of chapters 2, 3, and appendixes 1 and 3, Charles Darwin collected or made field notes on an impressive list of living bird species (approximately 180), groups, and ornithological topics during the voyage of the *Beagle*. He would of course have observed a great many more species in the wild than were collected as specimens or documented as closely observed by him. Many of his resulting written accounts of collected species included descriptions of the distribution, appearance, vocalizations, habits, nesting, diets, ecology, and behaviour of birds that he saw living in the wild. From his very first encounter with land birds beyond England (he would have seen seabirds while initially at sea), on the Cape Verde Islands in the Atlantic Ocean, CD would refer to bird species familiar to him in Britain in making his comparative observations with those new and exotic to him.

On the South American mainland, Darwin got down to some serious bird watching, although his unfamiliarity with the endemic bird groups there led to a few identification problems for him, such as recording seeing a crane and bee-eaters, birds that do not occur there. Although he was a keen student of beetles from his youth, his notes on zoology as a whole indicated to one authoritative Darwinian author that during his *Beagle* voyage, CD "was possibly even fonder of birds than of the beetles."[3] Although Darwin was expected to obtain bird specimens and record their collecting locations, he was presumably not "officially" required or expected to record observations on their biology, ecology, or behaviour. That he did so has to be seen as reflecting his personal interests and inquiring ornithological mind.

In South America, Darwin did make some insightful avian distributional—and early and pioneering, ecological, and behavioural—observations. He gave some particularly thoughtful considerations to the rheas, raptorial birds of prey, vultures, skimmers, waterfowl, seedsnipe, hummingbirds, cinclodes, ovenbirds, tapaculos, mockingbirds, American blackbirds, and species of a good number of other avian groups on that continent. His first publication upon returning to England recited observations and thoughts on the two flightless rheas. One rhea species was previously formally known, and the other, although noted to exist in a travel book, was first collected for the biological sciences by Darwin on the voyage and thus first established formally as a new, and the second, species of rhea.

In the Falkland Islands and the Galapagos Archipelago, Darwin observed the remarkable tameness of birds, most notably of the Galapagos Dove, and collected similarly tame mockingbirds and finches from various islands.

He shortly thereafter gave highly significant thought to what the various Galapagos island forms of mockingbirds, but not of the finches, might indicate with respect to the origin of species. Some observations made in his field notebooks, as opposed to appearing in his publications, reflect the fact that he was formulating significant and sophisticated initial insights into the zoogeography and broad ecology of the islands of the Galapagos Archipelago and its avifauna.

It is a little surprising that Darwin did not collect a specimen of the Galapagos Penguin (*Spheniscus mendiculus*), which is endemic to that archipelago and is quite distinctly different to the Magellanic Penguin (*Spheniscus magellanicus*) that he had observed on the Falkland Islands and, presumably, on the South American coast. As he failed to make any mention of the Galapagos Penguin in his notes and publications, it is probable that he and others on the *Beagle* did not happen to see it, despite the fact that it does occur on the northern coast of Indefatigable (or Santa Cruz) Island, which the *Beagle* visited, although most live on the western islands within the archipelago. This interpretation of this negative result of his visit is supported by the fact that although CD specifically noted the kinds of birds that he saw on the coast of the islands, in his original ornithological notes at the time, he did not mention a penguin or the Flightless Cormorant.[4] The Galapagos Penguin was not to be described as a new species made known to science until 36 years after Charles visited the Galapagos Islands.[5]

The results of Darwin's bird collecting, observations, and thinking during the voyage of the *Beagle* were presented in his 1839 *Journal and Remarks, 1832–1836*, and his *Journal of Researches into the Geology and Natural History of the Various Countries Visited by H.M.S. Beagle*, and in the volume on birds of the 1838–1841, *The Zoology of the Voyage of H. M. S. Beagle*, coauthored with John Gould;[6] see figure 8.1. A good deal of his field bird observations in the former publications were repeated in the latter one.

Darwin noted in *The Zoology of the Beagle* that although the marine frigatebirds apparently never touch the sea, they nevertheless have their toes webbed to some extent.[7] He was to discuss this obvious anomaly almost four decades later in his *The Origin of Species*. In the first edition of his *Origin*, he also observed that "He who believes that each being has been created as we now see it, must occasionally have felt surprise when he has met with an animal having habits and structure not at all in agreement. What can be plainer than that the webbed feet of ducks and geese are formed for swimming? Yet there are upland geese with webbed feet which rarely or never go near the water."[8] The observation of a small thing like the redundant webbing between toes of birds that no longer habitually swim was to have a large impact on the young Charles Darwin, with respect to the morphological modification of animal species as a result of changes in their biology and ecology over geological time.

Figure 8.1 A portrait of John Gould at the age of 45, by T. H. Maguire, 1849.

Darwin used the word "created" several times in his works, and of this he wrote to Joseph Hooker in March 1863: "I have long regretted that I truckled to public opinion, and used the Pentateuchal [*sic*] term of 'creation', by which I really meant 'appeared' by some wholly unknown process."[9] An authoritative student of Darwin interpreted this to mean that he was applying this regret to his published work back to "the *Voyage of the Beagle* and his dawning hypothesis of those early years."[10] Darwin's documented field observations on living wild bird habits and behaviour during the voyage of the *Beagle* are impressive, particularly so when right up to the publication of *The Origin of Species,*[11] ornithology "still largely consisted of shooting [birds] and collecting [their] skins."[12] Darwin, unlike most collectors of his time, often noted food items eaten by birds he shot as a result of their dissection, something a good number of modern collectors inexplicably, and inexcusably, fail to do.

CHARLES DARWIN AND JOHN GOULD

Following three and a half months spent working at Cambridge after his voyage, Darwin moved to London on 6 March 1837, and wasted no time in meeting John Gould during his very next week there. Gould had by then given some of his time over two preceding months to examining, describing, and naming Charles's Galapagos Islands bird specimens. Of particular value and significance was Gould's appreciation of most of Charles's collection of small Galapagos songbirds as representing 13 member species of a single new group of finches. Gould's taxonomic views enabled CD to perceive the finches as a single radiation of closely related species found throughout the Galapagos Islands, Gould having apparently pointed out that among them, "The bill appears to form only a secondary character."[13] The generally similar plumages contrasted by the obvious diversity of beak sizes and shapes exhibited among what we now know as Darwin's finches had, however, completely misled Charles into thinking that they must have represented a number of distinct bird families, until corrected by Gould; see "The Significance of Darwin's Finches to Darwin's Thinking" following.

As a result of Gould's advice, Darwin could appreciate that the finches reflected the very same evolutionary story that he had come to strongly suspect that the mockingbirds of that archipelago had indicated to him (see the following). Thus, contrary to widespread popular misconception, Darwin did *not* appreciate that the finches told a clear story of speciation *during the voyage*, and indeed he did not do so back in England until he had partly developed his thinking about the mutability of species. Only then was he able to appreciate their great significance. Contrary to what he then thought, no species of Galapagos finch that he had collected on the four islands he visited is (at least not now) restricted to a single island within the archipelago. Indeed, up to ten finch species are now known to coexist on a single island, with the larger islands unsurprisingly supporting more species than the smaller ones.[14]

Darwin's immediate post-*Beagle* relationship with John Gould "the bird man" was extremely important in enhancing his appreciation of ornithological taxonomic issues (if to a limited extent) and writing; and he was clearly appreciative of Gould's timely, if hastily provided, expertise and contributions to *The Zoology of the Voyage of the Beagle*. Gould's contribution was rushed, and not quite completed, because, as was previously noted, he was preparing for his prolonged expedition to Australia. Following his departure, Darwin added to Gould's text where he deemed it necessary, often substantially, and sought the help of George Gray with outstanding taxonomic and nomenclatural issues as detailed in chapter 3.

Although John Gould was a more experienced and much better avian taxonomist than Darwin was ever to become, Darwin was nevertheless quite a competent one; and he did express opinions contradictory to those of Gould, and on at least one occasion proved to be correct. After all, he had managed to classify many of the birds seen and collected during the *Beagle* years to order, family, genus, and even to species level with no more than his expertise of British and other birds gained prior to his leaving England, the doubtless limited knowledge of his fellow voyagers, and the literature available to him aboard the vessel.[15]

It was disappointing to Darwin that John Gould failed to commit to his views on the origin of species by natural selection in his subsequent high-profile bird books. In fact, Gould went to some lengths to avoid doing so, and not merely sitting on the fence in this regard. In the introduction to his great 1861 work on the hummingbirds, printed two years after *The Origin of Species* appeared, Gould wrote of those lovely birds, who "would not pause, admire, and turn his mind with reverence towards the Almighty Creator, the wonders of whose hand we at every step discover?"[16] Additional Gould text, some of it stating that the beauty of hummingbirds had no purpose but ornament and was in no other sense useful, impressed the long-serving librarian at The Royal Society, Charles Weld. In reviewing the hummingbird monograph for a magazine, Weld thus enthusiastically noted Gould's "strong antagonistic bearing to the Darwinian theory."[17] He was not alone in enjoying John Gould's reprobation of Darwin's evolutionary view of the plumage of adult male hummingbirds.

Upon reading the introduction to the Gould hummingbird book, in the copy presented to him by its author, Darwin responded in writing to Gould on 6 October 1861. He thanked Gould and pointed out the problem that the numerous cases of apparent variation or "races," probably subspecies, of doubtful species of hummingbird represent with regard to the difficulty of defining species.[18] He later went on to use Gould's hummingbird monograph to discuss and illustrate his natural and sexual selection theories, most notably in his *The Descent of Man*. Darwin also cited, perhaps with tongue in cheek, Gould's magnificent paintings of the hummingbirds in applying his analogy of artificial selection, observing that just as it has brought about the strikingly different breeds of domestic pigeon, so sexual selection by females has brought about the diverse beauty of the adult males of species within the hummingbird family.[19]

At the end of April 1864, the great ornithologist Alfred Newton wrote to his brother Edward about John Gould, five years after the publication of *The Origin of Species*: "It is most amusing to see how anxious he is to avoid committing himself about Darwin's theory. Of course, he does not care a rap whether it is true or not—but he is dreadfully afraid that by prematurely espousing it he might lose some subscribers, though he acknowledged to me the other day he thought it would be generally adopted before long." A great authority on John Gould observed of him "in all the Gould letters I have read, he rarely makes any comment on postulates, politics or personalities."[20] Clearly, Gould was a sharply focused entrepreneurial ornithologist and publisher who saw rocking any boats in evolutionary waters as poor, if not dangerous, politics and bad business with respect to his all-important contacts with conservative wealthy people in high places.[21]

John Gould as an Anti-Darwinian

In his recent fascinating book *Charles Darwin and Victorian Visual Culture*, the eloquent Jonathan Smith mounts an extraordinarily imaginative, partly contradictory, hypothesis under his subheading "*The Birds of Great Britain* as a visual response to Darwinism." This refers to John Gould's book published during 1862 to 1873. In essence, Smith argues that in showing many pairs of British birds at their nest with eggs or young, but supposedly including little sign of any "violence and competition," Gould "sought to offer a picture of the natural world very different from the one offered by Darwin" by presenting such "domestic" or "family" scenes.

Although Gould makes it clear in his text that he sees the hand of God the Creator at work in nature, it is difficult to follow the textual contortions and contradictions that Smith presents in drawing his very long bow. He writes of *The Birds of Great Britain* as being "a celebration of avian family values, of domesticity and parental nurturing. Numerous plates depict nestlings being fed. Others show parent birds protecting, incubating, or simply hovering over their offspring. The depiction of a male and female on each plate here creates a nuclear family, usually with an ornithological equivalent of 'separate spheres'—the female attending directly to the young while the male, having obtained the food, stands by."[22] It is an original and novel interpretation of Gould's illustrations and his, entirely hypothetical, motivations for their content as Smith imagines them to be; but is it sufficiently critical of all pertinent facts? The relationship between Charles Darwin and John Gould as ornithological colleagues is highly germane to the context of this book. However, I deal with other pertinent, and in my view controversial, issues raised by Smith's interpretations of Gould's work as anti-Darwin propaganda elsewhere to limit distraction here; see appendix 4.

John Gould's participation and expertise during the writing of the bird text for *The Zoology of the Voyage of the Beagle* was of great value to Darwin, as were his broad ornithological knowledge and his publications. Finely illustrated large bird books by Gould that were available to Darwin during the production of one of more of his own books included text and visual material of value to him and his work.[23]

Gould, having departed London on 16 May 1838, and reached Hobart in Tasmania on 18 September, departed Sydney on 9 April 1840, after nearly 20 months spent in Australia studying the birds there for his multivolume work on the avifauna of that continent. He reached England again on 18 August 1840. From the day of his return, Gould had a great deal to keep him fully occupied with the production of his own publications, on mammals as well as birds, and their marketing. Because Darwin's need for advice on bird taxonomy was by then far more limited, contact between the two men appears to have subsequently been infrequent, and as social as it was scientific. Gould went on to subsequently publish several other works but not in time for Darwin to use them for his own book contents.

CHARLES DARWIN AND ALFRED RUSSEL WALLACE

A fundamental error that Darwin made while in the Galapagos Archipelago was his failure to record from which particular island each of some of his bird specimens was collected. This, as is recited in chapter 3, was to prove critically unfortunate to him and to John Gould (and subsequently to others) in their attempts to sort and define the various species of Galapagos finches and their distributions for their volume on birds for *The Zoology of the Voyage of the Beagle*. Darwin could not, however, have been expected to then foresee that each of such similar, recently formed, closely adjacent, volcanic islands, located as an isolated, mid-ocean archipelago, might each house a distinctive species or more of a single family of similar and closely related small, volant songbirds.

Alfred Russel Wallace, on the other hand, was likely to have been well aware of the potential seriousness of inadequately recording specific collecting locations among adjacent islands that he collected from; see colour figure 8.2. He would presumably have noted this problem as having been detailed as a serious shortcoming in *The Zoology of the Voyage of the Beagle*.[24] Indeed, he is documented as writing "From my first arrival in the East I had determined to keep a complete set of certain groups from every island or distinct locality which I visited for my own study on my return home, as I felt sure they would afford me very valuable materials for working out the geographical distribution of animals in the archipelago, and also throw light on various other problems."[25]

Wallace departed England aboard ship bound for Singapore on 4 March 1854, and was to subsequently spend the eight-year period of April 1854 to April 1862 travelling extensively about the, then, Malay (now Indonesian) Archipelago. He explored from Singapore in the west to the Aru Islands and to Waigeo Island off New Guinea as his most easterly location.[26] He is reported to have collected some 310 mammals, 100 reptiles, 8,050 birds, 7,500 shells, almost 110,000 insects and a total of some 126,000 natural history specimens.[27] For clarity, Wallace, at the start of the Preface to his *The Malay Archipelago*, states that he found himself upon his return to England surrounded by "nearly three thousand bird-skins, of about a thousand species."[28] This figure refers to the specimens that he retained for his personal studies; the other 5,050 specimens having been disposed of for profit by his agent Samuel Stevens.

As Wallace was meticulous in documenting the specific islands from which his zoological specimens were collected, he was able to record, as he predicted, a great deal of fundamental importance with respect to the zoogeography of the area. His extensive data, for birds and other groups, enabled him to contribute a great deal of original thinking about the significance of the geographical distributions of animal life across that vast Malay Archipelago of, seemingly, countless islands—in fact, some 13,000, but of which he of course visited but only a tiny fraction.

By far Wallace's best known and most significant single finding in this regard was the defining of what was to become widely known as a major zoogeographical barrier. This he plotted as a line following the north–south aligned deep-water boundary between several close islands of the, now, central Indonesian Archipelago. Their closeness notwithstanding, these islands have the discrete Indo-Oriental and the Australasian faunas predominating to the west and to the east of his line of demarcation, respectively.[29] That a broad ecological difference, most obvious in climate and vegetation, existed either side of an imagined line in this very area was initially noted by Salomon Müller.[30] A clear, but not exclusive, geographical dichotomy of bird life in the area, of what was to subsequently become called "Wallace's Line," had, however, been noted in a broad and more general way before Wallace—by fellow Englishman and ornithologist Philip Sclater.[31] It was an important finding for Darwin to learn of, as a far better-detailed and confirmed phenomenon, in correspondence with and also in a paper published by Wallace.[32]

In his personal "Species Notebook," complied during 1855 to 1859,[33] Alfred Wallace mentioned the Galapagos Islands, presumably with reference to Darwin's specimen collecting there, and therein correlated reproductive isolation with speciation.[34] While still travelling among the eastern islands, Wallace published a number of significant papers on the distribution, morphology, and habits of birds of various parts of the Malay Archipelago. These, together with the ornithological content of letters that Wallace sent to Darwin, were available for Charles to consider and discuss in his own books.

Of even greater value to Darwin's, more theoretical, considerations were Wallace's insightful thoughts and discussions on natural and sexual selection. Indeed, Darwin found Wallace's, often frustratingly contrary to his own, views on his original and controversial theory of sexual selection most challenging. Nevertheless, they were valuable in providing a "sounding board" that enabled CD to see and thus address potentially telling difficult facts and criticisms. These trying but helpful demands on Darwin, to consider and deal with Wallace's contradictory views and facts by having to provide more convincing evidence and argument for his sexual selection theory, were to continue long after Wallace returned to England.[35] Darwin and Wallace remained firm and mutually respectful friends and close colleagues, notwithstanding strong differences of opinion over certain aspects of each other's writings and most particularly those by Darwin on sexual selection and the extent of natural selection on human evolution. Their correspondence and personal meetings were of much significance to the ornithological thinking behind, and content of, Darwin's published works.

BIRD STUDIES FROM DOWN HOUSE

During Darwin's static post-voyaging life, initially and briefly in Cambridge and London and thereafter for 40 years in Down House in Kent, he found himself studying and experimenting on several bird species in detail. The same was done with other animal groups, notably domesticated mammals, barnacles, and earthworms; and plant groups, particularly carnivorous plants, orchids, climbing, and other plants. Additional work was progressing with geology in general and volcanos and coral reefs especially. He observed, read about, and subsequently wrote to correspondents about, recalled, compiled, ordered, and synthesized a truly amazing number of facts about an impressive number of bird species, genera, other avian taxa, and ornithological topics in his garden and study (colour figures 4.1, 8.3, 8.4) over many years.

The abundant footnotes to the 213 pages of the four chapters dedicated to birds in *The Descent of Man* alone include numerous references to literature consulted by Darwin. I cannot review the cited references in all of his four bird chapters here, but a rough count indicates that he cites some 82 different publications and/or correspondents in his first bird chapter, and repeats of 26 of these references (e.g., 15 citations of Jerdon's 1862–1864 book *The Birds of India*) and adds a further 85 to the original 82 citations: a total of 168 citations for the first of those bird chapters alone.

Charles Darwin's expertise as a zoological taxonomist was in large part limited to the barnacles[36] and also, for want of a better way of putting it, to the breeds of the domestic pigeon.[37] He lacked the appropriate specialized knowledge, time, or ambition to gain such expertise in other animal groups. For this, and for plant taxonomy, he therefore sensibly relied heavily on the knowledge and advice of his highly qualified, specialist friends and colleagues. For bird taxonomy, these included John Gould, Philip Sclater, George Gray, and Tom Eyton. Three of his more significant providers of information about living birds were Alfred Wallace; Edward Blyth, particularly on the Indian avifauna for well in excess of a decade; and Leonard Jenyns. Other ornithologists involved are mentioned in chapter 4.

Darwin's major interest in zoology was in the diversity of life forms, variation within them, and the evolutionary processes that brought about species, and the morphology of the sexes and also of the various age classes within them. With regard to the latter, he found the study of bird courtship and associated plumages of the species highly instructive, fascinating, and enjoyable; avian taxonomy being something he used or applied where necessary but did not otherwise engage with. Thus, the content of appendix 1 shows that in a number of instances, CD used different common and/or scientific names for the same species, even within a single publication (e.g., pin-tail duck [*Anas acuta*] on page 599 and pintail duck [*Querquedula acuta*] on page 632 in *The Descent of Man*), but this is hardly surprising given the sum of ornithological literature that he pulled together into his own substantial volumes of complex texts and diverse subject matter.

As noted previously, Darwin was not without firm avian taxonomic opinions. For example, in a letter to Leonard Jenyns of 18 October 1846, he states his interest in the subject having read "the absurdly opposite conclusions of Gloger & Brehm; the one [Brehm] making half-a dozen species out of every common bird & the other [Gloger] turning so many reputed species into one." This observation resulted from him having discussed with Andrew Smith getting together "some hundreds of specimens of larks & sparrows from all parts of Great Britain & see whether with finest measurements he cd [could] detect any proportional variations in beaks or limbs &ct."[38]

Appearing in Darwin's personal letters is the novel idea that "I have sometimes thought that the progenitor of the whole class [Aves, or birds] must have been a crested animal." This notion was stimulated by him observing that so many domesticated birds produce races or breeds "with a tuft or with reversed feathers on their heads"; and the finding of a substantially crested, wild nesting, adult Common Blackbird producing a brood with two crested offspring.[39] Unfortunately, the vast majority of fossil bird remains can tell us little or nothing about what their head feathering consisted of in life.

The Origin of Species Book and Ornithology

The ornithological content of *The Origin of Species* is not inconsiderable, if largely dispersed, throughout the book. The 10 pages on "Breeds of the Domestic Pigeon, Their Differences and Origins" in the first chapter, titled "Variation Under Domestication" (although also briefly mentioned, along with other domesticated birds, on numerous other pages), proved but a mere precursor to two entire chapters on the domestic pigeon in volume 1 of his subsequent book, *The Variation of Animals and Plants Under Domestication*. Similarly, a mere three-page treatment of "Sexual Selection" in the fourth chapter of *The Origin*, citing mostly avian examples, is entirely superseded by four chapters on sexual selection in birds within *The Descent of Man*. The parasitic instincts of the Common Cuckoo and American cowbirds are given six pages of text, and the dispersal of plant seeds and minute animals by birds are given some six pages, in *The Origin of Species*.[40] Otherwise, birds are mentioned as examples during various and numerous discussions of more general topics, represented by approximately 240 entries in appendix 1.

Of course the ultimately revolutionary *The Origin of Species* book resulted in numerous naturalists and biologists looking at birds in a new and bright light. This was thrown on them with respect to their relationships within the group, or class (Aves), as a whole and with how the present living genera and species might have evolved with Darwin's new evolutionary approach in mind. Hot on the heels of the publication of *The Origin of Species*, indeed during the same year, was a paper ironically written by the ecclesiastic Canon Henry Tristram. In this, he observed that the larks and chats of Algeria appear to be representatives of European species that have adapted in their plumage colouration and structure to their desert habitats, and thus well illustrated Darwin's new theory of natural selection.[41] Unfortunately, he subsequently changed his, perfectly reasonable, interpretation of what these birds demonstrated and he became, as a Christian, a vigorous opponent of Darwin's theory.[42]

The first biologist to present an avian evolutionary, or genealogical, tree based on Darwin's origin of species concept was Ernst Haeckel.[43] This was followed by Thomas Huxley, as close a friend to Darwin as anyone, who wrote "On the Classification of Birds."[44] This involved a comparative anatomical review of the major bird groups but did not suggest a comprehensive phylogeny of the birds known at the time. These initial attempts have been followed by numerous classifications, as opposed to phylogenetic trees, of parts of or the entire class Aves over the past century and a half. They have culminated in lists and esoteric diagrammatic phylogenetic studies of them based on the results of highly sophisticated genetic laboratory techniques applied to numerous genera and species of the now known approximately 10,500 living bird species (of which over 350 have been described as new ones since 1946).[45]

A date as recent as 12 December 2014 has been described as "a red-letter day in the history of ornithology."[46] This is because that date saw the simultaneous publication of some 28 scientific papers on the taxonomy of the class Aves based on genomic data sets from 48 species from 32 of 35 recently proposed bird orders.[47] The total genetic work involved was undertaken by over 200 scientists based at 80 laboratories in 20 countries. The result is said to be "a nearly complete understanding of relationships among the world's major groups and the timing of the major events in their evolution."[48] The results are of course extremely complex, but the evolutionary processes that Darwin and Wallace elucidated provided the foundation for our understanding of newly affirmed and many surprising new results of this genetic research.

Suffice it to note here that major surprises include that the vast majority of living birds form three discrete subgroups: one vast group (the Neoaves) separate from two distinct small groups—one consisting of the landfowl (Galliformes) and waterfowl (Anseriformes) and the other of the ratites, including the tinamous, constituting the Palaeognathae.[49] The enormous Neoaves group surprisingly comprises only two fundamental subgroups: the Columbea, which as it name implies includes the pigeons as well as the mesites of Madagascar, the sandgouse, flamingos, and grebes; and the Passerea, which includes all other birds. The order Psittaciformes consisting of all parrots is affirmed to be the non-passerine group closest to the perching bird, or passerine, order Passeriformes.[50] Similarly, it is confirmed that the falcon family Falconidae, now exclusively in the order Falconiformes, is closest to the parrots and passerine birds and is far removed from all other birds of prey (of the order Accipitriformes). Several other resulting perceptions of the closest relationships between several bird groups are also most novel.[51] Charles Darwin would surely have revelled in such advances in our ever-increasingly accurate understanding of the evolution of the birds of the world; underscoring within the science of ornithology the application of his origin of species by natural selection principle.

There can be no doubt that limited though the bird content of *The Origin of Species* is, much of it was novel and original and stimulated further ornithological considerations and study; but the far greater bird content of the subsequent *The Descent of Man* did very much more so. Incredibly, it has recently been estimated that in excess of 380,000 ornithological publications have appeared since the 1859 release of *The Origin of Species*.[52] It would be interesting to know how many thousands, if not tens of thousands, of times Darwin's publications has been cited within these, particularly earlier, works for their ornithological content.

The Significance of Darwin's Finches to Darwin's Thinking

Today we know that the Galapagos, or Darwin's, finches are not finches at all. They are actually tanagers that, having reached the Galapagos Archipelago, adapted to the harsh island habitats over millennia to become superficially very finch-like in appearance. This represents as fine an example as any as to how we can now understand, through the evolutionary processes hypothesized by Charles Darwin and Alfred Wallace, such a scenario could come about. They also provide, as individual species and as a group, a fine example of convergent evolution with some true finches found on islands elsewhere on the planet (e.g., the fodies of the genus *Foudia* of the family Ploceidae in the western Indian Ocean).[53]

The drab little "finches" collected among the islands of the Galapagos Archipelago did eventually become of some significance to the development of Darwin's thinking with respect to the origin of species. This was not to be the case, however, until the late 1830s—and thus well after the voyage of the *Beagle*. The fact is that, contrary to doggedly persistent legend, Darwin remained entirely ignorant of the significance of the Galapagos finches during his voyaging. They did not in any way, shape, or form provide Darwin with an epiphany with respect to natural selection or the origin of species while he was in the Galapagos Islands or during the remainder of his time aboard the *Beagle* (whereas the mockingbirds did; see following). It might also have surprised many to learn, under "Galapagos Finches and Mockingbirds" in chapter 4, that not a single Galapagos finch, or those finches as a group, were so much as mentioned in *The Origin of Species*.

There are far too many publications that have contributed to the legend about some level of significance of the Galapagos finches to Darwin's original thinking about the evolution of species *while he was voyaging* to mention here. Ironically, the several printings of Darwin's second edition of his *Voyage of the Beagle* doubtless misled some unwary readers because text written in hindsight back in England expressing his views on the finches appear under subheadings such as "October 8th" under the page heading "1835" therein.[54] By incorporating things learnt about his *Beagle* birds subsequent to the voyage within a book following the chronology of the voyage, Darwin left room for some confusion (see the following).

A relatively recent high profile and influential publication (e.g., repeated all-but verbatim by Rice in 2000[55]) not only implied that Darwin correlated bill shape of various finch populations to specific differing diets of "nuts and seeds ... insects ... [and] fruits and flowers," but it also stated that "there was even a bird that had learned how to use a cactus spine to probe grubs out of holes."[56] Darwin made no observation while in the Galapagos Islands about diverse finch diets: To the contrary, he saw all of the species as simply eating "seeds" together. He failed to attribute any significance in the diversity of bill size and shape of the various forms of the birds to their food and feeding ecology; although he did record seeing finches eat bits of succulent leaves, possibly for their water content, and one eating the flesh and picking the flower of a cactus (colour figures 3.7, 3.10). Nor did he allude to what is now known as the Woodpecker Finch using a spine as a tool to feed on wood-dwelling grubs. In fact, the tool-using behaviour of the latter bird was not discovered until 84 years after Darwin visited the Galapagos.[57]

Another prominent recent example of the perpetuation of the legend appears in a major bibliography of Charles Darwin. In this work, statements are attributed to CD's *Journal of Researches* text without it being pointed out that they were made by him with the benefit of hindsight, that is, after the *Beagle* voyage. It is stated that CD tells his readers of the Galapagos finches that "What differentiated one from another was the shape of the beak, and there was a direct correlation between beak shape and diet. There was the heavy beak of *Geospiza Magnirostris* [Large Ground Finch], which fed on large seeds; there was the longer, curved beak of *Camarhynchus pallidus* [Woodpecker Finch], reminiscent of the woodpecker's and signifying its insectivorous, tree-climbing habits (it also shared with its cousin *C. heliobates* [Mangrove Finch], the habit of using a twig or cactus spine to poke hidden insects out into the open); there was the narrow pointed beak, like a warbler's, of *Certhidea olivacea* [Green Warbler-Finch], which happily pecked up and ate the tiniest insects." These are the words of the biographer Peter Brent[58] and not of Charles Darwin, and they contain several significant errors: Darwin did not observe any such correlation between finch bill shapes and sizes and their diets; he did not allude to the Large Ground Finch feeding on large seeds; the beak of the Woodpecker Finch is not in any way reminiscent of the beak of any woodpecker, and it cannot, in and of itself, signify any tree-climbing habit; and there is no record of CD observing or recording the Warbler Finch eating insects of any size.

Another influential book adds to the Galapagos finch legend by erroneously stating that, having examined specimens of them, Darwin "asked himself why these finches, which are broadly similar, vary enormously in bill size and shape. He concluded that the birds had developed different diets according to the food available on their islands, and by natural selection had adapted accordingly."[59]

Great Britain's Royal Mail further enhanced this misleading association between the great naturalist and the finches by issuing a 26-pence postage stamp in 1982 with a head of a Warbler Finch and a Large Ground Finch either side of a portrait and signature of an (inappropriately) elderly Charles Darwin; see colour figure 8.5. It would arguably have been more appropriate to have had domestic pigeons frame a portrait of him as a younger man, such as he appears in colour figure 1.1.

Charles had, well after sailing from the Galapagos, come to an inkling that differences in the appearances of the mockingbirds from several different islands there suggested something of great importance about the evolution of species. Having noticed their difference, he had labelled his mockingbird specimens as coming from specific islands, unlike his finch specimens. The fact that Darwin referred to the island forms of mockingbird as varieties rather than species has been emphasized as significant by some authors. It is, however, quite irrelevant to the fundamentally crucial observation he was making about their divergence reflecting adaptation in island isolation. However, he gained no such insight from the small songbirds that at the time he wrongly considered to consist of some finch species and members of several other different bird families. He made this error because the diversity among the, similarly plumaged, finches in their beaks was so marked that it blinded him to the reality that they represented a single group of extremely closely related species. This was because all of his previous experience, and that of most other naturalists of his time, indicated that members of closely related, small songbirds, particularly those of a single genus, typically have very similar beaks. The view has been expressed that "So anomalous is this condition that an ornithologist basing his classification upon the customary relationship between beak and plumage would unhesitatingly place Darwin's finches in at least six or seven genera, and perhaps even several subfamilies";[60] and this is what Darwin, quite understandably, did think upon first collecting them. It took John Gould's experience and insights, back in England, to make Darwin appreciate this, as is detailed under the heading

"Charles Darwin and John Gould" previously. Further insights into Darwin as an avian taxonomist can be found in his personal notebooks and correspondence, but these fall outside the scope of this work.

It was only at this point in time, in England, that Charles became aware of the important evolutionary history of speciation that these little finch-like birds reflected in their island distributions and discrete beak shapes and sizes. As if instantly illuminated by a blinding flash of light, the finches now mirrored the evolutionary principles that he had been groping towards in thinking about what the variable Galapagos mockingbirds (and tortoises) were indicating to him about speciation. Thus, in the corrected second edition of his *Journal of Researches*, Darwin could write so much more insightfully of the Galapagos finches; but even then, he did so in less than 30 short lines of text. The significance of his latter observation and of that group of birds to subsequent ornithology, right up to today, is emphasized by decades of intensive studies of, and a great deal of literature about, them.

Darwin's failure to expand on the finches as presenting such a fine example of his theory may have, at least in part, been due to the fact that John Gould's taxonomic perception of them as a morphologically diverse single group, notably in beak structure, of very closely related species was seriously doubted by most of his contemporary ornithologists. Much later, however, anatomical studies confirmed Gould's treatment of them as constituting a single group,[61] as have several subsequent genetic ones.[62]

The misleading legend about Darwin appreciating the evolutionary significance of the finches prior to Gould's taxonomic review of them was unfortunately generated by the erroneous histories of several authors, as was reviewed by Frank Sulloway.[63] This legend was also enhanced by David Lack's excellent and influential book in which he unfortunately opened his Preface with "Charles Darwin collected some dull-looking finches in the Galapagos Islands. They proved to be a new group of birds and, together with the giant tortoises and other Galapagos animals, they started a train of thought which culminated in the Origin of Species, and shook the world." The undiscerning reader might miss the fact that Lack clearly states elsewhere in his book that Darwin's appreciation of the importance of the finches was retrospective to his *Beagle* voyage. Moreover, in being titled *Darwin's Finches*, the book also potentially enhances a false close association between the finches and Darwin's voyage and original theory. In giving his book this title, Lack did note that the alternative "Galapagos Finches" was inappropriate because the Cocos Finch, undoubtedly a member of the group, lives quite remote from the Galapagos Islands.[64]

The finches of the Galapagos did not, then, inspire Darwin's initial thinking about evolution and the origins of species, nor did they do so subsequent to 1837 when he first became an evolutionist.[65] It was in July of that year that he commenced his notebooks on the "Transmutation of Species." It was instead with the development of his increasing understanding of evolutionary processes that he came to see, in retrospect, the graphic example of speciation that the finches represented. Even then, he failed to mention the Galapagos finches in his pertinent notebooks on speciation and related topics.[66]

Darwin outlined exactly just such a theoretical evolutionary modification of the beak of a bird, such as the Galapagos finches exhibit, in a letter to Charles Kingsley. He wrote from Down House on 29 April 1867: "When speaking of the formation for instance of a new sp. [species] of Bird with long beak Instead [*sic*] of saying, as I have sometimes incautiously done a bird suddenly appeared with a beak [particularly] longer than that of its fellows, I would now say that of all the birds annually born, some will have a beak a shade longer, & some a shade shorter, & that under conditions or habits of life favouring longer beak, all the individuals, with beaks a little longer would be more apt to survive than those with beaks shorter than average."[67] Remarkably, these are exactly the scenarios that Peter and Rosemary Grant and collaborators actually came to witness with respect to the fundamental change in bill morphology of wild Medium Ground Finches on Daphne Major island in the Galapagos. In this instance, it was brought about by drought conditions during but a few consecutive years, which reduced the species population from some 1,200 to a mere 180 individuals, rather than over the geological time spans than Darwin had in mind.[68] Although CD observed that the various finches ate seeds, he, unlike the Grants team of modern field biologists, could not appreciate that each species specializes in particular species, shapes, and sizes of them.

The Variation of Animals and Plants Under Domestication

Charles Darwin was more sensitive than any other person could possibly be to the furore that his fundamental theory behind *The Origin of Species* would bring about in society worldwide, particularly given his earlier ecclesiastical ambitions and his devoutly religious wife, associates, and other contemporaries. He was, however, also well aware that there existed a significant proportion of people globally that understood, or could be easily brought to understand, the power of artificial selection on domesticated animal and plant life. Thus, he saw that bringing about an appreciation in the mind of his readership, if only in a fundamental way, of what artificial selection had done and was continuing to do to domesticated life forms worldwide would be the ideal precursor to his theory of natural selection in nature. He wisely chose to use this highly significant appreciation of what this kind of selection had to date brought about in peoples' farmyards, pastures, gardens, greenhouses, pigeon lofts, aviaries, fish tanks, and the like to precede the introduction of his notion of natural selection in his *The Origin of Species*. Once he had opened this door ajar to understanding artificial selection in this book, he could go on to open it wide within his subsequent two volumes on *The Variation of Animals and Plants Under Domestication*.

Much of the major subject matter, if not the detailed content, in Darwin's somewhat daunting 1868 work, *The Variation of Animals and Plants*, originated as material he had gathered together as part of the evidence he intended to present in his work on the origin of species. As I have related, he suddenly found his quiet and settled world of ongoing research, thinking, and unpublished writing violently shaken to its foundations. This was caused by the arrival of a manuscript from Alfred Russel Wallace who had coincidentally come to appreciate the process of the origin of species through natural selection. Consequently, CD's usual careful and considered, tentative, pace had to be replaced by the absolutely frantic production of a brief outline of his long-gestating theory so that it might be read simultaneously with the manuscript by Wallace at a scientific meeting in London; see chapter 4.

After this intense and exhausting scare, Darwin clearly saw the need to get a polished presentation of the substantial evidence that he had accumulated for his natural selection theory published. It was intended to be but a "sketch," but it grew in the writing to what became his 1859 *The Origin of Species*. Notwithstanding its length, of some 500 pages in the first edition and 700 in the sixth, this book contained only a part of what CD had intended its content to be. Given his hasty need for a mere "sketch" of his theory and the evidence for it, he found that he could include only a single chapter, the first, on "Variation under domestication." Thus, his two volumes on *The Variation of Animals and Plants Under Domestication* appeared nine years after *The Origin of Species* and contained the results of a great deal more consideration of, and experimentation on, the subject of his opening chapter in *The Origin of Species*.

Darwin's studies of domesticated birds

Indicative of the kind of lengths Darwin went to in investigating variation in domesticated birds was a single paragraph note he published in the *Journal of Horticulture and Cottage Gardener* for December 1862. In this, he asks, "If any of your readers have kept Penguin Ducks [a breed of the domestic duck], and will have the kindness to observe one little point, and communicate the result, I should be greatly obliged. On examining the skeleton, I find that certain bones of the leg are longer than in the other breeds. I formerly kept these birds alive, and as far as I dare trust my memory, they could run considerably faster than other Ducks. Is this the case? It would, perhaps, be a good way to test their running powers to call the two kinds, when hungry, from a distance to their food, and see which arrived first."[69] He also asked similarly probing questions about various domesticated birds of many of his correspondents.

In his *Variation of Animals and Plants*, Darwin devotes two chapters, or 93 pages, to pigeons: the first, of 48 pages, on "Domestic Pigeons"; and the second, of 45 pages, titled "Pigeons—continued." A chapter of 51 pages deals with "Fowls"; and of the following 29-page chapter titled

"Ducks—Goose—Peacock—Turkey—Guinea-Fowl—Canary-Bird—Gold-Fish—Hive-Bees—Silk-Moths," 22 pages deal only with the birds indicated. Obviously, these are substantial texts about their various avian subjects, and although I cannot possibly detail all of their respective content, I provide here a summary and assessment of them. In completing this work, Darwin became an authority on domesticated birds in general, the pigeon in particular, and a world authority on their origins, variation, and the results of artificial selection on them.

Pliable pigeons

Charles Darwin could see the great service that demonstrating artificial selection on the variation inherent in domesticated animals first, before presenting his theory of natural selection on such variation in nature, would do for his cause. He also appreciated that the pigeon clearly provided the best domesticated species available to him for that masterful purpose, given its long history in association with humans and its resulting highly distinctive breeds. The remarkable extent to which the various more distinct breeds differed in external appearance was, as noted previously, so great that Darwin's generation of the very people that had brought them about, the pigeon fanciers, could not believe that they all derived from a single ancestor—the Rock Dove.

Going to great lengths and "tediously minute" details, Darwin described the external appearance and body proportions of the various breeds of domestic pigeon, resulting from intentional artificial selection by breeders; see "The Domesticated Pigeon" in chapter 5. He compared them to other wild pigeon species in general and to the Rock Dove in particular. Charles cultivated a wide British and international circle of pigeon-wise correspondence and barraged them with technical questions in his quest for pertinent information. In footnotes to the second page of his first pigeon chapter alone, he thanks 18 different men for their help in this way.[70] By keeping and breeding numerous breeds of pigeon over the three-year period 1855 to 1858, he systematically preserved their skeletons as well as the skins of most. He also obtained the same for additional breeds from correspondence around the globe. In this way, he was able to examine, compare, and then demonstrate to his readers the remarkable differences in skeletal structures that resulted from unintentional artificial selection: internal anatomy acquired as a by-product of intentional selection for externally visible characters.

The result of Darwin's slow and methodical work with pigeons in loft, laboratory, museum, and study (and also kitchens and outhouses, stinking of decaying pigeon corpses) was a 93-page argument, constituting two chapters of his *The Variation of Animals and Plants*. So comprehensive and convincing was this that there could never again be any doubt about the Rock

Dove being the one and only ancestor of all domestic pigeon breeds. One most significantly telling point that Charles clearly enunciated was that the crossing, or "hybridisation," of any two breeds would sooner or later result in at least the occasional individual showing the species-specific and characteristic wing, tail, or rump plumage traits of the ancestral Rock Dove. It was an extraordinary example of carefully planned, long-term, extensive and intensive ornithological research involving the application of comparative external and anatomical morphological and evolutionary studies. Such an approach was to be eagerly emulated by countless students of wild bird morphology and avian taxonomy over subsequent generations of ornithologists, and of course also by zoologists studying other animals.

There was nothing modest about Darwin's fundamental contributions to our understanding of artificial, and thus also natural, selection through his masterful use of the lowly Rock Dove and its domesticated descendent breeds. Although, as Darwin freely admitted, the case that he argued for artificial selection in pigeons often contained details that were tedious, the result was an absolute triumph. It provided a relatively digestible entrée for his subsequently elegantly served main course on natural selection.

As the great New World field and theoretical ornithologist Alexander Skutch wrote, "Certainly, it is as proper to speak of Darwin's pigeons as of Darwin's finches. Of all the contributions the study of pigeons has made to science, none is more important than its influence on Darwin's thought about evolution, which has so greatly changed our views of our world and ourselves."[71] In fact, it is far more so. Although Darwin described the Galapagos finches in broad terms, he wrote extremely little about the significance of them to his theory of natural selection and the origin of species: nothing in the first edition of his *Journal of Researches* and only a single vague sentence in the second edition,[72] nothing in the various editions of *The Origin of Species*, and nothing in his *The Descent of Man*. However, he wrote ten pages in *The Origin of Species* and a very great deal more in his *The Variation of Animals and Plants* about domesticated pigeons, as well as on numerous pages in several of his other books.

Alexander Skutch was far from alone in appreciating the significance of CD's use of the domesticated pigeon to get across the power of selection, be it artificial or natural. For example, upon the retirement of Richard Owen as Superintendent of the Natural History Departments, William Flower became the first Director of the Natural History Museum in 1884, and he quickly added an exhibition of various breeds of the domestic pigeon to the public galleries to highlight the great significance of Darwin's work on them.[73]

Darwin's published ornithological work reflects his global voyaging and the diversity of birds that he personally observed, documented, and subsequently studied in the literature, museum, and laboratory. But no single species contributed more to the intellectual formulation of and evidence for his major biological theories than the countless descendants of the modest Rock Dove under domestication.

In reviewing all known hybrids between any two wild pigeon species, Eugene McCarthy[74] suggested that "Darwin seems to have little actual knowledge of the history of pigeon breeding and his personal experience was minimal." He notes that Darwin kept and bred pigeons for three years and expressed the view that "Such an approach is more that of a collector or enthusiast than of a dedicated breeder." McCarthy fails to note, however, that the reproductive output of the species, which Darwin noted "can be propagated in great numbers and at a very quick rate,"[75] would provide significant experience over three years: not to mention that CD read about and communicated long and widely with the expert pigeon breeders of his time. Darwin also continued to study skeletal and skin specimens of domestic pigeons long after keeping them. Thus, McCarthy's view seems as inexplicable as it is unjustified, for it is surely possible for even an extraordinarily busy but very great biological scientist to be a "dedicated" breeder of pigeons for even a single year, let alone three consecutive ones.

I also find McCarthy's implied assertion that Darwin did not consider the possibility that the hybridization of two distinct breeds of domestic pigeon could not have influenced the creation of a new breed[76] highly questionable. Darwin wrote, "In some few instances it is credible, though for several reasons not probable, that well-marked races have been formed by crossing; for instance, a barb might perhaps have been formed by a cross between a long-beaked carrier, having large eye-wattles, and some short-beaked pigeon. That many races have been in some degree modified by crossing, and that certain varieties which are distinguished only by peculiar tints have arisen from crosses between differently-coloured varieties, may be admitted as almost certain."[77] McCarthy also implies that Darwin was remise in failing to cite older pigeon literature with specific reference to that by John Moore;[78] but a glance at the index to his *The Variation of Animals and Plants Under Domestication* shows five separate citations to that particular work.

The immensely diverse domesticated pigeon, much beloved by Charles Darwin, is mentioned in the last paragraph of volume 2 of his *The Variation of Animals and Plants* in questioning any influence of a divine creator by him writing, "Did He ordain that the crop and tail feathers of the pigeon should vary in order that the fancier might make his grotesque pouter and fantail breeds?" He couched his words as a rhetorical question, but he doubtless did so with a resounding "no!" ringing in his head.

Red canary or red herring?

In his interesting book *The Red Canary*, about how the red form of the domesticated canary came about, Tim Birkhead notes that Charles and Emma Darwin obtained a caged canary in October 1850 to amuse and comfort their terribly ill 10-year-old daughter Annie. Apparently there was a traditional belief in Europe that the close proximity of small birds, particularly red ones, to a sick child could cure it "by absorbing its sickness,"[79] albeit hard to imagine that this belief played any role in Charles Darwin's thinking. Annie tragically died of consumption in June 1851. She was Charles's favourite child, and her death was a most terrible blow to him and did nothing to improve his rapidly diminishing religious faith. Birkhead goes on to suggest that when Darwin wrote his *The Variation of Animals and Plants*, published 17 years after Annie's death, "he could barely bring himself to confront the canary, glossing over it is a single page and making no effort to reconstruct its history as he had for pigeons." It is possible that this interpretation is, I think for the following reasons, perhaps more romantic than it is realistic.

In the same chapter of *The Variation of Animals and Plants*, Darwin dedicated 11 pages to the common duck, three to the goose, two to the peacock, two to the turkey, half a page to the guineafowl, and the single page to the canary. These respective text lengths in fact largely reflect the relative significance and history of domestication of the various species concerned. Darwin noted that the canary "has been recently domesticated, namely within the last 350 years," this being far less time than the other species—and thus there was, inevitably, less evidence and far less to say about the species. Moreover, mention is made of the domesticated canary by Darwin at least eight additional times in *The Variation of Animals and Plants*, and also in eight different places in the subsequent *The Descent of Man* (see appendix 1).

Birkhead went on to write, "it was worse than this" because Darwin "got his canaries muddled up" and, "in his haste to be done with them [hardly something CD was known for!], referred to a feather-footed variety. There are feather-footed fowl and feather-footed pigeons, but no such canary. The canary, it seems, was Darwin's *bête noire* [black beast]." Birkhead notes that CD was led to refer to "feather-footed" or "feather-legged" canaries because he had underlined reference to them in his personal copy of Brent's 1864 book *The Canary, British Finches, and Some Other Birds*. The English edition of a work by Hervieux[80] had apparently misled Brent, in which "duvet" was mistranslated as "rough-footed"[81] and that Brent termed feather-footed and feather-legged. But Brent claimed to have seen a feather-footed canary,[82] an individual of a cultivated form he states to then be very scarce "if the breed is not altogether lost." In view of this, Darwin can hardly be blamed for any error should feather-footed canaries never have existed. At the time, Charles simply used Brent's mention of feathered-footed canaries in good faith, just as he did numerous other widely used contemporary references about birds.

Darwin was of course dependent on what he perceived at the time to be reliable sources of information, be they published or personal communication from people he saw as expert colleagues. In his introduction to *The Origin of Species*, he writes "No doubt errors will have crept in, though I hope I have always been cautious in trusting to good authorities alone." No less an ornithologist than the great Alfred Newton, assisted by Hans Gadow, also continued to include mention of the feather-footed canary, in his classic standard ornithology *A Dictionary of Birds*, both author and publication being much lauded by Tim Birkhead.[83] A quarter of a century after Darwin did so, Newton specifically mentioned among "the fancy" of the canary a variety with "feathered feet."[84]

It has been observed of his domestic bird specimens that "Given their vital role in his work and the very personal nature of the collection, Darwin's domestic birds are arguably more significant than the birds of the *Beagle*." Also that "for it is perhaps the domestic birds that should be regarded as the true ornithological icons of Darwin's evolutionary thought."[85]

The Descent of Man

As previously noted, Darwin was no specialist in avian taxonomy and nomenclature, and he was only too well aware of his limitations in that field of study. He noted that he therefore had his four chapters on birds in his *The Descent of Man* read by the ornithologist Philip Sclater, who primarily helped him with the nomenclature applied, but not without some errors remaining. Certainly Sclater also overlooked several mistakes Darwin made concerning the biology of some birds, which is perhaps understandable given that Sclater was a London-based taxonomist at the Zoological Society and not an avian field biologist.

Most of what is written about a great diversity of bird species in *The Descent of Man* relates to individual variation, sexual and age dimorphism relative to nest type and site, courtship and associated nuptial plumage characters, protective plumage colouration, and, very predominantly, sexual selection between males but mostly by females on males. The four chapters on birds therein are by far the most important of Darwin's ornithological contributions on wild, and a few domestic, birds. Of particular interest are Darwin's novel and insightful observations and interpretations of the plumage characters of adult male peacocks, Great Argus and other pheasant species, their use and significance to females in courtship and mate choice, and their evolutionary origins—detailed under the section "Secondary Sexual Characters" in chapter 6. These eventually stimulated recent novel and exciting experimental research. Results of this research supported Darwin's contention that female birds could and do discern subtle differences in the quality of male nuptial adornment and vigorously assess them in their critical mate selection behaviour.

Darwin discussed the vocalizations of birds and associated physical adaptations within the context of sexual selection in *The Descent of Man*, although at the time unaware of the significance of territoriality. His descriptions of feathers grossly modified in males to produce instrumental sounds for courting through sexual selection by females would have greatly surprised and intrigued most of his contemporary readers. As his ideas about pheasant nuptial plumage have encouraged recent experimental research, so his highlighting the fact that mechanical sound produced by feathers evolved for that purpose in males through female sexual selection has been followed by recent field research. Such studies have applied modern technology to record what rapid avian courtship display movements the human eye could not possibly see, resulting in a far better understanding of their function and evolution.[86]

In the light of the limited knowledge of the subject at the time, and the space available within his book for it, Darwin reviewed the courtship displays of birds and the plumage traits presented by them. In doing so he particularly mentioned pheasants and related birds: hummingbirds, bellbirds, bowerbirds, and birds of paradise, among others. Many others of his specific examples stimulated subsequent generations of ornithologists to study one or more of the species involved in far greater detail, resulting in great strides forward in our understanding of courtship, sexual selection, and related aspects of avian biology. Darwin was a pioneer of the study of animal behaviour, known today as ethology, with particular reference to both wild and domesticated birds. Yet the widespread serious study of this subject did not occur, in its infancy, until as recently as the 1910s–1920s.[87]

CHARLES DARWIN AS AN ORNITHOLOGIST

Taking the year 1900 as being at the peak of Charles Darwin's major works being published with significant ornithological content, the second edition of *The Descent of Man* appearing in 1901, I can document that he makes mention of some 600 different bird species, other ornithological topics aside. Given that the number of bird species known to science in 1935 was carefully considered to be about 8,500[88] and that a significant number were first described between 1900 and 1935, Darwin probably included mention of approximately 5–10 percent of the then known bird species in his publications. He also mentioned at least 1,700 discrete ornithological subjects in his books, including species; but this indicates nothing about the extent of his ornithological writings, which for some species and topics was considerable; see appendix 1.

Darwin's Ornithological Interest and Knowledge

Darwin's real interest in ornithology was expressed in the clearest possible terms by him prior to his teenage years. When Darwin's son Francis was young, he shared his interest in, and pleasure of, birds with the boy by making a game of the colour plates in the book *Illustrations of the Zoology of South Africa* by Andrew Smith (whom he met when the *Beagle* called into the Cape of Good Hope, or Cape Town, in June 1836). Father and son would enthusiastically flip through the plates of birds with the game being that each alternate picture "belonged" to one or the other of them. Francis clearly recalled that "after a series of dull thrush-like birds had been calmly shared between my father and myself, the agony of seeing a magnificent green and purple one fall to his lot. I am sure he tried to cheat himself, but this was not always possible."[89] His father's interest in birds persisted throughout his entire life, as is reflected in him lastly publishing a paper on the brood parasitism of American cowbirds in 1881, a few months before his death (see the following). But the fact is that the demands on his mind for the broad spheres of biology and geology and associated fields of study that he undertook did not permit him the luxury of much ornithology after the voyage of the *Beagle* and the writing of its zoological results; that is, not beyond the precious little ornithology in his *The Origin of Species*, the highly significant work on domesticated pigeons in *The Variation of Animals and Plants*, and the considerable bird content of *The Descent of Man*. He was simply not able to be a specialist ornithologist and, as such, he did not subscribe to the *Ibis*, the journal of the British Ornithologists Union and possibly the most scientifically significant of ornithological publications of his time,[90] let alone those published overseas.

It has been suggested that the extent of Darwin's ornithological expertise has been exaggerated in the literature.[91] Be that as it may, the previous chapters herein indicate that his ornithological knowledge and contributions were far from inconsiderable. A letter that Darwin wrote to William Darwin Fox from Maldonado on 23 May 1833, in part reads thus: "You ask me about Ornithology; my labours in it are very simple.—I have taught my servant to shoot & skin birds, & I give him money.—I have only taken one bird which has much interested me: I dare say it is as common as a cock sparrow, but it appears to me as if all the Orders had said 'let us go snacks in making a specimen'."[92] This statement, made after less than a third of CD's voyage was completed, led one author to suggest that this reveals Darwin's "true attitude to birds"[93] at that time. This is one individual's interpretation. Another interpretation is that by "labours," Darwin simply meant just that, quite probably implying that his geological, botanical, and zoological work dictated that he saw fit to leave the collecting of birds to Syms Covington. Dealing with killing and preserving dead birds was all that Covington did, making no observations of live ones as did Darwin. This cannot justifiably be taken to mean that Darwin was not interested in, lacked knowledge of, or was not observing and thinking about, birds. Indeed, he did of course publish many of his observations of and thoughts about living wild birds made during his voyage.

Charles's statement that he had, to that point in time, taken only one bird of particular interest to him (a Least Seedsnipe [*Thinocorus rumicivorus*]) could well do no more than reflect that he found the species concerned a particularly challenging taxonomic puzzle. He wrote to John Henslow at Cambridge describing this bird as striking him as a mixture of a "lark pidgeon [*sic*] and a snipe"; and, in typical self-deprecation, he wrote that he supposed it would turn out to be some well-known bird. Given that Charles was genuinely confused by the structure and appearance of this intriguing bird, which is indeed a member of the taxonomically odd and thus particularly interesting group consisting of the four seedsnipe species, the view of W. (Bill) R. P. Bourne[94] that "its importance seems exaggerated" does, in the context of the circumstances, time, place, and evidence, seem unreasonable.

A recent assessment of Charles Darwin as an ornithologist during the voyage of the *Beagle*, assuming that he "had all the tools needed in the field by an ornithologist of his day," observed that he would not have encountered all British species "especially those not commonly shot on hunts and offered at markets, such as nocturnal birds and rare passerines": but no factual evidence actually exists to suggest that he was not necessarily familiar with numerous British species that he had not shot as "game." Moreover, there can have been but a handful of his contemporaries that had "acquired thorough identification skills for all British birds" and with meaningful knowledge of many "rare passerines" in Britain. This seems an unreasonable expectation of a 22-year-old university medical dropout amateur general naturalist of that era. So do the statements that "even worse were Darwin's identification skills for the different plumages of seabirds and waders which he *sometimes* attributed to the wrong genus" and that "He did not know the scientific names of *all* British birds and had no special knowledge of the *scientific names of exotic species*, lacking the necessary years of training in the field as well as in reference collections"[95] (emphasis mine).

This criticism is perhaps unreasonable, because very few ornithologists of that time, and for long afterwards, would have acquired such specific skills and knowledge. Similarly, the critical observation that he failed to identify the Short-eared Owl when he encountered it on the Falkland Islands[96] seems harsh. He would not, after all, have expected to find this British species on an isolated island group on the other side of the globe and would surely, as a careful scientist, have therefore been appropriately reluctant to apply a specific name to it (other than the "Owl" that he did[97]) until able to identify it with comparative material at hand once back in England.

The preceding recent criticisms of Darwin notwithstanding, it is recorded that Professor John Henslow saw him prior to his departures on the *Beagle* as not "a finished Naturalist, but as amply qualified for collecting, observing, & noting anything new to be noted in Natural History."[98] Moreover, a recent author on Darwin considered that "he was actually an extraordinarily well-trained natural scientist."[99] Another believed that "Charles Darwin was as prepared to be the naturalist of the 'Beagle' as any contemporary university graduate could have been."[100]

Frank Steinheimer follows his criticisms of Darwin, however, with the observation that he "did make some excellent ornithological observations and carried out innovative experiments, noting down every minute detail of his research" and that "he was also good at asking the right questions, many of which he could then answer himself during the voyage through his own field studies."[101] A statement by the latter author that Darwin "showed no great interest" in any birds from the Galapagos Islands is, accepting that "great interest"[102] is subjective, contradicted by CD's notes on the Galapagos Hawk, mockingbirds, Galapagos Dove, Galapagos Crake, and the finches (at least under his introduction to the genus *Geospiza* where he alludes to their ecology and habits) in *The Zoology of the Voyage of H. M. S. Beagle*. When viewed in the context of other demands on his time and attention among the islands, not least of which was his geological work, he clearly did show interest in the birds there.

As a bird collector, Darwin was ahead of his time both in recording the food items found within his bird specimens and in making ecological and behavioural notes on, and comparative observations between, wild species. Given this, it is difficult to see how it "was mainly Gould who made Darwin's collection *and notes* into a significant contribution to ornithology"[103] (emphasis mine), particularly as it was Darwin who completed the volume after Gould's hasty departure for Australia, with some help from George Gray, and who contributed absolutely all of the numerous field observations therein.

Following the previous criticisms of Darwin, Frank Steinheimer concludes with "Darwin's contribution to ornithology [during the *Beagle* voyage] was not insignificant"; and "birds were not Darwin's first love, and that the ornithological knowledge gained in the "Beagle" voyage do [*sic*] not seem to have been indispensable for his evolution theory, although he did use many examples from captive birds."[104] That birds were not his first love is well documented and evident, and he never claimed them to be so. His youthful and subsequently long-lasting great interest in entomology and his exponentially increasing greater interest in geology and other topics during his voyage are a matter of record. The fact that *The Origin of Species* contains little expressing ornithological knowledge gained during the voyage of the *Beagle* cannot lead to the assumption that it was not significant to CD's theoretical considerations. Certainly his remarks about several South American birds, the extent of bird species endemism on the Galapagos Islands, and his thinking about the distinctive island-isolated forms of the mockingbirds within that archipelago while still aboard the *Beagle*, indicate otherwise. Beyond the ornithology of the *Beagle* voyage, Steinheimer did, however, concede that "For

Darwin, then, ornithology, especially of captive breeds, provided essential evidence for his theory of evolution."[105]

It has been stated that Darwin, once back in England with his specimens, was "stunned" to learn from John Gould that bird species endemic to the Galapagos "were closely related to species of the American mainland."[106] This particular observation is surely erroneous because Darwin had previously observed species of mockingbird and dove, if not also of additional species such as of heron, flamingo, pelican, duck, hawk, flycatcher, and warbler, in South America that would so obviously have struck him as being extremely closely related to those on the islands—as also would, of course, have the "iguanas" that he noted as having observed other species of on the South American mainland.[107] Indeed, he wrote in his bird notes while in the archipelago "the Ornithology [of the Galapagos] to my eyes resemble that of the temperate parts of that Continent [South America]."[108]

Darwin's anthropomorphism

In writing about birds, Darwin occasionally used anthropomorphic language, notably more often in his earlier publications, particularly in describing or discussing their behaviour. On St Paul's Rocks in the Atlantic, very early in the voyage of the *Beagle*, he found seabirds to be so tame that he described them as being of "stupid disposition." As Charles was to subsequently observe such tameness on many other islands lacking bird predators about the globe, and so was to come to understand the reason for it, he came to see such tameness not as stupidity but as inevitable adaptive behaviour. He wrote of the South American Red-winged Tinamou as being a "very silly bird" because it could be easily approached on horseback and killed with a stick. It was to be a good number of years hence that he could come to appreciate that this was because the bird may have evolved in the absence of horse-like predators and people.

In 1839, Darwin noted of the terrestrial Moustached Turca, which hops on long legs while holding its tail vertically erect above its back, that it is "ashamed of itself" and is "aware of its most ridiculous figure." At about the same time he noted, in his *The Zoology of the Beagle* volume on birds, Black Vultures soaring in circles at great height "for sport" or perhaps pair formation. He also found caracaras "disgusting," "crafty," "cowardly," and watching resting people with an "evil eye."

In 1901, he accounted for European Robins singing in autumn by writing "nothing is more common than for animals to take pleasure in practising whatever instinct they follow at other times for some real good" or that they should sing outside the courtship season "for their own amusement."[109] In fact, the autumnal singing by these robins is territorial in function.[110] He writes of bowerbirds' bowers being where both sexes assemble to "amuse themselves" and pay court;[111] and that male birds in general "take delight in displaying their beauty."[112] In fact, male bowerbirds build bowers as part of their dominance and sexual displays and as a traditional site at which they exhibit bower quality and bower decorations and attract and court visiting females.[113]

Given the considerable literature that Darwin published about birds, the extent of his anthropomorphic content is minimal and is, in any event, understandable given its commonplace usage and the lack of appreciation of animal behavioural motivation at that time. Moreover, some of it could have been included under artistic license, to liven his prose. Ironically, it is suggested that Darwin can be seen as the father of the modern study of animal behaviour because of his books *The Descent of Man* (1871) and *The Expression of the Emotions in Man and Animals* (1872).[114] But it was to be about a half a century into the future before ethology became firmly established as a biological science, primarily through the work of Konrad Lorenz and Niko Tinbergen.[115] It is a moot point, but one wonders if Darwin might ever have considered adding the word "other" to the title of his book, to thus make it read *The Expression of the Emotions in Man and other Animals.*

WAS CHARLES DARWIN AN ORNITHOLOGIST?

Although Darwin never professed to be a specialized ornithologist, and occasionally disparaged his own efforts at species identification with rare humour, he was as broadly accomplished in the field as were the vast majority of his contemporaries. Although admitting to the "extreme complexity" of the subject and noting that it remained in much need of the attention of "some competent ornithologist," CD used the literature available to him to attempt a review of the "classes of cases, under which the differences and resemblances between the plumage of the young and the old, in both sexes or in one sex alone [of birds], may be grouped." He did not, I am sure, mean to imply that he was not an ornithologist or was an incompetent one, but rather that as a biologist, zoologist, entomologist, botanist, geologist, vulcanologist, anthropologist, ethologist, and more: he was not an exclusive and specialist ornithologist. Thus, he could only take his studies of the significance of variation in the plumages of birds relative to sex, age, seasonality, and moult so far, given his other work. But at a time when ornithology was primarily a matter of shooting and preserving birds, taxonomic descriptions, nomenclature, and the observation of only the most basic, if any, bird biology, Darwin certainly made his mark. He did so with comparative and theoretical considerations of bird morphology, anatomy, ecology, behaviour, and natural and sexual selection based on a synthesis of personal experience, experimentation, and available knowledge—beyond the competence of numerous contemporary "ornithologists."

In 1879, a letter of statement was written to the great American ornithologist Elliot Coues and signed by 38 "brother ornithologists" in England to express their appreciation of his American ornithological bibliography. In adding his name to ornithological luminaries including H. E. Dresser, John Gould, Alfred Newton, Osbert Salvin, Philip L. Sclater, Henry Seebohm, Richard Bowdler Sharpe, W. B. Tegetmeier, and Alfred R. Wallace, Charles Darwin clearly comfortably saw himself as an ornithologist,[116] as did his peers—and he and they were perfectly justified in doing so. Given his, now thoroughly documented, moral character and the nature of the document concerned, he would not have signed it unless he sincerely considered himself qualified under the title of ornithologist. He did not, however, see himself as proficient as some or all of these other men. For example, when Wallace mentioned his intention to write a paper about the colouration of birds relative to their nesting sites and nests, Darwin referred to Wallace's "greater knowledge of ornithology" in a letter of 29 April 1867.[117] Charles Darwin was as modest an ornithologist as he was modest in most things.

At the start of their impressive book *Ten Thousand Birds: Ornithology Since Darwin*, Tim Birkhead and colleagues wrote "Darwin made so many perceptive observations and comments on birds that inspired a number of pioneers to test his ideas"; and "Darwin was more than an ornithologist—he was too broad for that—but he had good credentials as an ornithologist because he raised birds and he wrote extensively about their biology. Many of today's ideas have their genesis in his writings." They went on to write of him at the end of their book: "Since Darwin's day, ornithologists have been at the cutting edge of many theoretical developments (Mendelian, population, and molecular genetics, the Modern Synthesis, the new systematics, game theory, optimal foraging theory, population dynamics, life history theory) and have employed mathematical models as a basis for generating testable hypotheses in disparate subjects (including sexual selection, island biogeography, optimal foraging, optimal clutch size, life history trade-offs, population genetics, flight energetics and aerodynamic models, metabolic models with respect to energy budgets and energy expenditure, and phylogenetic models)."[118] The scholarly book by Birkhead and colleagues is concluded with a list of 500 of the most eminent ornithologists of all time; and Charles Darwin is, quite correctly, included in that list—for he most certainly fulfilled his schoolboy ambition to become an ornithologist, and one of significance at that.

Darwin's last published ornithological contribution appeared in 1881, his penultimate year of life. This was a brief paper titled "The Parasitic Habits of Molothrus" for the journal *Nature*.[119] It was stimulated by information received from a correspondent in Lima, sent in September of 1881, who observed of captive and wild Shiny Cowbirds (*Molothrus bonariensis*) that their eggs are like those of two of three host species in size and colour; that generally one cowbird egg is laid in a host nest, but up to six may be; that the nestling cowbird typically but does not invariably eject nestlings of its foster parents; and that he once found two cowbirds in the one foster parents' nest from which they had evicted the hosts' nestlings. In captivity, a female cowbird started laying at two years old, laying six eggs each "clutch" at intervals of four full days but more later in the season. Thus, the cowbird lays each egg at atypically long intervals, as does the similarly parasitic Common Cuckoo.

Darwin suggested that the female Common Cuckoo lays eggs in the nests of other species as a brood parasite "owing to her habit of laying them at intervals of two or three days," as it would be disadvantageous for her to have young of different ages and eggs in the same nest together at the same time. But he pointed out that this latter situation does occur "with the non-parasitic North American cuckoo"; and that if this had not been so, it "might have been argued that the habit of the Common Cuckoo to lay her eggs at much longer intervals of time than do most other birds, was an adaptation to give her time to search for foster parents." This led CD to note that there is "some close connection between parasitism and the laying of eggs at considerable intervals of time";[120] see the "Cuckoo Instincts" section of chapter 4. We now know that many and varied non-parasitic bird species lay their eggs at longer intervals and that one or more of a range of explanations might apply in the case of any one species. For example, the eggs of the Common Cuckoo hatch relatively quickly, their embryonic development being slightly advanced when laid, and so females would presumably need additional time between eggs laying to achieve this embryonic advancement.[121]

Six more scientific papers were subsequently published under Charles Darwin's authorship, on subjects other than ornithology, the last appearing in 1883. It would be difficult to argue with the proposition that, given the standards of biological studies and normal hierarchical systems within tertiary institutions of his time, Charles Darwin's ornithological work would not have been acceptable for a higher, if not a doctorate, degree in biology. Indeed, his research and findings on domesticated birds in general, if not that on pigeons alone, should very probably have qualified for such a degree given the academic standards of the time: so would his work on barnacles, on geology, on coral reef formation, on earthworms, on entomology, on orchids, on climbing plants, on insectivorous plants, on sexual selection, on the expression of emotions, on variation of animals and plants under domestication, and so forth.

Within little time after the publication of *The Descent of Man*, ornithologists of the day would have considered Darwin a most accomplished and eminent colleague, notwithstanding his taxonomic shortcomings and relatively limited bird contributions. The magnitude of his studies beyond birds cannot, however, be seen to diminish his ornithological work in the context of his time and the quality of work of contemporary ornithologists. It is said that he had six million words appear in 19 major works, numerous scientific papers, and an unknown number of letters but of which 14,000 are preserved in archive today.[122]

Darwin was beyond doubt and any objective and fair debate an ornithologist, and one of relatively high ability for his time. Most of Darwin's contemporary ornithological colleagues were concerned with the description, depiction, and naming of bird species. He, on the other hand, also thought much about the life histories, ecology, mating systems, evolution of their plumage and other morphology, and behaviour of birds; and in asking questions of numerous fellow naturalists about these topics, he undoubtedly influenced many to consider them too. Judged in the light of the vast field of knowledge, investigation, and global electronic communication that is the science of ornithology today, he was of course, and inevitably, not without his flaws and limitations. It was recently appropriately noted that of the "Subjects of key observations at different stages of his career, birds played a vital role in Darwin's research and the development of his evolutionary ideas."[123] Charles Darwin's published work, reviewed in this volume with respect to its ornithological content, had a profound influence on ornithology, stimulating diverse theoretical developments, and it continues to provide inspiration to some students of bird life today. In addition, and significantly, "There is no question that of all Darwin's works, his evolution theory has had the biggest impact on ornithology,"[124] which I assume is taken to include sexual selection.

DOWN TO THE ABBEY

In later life, Darwin doubtless enjoyed watching and hearing birds as he regularly walked his private and much loved Sandwalk circuit, located beyond the bottom of the garden at Down House; see colour figure 8.4. Here and elsewhere about his home, wrote his son Francis, "He always found birds' nests even up to the last years of his life, and we, as children, considered that he had a special genius in this direction. In his quiet prowls he came across the less common birds, but I fancy he used to conceal it from me, as a little boy, because he observed the agony of mind which I endured at not having seen the siskin or goldfinch, or whatever it might have been."[125]

On 17 November 1877, Cambridge University, somewhat belatedly, awarded Charles an honorary LLD (doctorate of law)—long after he had been awarded many similar and greater honours from institutions beyond Britain. He attended the ceremony and thoroughly enjoyed himself in the presence of an adoring audience. Oxford University had offered him an honorary D.C.L. doctorate (Doctor of Civil Law) previously, in 1870; but as he declined to attend and receive it on campus, the university decided to withdraw the offer. By his action, Darwin effectively declined the Oxford degree.

In conversation with two atheist philosopher guests at his home in September of 1881, in the presence of the Reverend Brodie Innes, Darwin claimed to be an agnostic, while agreeing with them that no evidence supported Christianity;[126] but he probably died, to all intents and purposes, an atheist.

During February 1882, Charles suffered pain about his heart and his pulse became irregular. His physically active life was over. Weather permitting, he sat in the garden orchard of his home with Emma, and he expressed enjoyment at seeing the crocuses and hearing the singing of birds while doing so.[127]

Charles Darwin died on 19 April 1882, having not been knighted by a politically and religiously conservative and invertebrate British establishment, government, and church (whereas three of his sons were to be knighted). Being profoundly hypocritical and pragmatic, they could see the political advantages of agreeing to having the body of this perceived antichrist placed in Westminster Abbey. Such an arrangement was explicitly contrary to Darwin's most clearly expressed firm wishes, those of his family, and of the Downe community. John Lewis, the carpenter of Downe, had already made his plain and simple rustic coffin, as Charles had requested it. His pain at seeing the remains of his friend and neighbour being lost to the village and its graveyard is clear in his protestation: "I made his coffin just the way he wanted it; all rough, just as it left the bench, no polish, no nothin'. But when they agreed to send him to Westminster . . . my coffin wasn't wanted, and they sent it back. This other one you could see to shave in."[128] Darwin had been long prone to not shaving; and it appears that the coffin for the Westminster funeral was in fact of unpolished oak.[129]

Unfortunately this politically motivated funeral conspiracy included several of Darwin's closest biologist friends and his cousin Frank Galton. They doubtless saw his mortal remains so placed as an appropriate acknowledgment of the man's greatness and the significance of his life to British and world science—but also as a scientific-political coup. As a result of the lobbying by these men of influence and associated favourable London press support, notably by the *Evening Standard*, 20 members of parliament signed a letter to the Dean of Westminster making the case for Darwin's remains to be interred in the Abby on 26 April 1882. In doing so, the pallbearers included Darwin's close friends and colleagues Thomas Huxley; Joseph Hooker; Alfred Wallace; William Spottiswoode, the President of the Royal Society; John Lubbock, 1st Baron of Avesbury; Reverend Canon Farrar; James Russell Lowell, the American Minister to the Court of St. James's; Edward Stanley, 15th Earl of Derby; William Cavendish, 7th Duke of Devonshire; and George Campbell, 9th Duke of Argyll. The stone slab covering his resting place beneath the floor of the abbey blandly reads, just as he requested that it should, "Charles Robert Darwin" together with his date of birth and death.

Charles Darwin had wanted to be buried in the small St. Mary's Church graveyard in Downe, beside his infant children and his much beloved elder brother Erasmus, who had died only seven months previously and, more importantly, where his dear wife Emma could eventually be buried beside him. Because of this, Emma was against the Westminster burial, until convinced by her eldest son William that his father would have graciously accepted this acknowledgement of his achievements.[130] Emma did not attend Westminster Abbey for his internment, a week after his death.

There can surely be no doubt that Charles would far rather have thought of the wings of wild birds fluttering over his, and Emma's, grave in the pleasantly foliaged, breezy, graveyard of Downe Village than the static, dusty, stone wings of chiselled angels in the dark, still, and, but for a single diurnal, socially monogamous and superstitious, vertebrate species, lifeless recesses of Westminster Abby. He was not let down to rest in Downe, as he should have been, but was let down by some close, doubtless well meaning, friends and by his government, and thus his nation. An appropriate observation made in *The Times* newspaper about the corpse of Charles Darwin was that "The Abbey needed it more than it needed the Abbey."

On 9 June 1889, a larger-than-life marble statue of Charles Darwin, by sculptor Joseph E. Boehm, was placed at the top of the main flight of stairs in the great entrance hall of the Natural History Museum in London. A public appeal for this statue raised £4,500 Sterling, a remarkable 2,296 subscriptions coming from Sweden alone.[131]

Emma Darwin died at Down House on 29 September 1896, 14 years after her revered husband, knowing she would be buried in St. Mary's Church graveyard in Downe Village alongside Charles's older brother Erasmus—all too far removed from her dearly beloved, agnostic, Charles.[132] Emma Darwin's biography is concluded with the appropriate "Suddenly from the trees above Emma's quiet grave a blackbird sings and sings, a pure, plangent melody that fills the air. It is a requiem she would have loved."[133] So too would have her lifelong devoted husband, companion, friend, and onetime open-minded and observant, globally voyaging, young naturalist aboard the HMS *Beagle*, the name of which ship is forever synonymous with that of the ornithologist Charles Robert Darwin (colour figure 8.6).

NOTES

1. F. Darwin 1950: 22.
2. Darwin 1881b.
3. Keynes 1997: 462.
4. Barlow 1963: 265; Grant and Estes 2009: 159–60.
5. Sundevall 1871.
6. Darwin 1839a,b; Gould and Darwin 1838–1841.
7. Gould and Darwin 1838–1841: 146.
8. Darwin 1859: 185.

9. F. Darwin 1887, vol. 3: 18.
10. Barlow 1963: 207.
11. Darwin 1859.
12. Mearns and Mearns 1998: 253.
13. S. J. Gould 1985: 357.
14. Grant 1986: 53, 62.
15. Sulloway 1982b: 10.
16. Gould 1849–1861: 25.
17. Weld 1862: 458.
18. Burkhardt et al. 1985–2014, vol. 9: 295.
19. Barrett and Freeman 1988, vol. 22: 402–3.
20. Sauer 1982: 368.
21. Tree 1991: 61.
22. J. Smith 2006: 99–114.
23. I.e., Gould 1830–1831; 1832–1837; 1833–1835; 1836–1838; 1837–1838a,b,c; 1840–1848; 1844–1850; 1848a,b; 1850–1883; 1851–1869; 1852–1854; 1855; 1858–1875; 1861; 1862–1873.
24. Gould and Darwin 1838–1841: 99.
25. In Quammen 1996: 82.
26. *Pace* Quammen 1996: 86.
27. van Oosterzee1997: 181; Mearns and Mearns 1998: 319.
28. Wallace 1883: vii.
29. Wallace 1860, 1869, 1876, 1880; van Oosterzee 1997; Benton 2013.
30. Müller 1846.
31. Sclater 1858.
32. Wallace 1860.
33. Cf. McKinney 1972.
34. Benton 2013: 82, 86, 89.
35. Darwin 1859, 1871, 1901; Benton 2013.
36. Darwin 1851a,b; 1854a,b.
37. Darwin 1859, 1868a.
38. Burkhardt et al. 1985–2014, vol. 3: 354.
39. Burkhardt et al. 1985–2014, vol. 13: 233, 238.
40. Darwin 1902: 330–35 and 509–14, respectively.
41. Tristram 1859.
42. Birkhead and Gallivan 2012: 900.
43. Haeckel 1866.
44. Huxley 1867.
45. E.g., Sibley and Ahlquist 1990; Fjeldså 2013; Gill and Donsker 2015.
46. Joseph and Buchanan 2015: 1.
47. Zhang et al. 2014; Jarvis et al. 2014, and references therein.
48. Joseph and Buchanan 2015: 1.
49. J. V. Smith et al. 2013; Jarvis et al. 2014; Lee and Worthy 2014; A. J. Baker et al. 2014.
50. Cf. Hackett et al. 2008.
51. See Hackett et al. 2008; Suh et al. 2011; Fjeldså 2013; Jarvis et al. 2014; Joseph and Buchanan 2015.
52. Birkhead et al. 2014: vii.
53. Crook 1961; Frith 1976.
54. Darwin 1845, 1860, 1889.

Figure 1.1 Charles Darwin at the age of 31. Watercolor on paper by George Richmond, 1840, © Darwin Heirlooms trust, courtesy of the English Heritage Photo Library.

Figure 1.2 The HMS *Beagle*, at anchor in Sydney Cove, Port Jackson, now Sydney Harbour, Australia, as it was in January 1836. Painting by, copyright, and courtesy of marine artist Frank Allen of Sydney.

Figure 2.1 A mounted Gaucho hunting a rhea with the bolas. From the front cover decoration of *A Naturalist's Voyage* (Darwin 1889).

Figure 2.2 The Lesser Rhea; by Gould, from Plate 47 in *The Zoology of the Voyage of H. M. S. Beagle. Part III: Birds*, by Gould and Darwin of 1838–1841.

Figure 2.3 Black-faced Ibis, El Chaltén, Argentina. Photograph by C. B. Frith.

Figure 2.4 A female Fuegian Steamer Duck with ducklings; Stanley, Falkland Islands. Photograph by C. B. Frith.

Figure 2.5 An adult male Magellanic Woodpecker, Fitzroy National Park, Patagonia, Argentina. Photograph by C. B. Frith.

Figure 2.6 An adult male Long-tailed Sylph hummingbird, Tandayapa Lodge, Ecuador. Photograph by C. B. Frith.

Figure 2.7 The Galapagos Crake; by Gould, from Plate 49 in *The Zoology of the Voyage of the H. M. S. Beagle, Part III: Birds*, by Gould and Darwin of 1838–1841.

Figure 2.9 An adult Lava Gull, Santa Cruz, Galapagos Islands. Photograph by C. B. Frith.

Figure 3.1 A Black Vulture, Manu River, Peru. Photograph by C. B. Frith.

Figure 3.2 A Striated Caracara, East Falkland, Falkland Islands. Photograph by C. B. Frith.

Figure 3.3 Galapagos Hawk. Adult to left, immature to right; by Gould, from Plate 2 in *The Zoology of the Voyage of H. M. S. Beagle. Part III: Birds*, by Gould and Darwin, 1838–1841.

Figure 3.4 The Floreana Mockingbird; by Gould from Plate 16 in *The Zoology of the Voyage of H. M. S. Beagle. Part III: Birds*, by Gould and Darwin, 1838–1841.

Figure 3.5 The San Cristobal Mockingbird; by Gould, from Plate 17 in *The Zoology of the Voyage of H. M. S. Beagle. Part III: Birds*, by Gould and Darwin, 1838–1841.

Figure 3.6 A Seaside Cinclodes; by Gould, from Plate 20 in *The Zoology of the Voyage of H. M. S. Beagle. Part III: Birds*, by Gould and Darwin, 1838–1841.

Figure 3.7 Medium Ground-Finch eating part of a succulent coastal plant; Santa Cruz, Galapagos Islands. Photograph by C. B. Frith.

Figure 3.8 Large Ground-Finch; by Gould, from Plate 36 in *The Zoology of the Voyage of H. M. S. Beagle. Part III: Birds*, by Gould and Darwin, 1838–1841.

Figure 3.9 Small Tree-Finch; by Gould, from Plate 39 in *The Zoology of the Voyage of H. M. S. Beagle. Part III: Birds*, by Gould and Darwin, 1838–1841.

Figure 3.10 Common Cactus-Finch; by Gould, from Plate 42 in *The Zoology of the Voyage of H. M. S. Beagle. Part III: Birds*, by Gould and Darwin, 1838–1841.

Figure 3.11 Warbler Finch; by Gould, from Plate 44 in *The Zoology of the Voyage of H. M. S. Beagle. Part III: Birds*, by Gould and Darwin, 1838–1841.

Figure 3.12 A Galapagos Dove, Santa Cruz, Galapagos Islands. Photograph by C. B. Frith.

Figure 3.13 A Cape Petrel, Australian Subantarctic Islands. Photograph by C. B. Frith.

Figure 4.1 Down House, home of Charles Darwin, seen from the garden; Downe, Kent, England. Photograph by C. B. Frith.

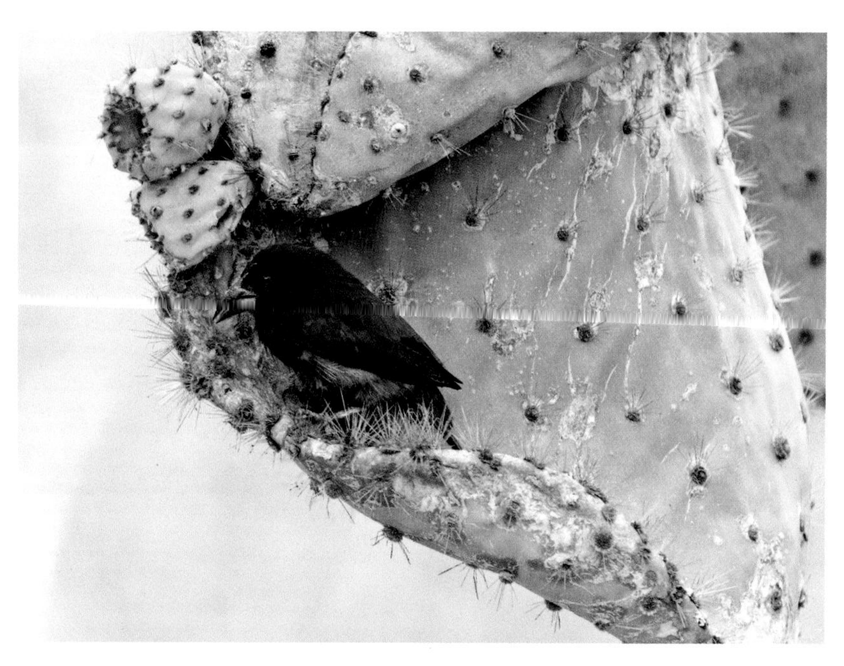

Figure 4.3 A Common Cactus Finch, Santiago Island, Galapagos Archipelago. Photograph by C. B. Frith.

Figure 5.2 Two Rock Doves, captive in England. Photograph by C. B. Frith.

Figure 6.2 An African Wattled Lapwing (*Vanellus senegallus*) wears a spur at the anterior bend of each wing. South Africa. Photograph by C. B. Frith.

Figure 6.4 An adult male Amazonian Umbrellabird. From Temminck and Laugier de Chartrouse, 1820–1839, p. 48.

Figure 6.6 An adult male Satin Bowerbird (*Ptilonorhynchus violaceus*) courts a female in his avenue bower on the southern Atherton Tablelands, north Queensland, Australia. Photograph by C. B. Frith.

Figure 6.7 Adults of the South American bellbirds. Top left, male Three-wattled Bellbird (*Procnias tricarunculata*; female similar to that of the White Bellbird); top right, male and female White Bellbird (*P. alba*); bottom left, male and female Bearded Bellbird (*P. averano*); bottom right, male and female Bare-throated Bellbird (*P. nudicollis*)—showing marked differences between males and similarities between the (green) females. From *The Cotingas* by David Snow, 1982, painted by Martin Woodcock. Courtesy of and © The Trustees of the Natural History Museum, London.

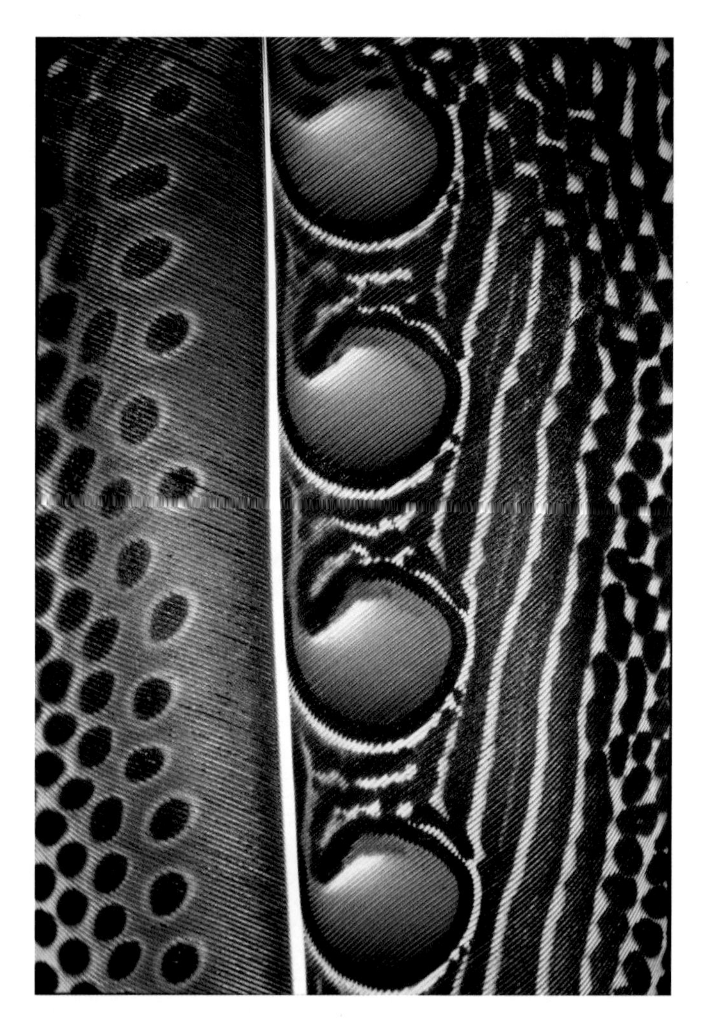

Figure 6.19 Portion near summit of one of the secondary wing feathers of an adult male Great Argus pheasant, as was illustrated by Darwin in black and white, to show "perfect ball-and-socket ocelli." Photograph by C. B. Frith.

Figure 7.1 An adult male Southern Cassowary (*Casuarius casuarius*) stands over his nest depression and clutch of eggs, which he alone incubates and raises the offspring that hatch; North Queensland, Australia. Photograph by C. B. Frith.

Figure 7.3 An adult Secretarybird, a bird of prey uniquely adapted for killing reptiles including venomous snakes; Kgalagardi Transfrontier Park, South Africa. Photograph by C. B. Frith.

Figure 8.2 Alfred Russel Wallace, with two freshly killed King Birds of Paradise in the Aru Islands in 1857; based on a 1950s painting by Russian artist Victor Evstafieff.

Figure 8.3 Charles Darwin's study in his home, Down House, Kent, England. Photograph by C. B. Frith.

Figure 8.4 The Sandwalk at the foot of the Down House garden, where Charles Darwin regularly exercised his body and mind. Photograph by C. B. Frith.

Figure 8.5 A 1982 British postage stamp depicting Charles Darwin and a Warbler Finch and a Large Ground Finch of the Galapagos Islands. Royal Mail, UK.

Figure 8.6 Charles Darwin at Down House, when 72 years of age. Photograph by Elliott and Fry.

Figure A.1 Satin Bowerbirds at a bower of the species. From *The Birds of Australia* by Gould, 1840–1848, volume 4: double page Plate 10.

Figure A.2 Two Spotted Bowerbirds at a bower of the species. From *The Birds of Australia* by Gould, 1840–1848, volume 4: double page Plate 8.

55. Rice 2000: 235.
56. Moorehead 1969: 162–3.
57. Gifford 1919.
58. Brent 1981: 199.
59. Mearns and Mearns 1998: 28.
60. Lack 1947: 14; Sulloway 1982b: 10.
61. Sushkin 1925, 1929; Sulloway 1982b, and references therein.
62. Petren et al. 1999, 2005; Abzhanov et al. 2006.
63. Sulloway 1982b.
64. Lack 1947.
65. See Gould and Darwin 1838–1841; Darwin 1839a, 1859, 1889, 1902.
66. Sulloway 1982b: 32.
67. Burkhardt et al. 1985–2014, vol. 15: 299.
68. Weiner 1994, and references therein; Grant and Estes 2009: 148–51.
69. Darwin 1862.
70. Darwin 1868a: 132.
71. Skutch 1991: 111.
72. Darwin 1845; 1889: 380.
73. Stearn 1998: 185.
74. McCarthy 2006: 159.
75. Darwin 1902: 47.
76. McCarthy 2006: 159.
77. Darwin 1868a: 188–9.
78. Moore 1735.
79. Birkhead 2003.
80. Hervieux 1719.
81. Robson and Lewer 1911: 26.
82. Birkhead 2003: 245.
83. In Birkhead and Gallivan 2012.
84. Newton 1893–1896: 71.
85. Cooper 2010: 28.
86. Snow 2004; Kirwan and Green 2011, and references therein.
87. E.g., Heinroth 1910, 1924; Heinroth and Heinroth 1924–1928.
88. Mayr 1935.
89. In Keynes 2001: 93.
90. Burkhardt et al. 1985–2014, vol. 15: 488.
91. Sulloway 1982b.
92. Burkhardt et al. 1985: 316.
93. Bourne 1992: 31.
94. Bourne 1992: 32.
95. Cf. Steinheimer 2004: 301–2.
96. Steinheimer 2004: 302.
97. Keynes 2000: 160.
98. Barlow 1967: 30.
99. Thomson 1995: 144.
100. Porter 1985: 977, in Steinheimer 2004: 302.
101. Steinheimer 2004: 307.
102. Steinheimer 2004: 309.
103. Cf. Steinheimer 2004: 311.
104. Steinheimer 2004: 313, 316.

105. Steinheimer 2004: 317.
106. Quammen 1996: 228–9.
107. Cf. Keynes 2000: 295.
108. Barlow 1963: 265.
109. Darwin 1901: 566–7.
110. Lack 1976: 35.
111. Darwin 1901: 583.
112. Darwin 1901: 601.
113. Frith and Frith 2004, 2008, 2009b.
114. Birkhead et al. 2014: 248.
115. Nisbett 1976; Kruuk 2003.
116. Sauer 1982: 195–6.
117. Burkhardt et al. 1985–2014, vol. 15: 240.
118. Birkhead et al. 2014: ix, xi, 428.
119. Darwin 1881b.
120. Darwin 1881b: 51–2.
121. N. B. Davies 2000: 52.
122. Jones 2008: 3.
123. Cooper 2010: 23.
124. Cf. Stresemann 1951, in Steinheimer 2004: 309.
125. F. Darwin 1950: 79.
126. Healey 2001: 317.
127. Browne 2002: 495.
128. Moore 1982: 102.
129. Cf. Brent 1981: 520.
130. Clements 2009: 137.
131. Stearn 1998: 73.
132. Cf. Browne 2002: 484.
133. Healey 2001: 355.

APPENDIX 1

The Complete Published Ornithology of Charles Darwin

This compilation includes all ornithological entries within those published works written by Charles Darwin (CD) containing such matter. Reference to the location of each is provided by the pertinent pagination of each publication containing ornithology using the following abbreviations, to save significant space, in **bold typeface**:

1837 = Notes on *Rhea americana* and *Rhea darwinii* (see References).

L = *Life*, 1838: in Secord 2010: 351–354 (see References).

Z = *The Zoology of the Voyage of H. M. S. Beagle, under the Command of Captain Fitzroy, R. N., During the Years 1832 to 1836. Part III: Birds*. 1838–1841, Smith, Elder & Co., London. By Gould, J. and Darwin, C.

J = *Journal of Researches into the Geology and Natural History of the Various Countries Visited by H. M. S. Beagle*. 1839, first edition. Henry Colburn, London (Darwin 1839b in References).

1839 = *Questions About the Breeding of Animals*. Stewart & Murray, London (Darwin 1839c in References).

1849 = Section VI: Geology. In Herschel, J. F. W. ed., *A manual of scientific enquiry: Prepared for the Use of Her Majesty's Navy and Adapted for Travellers in General*. London: John Murray, pp. 156–195.

1858 = Darwin, C. R. and Wallace A. R. 1858. On the tendency of species to form varieties; and on the perpetuation of varieties and species by natural means of selection. *Journal of the Proceedings of the Linnean Society of London. Zoology*, 3, 46–50.

V1 = *The Variation of Animals and Plants Under Domestication*. Volume 1. John Murray, London 1868 (Darwin 1868a in References).

V2 = *The Variation of Animals and Plants Under Domestication*. Volume 2. John Murray, London 1868 (Darwin 1868b in References).

F = *The Effects of Cross and Self Fertilisation in the Vegetable Kingdom*. John Murray, London 1876.

1881 = The Parasitic Habits of Molothrus. *Nature. A Weekly Illustrated Journal of Science*, **25**, 51–52 (Darwin 1881b in References).

J2 = A Naturalist's Voyage: *Journal of Researches into the Geology and Natural History of the Various Countries Visited by H. M. S. "Beagle", Under the Command of Capt. FitzRoy, R. N.* 1889, new edition, John Murray, London (Darwin 1889 in References).

D = *The Descent of Man and Selection in Relation to Sex*. John Murray, London 1901, second edition (first published 1871).

O = *The Origin of Species by Means of Natural Selection or the Preservation of Favoured Races in the Struggle for Life*. December 1902 reprint with "Additions and Corrections" and "An Historical Sketch," John Murray, London, sixth edition (first edition was published in 1859). The sixth edition, first published by Murray in 1872, was the last that Darwin personally contributed to and is thus taken as his definitive text.

E = *The Expression of the Emotions in Man and Animals*. Edited by Francis Darwin. John Murray, London, 1904, popular edition (first published 1872).

A = *Charles Darwin's Autobiography*. Edited by Sir Francis Darwin. Henry Schuman, New York. (Charles Darwin wrote his "Autobiography" in 1876, for his family's eyes only, but it was subsequently edited and appeared in this book (Darwin, F. 1950 in References).

The listing of the ornithological content of Darwin's publications is most appropriately done here in avian systematic order, while alphabetical access to many of them is readily available in the comprehensive *Index of chapters*. Text following CD's actual word(s) or names, which appear in **bold typeface** or ***bold italic typeface*** with upper and lower case as appearing in his works, provides present common and/or scientific name(s) and the briefest summary of what he wrote about each subject. What CD wrote might (relatively rarely) have proved incorrect in the light of present-day knowledge (sometimes noted within parenthesis, where any material added by me appears). Words in square parenthesis are those not used but meant or implied by CD in the context of what he did write. Any "See chapter . . ." reference implies that the bird or topic concerned is mentioned in the chapter indicated.

Although a good number of "lists of the birds of the world" have appeared since the revolutionary one of Sibley and Monroe (1990), the comprehensive index of which did make it eminently suited to the complex needs involved in compiling an appendix like this, it is simply too outdated to use here. The ornithologically groundbreaking nature of the Sibley and Monroe volume, based on Sibley and Ahlquist's (1990) revolutionary genetic revision of the Class Aves, ensured that it stimulated much avian taxonomic revision based on more extensive and sophisticated genetic work. Ever since the first attempt at a broad classification of the birds by Carl Linnaeus (1758), the order of the numerous taxa have been changed by taxonomist and this will long continue to be the case—thus, I am not as concerned about which recent taxonomic order I use here as I am about the taxa being findable within this work, which the *Index to chapters* at the back of this volume assists.

Where bird names given by CD remain in use today, nothing is added; but if the scientific and/or common names used today are different (as is mostly the case), then these are given in normal italics (i.e., not bold typeface) and are taken from the International Ornithologists' Union's (IOU) World Bird List, version 5.2 (Gill and Donsker 2015) as are the taxonomy and systematic order in which the entries appear. The common names used will perhaps not please everyone, given the policies applied by the IOU and International Ornithological Committee (IOC), but this would be the case no matter what single list be followed. As they are predominantly long disused, I do not list every family and/or sub-family names under which species are listed in *The Zoology of the Voyage of H. M. S. Beagle* (**Z**), but I do provide the order and family name of Gill and Donsker (2015; see http://www.worldbirdnames.org) in **BOLD UPPERCASE TYPEFACE**.

I stress that the following listing includes only ornithological content of Darwin's publications that could be conveniently included under taxa or subject and that a very great deal of discussion of these or of other avian matters could not be included in this appendix. Much of the latter kind of material is, however, touched on, if not dealt with quite fully, in the appropriate chapters of this volume. It must be understood that limited space has not always permitted me to indicate the context in which CD included some ornithological content detailed following. That I have found it necessary to keep these entries as brief as possible emphasizes the great extent of Darwin's published ornithological considerations.

Darwin wrote exhaustively about the domestic pigeon (*Columbia livia*) in two major chapters of volume 1 of *The Variation of Animals and Plants Under Domestication* as well as a chapter on the domesticated fowl (*Gallus gallus*) and a fourth titled "Ducks—goose—peacock—turkey—guinea-fowl—canary bird—gold-fish—hive-bees—silk-moths" with a dozen or so mentions of birds on preceding pages—and the significant ornithological content of these is covered in chapter 5. In volume 2 of *Variations*, however, there is a great number of predominantly brief mentions of (mostly domesticated) birds. Most bird observations in the latter volume refer to domestic pigeons and their various breeds, fowls, turkeys, peafowl, and ducks and are far too numerous to treat as separate entries here. I do, however, provide page numbers for wherever mention of these domesticated species is made under the applicable headings (i.e. **fowl(s)** or **poultry**; **peafowl**, **peacock,** or **peahen**; **turkey(s)**; **goose** or **geese**; **duck(s)** and **pigeon(s)** and/or **rock pigeon** or ***C. livia***.). Suffice it to say that they are mentioned in his discussions of many and varied aspects of variation under domestication, including within the four chapters dedicated to birds. Because he presented the great weight of evidence obtained from these birds under numerous subheadings, covering many different aspects of his general theme of variation, many brief observations are repeated by him.

The words "pigeons" or "poultry" or both together appear on numerous pages and sometimes several times on a single page of *Variation of Animals and Plants*. This work emphasizes just how extensive was the knowledge that CD had gained about pigeons, fowls, and other domesticated birds, as a result of personal observations and experiments, the literature available to him, and his copious correspondence with numerous authorities, and that he applied so effectively as evidence supporting his theories.

Darwin wrote extensively about birds in the four chapters of his *The Descent of Man* that are dedicated to them, and although all of the species and topics he included appear following, more extensive treatment of some of his ornithology therein appears in chapter 6. All bird species mentioned in his published work appear following. It is inevitable, given the amount and density of literature involved, that I will have missed some ornithological content in this analysis, particularly as CD often alluded to a bird species or topic most briefly in passing. I would much appreciate being made aware of any such oversights.

Mention of Wallace refers to the English biologist and evolutionist Alfred Russel Wallace; of Gould to English ornithologist and bird artist John Gould; and of Audubon to the American ornithologist and bird artist John James Audubon.

A cursory count shows that Darwin mentioned over 600 bird species in his publications, plus numerous higher avian taxa and subjects. There are some 1,700 ornithological subjects listed (including synonyms), and these are shown following in **bold typeface** to the extreme left of the page—but these hardly reflect the total number because many have multiple entries (e.g., under **hummingbirds** 19, **pigeons** 47, and **bird[s]** 115). These statistics can, however, give no adequate measure of Charles Darwin's published ornithology, as they cannot indicate the length of many of his discussions about birds as contained within *The Zoology of the Voyage of H. M. S. Beagle, Part III: Birds*; the four bird chapters in *The Variation of Animals and Plants Under Domestication*; and the four bird chapters in *The Descent of Man*, totalling 613 pages dedicated to birds. This excludes any bird material in the substantial remaining pagination of these works and in his entire *Journal of Researches . . .*, *Origin of Species . . .* and *The Expression of the Emotions in Man and Animals*.

FOSSIL BIRDS

Archeopteryx = *A. lithographica*, an Upper Jurassic bird/reptile fossil found in Germany in 1861 showing that birds did not suddenly appear during the Eocene, **O**: 44344–4. See chapter 4:

That "strange Secondary bird, with a long lizard-like tail" that serves "to connect several of the great vertebrate classes more or less closely," **D**: 245.

NON-PASSERIFORMES

DINORNITHIDAE—Moas

Deinornis = *Dinornis*, an extinct genus of flightless New Zealand Moas, mentioned as appearing to have "replaced (in the ecological sense) mammiferous quadrupeds, in the same manner as the reptiles still do on the Galapagos archipelago," **J2**: 427–428.

gigantic birds of New Zealand = unspecified extinct flightless Moas, as an example of "this wonderful relationship in the same continent between the dead and the living" [i.e., living **Apteryx** or kiwis], **O**: 485.

Order: Cursores = a disused taxon containing the ratites and waders, **Z**: 120.

They include few species with "strongly-marked sexual differences" except among the bustards, **D**: 339.

TINAMIFORMES

TINAMIIDAE—Tinamous

Tinamus = tinamous; see **partridge**, **J**: 131.

Rhynchotus rufescens = Red-winged Tinamou, locations, habitat, habits, call, rising in flight like partridges, **Z**: 120. See chapter 2.

Tinamus rufescens = as a synonym of **Rhynchotus rufescens**, **Z**: 120.

Rhynchotus fasciatus = as a synonym of **Rhynchotus rufescens**, **Z**: 120.

Cryptura Guaza = as a synonym of **Rhynchotus rufescens**, **Z**: 120.

Crypturus rufescens = as a synonym of **Rhynchotus rufescens**, **Z**: 120.

Nothura perdicaria = Chilean Tinamou *Nothoprocta perdicaria*, like **N. major** in looks and habits but larger with longer beak, distribution, habits, call, flight, egg (compared with that of **N. minor**); **Z**: 199–220.

Crypturus perdicarius = as a synonym of **Nothura perdicaria**, **Z**: 119.

Northura minor = Lesser Nothura *Nothura minor*, specimen from Bahia Blanca, habitats, habits, egg, **Z**: 119.

Tinamus minor = as a synonym of **Northura minor**, **Z**: 119.

Northura major = Spotted Nothura *Nothura maculosa*, common on northern shore of La Plata, habits, egg, call, capturing them; **Z**: 119.

Tinamus major = as a synonym of **Northura major**, **Z**: 119.

Eudromia elegans = Elegant Crested-Tinamou *Eudromia elegans*, **J**: 131. See **partridge**.

Partridges(s) = Red-winged Tinamou *Rhynchotus rufescens*, in great numbers several days ride from Maldonado, habits, killing and catching them from horseback; **J**: 51. See chapter 2.

An unidentified species of tinamou attacked by a **Carrancha**, **J**: 65.

Three species, "two species of Tinamus, and *Eudromia elegans* (Elegant Crested-Tinamou)," briefly alluded to as being predated by a fox; **J**: 131.

Unspecified "partridges" briefly mentioned as suffering in drought, **J**: 156.

The same species of some partridges occur of both sides of the Andes, **J**: 400.

Red-winged Tinamous take flight like (true) partridges, **Z**: 120. See chapter 2.

STRUTHIONIFORMES

Ostrich order = Struthioniformes, mentioned prior to discussing some members (cassowaries and emus); **D**: 729.

STRUTHIONIDAE—Ostriches

Ostrich family = unspecified but probably including at least the rheas and Emu, most or all of which lack the oil gland typical of most birds; **V1**: 147 footnote.

ostrich-tribe = "evidently a widely-diffused remnant of a larger group," and a group that serves "to connect several of the great vertebrate classes more or less closely"; **D**: 245.

Ostrich = Common Ostrich *Struthio camelus*, although not named as such, CD writes, in an account of the Greater Rhea, "I may add, also, that it is believed in Africa, that two females lay in one nest." **J**: 107.

Briefly alluded to in noting distributions of the two rhea species, **O**: 496.

African ostrich = Common Ostrich *Struthio camelus*; males are larger, more colourful, and better plumed than females, but only males incubate; **D**: 730.

Lays 25 to 30 eggs [a clutch] in the wild, but captives in the south of France lay only 12 to 15; **V2**: 156.

RHEIFORMES

RHEIDAE—Rheas

ostriches (*Struthio Rhea*) or **ostrich(s)** = the Greater Rhea *Rhea americana*, discussed in **J**: 48, 50, 74, 90, 93, 105–109, 131, 137, 139, 177, 198, 400, and **1837** and **1881** papers. See chapters 2 and 3.

The species lays clutches of "a score" yet may be less numerous than the **condor**, which lays only "a couple of eggs"; **O**: 81.

Rhea Americana = Greater Rhea *Rhea americana*, distribution, habits, food, how hunted by Gauchos, swimming, sexual dimorphism, call, nest, eggs, male parental care of multiple females' eggs, male defence of nest/young; **Z**: 120–123; **1837** and **1881** papers. See chapters 2 and 3.

Compared to **Rhea darwinii**, **Z**: 123–124. See chapter 2.

Males are larger, stronger, and swifter than females but provide all parental care; **D**: 731 footnote.

Rhea darwinii = Lesser (Darwin's) Rhea *Rhea pennata*, described and compared with Greater Rhea, illustrated, distribution, eggs, swimming; in **Z**: 123–125; **1837** and **1881** papers. **J**: 475. See chapter 3.

Males are larger, stronger, and swifter than females but provide all parental care; **D**: 731 footnote. See colour figure 2.2, chapter 2.

Rhea pennata = see **Rhea darwinii**, **Z**: 125. See colour figure 2.2.

Avestruz Petise = Lesser (Darwin's) Rhea, *Rhea pennata*; **J**: 108–110; **Z**: 123–125. See chapters 2 and 3 and colour figure 2.2.

ostrich feathers = Lesser Rhea *Rhea pennata*, bartered for with natives of the Strait of Magellan; **J2**: 232. See colour figure 2.2.

Ostrich(es) = unspecified but for both rheas and the ostrich, in noting they inhabit continents where exposed to dangers yet are flightless (but kick in defence). CD believes their progenitor was of the size and habits of "the bustard," **O**: 167–168. See chapter 4.

Wing use, **O**: 218. See chapter 4.

CD responds to the question "why has not the ostrich acquired the power of flight?" with because it would require an "enormous supply of food"; **O**: 281. See chapter 4.

Females lay eggs in several nests of males, and the reason for this is discussed; **O**: 335–336.

Their wings are considered "rudimentary" and "almost useless . . ., which serve merely as sails"; **O**: 621.

Their eyesight is good, **Z**: 125.

Young are "covered with longitudinally striped down," **D**: 708.

American ostrich = unspecified but probably refers to both rheas, laying many eggs out of nests; **O**: 336.

Rhea (American ostrich) = referring to both Rhea species in noting their exclusive geographical distributions and the lack of the African ostriches and Australian Emu on South America, **O**: 496.

CASUARIIFORMES

CASUARIIDAE—Cassowaries

Cassowary = unspecified, of the genus *Casuarius*, erect their feathers in anger; **E**: 99.

common cassowary (Casuarius galeatus) = Southern Cassowary *Casuarius casuarius*, males are smaller and less colourful than females and perform all incubation; females are pugnacious in the breeding season when their bare face and neck becomes most colourful; **D**: 729. See chapter 7 and colour figure 7.1.

mooruk (*Casuarinus Bennetti*) = Dwarf Cassowary *Casuarius bennetti*, from New Ireland (where introduced), frequently bred in European menageries; **V2**: 156.

DROMAIIDAE

emus (*Dromæus irroratus*) = *Dromaius novaehollandiae*, the female is considerably larger than the male, more vocal and pugnacious, and males perform all parental duties; **D**: 730.

emu(s) = *Dromaius novaehollandiae,* CD noted birds were observed swimming the Murrumbidgee River, Australia, by Captain Sturt; **J**: 106.

While in the Blue Mountains, Australia, CD noted the Emu "is banished at a long distance"; **J**: 525.

Briefly alluded to in noting distributions of the two rhea species, **O**: 496. See chapter 2.

In captivity, only the male attended the nest, **Z**: 122.

Emu used as name for rheas by Dobrizhoffer, **Z**: 125.

APTERYGIFORMES

APTERYGIDAE—Kiwis

Apteryx = the genus of flightless kiwis of New Zealand mentioned in a review of conflicting opinions published by Professor Richard Owen about the origin of species, **O**: xxv–xxvi.

Loss of flight in them, **O**: 210.

CD considers their wings to be useless and thus truly redundant, **O**: 622. See also chapter 4.

ANSERIFORMES

ANHIMIDAE—Screamers

Palamedea cornuta = Horned Screamer *Anhima cornuta*, formidable wing spurs worn by both sexes, illustrated; **D**: 558. See chapter 6 and colour figure 6.2.

ANATIDAE—Ducks, Geese, and Swans

Anatidae = waterfowl, the meaning of differences in the trachea between the sexes within the family is not understood; **D**: 572.

Dendrocygna = see **Indian goose**.

Musk duck (*Dendrocygna viduata*) = White-faced Whistling Duck, perches and roosts in trees; **V1**: 181–182.

Dendrocygna viduata = White-faced Whistling Duck, easily tamed, is kept by Guianan Indians, but "absolutely refuses to bred in captivity"; **V2**: 157.

Chinese geese or **goose (*A. cygnoides*)** = Swan Goose *Anser cygnoides*, the results of hybridization with **common geese** [in domestication] are highly distinctive; **O**: 373–374, **D**: 632–3.

The knob on the beak, much larger in males than females, may be attractive to females; **D**: 650.

CD notes that the wild parent form was said to remain unknown, **V1**: 237.

Grey-lag Goose (*A. ferus*) = Greylag Goose *Anser anser*, most judges consider this the ancestor of the domestic goose; **V1**: 287. See chapter 5.

Unlike in domestic geese, the older ganders are not white feathered; **V1**: 288.

goose = domestic or Greylag Goose *Anser anser*, CD notes the lack of many breeds of it; **O**: 48.

In the "Sebastopol goose the scapular feathers are greatly elongated, curled, or even spirally twisted, with the margins plumose"; **D**: 589.

Said to have run wild in La Plata, **V1**: 190.

Varies more than almost any wild bird, so it cannot be affiliated with certainty to any natural species; **V2**: 254.

goose or **geese** = domestic or Greylag Goose *Anser anser*, **V1**: 274, 287–290; **V2**: 104, 112, 161–162, 199, 204, 209, 233, 254, 258, 287, 297, 332, 392. See chapter 5.

common goose/geese = Greylag Goose *Anser anser*, describes beak use in foraging; **O**: 288. See chapter 4.

Hybridizes with Chinese Goose under domestication with highly distinctive results, **O**: 373–4, **D**: 632–633.

goose-quill = unspecified but likely of domestic goose *Anser anser*, likened to quills of a porcupine's tail; **E**: 94.

white-fronted goose (*A. albifrons*) = Greater White-fronted Goose *Anser albifrons*, the trachea and feathers at the beak base of the domestic goose indicate past crossing with this species; **V1**: 288.

snow-goose (*Anser hyperboreus*) = Snow Goose *Chen caerulescens*, the plumage becomes white only at maturity; **D**: 751.

Canada geese or **goose** = *Branta canadensis*, one successfully hybridized with a **Bernicle gander** in domestication; **D**: 632.

Canada geese (*Anser canadensis*) = *Branta canadensis*, Audubon observed that within flocks, those previously mated renewed courtship early while others spent hours every day doing so until satisfied with their choice **D**: 635.

Wild geese (*Anser Canadensis*) = Canada Goose *Branta canadensis*. Audubon kept some for over eight years, but they would not mate; **V2**: 157.

Bernicle gander = Brant Goose *Branta bernicla*, one successfully hybridized with a **Canada goose** in domestication; **D**: 632.

black swan = *Cygnus atratus*; both sexes are black throughout, but this is contrasted with a red beak; **D**: 749.

A small domestic goose variety has structurally similar scapular feathers, **V1**: 289.

black-necked swan = *Cygnus melancoryphus*, seen on a pampus lake in September 1833; **J**: 133, 346.

CD cites "Pernety (1763)" who noted that whereas all the endemic birds on the Falklands were remarkably tame, the Black-necked Swan, a bird of passage there, was impossible to kill because it had learnt to fear humans elsewhere; **J**: 477. See chapter 2.

Is a "piebald" species in a group with both black and white species, **D**: 753.

common swan (*Cygnus olor*) = Mute Swan, young do not replace their dark plumage with white until 18 to 24 months old; **D**: 736.

Cygnus immutabilis = Mute Swan *Cygnus olor*, the name was applied to the Polish population in which young are reported to always be white; **D**: 737 footnote.

wild swan (*Cygnus ferus*) = Whooper Swan *Cygnus cygnus*, the trachea is more deeply embedded in the sternum in adult males than in females or young males; **D**: 527.

Swan(s) = unspecified but doubtless Mute Swan *Cygnus olor*, threat display described and illustrated; **E**: 99–101. See chapter 7 and figure 7.2.

Unspecified, young are slate coloured, whereas adults are white; **D**: 734.

Unspecified, it is probable that the complete or partial blackness of some species is the result of sexual selection "accompanied by equal transmission to both sexes"; **D**: 749.

Unspecified, sexual selection has brought about both white and black species; **D**: 752.

Unspecified, tail feathers are unusually numerous and apt to vary in number; **V1**: 158.

Micropterus brachypterus = Fuegian Steamer Duck *Tachyeres pteneres* (see **Loggerheaded duck)**, locomotion, habits, food, calls; **Z**: 136. See colour figure 2.4.

Anas brachytera = as a synonym of **Micropterus brachypterus**, **Z**: 136.

logger-headed duck or goose *Anas brachyptera* = Fuegian Steamer Duck *Tachyeres pteneres* (notwithstanding the specific name CD gave, which in fact refers to the flying Falkland Steamer Duck [*T. brachypterus*] because he describes its flightlessness), described and observed; **J**: 257–258; **O**: 167. See chapter **2**.

Wing use, **O**: 218, 648. See chapters 2 and 4.

In discussing evolution of the bird wing and compared with penguins, **O**: 443.

Weight of this bird, **Z**: 136. See chapter 2.

Merganetta armata = Torrent Duck, bill structure (and whale baleen) mentioned in discussing adaptation to sifting mud and water; **O**: 287. But see also chapters 2 and 4.

spur-winged goose (*Plectropterus gambensis*) = males have much larger spurs than females and use them in fighting and in defence of young, **D**: 558. See chapter 6.

Indian goose (*Sarkidiornis melanotus*) = Knob-Billed Duck *Sarkidiornis melanotos*, the young closely resemble mature birds of the (now not so closely) allied genus ***Dendrocygna***; **D**: 709.

Egyptian Goose (*Chenalopex*) = *Alopochen aegyptiaca*, bill structure mentioned in discussing adaptation to sifting mud and water (and whale baleen); **O**: 288–289. See chapters 4 and 5.

The "bare obtuse knobs" on its wings probably show "first steps" of the true spurs in other species, **D**: 558.

Offspring from a cross with the domestic Penguin duck showed much influence of the latter, **V1**: 282; **V2**: 68.

The rate of change in breeding seasonality under domestication has been recorded, **V2**: 304.

Anser melanopterus = Andean Goose *Chloephaga melanoptera*, Captain FitzRoy purchased a skin at Valparaiso, illustrated; **Z**: 134.

upland goose/geese = *Chloephaga picta*, noted as tame on the Falklands; **J**: 476–477.

Hunters sometimes kill more in one day than can be carried home, **J**: 477.

Have webbed feet but "rarely go near water," **O:** 222–223, 250.

Anas leucoptera = Upland Goose *Chloephaga picta*, which CD referred to as "The upland species"; **J**: 257. See chapter 2.

Chloephaga magellanica = Upland Goose *Chloephaga picta*, locations, habitats, habits, nesting, food; **Z**: 134. See chapter 2.

Anas Magellanica = as a synonym of **Chloephaga magellanica**, **Z**: 134.

Bernicla leucoptera = as a synonym of **Chloephaga magellanica**, **Z**: 134.

Rock Goose or **Bernicla antarctica** = Kelp Goose *Chloephaga hybrida*, locations, habits; **Z**: 134–135.

Anas Antarctica = as a synonym of **Bernicla antarctica**, **Z**: 134.

The males are wholly white, doubtless through sexual selection; **D**: 750; **V1**: 288.

rock goose (*Anas Antarctica*) = Kelp Goose *Chloephaga hybrida*, CD notes it exclusively living on the "sea-beach" and found it "a common feature in the landscape"; **J**: 257. See chapter 2.

lowland goose = Ruddy-headed Goose *Chloephaga rubidiceps*, tame on Falkland Islands; **J**: 476–477.

Chloephaga – in two species, the sexes are indistinguishable; but in the other two, the sexes are so different that they might be mistaken for distinct species; **D**: 702.

geese or **goose** = unspecified, wounded birds on the Falkland Islands attacked by Striated Caracara *Phalcoboenus australis*; **J**: 67.

Small flocks seen on East Falkland Islands, **J**: 246.

On Falkland Islands, they nest on islets to avoid fox predation, but are tame toward people; **J**: 477.

Webbed feet are "formed" for swimming, yet webbed upland geese "rarely go near water"; **O**: 222, 646. See chapter 2.

Members of nocturnally migrating flocks call to one another, **D**: 563.

When many species are together in domestication, pairs of different species frequently form bonds; **D**: 632.

Mr. Dixon says there is not a species that is not domesticable to the extent of breeding under confinement, but CD thought this too bold a statement; **V2**: 157.

Sheldrake(s) = Common Shelduck *Tadorna tadorna*, foot patting in foraging for marine worms and also when hungry; **E**: 50. See chapter 7.

shield-drake (Tadorna vulpanser) = Common Shelduck *Tadorna tadorna*, CD notes a "remarkable attachment" between one and a **common duck** in discussing hybrids in domestication; **D**: 632.

New Zealand shieldrake (*Tadorna variegata*) = Paradise Shelduck, sexual dimorphism described, parental care briefly mentioned as is that young resemble adult males; **D**: 731 footnote.

musk-duck (*Cairina moschata*) = Muscovy Duck, males have "bloody fights" during breeding season; **D**: 554.

Fond of roosting and perching on trees or buildings, **V1**: 182.

Domestic birds resulting from hybridization with common duck wild in North America, Belgium, and near the Caspian Sea; **V1**: 190.

Aix sponsa = Wood Duck, bill structure (and whale baleen) mentioned in discussing adaptation to sifting mud and water; **O**: 287. See chapter 4.

mandarin Teal (*sic*) = Mandarin Duck *Aix galericulata*, a duck losing her stolen mate "remained disconsolate" (while another drake sedulously courted her) only to recognize her mate "with extreme joy" when he was returned three weeks later; **D**: 626.

Querquedula erythrorhyncha = Brazilian Teal *Amazonetta brasiliensis*, specimens obtained at Buenos Ayres and Straits of Magellan; **Z**: 135.

Anas erythrorhyncha = as a synonym of **Querquedula erythrorhyncha**, **Z**: 135.

wigeon (*Mareca penelope*) = Eurasian Wigeon *Anas penelope*, a male successfully hybridized with a **pintail duck** in domestication despite conspecific females being present; **D**: 632.

domestic duck = Mallard *Anas platyrhynchos*, leg bones heavier and wing bones lighter in proportion to entire skeleton than in wild Mallard, as adaptations to domestication; **O**: 12. See figure 5.10.

Beak structure, **O**: 310. See figure 5.10.

Rev. W. D. Fox informed CD that "so many mallards were shot" on a large pond that "only one [male] was left for every seven or eight females, yet unusually large broods were reared"; **D**: 339.

common duck = Mallard *Anas platyrhynchos*, bill structure (and whale baleen) mentioned in discussing adaptation to sifting mud and water; **O**: 287–288. See chapter 4 and figure 5.10.

The "male hisses, whilst the female utters a loud quack"; **D**: 572.

CD notes a "remarkable attachment" between one and a **shield-duck** in discussing hybrids in domestication, **D**: 632.

Domestic birds have become almost wild in Norfolk, and ones resulting from hybridization with Muscovy Ducks wild in North America, Belgium, and near the Caspian Sea; **V1**: 190.

common drake (*Anas boschas*) = Mallard *Anas platyrhynchos*; after breeding, the drake loses male plumage for three months, during which he assumes female plumage; **D**: 598–599.

duck = wild or domestic duck, Mallard *Anas platyrhynchos*; CD notes that fanciers he knew were convinced that each main domestic breed was descended from a distinctly different species; **O**: 33.

In discussing origin of whale baleen, as duck sifts mud and water with a lamellated beak; **O**: 285.

Wild ducks disperse seeds and small molluscs on their feet and legs, **O**: 537–538. See chapters 3 and 4.

Although it has wings, it is nearly incapable of flight in domestication; **O**: 648.

Well-marked sexual differences in British wild ducks in which males pair with a single female noted, **D**: 337–338.

The wild duck is strictly monogamous, the domestic duck highly polygamous; **D**: 339.

The green wing speculum, common to both sexes, is acquired early in life, whereas the curled tail feathers and other ornaments of the male are acquired later; **D**: 364.

Captives "recently descended from wild birds" fled from strange dogs and cats but ignored their owner's dog, **D**: 628.

A wild duck reared in captivity and paired to a drake **mallard** then reproduced with a **Pintail**, **D**: 633, 640.

Young of both sexes closely resemble adult females, whereas adult males differ conspicuously; **D**: 712.

The tufted domestic duck has a fleshy fibrous mass beneath the tuft of head feathers but no skull modification, **V1**: 266; but CD also noted they, like "nearly all crested fowls," show an "imperfectly ossified condition" of the skull associated with the crest; **V1**: 274. See chapter 5.

Brief discussion of "reversion," or the appearance of ancestral traits, in individuals at various ages; **V2**: 45–46.

duck(s) = the domesticated species and its various breeds, often with reference to the original wild species *Anas platyrhynchos* and *Cairina moschata*; **V1**: 190, 274, 276–287; **V2**: 25, 40, 45–46, 51, 68, 97, 199, 233, 262–264, 278, 287, 297–298, 304, 325, 332, 350, 371, 407–408, 414. See chapters 5 and 8.

Aylesbury duck = a form of **domestic duck** that has adult wings in a similar reduced condition to those of the Flightless Steamer Duck, **O**: 167; inheritance of hatchling period in England, **V2**: 25.

Rhynchaspis maculatus = Red Shoveler *Anas platalea*, CD received a skin from the Rio Plata; **Z**: 135.

shoveller = Red Shoveler *Anas platalea*, CD cites a published painting of same; **Z**: 135.

shoveller-duck = Northern Shoveler *Anas clypeata*, the lamellae of its beak compared to whale baleen and origins of these discussed; **O**: 286–289, 310. See chapter 4.

Querquedula creccoides = Yellow-billed Teal *Anas flavirostris*, specimens from Rio Plata and Straits of Magellan, closely allied to **Q. Carolinensis**; **Z**: 135–136.

Anas creccoides = as a synonym of **Querquedula creccoides**, **Z**: 135.

Paecilonitta bahamensis = Yellow-billed Pintail *Anas georgica*, a specimen from a Galapagos Islands salt-water lagoon; **Z**: 135.

Anas Bahamensis = as a synonym of **Paecilonitta bahamensis**, **Z**: 135.

Mareca Bahamensis = as a synonym of **Paecilonitta bahamensis**, **Z**: 135.

Dafila urophasianus = Yellow-billed Pintail *Anas georgica*, recorded at Bahia Blanca, Patagonia; **Z**: 135.

Anas urophasianus = as a synonym of **Dafila urophasianus**, **Z**: 135.

pin-tail duck (*Anas acuta*) = Northern Pintail, males loose nuptial plumage for six to eight weeks each year; **D**: 599. See chapter 8.

pintail duck or **Pintail (*Querquedula acuta*)** = Northern Pintail *Anas acuta*, a female successfully hybridized with a male **Wigeon** in domestication despite conspecific females being available to him; **D**: 632. See chapter 8.

A wild **mallard** duck reared in captivity and paired to a drake of her kind then paired and reproduced with a **Pintail**, **D**: 633, 640.

Querquedula Carolinensis = Eurasian Teal *Anas crecca*, see **Querquedula creccoides**; **Z**: 135–136.

Anas punctata = Hottentot Teal *Anas hottentota*, the bony enlargement of the trachea is "only a little more developed in the male than in the female"; **D**: 572.

Anas = unspecified, young fertile females may acquire the characters of the [adult] males; **D**: 704.

Harlequin duck = *Histrionicus histrionicus*, males take three years to acquire full adult plumage, but many breed in their second year; **D**: 738–739 footnotes.

black scoter-duck (Oidemia) = Common Scoter *Melanitta nigra*, males are black with a brightly coloured beak, and the females brown or mottled, doubtless through sexual selection; **D**: 749.

Long-tailed duck (*Harelda glacialis*) = *Clangula hyemalis*, certain females are more attractive to males than others and may be surrounded by six to eight courting males; **D**: 641.

Mergus cucullatus = Hooded Merganser *Lophodytes cucullatus*, sexual dimorphism with reference to that in the speculum and the age that this is acquired is discussed; **D**: 365.

Merganser = Common Merganser *Mergus merganser*, has "recurved teeth" in the beak for fishing; **O**: 288–289.

In males, "the enlarged portion of the trachea is furnished with an additional pair of muscles"; **D**: 572.

goosander (*Mergus merganser*) = Common Merganser, the adult male is conspicuously more colourful than females and has longer scapulars and secondaries, but the crest of females and young is longer than in males; **D**: 713–714, 753. See also chapter 2.

Merganser serrator = Red-breasted Merganser *Mergus serrator*, males "undergo a change of plumage, which assimilates them in some measure to the female [in appearance]"; **D**: 599.

Australian musk-duck (*Biziura lobata*) = Musk Duck, males emit an odour in summer, which may be retained throughout the year by some individuals; but Mr. Ramsay never shot a female in the breeding season with the odour; **D**: 549–550. See chapter 6.

Males are twice as large as females; **D**: 554.

duck-family = mentioned in discussing bill adaptations; **O:** 289, 310. See chapter 4.

No birds breed "with such complete facility under confinement as members of this group," **V2**: 157.

Duck tribe = only one species had bred at a French facility prior to many being bred in England; **V2**: 149.

ducks = unspecified, webbed feet are "formed" for swimming; **O**: 222.

Females may partly assume secondary male characters; **V2**: 51.

wild fowl = mentioned in briefly alluding to a lake "swarming with wild fowl," **J**: 133.

Females are generally similar to those of allied species, whereas the males differ dramatically; **D**: 717.

waterfowl or water-fowl = unspecified, "particularly numerous" on east Falkland Islands yet less so than recorded by "old navigators"; **J**: 256.

Merely states it is difficult to know if there are any new species on Galapagos Islands; **J**: 461.

Noted as tame on the Falkland Islands; **J**: 477.

Unspecified, nocturnally migrating flock members call to one another; **D**: 563.

Harsh trumpet-like cry of some "pleasing to the females," **D**: 580.

Order: Palmipedes = a disused taxon that included the waterfowl and some other web-footed swimming and nonswimming groups; **Z**: 134.

GALLIFORMES

Razores or **Rasores** = a disused taxon that included the Galliformes, pigeons, tinamous; anatomical characters confirm affinity of seedsnipe to this group and to the **Grallatores**; **Z**: 118, 122.

MEGAPODIIDAE—**Megapodes**
Talegalla lathami = Australian Brushturkey *Alectura lathami*, a footnote to **Rhea americana** notes that multiple female Brush-Turkeys lay in one male's nest mound; **Z**: 122.

CRACIDAE—**Chachalacas, Curassows, and Guans**
Guans or Cracidæ = are easily tamed but are very shy breeders in the UK (although various species were previously bred in Holland) and in South America where kept by native Indians; **V2**: 156.

Chamæpetes unicolor = Black Guan, in males the first primary arches toward the tip and is much more attenuated than in females; **D**: 577.

Penelope nigra = Highland Guan *Penelopina nigra*, Mr. Salvin observed a male flying downward with outstretched wings "gave forth a kind of crashing rushing noise" like the falling of a tree; **D**: 577.

black curassow = *Crax alector*; while covered with down, the young are longitudinally striped on the back, but this is lost with maturity; and CD discusses why domestic fowl chicks are not so marked; **V1**: 249.

NUMIDIDAE—**Guineafowl**
guinea-fowl = Helmeted Guineafowl *Numida meleagris*, a flock of 50 or 60 on St. Jago Island; **J**: 3. See chapter 2.

CD notes unspecified "guinea-fowl" imported to Ascension Island, Atlantic Ocean, from the Cape Verde Islands; **J**: 587.

The contrast in sexual dimorphism between this and the polygamous **peacock** or **pheasant** noted; **D**: 338.

It is "strictly monogamous; but Mr. Fox finds his birds succeed best when he keeps one cock to two or three hens"; **D**: 339.

The sexes are indistinguishable and the plumage could have been acquired through sexual selection by males and then "transmitted to both sexes," **D**: 719.

Domesticated birds had become wild on Ascension Island and in Jamaica, **V1**: 190.

Feral birds in West Indies vary more than in the domesticated state; **V2**: 33.

A native of Africa, it nevertheless lays many eggs in the damp and cool UK; **V2**: 161.

See Numida ptilorhynca.

Numida ptilorhynca = Helmeted Guineafowl *Numida meleagris*, domestic guineafowl believed to have descended from the species defined by this disused species name; **V1**: 294. See chapter 5.

guinea-fowl(s) = unspecified, their plumage presents a good example of white spots surrounded by darker zones, in discussing the origins of plumage ocelli; **D**: 654–655.

None of the species in which plumage colours have been largely transferred from males to females are brilliantly coloured; **D**: 720.

Has not varied at all under domestication, or only in colour; **V2**: 287.

PHASIANIDAE—**Pheasants and allies**

wild turkey-cock = Wild Turkey *Meleagris gallopavo*, CD considered "the tuft of hair on the breast" [actually modified feathers] of no use and doubtfully attractive to females (he was wrong); **O**: 110. See chapter 4.

Are ready to fight whenever they meet; **D**: 560.

Scrape their wings against the ground; **D**: 573, 581. See chapter 6.

Males swell and intensify colours of appendages in display; **D**: 584.

Small barbless feathers on its breast "appear like bristles," **D**: 587.

In an extinct variety, "the top-knot consisted of bare quills surmounted with plumes of down"; **D**: 589.

Brief description of male display as "superb but grotesque to human eyes," **D**: 601.

Young turkey cocks "in fighting always seize hold of each other's wattles," **D**: 614.

Females apparently more numerous than males, suggesting to CD the former "would have been led" to compete to court the latter as with **peahen, Indian Turnix**, and some **grouse**; **D**: 733.

wild turkey = *Meleagris gallopavo*, when the female gives her morning call, the male answers by a note other than the gobbling given in display; **D**: 573.

Audubon raised one that would run away from strange dogs but not his own, even when it escaped into the wild and was retrieved by its canine "friend"; **D**: 628.

Domesticated females prefer to be mated by wild males than by fellow captives; **D**: 638.

Audubon states older females usually make first advances to mating; **D**: 640.

In two-year-old males, the tuft of breast "bristles" is four inches long but is hardly apparent in females of that age; **D**: 705.

Females differ from males only in their colours being much duller; **D**: 719.

Turkey = unspecified but Wild or domestic Turkey *Meleagris gallopavo*, CD noting that the bare head of males is similar to that of the "**vulture**," which is generally seen as adapted to "wallowing in putridity"; but that caution is required as the turkey is "clean-feeding"; **O**: 248.

Young will run from mother giving alarm call to hide, thus allowing her to escape; **O**: 329; **V1**: 181.

Hard fruit seeds pass uninjured through the digestive organs; **O**: 510.

Domesticated birds once almost feral on banks of the Parana River; **V1**: 190.

Mexican [Turkey] **(*Meleagris Mexicana*)** = CD notes this as being thought as a distinct species that gave rise to the domestic bird (but in fact is the nominate subspecies *Meleagris g. gallopavo* of the Wild Turkey); **V1**: 292. See chapter 5.

turkey(s) = the domesticated *Meleagris gallopavo* and its forms; **V1**: 190, 292–294, 161, 199, 233, 258, 262–263, 278, 287, 413. See chapter 5.

Tetrao umbellus = Ruffed Grouse *Bonasa umbellus*, CD cites an author who states that fights among males before females are a "sham" designed to show themselves to best advantage; **D**: 562.

Displaying males rapidly strike their wings together above the back and not against the sides as Audubon thought, **D**: 573–574.

Breed at Lord Derby's UK aviaries; **V2**: 156.

capercailzie = Western Capercaillie *Tetrao urogallus*, males of this polygamous species "differ greatly from the females" (see **black-cock**); **D**: 338.

Scandinavian broods produce more males than females (see **black-cock**); **D**: 382.

Have "appointed places" for fighting and courting; **D**: 556.

CD told that females sometimes steal away with a young male too timid to approach older rivals; **D**: 561.

Males sometimes hold traditional leks during autumn; **D**: 567.

Females are "charmed" by the call of males; **D**: 573.

Leks of these "polygamists" active from end of March to mid or late May [presumably in the UK]; **D**: 617–618.

Females "flit round the male" as he displays and solicit his attention; **D**: 640.

The black male plumage would be too dangerous for the female to wear while incubating; **D**: 676.

Males are black with crimson bare skin over the eyes, and the females brown or mottled, this doubtless being the result of sexual selection; **D**: 749.

Bred in London Zoo, and without difficulty in Norway and Russia; **V2**: 156.

Tetrao urogalloides = Black-billed Capercaillie *Tetrao urogalloides*, "follow nearly the same [courting] habits" as **black-cock** and ***Tetrao phasianellus***; **D**: 617.

black-cock or **blackcock** = Black Grouse *Lyrurus tetrix*, males of this polygamous species "differ greatly from the females" (see **capercailzie**); **D**: 338.

Scandinavian broods produce more males than females (see **capercailzie)**; **D**: 382.

Have "appointed places" for fighting and courting, and courtship display described; **D**: 556.

The "*spel*" vocalization of males serves as a call to females; **D**: 573.

Moult only once in the year; **D**: 597.

In Germany and Scandinavia, traditional leks are active mid-March into May involving 40 or 50 birds, which are "polygamists"; **D**: 617–618.

Males are black with crimson bare skin over the eyes, and the females brown or mottled, doubtless through sexual selection; **D**: 749.

black game = Black Grouse = *Lyrurus tetrix*, CD cites but one record of lekking males in Scotland, despite this being well known in Germany and Scandinavia; **D**: 620.

black grouse or **black-grouse** = *Lyrurus tetrix*, 18 cases of hybridization with the **pheasant** in Great Britain noted; **D**: 631.

The sexes differ greatly but build an open nest like Willow Ptarmigan in which the sexes are similar; **D**: 694.

Young resemble the young as well as the old of certain other species, for example, **red-grouse** or ***T. scoticus***; **D**: 709.

Female and young "resemble pretty closely" both sexes and young of Willow Ptarmigan, so CD assumed common descent from an ancestor in which both sexes were coloured like **red-grouse**; **D**: 719–720. See chapter 7.

Can the very slight differences in tints and markings between females, and those of the **red-grouse** serve as a protection; **D**: 723.

While covered with down, the young are longitudinally striped on the back but loose this with maturity, and CD discusses why chicks of domestic fowl are not so marked; **V1**: 249.

Tetrao tetrix = Black Grouse = *Lyrurus tetrix*, bred in captivity in Norway; **V2**: 156.

pheasant-grouse = a loose term applied to cover **black-cock**, ***Tetrao phasianellus***, and ***Tetrao urogalloides*** in considering them all "polygamists"; **D**: 618.

grouse (*Tetrao urophasianus*) = Sage Grouse *Centrocercus urophasianus*, a courting male has his "bare yellow œsophagus inflated to a prodigious size, fully half as large as the body"; **D**: 569–570.

Tetrao phasianellus = Sharp-tailed Grouse *Tympanuchus phasianellus*, lek display described; **D**: 580.

Their "dances" (i.e., courtship display season) "last for a month or more," **D**: 617.

grouse = unspecified, females of certain kinds apparently more numerous than males, suggesting to CD the former "would have been led" to compete to court the latter as with **peahen**, **wild turkey**, and some **Indian Turnix**; **D**: 733.

Tetrao cupido = Greater Prairie-Chicken *Tympanuchus cupido*, the courtship and the selection by females of competing males described; **D**: 562. See figure 6.3.

Males inflate bare orange sacs on their neck, which amplify vocalizations, illustrated; **D**: 568. See chapter 6; see figure 6.3.

Bred in captivity in North America; **V2**: 156.

alpine ptarmigan = Rock Ptarmigan *Lagopus muta*, adaptive white winter plumage noted; **O**: 103.

Red Grouse, red grouse, or **red-grouse** = Willow Ptarmigan *Lagopus lagopus*, mentioned in a review of conflicting opinions about the origin of species; **O**: xxv–xxvi.

CD asks is this bird in Britain a "race of a Norwegian species" or "an undoubted species peculiar to Great Britain" and what distance between two such forms suffices for species recognition; **O**: 59–60.

Wear adaptive plumage colour like that of heather; **O**: 103.

The sexes of this monogamous species "differ very little" (see **ptarmigan**); **D**: 338.

Can the very slight differences in tints and markings between females and those of the **black-grouse** serve as a protection?; **D**: 723.

red grouse (*Tetrao scoticus*) = Willow Ptarmigan *Lagopus lagopus,* the sexes are similar but build an open nest like Black Grouse in which the sexes are greatly dissimilar; **D**: 694. See chapter 7.

See **black-grouse**.

Tetrao Scoticus = Willow Ptarmigan *Lagopus lagopus*, bred in captivity in Ireland; **V2**: 156.

grouse tribe = mentioned in discussing sexual dimorphism in polygynous and monogamous species; **D**: 338.

grouse = CD found the flight of seedsnipe (Thinocoridae) grouse-like; **Z**: 117.

Stock of unspecified grouse on estates depends on the destruction of "vermin" [i.e., predators]; **O**: 84.

Numbers are kept in check largely by birds of prey; **O**: 103–104.

Flight of **Attagis gayii** is grouse-like; **Z**: 117.

Unspecified, cocks are ready to fight whenever they meet; and when they congregate at an "appointed spot" and fight, they are generally attended by females; **D**: 560.

Some kinds scrape their wings against the ground and "thus produce a buzzing sound," **D**: 537. See chapter 6.

Unspecified, males sometimes display in absence of females and "take delight" in it; **D**: 600.

Unspecified, species within a genus may include those with identical or very different sexes; **D**: 702.

Unspecified, many bred in captivity in the UK; **V2**: 156.

Their fine leg down is unlike the primary feather-like ones on feet of some domestic fowl; **V2**: 323.

ptarmigan(s) = three species of the genus *Lagopus*, of which CD would have been most familiar with *L. lagopus* and/or *L. mutus*, mentioned in passing in comparing them with the two *Attagis* seedsnipe in being similar "in almost every respect to ptarmigans"; **J**: 111; **Z**: 117.

The sexes of this monogamous species "differ very little" (see **red grouse**); **D**: 338.

In a species called "Dal-ripa," more males than females attend the courting site, possibly because more females are predated than are males; **D**: 382.

As an example of a species in which summer and winter plumages serve as (cryptic) protection; **D**: 595.

Moult "twice or even thrice in the year," **D**: 597, 705.

Unspecified in Scandinavia, being monogamous yet holding "nuptial assemblages" during mid-March to mid-May; **D**: 619.

Their summer and winter plumages are cryptic, and they do "suffer greatly from birds of prey"; **D**: 723.

red-legged partridge (Caccabis rufa) = *Alectoris rufa*, as potential plant disperser by carrying seeds in mud stuck to the legs, as observed by CD, **O**. 512. See chapters 3 and 4.

red-legged partridge (*P. rubra*) = *Alectoris rufa*, captive bred in France; **V2**: 156 footnote.

Indian partridge (*Ortygornis gularis*) = Swamp Francolin *Francolinus gularis*, almost every male has a breast scarred by the spurs of rival males; **D**: 556.

partridge = Grey Partridge *Perdix perdix*, CD noting that **guinea-fowl** avoided him "like partridges on a rainy day in September"; **J**: 4.

Compared with Least Seedsnipe (*Thinocorus rumicivorus*) in their flock flight behaviour; **J**: 111.

"Partridges and pheasants are tolerably abundant" on St. Helena, South Atlantic, in July 1836; **J**: 584.

Take to the wing in a flock and soar like **Tinochorus rumicivorus**; **Z**: 118.

Stock of unspecified partridges on estates depends on the destruction of "vermin" [i.e., predators]; **O**: 84.

Unspecified, but likely Grey Partridge *Perdix perdix*, mentioned in noting the Corn Crake being nearly as terrestrial as the partridge, in discussing toes adapted for aquatic life; **O**: 222.

The contrast in sexual dimorphism between this and the polygamous **peacock** or **pheasant** noted; **D**: 338.

CD considered the species "in considerable excess in the south of England and Scotland"; **D**: 382.

A man that shot males of pairs found that their surviving mate always found a fresh partner; **D**: 622.

Always seen during spring in pairs or small parties, not alone, sometimes of same sex; **D**: 623.

Sometimes "live in triplets"; **D**: 624.

Captives distinguish different individual people and exhibit affection to them; **D**: 627.

Females differ from males only in having a smaller red breast mark; **D**: 719.

English partridge = Grey Partridge *Perdix perdix*, black mark on breast is like that of **Tinochorus rumicivorus**; **Z**: 118.

Partridge(s) = unspecified, none of the species in which plumage colours have been largely transferred from males to females are brilliantly coloured; **D**: 720.

Are they, as now coloured, better protected than if resembling quails?; **D**: 723.

Are difficult to see when crouched on the ground; **D**: 746.

Young will run from mother giving alarm call to hide, thus allowing mother to escape; **V1**: 181.

Females may partly assume secondary male characters; **V2**: 51.

quail(s) = of unknown species, mentioned as being similar to seedsnipe and closely related to them (which they are not); **J**: 110–111.

Unspecified, but doubtless Common Quail (*Coturnix coturnix*), mentioned in noting the Corn Crake being nearly as terrestrial as the quail, in discussing toes adapted for aquatic life; **O**: 222.

Have "a close affinity" with **Tinochorus**; **Z**: 118.

Unspecified, adpress, or sleek, their feathers in fright; **E**: 100.

None of the species in which plumage colours have been largely transferred from males to females are brilliantly coloured; **D**: 720.

tree-partridges (Arboricola) = of the genus *Arborophila*; geographically, males have undergone intraspecific morphological change into distinguishable forms, but females and young have not; **D**: 714.

Galloperdix = the genus of the three Spurfowl species, males typically have two leg spurs but the females only one; **D**: 557–558, 685. See chapter 6.

Blood-pheasants (*Ithaginis cruentus*) = Blood Pheasant, birds may have up to five spurs per leg; **D**: 557.

Tragopan pheasant (*Ceriornis Temminckii*) = Temminck's Tragopan *Tragopan temminckii*, displaying males swell appendages into two head horns and a large throat lappet of intensified blue; **D**: 584.

Males display in a similar way to **Polyplectron chinquis**; **D**: 605.

tragopan or **tragopan pheasants** = species of *Tragopan*; in asking of Mr. Bartlett if the male is polygamous, CD was struck by the reply "I do not know, but should think so from his splendid colours"; **D**: 339.

CD states no one can doubt the beauty gained by the male tragopan in distending his blue wattles in display; **D**: 614–615.

Unspecified, faint white lines in females represent the white spots in males; **D**: 653.

The characteristic plumage of the [adult] males is essentially similar to that of guineafowl; **D**: 719.

Lophophorus = unspecified, the genus of the three monal pheasants, thought to be polygamous, with females fighting each other in the presence of a male; **D**: 640.

game-cock, **game cock, cock**, **domestic cock** = cockerel of domesticated Red Junglefowl *Gallus gallus*, briefly compared to placid species; **O**: 34.

A spurless cock is unlikely to leave many offspring; **O**: 108.

Artificial selection for victory between fighting males; **O**: 277; **D**: 325.

Erection of head feathers in the "cock-pit" long recognized as "a sign of cowardice;" **E**: 100.

CD reports having heard of individuals feeding blind companion(s); **D**: 157.

Male **ruffs** fight like them; **D**: 552. See figure 6.1

A case of a game-cock killing a **kite** cited and another of one fighting on in the cockpit with its broken legs splinted until it was killed; **D**: 555.

The "cock clucks to the hen, and the hen to her chickens, when a dainty morsel is found"; **D**: 563.

Cocks crow in triumph over a defeated rival; **D**: 563.

Cock-fighters "dubb" their cocks by trimming their hackles and "cut off the comb and gills" so their opponents lack these points of purchase when fighting; **D**: 614.

Feather structure and pigmentation analogous to the ocelli of **peacocks** described; **D**: 656. See chapter 6.

Pedigrees of famous strains were kept and extended back for a century; **V2**: 3.

No less than nine sub-varieties kept and named in the Philippines; **V2**: 210.

Selection for morphological traits by people plus for aggression and fighting vigour in the cock-pit; **V2**: 225.

game-fowl = cockerel of domesticated Red Junglefowl *Gallus gallus*, are ready to fight whenever they meet; **D**: 560.

Independent of artificial selection in some "sub-breeds," females are very different, whereas males can hardly be distinguished; **D**: 718.

Game fowl = the extremely close resemblance between it and ***Gallus bankiva*** (= *G. g. bankiva*), **V**. 230. See chapter 5.

common hen = female domestic chicken *Gallus gallus*, description of aggressive display in defence of chicks; **E**: 99–100.

Some old hens acquire male external characters, voice, and aggression; **D**: 31.

The vocalizations of "joy" at egg laying described; **D**: 563.

chicken(s) = domestic chickens *Gallus gallus*, difficult to understand how unhatched birds instinctively break their egg shell; O: **296**.

Young have lost "wholly by habit, that fear of dog and cat which no doubt was originally instinctive in them" but respond to the call of mother hen for other dangers; **O**: 329.

A few hours after hatching, chicks pick up food, initially stimulated by hearing mother foraging; **E**: 49–50. See chapter 7.

bantam = domestic fowl *Gallus gallus* briefly compared to game-cock; **O**: 34.

Artificial selection of; **O**: 109.

poultry = briefest mention of domestic poultry on a New Zealand farm in December 1835; **J**: 507.

Briefly alludes to numerous domestic forms; **O**: 64.

Natural selection on eggs and the colour of chicks; **O**: 105.

In discussing relative proportionate body-part growth, CD states, "a large tuft of feathers on the head is generally accompanied by a diminished comb, and a large beard by diminished wattles"; **O**: 182.

As people enhance "beauty' in breeds, so wild female birds select for more attractive males; **D**: 326.

Females "prefer the more vigorous and lively males"; **D**: 330.

If both adult sexes vary late in life, the young will be left unaffected; **D**: 761.

common fowl or **fowl(s)** = domestic chicken *Gallus gallus*, found running wild on Ascension Island, South Atlantic, in July 1836; **J**: 587.

That natural instincts may be lost under domestication is shown by "breeds of fowls which very rarely or never become 'broody'" (or incubate eggs); **O**: 328.

In several breeds, the chicks, individuals in "first true plumage," and adults differ greatly from one another "as well as from their common parent-form, the ***Gallus bankiva***; and these characters are faithfully transmitted by each breed to their offspring at the corresponding periods of life"; **D**: 353.

In most breeds, the "characters proper to each sex are transmitted to the same sex alone"; **D**: 355.

With the domestic breeds, the inheritance of characters by one or both sexes seems determined by the period at which the characters are acquired. Thus, in those in which the adult males differ greatly in colour from females, "as well as from the wild parent-species," they also differ from young males—"so that the newly-acquired characters must have appeared at a rather late period of life." This is followed by further examples, and some that contradict them; **D**: 368.

Of 1,001 chicks of "a highly-bred stock of Cochins reared over eight years," 487 were male; **D**: 381–382. See chapter 6.

In certain breeds, "the feathers are plumose, with some tendency in the shafts to be naked"; **D**: 589.

Under domestication and "high feeding," birds' natural instincts may be corrupted; **D**: 634.

Courtship and female choice of mate in domestication described and discussed; **D**: 635–636. See also chapter 6.

CD discusses the feather shape and pigmentation of hackles, top-knots, and "correlated" plumage in various breeds; **D**: 650–651.

The process by which zoned feathers of mongrels from different coloured fowls originate is not complex; **D**: 655.

Feathers may have separated or decomposed barbs for a significant part of their length; **D**: 656.

The inheritance of characters between the sexes briefly discussed; **D**: 680–681.

An extinct German breed had hens with leg spurs, and although good egg layers, their spurs prevented successful incubation; **D**: 685.

Of the numerous breeds the sexes typically differ in plumage, and exceptions are thus notable; and CD predicts this would hold true of other species if so diversified under domestication; **D**: 701.

Young of both sexes closely resemble adult females, as in many birds with beautiful adult males; **D**: 711.

The sexes differ to an extreme degree; **D**: 718.

Young will run from mother giving alarm call to hide, thus allowing her to escape; **V1**: 181.

Have become feral in South America, perhaps West Africa, and on several islands; **V1**: 190.

Discussing "reversion," or the appearance of ancestral traits, in individuals at various ages; **V2**: 38–40, 44–45.

fowl(s) or **poultry** = the domesticated species *Gallus gallus* and its various breeds; **V1**: 37, 73, 160, 181, 225–275; **V2**: 3, 14, 22, 29, 31, 33, 35, 44–45, 51–55, 68–69, 74, 76, 92–93, 95–97, 100–101, 104, 109, 112, 117, 124–125, 131, 161–162, 189, 196–199, 202, 207, 209–210, 225, 235, 238–240, 242, 245–246, 257, 269, 278, 286–287, 289, 297–298, 304, 306, 315, 317–318, 322, 325, 332–333, 336, 349–350, 354, 369, 373, 391–393, 401, 407–408, 410, 412–415, 428. See chapters 5, 7, and figures 5.6–5.9.

Dorking fowl = given as an example of a lack of artificial selection on a body part (in this case the comb), resulting in a lack of uniformity of that part; **O**: 187.

Result of crossing with common four-toed breeds; **V2**: 14.

Sebright bantam = mentioned with regard to its characters being artificially selected for; **D**: 326; **V2**: 22.

Spanish cock = to the human eye, its glossy black plumage "is much enhanced by his white face and crimson comb"; **D**: 614. See figure 5.6.

Spanish fowl = the male has an immense upright comb, and the female has one many times larger than the parent species but is not upright; **D**: 681–682. See figure 5.6.

Shanghai cock = a quarrelsome hen was "subdued by the gentle courtship" of one; **D**: 636.

pencilled Hamburghs = the sexes differ greatly and also from the original **Gallus bankiva**; **D**: 681.

Gallus bankiva = Red Junglefowl *Gallus g. gallus* (Javan subspecies), "when reared in India under a hen, are at first excessively wild"; **O**: 329. See chapters 4 and 5.

Wild males retain their neck-hackles for nine or 10 months of the year; **D**: 598.

Mentioned in discussing it compared to the domestic and other *Gallus* fowl; **V1**: 226, 233–246, 248, 251, 254, 257–261, 271–273. See figure 5.7.

See **common fowl** or **fowl** and **pencilled Hamburghs**.

Gallus = examination of 500 eggs of various crosses between three species and their hybrids found the majority fertile, but all but 12 died as embryos or chicks; **O**: 387–388.

The genus inhabits the base of the Himalaya, is replaced higher by **Gallophasis** (*Lophura*) and still higher by **Phasianus**; **V1**: 237.

It is as improbable that [native species of] *Gallus* should inhabit South America as that hummingbirds should inhabit the Old World; **V1**: 237.

From the character of other gallinaceous African birds, it is not improbable *Gallus* is African; **V1**: 237.

The "topknot" of the domestic Polish fowls is a new male character in this genus; **V2**: 74. See also chapter 5.

No one has compared the fertility of *G. bankiva* or the domestic fowl crossed with another distinct *Gallus* species; **V2**: 109.

Gallus Sonnerati = Grey Junglefowl *G. sonneratii*, distribution, description, voice, hybrids; **V1**: 233–234.

The barbs and barbules of its scapular feathers blend together to "form thin horny plates of the same nature with the shaft"; **V1**: 289.

Gallus Stanleyi = Sri Lanka Junglefowl *G. lafayettii*, cocks fight to the death defending their harem; **D**: 555–556.

Voice, hybrids with domestic fowl, **V1**: 234.

Gallus varius* (or *furcatus*)** = Green Junglefowl, distribution, description, hybrids with ***G. Sonnerati and ***G. Stanleyi***; **V1**: 234–235.

Gallus œneus and ***G. Temminckii*** = disused names once applied to ***Gallus varius*** hybrids with domestic fowls; **V1**: 234–235.

Gallus giganteus = a name CD notes as being wrongly applied to a hybrid form of tame fowl; **V1**: 235.

Gallus Turcicus = a disused name wrongly applied to the pencilled Hamburgh domesticated breed; **V1**: 247.

Gallophasis = the genus **Gallus** inhabits the base of the Himalaya, is replaced higher by **Gallophasis** (*Lophura*) and still higher by **Phasianus**; **V1**: 237.

Kalij-pheasant = Kalij Pheasant *Lophura leucomelanos*, males make a "singular drumming noise" with the wings; **D**: 574.

Kalij pheasants (Gallophasis) = *Lophura leucomelanos*, geographically males have undergone intraspecific morphological change into distinguishable forms but not the females and young; **D**: 714.

silver-pheasant = *Lophura nycthemera*, a male triumphant over his rivals in mating females was immediately dominated by a rival upon his plumage being spoiled; **D**: 639. See chapter 6.

Females are generally similar to those of allied species, whereas the males differ dramatically; **D**: 717.

The males are partially white, doubtless through sexual selection; **D**: 750.

small fire-backed pheasant (*Euplocamus erythropthalmus*) = Hoogerwerf's Pheasant *Lophura hoogerswerfi*, females have leg spurs as well as males; **D**: 557.

Acomus = now part of the genus *Lophura* in which females possess well-developed leg-spurs; **D**: 685.

eared pheasant (*Crossoptilon auritum*) = Blue Eared-Pheasant, is an exception to pheasants in which "the males differ conspicuously from the females, and they acquire their ornaments at a rather late period of life," as both sexes exhibit ornate characters acquired early in life (see **Pheasants**); **D**: 364–365, 702, 720–721.

Mr. Bartlett informs CD that males of this dull pheasant do not display plumage to females; **D**: 609.

Sexual dimorphism in "tail" length discussed; **D**: 688.

Cheer pheasant (*Phasianus wallichii*) = Cheer Pheasant *Catreus wallichii*, males of this dull pheasant do not display plumage to females; **D**: 609.

The sexes closely resemble each other, and their colours are dull; **D**: 720–721.

Sœmmerring's pheasant = Copper Pheasant *Syrmaticus soemmerringii*, describes gross sexual dimorphism in tail length and the influence on this by hybridization; **D**: 678–679.

Sexual dimorphism in tail length discussed; **D**: 689.

Reeve's pheasant = Reeves's Pheasant *Syrmaticus reevesii*, sexual dimorphism in tail length discussed; **D**: 689.

Phasianus colchicus = Common Pheasant, hybrid with ***P. torquatus*** fertile; **O**: 373.

Phasianus = the genus **Gallus** inhabits the base of the Himalaya, is replaced higher by **Gallophasis** (*Lophura*) and still higher by **Phasianus**; **V1**: 237.

No one has compared the fertility of a species of this genus crossed with the domestic fowl or *Gallus bankiva*; **V2**: 109.

Pheasant(s) = Common Pheasant *Phasianus colchicus*, young reared under a hen in England are at first excessively wild; **O**: 329. See chapter 4.

CD notes the contrast in sexual dimorphism between this and the monogamous **guinea-fowl** or **partridge**; **D**: 338.

CD notes that young males "become brightly coloured in the autumn of their first year"; **D**: 346.

Males have acquired their bright [adult] plumage through sexual selection; **D**: 744.

Young will run from mother giving alarm call to hide, thus allowing her to escape; **V1**: 181.

Offspring of tame cocks crossed with domestic fowl show "extraordinary wildness," **V2**: 45

Females may partly assume secondary male characters; **V2**: 51.

Lays 18 to 20 eggs [as a clutch] in the wild but seldom more than 10 in confinement; **V2**: 155–156.

See **partridges**; **J**: 584.

common pheasant = *Phasianus colchicus*, sexual dimorphism in tail length and influence on this by hybridization; **D**: 678–679.

Sexual dimorphism in tail length discussed; **D**: 688.

Young of both sexes closely resemble adult females, whereas adult males differ conspicuously; **D**: 712.

Females are generally similar to those of allied species, whereas the males differ dramatically; **D**: 717.

Do the slight differences between females and those of **Japan** and **gold pheasants** serve as protection, or might not their plumages have been exchanged with impunity?; **D**: 723.

Hybrid offspring with fowls are considerably larger than either progenitor; **V2**: 125.

P. torquatus = Common Pheasant (Chinese subspecies *P. c. torquatus*), hybrids with **Phasianus colchicus** are fertile; **O**: 373.

Japan pheasant = Green Pheasant *Phasianus versicolor*, females are generally similar to those of allied species, whereas the males differ dramatically; **D**: 717.

Do the slight differences between females and those of **common** and **gold pheasants** serve as protection, or might not their plumages have been exchanged with impunity?; **D**: 723.

Gold pheasant or **gold-pheasant** = Golden Pheasant *Chrysolophus pictus*; in courtship display, males twist their erected plumage, body, and tail toward the female to best show it off; **D**: 603.

Females are generally similar to those of allied species, whereas the males differ dramatically; **D**: 717.

Do the slight differences between females and those of **Japan** and **common pheasants** serve as protection, or might not their plumages have been exchanged with impunity?; **D**: 723.

Bird fanciers pull head or neck feathers from young to sex them by the underfeather colour; **D**: 738.

Males can be distinguished from females at about three months old but do not acquire full adult plumage until the "end of the September in the following year"; **D**: 738 footnote.

There is a close relationship in colour and structure between their head and loin plumes; **V1**: 275.

Hybrid offspring with fowls are considerably larger than either progenitor; **V2**: 125.

Amherst pheasant = Lady Amherst's Pheasant *Chrysolophus amherstiae*; in courtship, display males twist their erected plumage, body, and tail towards the female to best show it off; **D**: 603.

Females are generally similar to those of allied species, whereas the males differ dramatically; **D**: 717.

There is a close relationship in colour and structure between their head and loin plumes; **V1**: 275.

Pheasant(s) = unspecified, "the sexes differ to an extreme degree" (**D**: 718), "and they acquire their ornaments at a rather late period of life" (see **eared pheasant**); **D**: 364, 702.

Birds reared from large numbers of wild-laid eggs produced four or five males to each female; **D**: 382.

Head covered with velvety down; **D**: 584.

Males of so many [species] display their plumage before the females; **D**: 609.

CD notes 18 cases of hybridization with the **black grouse** in Great Britain; **D**: 631.

In hybridizing with common chickens, male pheasants prefer older hens regardless of colour; **D**: 641.

Females of various species have long tails although exposed on their ground nests; **D**: 687.

Function and origins of sexual dimorphism in tail length discussed; **D**: 688–689.

None of the species in which plumage colours have been largely transferred from males to females are brilliantly coloured; **D**: 720.

When a cock crosses with a domestic fowl, its characters dominate in offspring; **V2**: 68.

Hybrid offspring with fowls and between distinct pheasant species are considerably larger than either progenitor; **V2**: 125, 131.

Polyplectron = the eight Peacock-Pheasants of this genus may have two or more spurs per leg; **D**: 557.

They so resemble peafowl in appearance, voice, and habits that they are sometimes called "peacock-pheasants"; courtship described and great length and pigmentation of tail-coverts described; **D**: 658–659.

The ocelli of their plumage described, illustrated, and their origins discussed; **D**: 660–661. See chapter 6.

Females of the several species dimly exhibit the ocelli of adult male plumage; **D**: 718–719.

Polyplectron chinquis = Grey Peacock-Pheasant *Polyplectron bicalcaratum*, males court not by standing in front of the female to erect plumage, as does a peacock, but present their tail obliquely and "lowering the expanded wing on the same side, and raising that on the opposite side," illustrated; **D**: 604–605. See figure 6.10.

Describes plumage detail with reference to ocelli, illustrated; **D**: 658–660. See chapter 6 and figure 6.10.

P. malaccense = Malayan Peacock-Pheasant *P. malacense*, plumage and ocelli described and illustrated in discussing ocelli generally; **D**: 659–660. See chapter 6.

P. hardwickii = Malayan Peacock-Pheasant *P. malacense*, has a peculiar topknot somewhat like the **Java peacock**; **D**: 658–659.

P. napoleonis = Palawan Peacock-Pheasant *Polyplectron napoleonis*, plumage and ocelli described in discussing ocelli generally; **D**: 658, 660. See chapter 6.

Argus pheasant = Great Argus *Argusianus argus*, the tail is greatly increased in length and "its body not larger than that of a fowl; yet the length from the end of the beak to the extremity of the tail is no less than five feet three inches, and that of the beautifully ocellated secondary wing-feathers nearly three feet"; **D**: 585–586. See chapter 6 and figure 6.11.

Male and its plumage detail and courtship display described and illustrated; **D:** 605–609. See chapter 6 and figure 6.11.

The vast secondaries of adult males render the birds almost flightless and thus easy prey; **D**: 613, 750. See figure 6.11.

Females may attend to each detail of the male's plumage during courtship; **D**: 643. See chapter 8. See figure 6.11.

Faint white line-like markings in females represent the spots in male plumage; **D**: 653–654.

The striking male plumage ocelli patterns described, illustrated, and their origins discussed; **D**: 661–673. See figures 6.11–6.19.

CD presents a detailed, and illustrated, description and discussion of the ocelli on wing feathers and their origin; **D**: 661–673. See chapters 5, 6, and figures 6.11–6.19.

They [males at least] do not cast their plumes during the winter; **D**: 705.

Gradations in morphology between males of allied species indicate the nature of steps through which they have passed and explain how characters such as the feather ocelli originated; CD finds no natural history fact more wonderful than that females appreciate the "ball-and-socket" ocelli and other plumage markings of males [and, by implication, have brought them about]; **D**: 942. See chapters 5 and 6.

peacock(s) = Indian Peafowl *Pavo cristatus*, lack of many domesticated forms noted; **O**: 48. See chapter 5.

A pied peacock was "eminently attractive" to females; **O**: 109.

Notes the contrast in sexual dimorphism between this and monogamous **guinea-fowl** or **partridge**; **D**: 338.

The sexes differ conspicuously (**D**: 718) but share the "head-crest," which is acquired early in life by both sexes (unlike the nuptial plumage of males); **D**: 364.

Males fight fiercely, including in extensive flight; **D**: 557.

Male **Menura Alberti** raise and spread their tail like a peacock; **D**: 568.

Males "rattle their quills together"; **D**: 537. See chapter 6.

Tail-coverts are elongated and tail bones modified to support them in adult males; **D**: 585–586.

Offer a striking instance of sexual dimorphism; **D**: 589.

Males expand and erect their "tail" transversely to their body in front of the female; **D**: 605.

Strutting and displaying males seem "the very emblem of pride and vanity"; **D**: 613.

Their long train of feathers doubtless render them easy prey to tigers; **D**: 613.

The striking male plumage ocelli, described, illustrated, and their origins discussed; **D**: 655–661. See chapters 5 and 6.

They [males at least] do not cast their plumes during the winter; **D**: 705.

Young of both sexes closely resemble adult females, as in many species with beautiful adult males; **D**: 711.

The "train" of males increases in beauty years after they are fully mature; **D**: 740. See chapter 7.

Males have acquired their bright [adult] plumage through sexual selection; **D**: 744.

Gradations in morphology between males of allied species indicate the nature of steps through which they have passed and explain how characters as the feather ocelli originated; **D**: 758.

It could not be supposed that males display their plumage to females for no purpose; **D**: 942.

Have become feral on Jamaica; **V1**: 190.

common peacock (*Pavo cristatus*) = Indian Peafowl, only the male wears spurs on the legs, but in the **Java Peacock** both sexes do so; **D**: 364–365.

Hybridized with **Javan Peacocks**; **V1**: 290.

pea-fowl = Indian Peafowl *Pavo cristatus*, CD quotes an author writing of 20 to 30 peacocks displaying to "gratified" females in forest; **D**: 601.

Females preferred an old pied male when visible, even when caged from their reach; **D**: 638, 648, 674.

Sir R. Heron states first advances to mating are always by the female; **D**: 640.

Function and origins of sexual dimorphism in "tail" length discussed; **D**: 688.

Indian peacocks (Pavo cristatus) = Indian Peafowl, the striking male plumage ocelli described, illustrated, and their origins discussed; **D**: 655–657.

Pavo nigripennis = Indian Peafowl *Pavo cristatus*, the name was applied to a black-winged peacock by Mr. Sclater as a distinct species, but it is but a form of *P. cristatus*; **V1**: 290–292. See chapter 5.

Javan peacock(s) (*Pavo muticus*) = Green Peafowl, the striking male plumage ocelli described, illustrated, and their origins discussed; **D**: 656, 658. See chapter 6.

Females possess well-developed leg spurs; **D**: 685.

Hybridized with **Indian peacocks**; **V1**: 290.

Java Peacock (*P. muticus*) = Green Peafowl, both sexes wear spurs on the legs; but in the **common peacock,** only the males do so; **D**: 364–365, 557.

peacock = unspecified, males sometimes display in absence of females and "take delight" in it; **D**: 600.

peahen = unspecified *Pavo* species, it need not be supposed that they admire each detail of a peacock's train but are struck by the general effect of their courtship display; **D**: 643.

The long tail of the peacock would be inconvenient and dangerous for peahens to have [e.g., to sit on her eggs—of the 4th edition of **O**: 241]; **D**: 676, 687.

Females are apparently more numerous than males, suggesting to CD the former "would have been led" to compete to court the latter, as with **Indian Turnix**, **wild turkey**, and some **grouse**; **D**: 733.

May partly assume secondary male characters; **V2**: 51.

Lay fewer eggs in the UK than in their native India; **V2**: 161.

peafowl, **peacock** or **peahen** = of the genus *Pavo*, **V1**: 290–292; **V2**: 112, 161, 235, 268, 287, 332. See chapter 6.

game or **game birds** = Darwin found seeing game and other wild birds a delight; **L**: 353. See chapter 1.

UK sporting dogs will not touch their bones, whereas other dogs will eagerly devour them; **V2**: 303.

gallinaceous birds or **order** = of the order Galliformes, briefly mentioned regarding differences in secondary sexual characters among males; **O**: 192.

Perhaps the running ability of **caracaras** "indicates an obscure relationship" with this group; **Z**: 18. See chapter 2.

Mentioned in noting the bill of **Pterophtochos albicollis** is like that of this group; **Z**: 152.

Males, especially of polygamous species, wear spurs for fighting, and a case of a **game-cock** killing a **kite** is cited, although even species lacking spurs engage in fierce conflicts; **D**: 555–556. See chapter 6.

Some species have bare-shafted head feathers terminating in a disc of vanes; **D**: 587.

Males of so many [species] display their plumage before the females; **D**: 609.

Many species thought to be polygamous; **D**: 640.

The comb of many species is highly ornamental and is colourful during courtship; **D**: 649.

The development of leg spurs in females has been checked through natural selection, as they are detrimental to successful nesting and incubation (unlike wing spurs, which persist in females of many species); **D**: 685.

Probably no one will dispute that dull ground-frequenting species acquired their colours as protection; **D**: 723.

Wallace's observation of certain species in the East led him to think that slight differences in female plumage between species are beneficial (for protection); **D**: 723.

Mr. Blyth remarked that these birds generally have a restricted range, as CD illustrates in India; **V1**: 237.

Males in many genera are barred or pencilled; **V1**: 255.

Judging from European species, CD thinks wild ***Gallus bankiva*** would use its legs and wings more than do domestic fowl; **V1**: 270.

Both sexes of some wild species have heads ornamented with hackle-like feathering, but there is often a difference in feather size and shape forming the crest; **V1**: 275.

Barred plumage is typical of the females of the various species; **V2**: 74.

Species of many genera show an "eminent capacity" for captive breeding; **V2**: 155.

Many species are spangled or pencilled [in plumage markings]; **V2**: 350.

Gallinaceae = of the order Galliformes, the habit of laying their eggs in nests of other birds "not very uncommon"; mentioned in context of unspecified "ostrich"; **O**: 335.

Exhibit "almost as strongly marked sexual differences as birds of paradise or hummingbirds"; some species being polygamous and others monogamous; **D**: 338.

In some species, the barbs of feathers are filamentous or plumose; **D**: 587.

Young of almost all species are "covered with longitudinally striped down"; **D**: 708.

Females of the various species resemble one another, whereas males differ dramatically; **D**: 717.

None of the species in which plumage colours have been largely transferred from males to females are brilliantly coloured; **D**: 720.

Differences generally distinguishing the sexes suggest some characters in domestic fowls have been transferred from the one sex to the other; **V1**: 255.

Species of many genera show an "eminent capacity" for captive breeding but with marked and inexplicable exceptions; **V2**: 155–156.

GAVIIFORMES

GAVIIDAE—Loons (or Divers)

Colymbus glacialis = Great Northern Loon *Gavia immer*, Mr. Blyth saw specimens of young that had anomalously assumed adult plumage; **D**: 736.

sea-mews (*Gavia*) = the loon genus in which the adult head and neck becomes grey or mottled during summer and white in winter in both adults and young; **D**: 751.

SPHENISCIFORMES

SPHENISCIDAE—Penguins

penguins = Magellanic Penguins *Spheniscus magellanicus*, surrounding the *Beagle* out of Rio and making strange noises like cattle bellowing on shore; **J**: 44.

penguin *Aptenodytes demersal* = Magellanic Penguin *Spheniscus magellanicus*, behaviour in going to sea; **J**: 256. See chapter 2.

jackass penguin = Magellanic Penguin *Spheniscus magellanicus*, habits on land and sea, calls; **J**: 256–257.

Spheniscus humboldtii = Humboldt Penguin *Spheniscis humboldti*, a specimen from near Valparaiso; **Z**:137.

penguin(s) = unspecified, wing use; **O**: 218.

CD notes the evolutionarily intermediate state of their wings; **O**: 442–443. See chapter 4.

CD does not consider their wings rudimentary, as they are "of high service, acting as a fin" and may therefore "represent the nascent state of the wing"; **O**: 622.

PROCELLARIIFORMES

DIOMEDEIDAE—Albatrosses

albatross = said by CD to very closely resemble the Southern Giant Petrel *Macronectes giganteus* in "habits and manner of flight"; **J**: 354; **Z**: 139.

HYDROBATIDAE—Storm Petrels

Thalassidroma oceanica = Wilson's Storm-Petrel *Oceanites oceanicus*, a specimen from Malanado, habits, nesting on Georgia where they arrive in September; **Z**: 141; **J**: 475. See chapter 3.

Procellaria oceanica = as a synonym of **Thalassidroma oceanica**; **Z**: 141.

PROCELLARIIDAE—Petrels, Shearwaters

***Procellaria gigantea*, or nelly (quebranta-huesos, or break-bones, of the Spaniards)** = Southern Giant Petrel *Macronectes giganteus*, a common bird of seas off southwestern South America. CD notes that "quebranta-huesos" properly refers to the Western Osprey *Pandion haliaetus*; **J**: 354; **Z**: 139.

Habitats, habits, food, nesting locations; **Z**: 139; **J**: 475. See chapter 3.

Procellaria glacialoïdes = Southern Fulmar *Fulmarus glacialoides*, on east and west coasts of South America, nesting sites and season in Georgia, tameness, habits; **Z**: 140–141; **J**: 475. See chapter 3.

Fulmar petrel = unspecified *Fulmarus* species, which "lays but one egg, yet it is believed to be the most numerous bird in the world" (see also **condor** and **ostrich**); **O**: 81–82.

Daption capensis or **Pintado** = Cape Petrel *Daption capense*, abundant throughout southern waters, habits, flight, tameness, calls, nesting on Georgia when **Procellaria glacialoides** does; **Z**: 140–141; **J**: 475. See chapter 3 and colour figure 3.13.

Procellaria Capensis = as a synonym of **Daption capensis**; **Z**: 140.

Prion = *Pachyptila*, mentioned in discussing bill structure and origin of whale baleen; **O**: 287.

Prion vittatus = Antarctic Prion *Pachyptila desolata*, distribution, wild and solitary, flight, nest location, burrows excavated, eggs; **Z**: 141.

Procellaria Vittata = as a synonym of **Prion vittatus**; **Z**: 141.

Puffinus cinereus = Grey Petrel *Procellaria cinerea*, flocking, calls, food; **J**: 354–355.

Distribution and locations collected, flocking, habits, flight, food, call; **Z**: 137–138. See chapters 2 and 3.

Procellaria puffinus = as a synonym of ***Puffinus cinereus***; **Z**: 137.

PELECANOIDIDAE—**Diving Petrels**

Pelecanoides garnotii = Peruvian Diving Petrel, a specimen from Iquique, Peru; **Z**: 139.

Puffinuria Garnotii = as a synonym of **Pelecanoides garnotii**; **Z**: 139.

Procellaria urinatrix = as a synonym of **Pelecanoides garnotii**; **Z**: 139.

Puffinuria Berardii = Common Diving Petrel *Pelecanoides urinatrix*, habitat, habits, its morphology auk-like; **J**: 355.

Pelecanoides berardi = Common Diving Petrel *Pelecanoides urinatrix*, distribution, habitats, habits; **Z**: 138–139; **J**: 475. See chapter 3.

Puffinuria Berardi = as a synonym of **Pelecanoides berardi**; **Z**: 138.

Puffinuria berardi = Common Diving Petrel *Pelecanoides urinatrix*, flight, swimming, and diving auk- or grebe-like; **O**: 221, 646.

petrels and albatross = unspecified, "exceedingly abundant" between latitudes 56° and 57° south of Cape Horn, Atlantic Ocean. CD assumes "the albatross, like the condor, is able to fast" for long periods; **J**: 190.

Both noted to continue to fly "as if the storm were their proper sphere"; **J**: 603.

Petrel(s) or **petrel(s)** = "These southern seas [off southwestern South America] are frequented by several species of Petrels"; **J**: 354–355.

Flight, swimming, and diving ability; **O**: 221, 223.

Their bill structure mentioned in discussing origin of whale baleen; **O**: 287.

Unspecified, sexual selection has brought about both white and black species; **D**: 752.

seafowl = On Ascension Island, South Atlantic, CD found a "whole plain was mottled; I now found they were seafowl, which were sleeping in such full confidence, that even in midday a man could walk up to, and seize hold of them" (possibly boobies or terns); **J**: 589.

Divers, and other fishing birds = feeding among kelp: "divers" would usually mean *Gavia* species of the Gaviidae (loons or divers), but as these do not occur in South America, CD must have used the word generically for aquatic diving birds (i.e., penguins, ducks, grebes, etc.); **J**: 305. See chapter 2.

A "diver" chased and killed by an Antarctic Giant Petrel *Macronectes giganteus*; **J**: 354. See chapter 2.

PODICIPEDIFORMES

PODICIPEDIDAE—Grebes

Podiceps rollandii = White-tufted Grebe *Rollandia rolland*, locations, habitats. Gould finds it as nearly related to ***Podiceps cornutus*** as ***P. auritus*** is to **P. kalipareus**; **Z**: 137.

Podiceps Rolland = as a synonym of **Podiceps rollandii**; **Z**: 137.

Podiceps chilensis = White-tufted Grebe *Rollandia rolland*, near Buenos Ayres and Tierra del Fuego, habitat, call; **Z**: 137.

Podiceps auritus = Horned Grebe, see **Podiceps kalipareus**; **Z**: 136.

See ***Podiceps cornutus*; Z**: 137.

Podiceps cornutus = Horned Grebe *Podiceps auritus*, as closely related to *P. rollandii* as *P. kalipareus* is to *P. auritus*; **Z**: 137.

Podiceps kalipareus = Silvery Grebe *Podiceps occipitalis*, locations, habitats, habits, in size and many characters like Horned Grebe *P. auritus*; **Z**: 136.

grebe(s) = unspecified, flight; swimming and diving of **Pelecanoides berardi** is grebe-like; **O**: 221.

Foot structure modified for swimming; **O**: 222.

PHOENICOPTERIFORMES

PHOENICOPTERIDAE—Flamingos

Flamingoes = Chilean Flamingo *Phoenicopterus chilensis*, frequent salt lakes; **J**: 77–78. See chapter 2.

Flamingo = unspecified, in captivity beat ground with feet when hungry (see **Sheldrake** and **Kagu**); **E**: 50.

Take "several years" to acquire their "perfect plumage"; **D**: 738 footnote.

PHAETHONTIFORMES

PHAETHONTIDAE—Tropicbirds

Tropic-birds = Red-tailed Tropicbird *Phaethon rubricauda*, when disturbed on their nests are said not to fly away, but "merely to stick out their feathers and scream'"; **E**: 99.

tropic-birds = unspecified, the plumage becomes white only at maturity; **D**: 751.

CICONIIFORMES

CICONIIDAE—Storks

Ibis tantalus = Wood Stork *Mycteria americana*, take four years to acquire "perfect plumage"; **D**: 738 footnote.

Immature second year birds sometimes breed; **D:** 739 footnote.

gaper (*Anastomus oscitans*) = Asian Openbill, young and adults of both sexes are dark in winter, but adults become white in summer; **D**: 741.

Its white plumage is a summer-only nuptial character, with immatures and adults being grey and black in winter; **D**: 751.

black stork = *Ciconia nigra*, males show "a well-marked sexual difference in the length and curvature of the bronchi"; **D**: 572.

Both sexes are black throughout but contrasted with a bright red beak; **D**: 749.

Xenorhynchus = the genus *Ephippiorhynchus*, which consists the Black-Necked *E. asiaticus* and Saddle-Billed *E. senegalensis* Storks, in which the iris colour differs between the sexes, are described; **D**: 649.

storks = unspecified, captives eating fish CD had placed seeds within to see how long and viable they remained in the birds' stomachs as potential plant dispersers; **O**: 510. See chapter 3.

When excited, they "make a loud clattering noise with their beaks"; **E**: 95.

It is probable that the complete or partial blackness of some species is the result of sexual selection "accompanied by equal transmission to both sexes"; **D**: 749.

Sexual selection has brought about both white and black species; **D**: 752.

Grallatores = once used for a long-disused combination of bird groups, predominantly of today's order Ciconiiformes, CD observing that their long unwebbed toes are "formed for walking over swamps and floating plants"; **O**: 222.

Anatomical characters confirm affinity of seedsnipe to this group and to the **Razores**; **Z**: 118.

Within them, "extremely few species differ sexually," but the **ruff** is an exception; **D**: 339

In some species, the sexes resemble each other "but in which the summer and winter plumage differ slightly in colour"; **D**: 594.

Order: Grallatores = a disused combination of bird groups, predominantly of today's Ciconiiformes; **Z**: 125.

PELICANIFORMES

THRESKIORNITHIDAE—Ibises, Spoonbills

Ibis malanops = Black-faced Ibis *Theristicus melanopis*, not uncommon in desert areas of Patagonia, eating grasshoppers, cicadae, small lizards, and even scorpions; **J**: 194. See chapter 2 and colour figure 2.3.

Theristicus melanops = Black-faced Ibis *Theristicus melanopis*, confused with an African species by CD; distribution, habits, call, food, nest site, egg; **Z**: 128–129. See chapter 2 and colour figure 2.3.
Ibis melanops = as a synonym of **Theristicus melanops**; **Z**: 128.

white ibis = American White Ibis *Eudocimus albus*, bare facial and throat skin changes to crimson for breeding; **D**: 593. See chapter 7.

scarlet ibis = *Eudocimus ruber*, adults are alike (their colour fading in captivity), but the young are brown; **D**: 734. See chapter 7.

Ibis (Falcinellus) ordi = White-Faced Ibis *Plegadis chihi*, a specimen from Rio Negro; numerous in large flocks on swampy plains between Bahia Blanca and Buenos Ayres, flight; **Z**: 129.
Tantalus Mexicanus = as a synonym of **Ibis (Falcinellus) ordi**; **Z**: 129.
Tantalus chalcopterus? = as a synonym of **Ibis (Falcinellus) ordi**; **Z**: 129.
Ibis Falcinellus = as a synonym of **Ibis (Falcinellus) ordi**; **Z**: 129.

ibises = unspecified, in some species the barbs of feathers are filamentous or plumose; **D**: 587.
Colouration of bare facial skin changes seasonally; **D**: 593.
Both sexes in several species acquired white plumage through sexual selection; **D**: 751.
Sexual selection has brought about both white and black species; **D**: 752.

spoonbill (*Platalea*) = Eurasian Spoonbill *P. leucorodia*, has a trachea convoluted into a figure of eight, "and yet this bird is mute" [not entirely true, as it produces "deep grunting or groaning" notes that are only audible over short distances; Hancock et al. 1992: 254]; **D**: 572 footnote.
Chinese females in spring of their second year resemble males of their first year and do not acquire the plumage of adult males until their third year; **D**: 703.

ARDEIDAE—Herons, Bitterns

dwarf bitterns (Ardetta) = *Ixobrychus* species, males acquire mature plumage with the "first moult," but females not before their third or fourth; **D**: 702–703.

night heron (*Ardea nycticorax*) = Black-crowned Night Heron *Nycticorax nycticorax*, a captive kept by Audubon hid from an approaching cat and gave a "most frightful" cry to deter it; **D**: 563.

Nycticorax americanus = Black-crowned Night Heron *Nycticorax nycticorax*, at Valparaiso, Chile; **Z**: 128.

Nycticorax violaceus = Yellow-crowned Night Heron *Nyctanassa violacea*, one young described from Galapagos Islands; **Z**: 128.

Ardea violacea = as a synonym of **Nycticorax violaceus**; **Z**: 128.

Ardea callocephala = as a synonym of **Nycticorax violaceus**; **Z**: 128.

Ardeola = the genus of squacco and pond herons, three species on separate continents are strikingly different in summer plumage but hardly distinguishable in winter; in two other species, both sexes retain a plumage year round similar to that worn by the previous three species in winter and as immatures, which is probably the ancestral plumage of the genus; **D**: 715. See chapter 7.

Herodias bubulcus = Western Cattle Egret *Bubulcus ibis*, see **heron(s)**.

Buphus coromandus or **egrets of India (*Buphus coromandus*)** = Western Cattle Egret *Bulbulcus ibis*, young and adults of both sexes are white in winter, with adults becoming golden-buff in summer, leading CD to conclude a progenitor acquired white plumage for nuptial purposes and transmitted it to their young, which was retained by the young but exchanged by adults for more colour; **D**: 741, 753–755. See chapter 7.

Ardea herodias = Great Blue Heron, coastal on the Galapagos Islands; **Z**: 128.

Audubon's brief description of courtship by male quoted by CD; **D**: 580.

Heron(s) = passing mention of them on Galapagos Islands and difficult to know if any new species there; **J**: 461.

Herons eating fish with seeds in them thus disperse plants; **O**: 539–510. See chapter 3.

In some species, barbs of feathers are filamentous or plumose; **D**: 587.

Colouration of bare facial skin changes seasonally; **D**: 593.

The slight seasonal difference in male from female plumage is doubtless for courtship; **D**: 595.

Elongated feathers on the back, neck, and crest in certain species (and specifically *Herodias bubulcus*) are acquired only during spring; **D**: 597.

Elegant plumes, long pendant feathers, crests, and the like worn during summer serve ornamental and nuptial purposes although common to both sexes; **D**: 704–705.

In many species young differ greatly from adults, and adult summer plumage is nuptial; **D**: 734.

Certain species are apparently (and are now known to be) dimorphic; **D**: 739 footnote.

The crest and plumes of some species increase in beauty for many years after they mature; **D**: 740. See chapter 7.

Egretta leuce = Great Egret *Ardea alba*, one collected at Maldonado, seen in Patagonia; **Z**: 128.

Ardea Leuce = as a synonym of **Egretta leuce**; **Z**: 128.

Ardea Egratta = as a synonym of **Egretta leuce**; **Z**: 128.

Ardea rufescens = Reddish Egret *Egretta rufescens*, the young are white, adults reddish; **D**: 754 footnote.

Ardea ludovicana = Tricolored Heron *Egretta tricolor*, take two years to acquire "perfect plumage"; **D**: 738 footnote.

Their crest and plumes increase in beauty for many years after they mature; **D**: 740.

Ardea cærulea = Little Blue Heron *Egretta caerulea*, adults are blue and young are white—but white, mottled, and blue birds may breed together; **D**: 739, 754 footnotes.

Ardea asha = Western Reef-Egret *Egretta gularis*, the young are white but adults are dark slate, leading CD to conclude a progenitor acquired white plumage for nuptial purposes and transmitted it to their young, which was retained by the young but exchanged by adults for more colour; **D**: 753–755.

Ardea gularis = Western Reef-Egret *Egretta gularis*, CD states it shows the opposite adult/young plumage dimorphism to that of ***Ardea asha***; **D**: 754.

egrets, unspecified, seen with "cranes" a day's ride from Rio de Janeiro; **J**: 22. See chapter 2.

The slight seasonal difference in male from female plumage is doubtless for courtship; **D**: 595.

Elegant plumes, long pendant feathers, crests, and the like worn during summer serve ornamental and nuptial purposes although common to both sexes; **D**: 704–705.

Both sexes in several species acquired white plumage through sexual selection; **D**: 751.

herons and egrets of North America and India = young of some species differ from both parents in being white; **D**: 741.

PELECANIDAE—Pelicans

Pelecanus onocrotalus = Great White Pelican, a rosy tint, with lemon-coloured marks on the breast, overspreads the whole plumage in spring; **D**: 599.

P. erythrorhynchus = American White Pelican *Pelecanus erythrorhynchos*, a thin horny crest on the beak is worn only during breeding seasons; **D**: 593–594.

blind pelican = American White Pelican *Pelecanus erythrorhynchos*, reported to be found "very fat" and "must have been well fed for a long time by its companions" (see also **Indian crows**); **D**: 157.

pelican(s) = unspecified, captives eating fish, CD placed seeds within to see how long and viable they remained in the birds' stomachs as potential plant dispersers; **O**: 510. See chapter 3.

They "fish in concert"; **D**: 154.

Males drive away weaker ones, snapping with beaks and beating with wings; **D**: 554. See chapter 6.

SULIFORMES

FREGATIDAE—Frigatebirds

Fregata aquila = Ascension Frigatebird, CD was informed they take hatchling sea turtles off beach sand—and he describes habits of Galapagos Island frigate birds (i.e., Great Frigatebirds *Fregata minor*); **Z**: 146. See chapter 2.

Pelecanus Aquilus = as a synonym of **Fregata aquila**; **Z**: 146.

frigate-bird(s) = unspecified (but Great and/or Lesser), nesting on (Cocos) Keeling Island; **J**: 544.

Have all toes rudimentarily webbed, but only once seen to alight on water; **O**, 222–223, 250. See chapter 2.

SULIDAE—Gannets, Boobies

booby = Brown Booby *Sula leucogaster*, on St Pauls Rocks, Atlantic Ocean; **J**: 9. See chapter 2.

gannets = unspecified boobies (Masked, Brown and/or Red-footed), nesting on (Cocos) Keeling Island; **J**: 544–545.

In most species, the plumage becomes white only at maturity; **D**: 751.

PHALACROCORACIDEA—Cormorants, Shags

Phalacrocorax carunculatus = Imperial Shag *Leucocarbo atriceps*, a specimen from Port St. Julian, Patagonia, was nest building in January; **Z**: 145.

Pelecanus carunculatus = as a synonym of **Phalacrocorax carunculatus**; **Z**: 145.

Phalacrocorax imperialis = as a synonym of **Phalacrocorax carunculatus**; **Z**: 145.

cormorants = Darwin gained pleasure seeing them flying home, in England; **L**: 354.

Of unknown species, feeding on large prawn-like crustacea; **J**: 18.

One killed by a Crested Caracara on the Falkland Islands; **J**: 67.

One "playing with a fish which it had caught" in a "wilfully cruel" way in the Falklands; **J**: 256; **D**: 566.

Feeding among kelp; **J**: 305. See chapter 2.

ACCIPITRIFORMES

CATHARTIDAE—New World Vultures

Vulturidae = the vulture family, with reference to the New World vultures; **Z**: 3.

Cathartes aura = Turkey Vulture, range, foods, and habits; **Z**: 8–9. See chapters 2 and 3.

Vultur jota = as a synonym of **Cathartes aura**; **Z**: 8.

Parties of 10 or more assemble on fallen logs and pair off before departing together; **D**: 635.

Vultur aura = Turkey Vulture *Cathartes aura*, its ability to smell; **J**: 222.

Only bird in mountains inland of Iquique, northern Chile; **J**: 444. See chapter **2**.

turkey buzzard *Vultur aura* = Turkey Vulture *Cathartes aura*, range, habitat, habits; **J**: 68, 346. See chapters 2 and 3.

Turkey-buzzard = common name for **Cathartes aura**, **Z**: 8–9. See chapter 2.

(*Cathartes jota*) = Black Vulture *Coragyps atratus* or Turkey Vulture Cathartes aura, CD cites Audubon observing the courtship movements and postures of males as being "extremely ludicrous"; **D**: 581.

Gallinazo = Black Vulture *Coragyps atratus*, range, habitat, habits; **J**: 64, 66. See chapter 3 and colour figure 3.1

Compared most briefly with Andean Condor as being at times gregarious; **J**: 221; **Z**: 4.

This bird "tame as poultry" at Lima, Peru; **J**: 450. See colour figure 3.1.

Common near Bahia Blanca; **Z**: 7. See colour figure 3.1.

Feed with caracaras on carrion; **Z**: 10. See chapter 2 and colour figure 3.1.

Black Vulture = common name of **Cathartes atratus**; **Z**: 7–8. See chapter 3 and colour figure 3.1.

Cathartes atratus = Black Vulture *Coragyps atratus*, distribution, habitats, association with humans, habits; **Z**: 7–8. See colour figure 3.1.

Cathartes urubu = as a synonym of **Cathartes atratus**; **Z**: 7.

Vultur atratus = as a synonym of **Cathartes atratus**; **Z**: 7.

Vultur jota = as a synonym of **Cathartes atratus**; **Z**: 7.

Carrion Crow = alternative common name of **Cathartes atratus** and **Cathartes aura**; **Z**: 7.

Condor(s) = Andean Condor *Vultur gryphus*, briefest mention of it being able to fast for long periods—see also **petrels and albatross**; **J**: 190.

Follow guanaco herds; **J**: 216.

An account of the range, habits, nesting, and growth of the bird, and the killing of it by people; **J**: 219–224. See chapter 2.

Said to feed on carcasses hidden beneath foliage by Pumas, and the only bird high in the Andes; **J**: 328, 394, respectively. See chapter 2.

Lays clutches of only "a couple of eggs" but may be more numerous than "the ostrich," which lays "a score"; **O**: 81. See chapter 2.

Age changes and adult sexual dimorphism in iris colour noted, as is the fleshy "crest or comb" of males; **D**: 649.

One laid a fertile egg(s) at London Zoo; **V2**: 154.

Sarcoramphus gryphus = Andean Condor *Vultur gryphus*, size and habits (as in **J**: 219–224); **Z**: 3–6.

Vultur gryphus = as a synonym of **Sarcoramphus gryphus**; **Z**: 3.

Sarcoramphus Condor = as a synonym of **Sarcoramphus gryphus**; **Z**: 3.

vulture(s) = unspecified New World species, CD noting that their bare head is generally seen as adapted to "wallowing in putridity"; but caution is required, as the head of the "clean-feeding" male **Turkey** is similarly naked; **O**: 247–248.

CD observes that they "roam far and wide high in the air, like marine birds over oceans"; and that three of four species are wholly or largely white and many others black; and their resulting conspicuousness may aid the sexes in finding each other in the breeding season; **D**: 752 footnote.

None had laid fertile eggs in zoos; **V2**: 153–154.

vultures = CD quotes Captain Cook as seeing vultures at Christmas Sound, Tierra del Fuego, and expressed "little doubt" they would have been Cathartes aura; **Z**: 15.

solitary vultures = unspecified, mentioned in passing; **J**: 93.

SAGITTARIIDAE—**Secretarybird**

Secretary-hawk (*Gypogeranus*) = Secretarybird *Sagittarius serpentarius*, CD notes this bird's "whole frame" is modified for killing snakes, and he hypothesizes it is "highly probable" that it would ruffle its feathers [as protection from bites] in attacking them; **E**: 111. See chapter 7 and colour figure 7.3.

PANDIONIDAE—**Ospreys**

Osprey = see ***Procellaria gigantea***; **J**: 354.

ACCIPITRIDAE—**Kites, Hawks, and Eagles**

Sub-family: Circinae = Accipitrinae of the Accipitridae, includes *Circus*; **Z**: 29.

Sub-family Buteoninae = Accipitrinae of the Accipitridae, includes **Craxirex** and **Buteo**; **Z**: 22.

Honey buzzard (*Pernis cristata*) = Crested Honey Buzzard *Pernis ptilonorhyncus*, geographical variability in crest size described; **D**: 646.

Aquila fusca = Tawny Eagle *Aquila rapax*, seen to mate in London Zoo; **V2**: 154.

golden eagle (*Aquila chrysaëtos*) = Golden Eagle *Aquila chrysaetos*, CD is informed that if one of a pair is killed, the survivor soon finds another mate; **D**: 622.

Falco albidus = Northern Goshawk *Accipiter gentilis*, a captive "returned to the dress of an earlier age"; **V2**: 158.

Circus aeruginosus = Western Marsh Harrier, CD quotes Gould as noting this "as perfect an analogue" to **Circus megaspilus**; **Z**: 31.

Circus megaspilus = Long-winged Harrier *Circus buffoni,* described and habitat noted; **Z**: 29–30. See ***Circus aeruginosus***.

Circus cyaneus = Hen Harrier, CD quotes Gould as noting this "as perfect an analogue" to **Circus cinerius**; **Z**: 31.

Circus cinerius = Cinereous Harrier *Circus cinereus*, specimen locations and habits noted; **Z**: 30–31. See ***Circus cyaneus***.

  Falco histrionicus = as a synonym of **Circus cinerius**; **Z**: 30.
  Circus histrionicus = as a synonym of **Circus cinerius**; **Z**: 30.

kite = Red Kite *Milvus milvus*, one killed by a **game-cock**; **D**: 555.

Kite (*Milvus niger*) = Black Kite *Milvus migrans*, CD notes one laid fertile egg(s) in London Zoo; **V2**: 154.

kite = unspecified; in flocks of domestic pigeons, white individuals first fall victims to the kite; **V2**: 229–230.

sea-eagle (*Falco ossifragus*) = White-tailed Eagle *Haliaeetus albicilla*, picks out black fowls to predate on the Irish west coast; **V2**: 230.

White-headed Eagle (*Falco leucocephalus*) = Bald Eagle *Haliaeetus leucocephalus*, known to breed in immature plumage; **D**: 739 footnote.

Haliæetus leucocephalus = Bald Eagle *Haliaeetus leucocephalus*, seen to mate in London Zoo; **V2**: 154.

fishing-eagles = unspecified, captives eating fish CD had placed plant seeds within to see how long and viable they remained in birds' stomachs with respect to them as plant dispersers; **O**: 510. See chapter 3.

Buteo erythronotus = Variable Hawk *Geranoaetus polysoma*, distribution and habits; **Z**: 26. See chapter 3.

  Haliaëtus erythronotus = as a synonym of **Buteo erythronotus**; **Z**: 26.
  Buteo tricolor = as a synonym of **Buteo erythronotus**; **Z**: 26.

Buteo varius = Variable Hawk *Geranoaetus polysoma*, described; **Z**: 27–28.

buzzard = the Galapagos Hawk *Buteo galapagoensis*, briefly mentioned as unique to the islands but with characters of *Polyborus* or Caracara; **J**: 461. See chapters 2 and 3, and colour figure 3.3.

Craxirex = described as a new genus for the Galapagos Hawk, now *Buteo galapagoensis*, the species described and illustrated, and habits noted; **Z**: 22–25. See colour figure 3.3.

Craxirex galapagoensis = the Galapagos Hawk *Buteo galapagoensis*, described and illustrated, and its habits noted; **Z**: 22–25. See chapter 3 and colour figure 3.3.

Polyborus galapagoensis = as a synonym of **Craxirex galapagoensis**; **Z**: 23.

Buteo ventralis = Rufous-tailed Hawk, described; **Z**: 26–27.

Buteo vulgaris = Common Buzzard *Buteo buteo*, seen to mate in London Zoo; **V2**: 154.

Buteo = genus for typical buzzards, compared with **Craxirex galapagoensis**; **Z**: 24. See chapter 3.

hawk(s) = unspecified, large species as predators of rabbits introduced to Falkland Islands; **J**: 249.

A species found in more open parts of Tierra del Fuego; **J**: 301.

CD wonders if hawks ever bring live prey to the nest because, in so doing, they may introduce at least small mammals to islands; **J**: 351. See chapter 4.

Listed among birds that were tame on the Galapagos Islands; **J**: 475; **V1**: 20.

Present as predators on the Falkland Islands, and yet other birds there tame to humans; **J**: 477.

Hawks "look out for tired birds" [i.e., migrants] and in eating them may disperse any seeds within them; **O**: 510. See chapter 2.

Ruffle their feathers and spread wings and tail when approached; **E**: 99.

CD cites the observation that of five nestlings in a nest, four were killed by a gamekeeper who shot the parents next day; but the following day, another adult pair were tending the nestling, and so they too were shot, only to be replaced by another pair feeding the nestling, one of which was shot; **D**: 624–625 footnote.

A long-winged and long-tailed breed of domestic pigeon had the appearance of a long-winged hawk; **V1**: 157.

As hawks are kept under almost free circumstances in their native lands, yet do not breed, leads CD to infer that this is not solely due to want of exercise; **V2**: 159.

People in France and Germany are advised to not keep white pigeons, as hawks abound; **V2**: 229.

OTIDIFORMES

OTIDIDAE—Bustards

bustard(s) = unspecified but no doubt the Great Bustard *Otis tarda*, thought to have been the kind of progenitor, in size and habits, of the rheas and ostrich; **O**: 168.

Only bustards show "strongly-marked sexual differences within the Cursores"; **D**: 339.

In certain species, the sexes differ in plumage; but females remain the same throughout the year, whereas males undergo a change in colour; **D**: 595.

With certain species, "the vernal moult is far from complete, some feathers being renewed, and some changed in colour," and some older males retain nuptial plumage throughout the year; **D**: 597.

Great, great English or **European bustard (*Otis tarda*)** = Great Bustard, said to be polygamous [which it is]; **D**: 339.

The "throat pouch" of males of this and at least four other bustard species is not used to hold water but to enhance vocalizations; **D**: 570. See chapter 6.

Throws itself into "indescribably odd attitudes whilst courting the female"; **D**: 581.

Otis = some species of this genus lack the oil gland typical of most birds; **V1**: 147 footnote.

Australian Bustard = *Ardeotis australis*, mentioned as being illustrated by Dr. Murie, in discussing the function of the "throat pouch" in that species; **D**: 570 footnote.

An allied Indian bustard (*Otis bengalensis*) = Bengal Florican *Houbaropsis bengalensis*, brief description of the males' leaping "flight" and subsequent display; **D**: 581. See chapter 6.

Indian bustard (*Sypheotides auritus*) = Lesser Florican *Sypheotides indicus*, has its primaries greatly acuminated, probably for noise production in display; **D**: 577.

The feathers forming the ear-tufts terminate in a disc; **D**: 587.

EURYPYGIFORMES

RHYNOCHETIDAE—Kagu

Kagu (*Rhinochetus jubatus*) = *Rhynochetos jubatus*, "when anxious to feed, beat the ground with their feet" (see also **flamingo** and **Sheldrake**); **E**: 50. See chapter 7.

GRUIFORMES

RALLIDAE—Rails, Crakes, and Coots

Zapornia notata = Speckled Rail *Coturnicops notatus*, specimen shot aboard the *Beagle* at Rio Plata, described and illustrated; **Z**: 132.

Crex lateralis = Rufous-sided Crake *Laterallus melanophaius*, recorded at Maldonado, Rio Plata, readily flies if disturbed; **Z**: 132.

Zapornia spilonota = Galapagos Crake *Laterallus spilonata*, described, illustrated, habitats, habits, call, clutch size; **Z**: 132–133. See colour figure 2.7.

water-rail = Galapagos Crake *Laterallus spilonata*, confined to damp region of Galapagos; **J**: 459, 473. See chapter 2 and colour figure 2.7.

Rallus phillipensis = Buff-banded Rail *Gallirallus philippensis*, common on Keeling or Cocos Atoll, Indian Ocean; **Z**: 133.

rail = Buff-banded Rail *Gallirallus philippensis*, observed by CD on Cocos Keeling Islands; **J**: 543.

Another "rail (*Porphyrio*)" CD recorded as shot near the summit of Ascension Island in July 1836 is said to be Allen's Gallinule *Pophyrio alleni* (Steinheimer 2004, appendix 9); **J**: 543.

landrail = Corn Crake *Crex crex*, "nearly as terrestrial as the quail or partridge," in discussing toes adapted for aquatic life; **O**: 222.

corncrakes = Corn Crakes *Crex crex*, with long toes but living in meadows and not swamps; **O**: 223.

Rallus ypecaha = Giant Wood Rail *Aramides ypecaha*, recorded at Buenos Ayres; **Z**: 133.

Crex melampyga = as a synonym of **Rallus ypecaha**; **Z**: 133.

Rallus sanguinolentus = Plumbeous Rail *Pardirallus sanguinolentus*, recorded at Valparaiso; **Z**: 133.

(Gallicrex cristatus) = Watercock *Gallicrex cinerea*, males are a third larger than females and are so pugnacious when breeding that people of Eastern Bengal keep them as fighting birds; **D**: 552. See chapter 6.

Porphyrio simplex = Allen's Gallinule *Porphyrio alleni*, described from Ascension Island, Atlantic Ocean; **Z**: 133–134. See **rail**.

Porphyrio = unspecified, captive bred in Sicily; **V2**: 156.

water-hen (*Gallinula nesiotis*) = Tristan Moorhen, CD found this island bird's loss of flying ability "equal to that of the domesticated duck but anatomically similar to the **European water-hen**"; **V1**: 287. See chapter 5.

Fulica galeata = Common Moorhen *Gallinula chloropus*, recorded on Conception Island, Chile; **Z**: 133.

Crex galeata = as a synonym of **Fulica galeata**; **Z**: 133.

Yahana proprement = as a synonym of **Fulica galeata**; **Z**: 133.

Gallinula galeata = as a synonym of **Fulica galeata**; **Z**: 133.

water-hen = Common Moorhen *Gallinula chloropus*, is nearly as aquatic as the Common Coot but lacks the adapted toes of the latter; **O**: 222.

common or **European water-hen (*Gallinula chloropus*)** = Common Moorhen, pairing males fight violently for females, striking one another with their feet; **D**: 552.

See **water-hen (*Gallinula nesiotis*)**.

Gallinula chloropus = Common Moorhen, captive bred in London Zoo; **V2**: 156.

Gallinula crassirostris = Spot-flanked Gallinule *Gallinula melanops*, collected on the banks of La Plata at Valparaiso; **Z**: 133.

Gallicrex cristatus = Red-knobbed Coot *Fulica cristata*, a "large red caruncle" develops on the head of males for breeding; **D**: 593.

coot = Eurasian Coot *Fulica atra*, has lobed toes adapted to aquatic life; **O**: 222.

spur-winged rails = unspecified, in some such species the spurs are of similar size in both sexes; **D**: 558. See chapter 6.

rail-like birds = unspecified, which undergo a double moult, have older males retain nuptial plumage throughout the year; **D**: 597.

PSOPHIIDAE—**Trumpeters**

Psophia = the genus for the three species of trumpeter, often kept about homes of Guianan Indians but "seldom or never known to breed"; **V2**: 157.

GRUIDAE—**Cranes**

cranes, in error, probably for storks, as no cranes live in South America; **J**: 22. See chapters 2 and 8.

cranes = "the cranes breed more readily [in captivity] than the other ["wader"] genera; **V2**: 156.

Grus antigone = Sarus Crane, has bred several times in captivity in Calcutta; **V2**: 156.

cranes (*Grus virgo*) = Demoiselle Crane, in both sexes the trachea penetrates the sternum but presents "certain sexual modifications"; **D**: 572.

Tetrapteryx paradisea = Blue Crane *Grus paradisea*, captive bred at Knowsley [UK]; **V2**: 156.

Grus montigresia = a misspelling of *G. montignesia* = Red-Crowned Crane *G. japonensis*, this crane has bred several times in the London Zoo and Paris; **V2**: 156.

Grus americanus = Whooping Crane *G. americana*, take four years to acquire their "perfect plumage" but breed prior to this; **D**: 738–739 footnotes.

Grus cineria = Common Crane *G. grus*, this crane has bred several times in London Zoo; **V2**: 156.

CHARADRIIFORMES

TURNICIDAE—Buttonquail

Turnix = the genus of buttonquail—in one section of these, the female is larger than the male, which is unusual among the **Gallinaceae**; and in most species, the female is the brighter sex; **D**: 726.

Females flock at the end of summer; **D**: 728.

Females have acquired their [adult] brighter colours through sexual selection; **D**: 744.

In all species of the sexually dimorphic Gallinaceae except *Turnix*, the male is "the most beautiful"; **V1**: 255.

Indian Turnix = Spotted Buttonquail *Turnix ocellatus*, females apparently more numerous than males, suggesting to CD the former "would have been led" to complete to court the latter as with peahen, wild turkey, and some grouse; **D**: 733.

Turnix taigoor = Barred Buttonquail *Turnix suscitator*, sexual dimorphism described, the female being more colourful, vocal, and pugnacious and thus used as fighting birds like **game-cocks**; **D**: 726.

CHIONIDIDAE—Sheathbills

Chionis alba = Snowy Sheathbill *C. albus*, encountered far at sea, food, flight; **Z**: 118–119.

Mentioned in passing as being closely related to seedsnipe (CD noted the taxonomy of sheathbills as tentative, and that they were no longer considered particularly close to seedsnipe); **J**: 111.

HAEMATOPODIDAE—Oystercatchers

Haematopus palliatus = Eurasian Oystercatcher *Haematopus ostralegus*, recorded at Rio Plata; **Z**: 128.

RECURVIROSTRIDAE—Stilts, Avocets

Himantopus nigricollis = Black-winged Stilt *Himantopus himantopus*, locations, habitats, habits, call; **Z**: 130.

Himantopus melanura = White-backed Stilt *Himantopus melanurus*, described as "a kind of plover" seen commonly in large flocks; **J**: 133. See chapter 2.

CHARADRIIDAE—**Plovers**

Charadriidae = the pelvis of this group in general agrees "perfectly" with that of **Tinochorus rumicivorus**; **Z**: 156.

pewits = Northern Lapwing *Vanellus vanellus*, compared with Southern Lapwing (see **teru-tero**); also collective word for lapwing species; **J**: 133.

common peewit (*Vanellus cristatus*) = Northern Lapwing *Vanellus vanellus*, the wing spur becomes more prominent in the breeding season, and males fight together; **D**: 558.

Vanellus cristatus = Northern Lapwing *Vanellus vanellus*, resemble **Philomachus cayanus** in habits; **Z**: 127.

Hoplopterus armatus = Blacksmith Lapwing *Vanellus armatus*, wing spurs do not increase in size for the breeding season (but they do fight like Northern Lapwings and drive off enemies); **D**: 560.

Lobivanellus lobatus = Masked Lapwing *Vanellus miles*, both sexes wear wing spurs, but they are much larger in males; **D**: 559–560.

Lobivanellus = once applied to several *Vanellus* species, a wing "tubercle" develops into a "short horny spur" during the breeding season; **D**: 558–559.

teru-tero (*Vanellus cayanensis*) = Southern Lapwing *Vanellus chilensis*, appearance, call, habits including distraction display and (highly edible) eggs briefly described; **J**: 130, 133. See chapter 2.

Philomachus cayanus = Southern Lapwing *Vanellus chilensis*, distribution, likened to ***Vanellus cristatus*** (= *V. vanellus*), habitat, habits, calls, distraction display, eggs; **Z**: 127.

Chardrius Cayanus = as a synonym of **Philomachus cayanus**; **Z**: 127.

Charadrius pluvialis = European Golden Plover *Pluvialis apricaria*, CD refers his reader to Macgillivray's *History of British Birds* as an example of young generally resembling the adult females of their kind; **D**: 740 footnote.

Charadrius virgininus = American Golden Plover *Pluvialis dominica*, common in large and small flocks on La Plata, also in Chile; **Z**: 126.

Charadrius marmoratus = as a synonym of **Charadrius virgininus**; **Z**: 126.

Hiaticula semipalmata = Common Ringed Plover *Charadrius hiaticula*, recorded on Galapagos; **Z**: 128.

CD refers his reader to Macgillivray's *History of British Birds* for an example of young generally resembling the adult females of their kind; **D**: 740 footnote.

Recurvirostra himantopus = as a synonym of **Hiaticula semipalmata**; **Z**: 130.

Tringa semipalmate = as a synonym of **Hiaticula semipalmata**; **Z**: 128.

Charadrius semipalmatus = as a synonym of **Hiaticula semipalmata**; **Z**: 128.

Hiaticula azarae = Kentish Plover *Charadrius alexandrinus*, locations, described; **Z**: 127.

Charadrius Azarae = as a synonym of **Hiaticula azarae**; **Z**: 127.

Charadrius collaris = as a synonym of **Hiaticula azarae**; **Z**: 127.

Hiaticula trifasciatus = Two-banded Plover *Charadrius falklandicus,* two specimens from Bahia Blanca; **Z**: 127.

Charadrius bifasciatus = as a synonym of **Hiaticula trifasciatus**; **Z**: 127.

Charadrius trifasciatus = as a synonym of **Hiaticula trifasciatus**; **Z**: 127.

Squatarola cincta = Rufous-chested Plover *Charadrius modestus*, locations, habitats, large flocks in fields; **Z**: 126.

Tringa Urvillii = as a synonym of **Squatarola cincta**; **Z**: 126.

Vanellus cinctus = as a synonym of **Squatarola cincta**; **Z**: 126.

Squatarola cincta = as a synonym of **Squatarola cincta**; **Z**: 126.

Charadrius rubecola = as a synonym of **Squatarola cincta**; **Z**: 126.

Squatarola fusca = Rufous-chested Plover *Charadrius modestus*, described from a Maldonado specimen; **Z**: 126–127.

plovers = unspecified; in certain species, wing spurs must be considered sexual characters; see **common peewit**; **D**: 558.

With certain species “the vernal moult is far from complete, some feathers being renewed and some changed in colour”; **D**: 597.

Some are difficult to see when crouched on the ground; **D**: 746.

dotterel plover (*Eudromias morinellus*) = Eurasian Dotterel *Charadrius morinellus*, females are larger and more colourful than males and appear to perform most, if not all, parental care; **D**: 729.

Oreophilus totanirostris = Tawny-throated Dotterel *Oreopholus ruficollis*, locations, habitat, habits, call; **Z**: 125–126.

water-birds = unspecified, CD noting that these on the Galapagos were not endemic, in contrast to most of the land birds, and that he was “able to get only eleven kinds” together with “waders”; **J2**: 380.

ROSTRATULIDAE—**Painted-snipes**

R. bengalensis = Greater Painted-snipe *Rostratula benghalensis*, the trachea is not convoluted in either sex, the young of both sexes are said to resemble the adult male, and males are believed (correctly) to incubate the clutch; **D**: 727–728. See **R. australis**. See chapter 7.

R. australis = Australian Painted-snipe *Rostratula australis*), the trachea is simple in males but with four convolutions before entering the lungs in females; **D**: 727. See *R. bengalensis*.

Rhynchaea semicollaris = South American Painted-snipe *Nycticryphes semicollaris*, recorded at Monte Video, Rio Plata, on swamps, habits like **Scolopax Gallinago**; **Z**: 131.

Painted Snipes (Rhynchæa) = the females of the three species are larger and brighter coloured than males, with **R. capensis** (= Greater Painted-snipe *Rostratula benghalensis*) illustrated; **D**: 727. See chapter 7.

Rhynchæa = painted snipe, females acquired their bright [adult] plumage through sexual selection; **D**: 744.

PEDIONOMIDAE—**Plains-wanderer**
Australian Plain-wanderer (*Pedionomus torquatus*) = Plains-wanderer, shows similar sexual dimorphism to **Turnix**; **D**: 726 footnote.

THINOCORIDAE—**Seedsnipes**
Attagis gayii = Rufous-bellied Seedsnipe *A. gayi*, habitat, habits, flight of a covey was grouse-like; CD saw the seedsnipe as the ecological equivalents of ptarmigan; **Z**: 117.

Attagis falklandica = White-bellied Seedsnipe *Attagis malouinus*, habitat, habits; **Z**: 117.
  Tetroa Falklandicus = as a synonym of **Attagis falklandica**; **Z**: 117.
  Coturnix Falklandica = as a synonym of **Attagis falklandica**; **Z**: 117.
  Perdix Falklandica = as a synonym of **Attagis falklandica**; **Z**: 117.
  Ortyx Falklandica = as a synonym of **Attagis falklandica**; **Z**: 117.

Tinochorus Eschscholtzii = Least Seedsnipe *Thinocorus rumicivorus*, range, habitats, appearance, structure, and habits are quail- and snipe-like; **J**: 110–111. See chapters 2, 3, and 8.

Tinochorus rumicivorus = Least Seedsnipe, in appearance and habits like both a wader and a gallinaceous species; distribution, habitats, habits (like partridges they only take wing as a flock); wing and tail structure like *Tringa hypoleucos*, nest location, food, anatomical notes by T. C. Eyton; **Z**: 117–118, 155–156; **J**: 475. See chapters 3 and 8.
Tinochorus Eschscholtzii = as a synonym of **Tinochorus rumicivorus**; **Z**: 117.

Two species of Attagis = two species of seedsnipe that CD considered to be "in almost every respect ptarmigans in their habits"; **J**: 111.

Tinochorus = one of the two seedsnipe genera, which T. C. Eyton believed to form a link between "the orders *Grallatores* and *Razores*" based on external morphology and anatomy; **Z**: 156.

SCOLOPACIDAE—Sandpipers, Snipes

woodcock = Eurasian Woodcock *Scolopax rusticola*, a leg of one had plant seeds in attached mud, which germinated and flowered, indicating it could disperse plants in this way; **O**: 512. See chapter 3.

woodcock = unspecified, are difficult to see when crouched on the ground; **D**: 746.

Are coloured for concealment but "with extreme elegance," leading CD to conclude that both natural and sexual selection were involved; **D**: 748.

jack-snipe = *Lymnocryptes minimus*, are coloured for concealment but "with extreme elegance," leading CD to conclude that both natural and sexual selection were involved; **D**: 748.

Solitary snipe (*Scolopax major*) = Solitary Snipe *Gallinago solitaria*, large numbers (contrary to name!) "assemble during dusk in a morass; and the same place is frequented for the same purpose [display] during successive years"; believed to be "polygamists"; **D**: 618.

Scolopax Gallinago = Common Snipe *Gallinago gallinago*, CD found habits of **Rhynchaea semicollaris** like those of this species; **Z**: 131.

English snipe = Common Snipe *Gallinago gallinago*, habits, flight, and drumming similar to **Scolopax (Telmatias) magellanicus**; **Z**: 131.

Scolopax Gallinago = Common Snipe *Gallinago gallinago*, its habits similar to those of **Rhynchaea semicollaris**, Monte Video, on swamps; **Z**: 131.

Scolopax Wilsonii = Common Snipe *Gallinago gallinago*, its aerial display, which causes "a switching noise," described; **D**: 576–577.

common snipe (*Scolopax gallinago*) = *Gallinago gallinago*, its aerial "drumming" display described and the tail feathers involved illustrated (also those of *S. frenata* and *S. javensis*); **D**: 575–576.

Snipe = South American Snipe *Gallinago paraguaiae*, rising and calling, near Mandetiba, north of Cape Frio, out of Rio de Janeiro on April 8, 1832; **J**: 22.

Unspecified, males fight together using their bills; **D**: 554.

Drumming of the tail [in flight display] "pleasing to the females"; **D**: 580.

Darwin describes his excitement at shooting his first snipe (Common Snipe *Gallinago gallinago*); **A**: 18.

Unspecified, CD killed 23 with 24 shots in South America; **A**: 81.

Scolopax (Telmatias) paraguaiae = South American Snipe *Gallinago paraguaiae*, at Valparaiso and Maldonado, Rio Plata; **Z**: 131.

Scolopax Paraguai = as a synonym of **Scolopax (Telmatias) paraguaiae**; **Z**: 131.

Scolopax Brasiliensis = as a synonym of **Scolopax (Telmatias) paraguaiae**; **Z**: 131.

Scolopax (Telmatias) magellanicus = South American Snipe *Gallinago paraguaiae*, specimens from Maldonado and East Falklands, habits, flight and drumming similar to "English snipe" (= Common Snipe *Gallinago gallinago*) when breeding; **Z**: 131.

Scolopax frenata = South American Snipe *Gallinago paraguaiae*, tail feathers involved in aerial "drumming" display, illustrated; **D**: 576.

Scolopax javensis = Pin-tailed Snipe *Gallinago stenura* or Swinhoe's Snipe *G. megala*, tail feathers involved in aerial "drumming" display illustrated; **D**: 576.

Snipe(s) = unspecified, mentioned in passing as being similar to seedsnipe; **J**: 110–111.

One seen on East Falkland Island in March 1834 was most likely South American Snipe *Gallinago paraguaiae*; **J**: 246.

The latter species described as tame on the Falkland Islands; **J**: 476.

CD observed snipe (most likely Pintail Snipe *Gallinago stenura*) on the Cocos Keeling Islands, April 1836; **J**: 543.

Migrant males arrive on the breeding grounds before females; **D**: 327.

In some species, the "sexes resemble each other, and do not change colour at any season"; **D**: 594.

Are difficult to see when crouched on the ground; **D**: 746.

Limosa hudsonica = Hudsonian Godwit *Limosa haemastica*, collected in Falkland Islands and Chiloe, from tidal mud bank flocks; **Z**: 19.

Scolopax Hudsonica = as a synonym of **Limosa hudsonica**; **Z**: 129.

Limosa lapponica = Bar-Tailed Godwit, females are larger and more colourful than males; **D**: 729.

Numenius brevirostris = Eskimo Curlew *Numenius borealis*, recorded at Buenos Ayres; **Z**: 129.

curlews = unspecified, in some species the "sexes resemble each other, and do not change colour at any season"; **D**: 594.

Numenius hudsonicus = Whimbrel *Numenius phaeopus*, very abundant on mud banks of Chiloe; **Z**: 129.

Totanus = *Tringa*, in some species the sexes resemble each other "but in which the summer and winter plumage differ slightly in colour"; **D**: 594.

Totanus melanolecos = Greater Yellowlegs *Tringa melanoleuca*, recorded at Rio Plata; **Z**: 130.

Scolopax melanoleaca = as a synonym of **Totanus melanolecos**; **Z**: 130.

Scolopax vociferus = as a synonym of **Totanus melanolecos**; **Z**: 130.

Totanus solitarius = as a synonym of **Totanus melanolecos**; **Z**: 130.

Totanus flavipes = Lesser Yellowlegs *Tringa flavipes*, recorded at Monte Video, Rio Plata; **Z**: 129.

Totanus macropterus = Solitary Sandpiper *Tringa solitaria*, recorded at Monte Video, Rio Plata; **Z**: 129.

Totanus = Wandering Tattler *Tringa incana*, which CD thought could perhaps be a related new species on the Galapagos Islands (but was not); **J**: 461.

Totanus fuliginosus = Wandering Tattler *Tringa incana*, described from Galapagos Islands; **Z**: 130.

Tringa hypoleucos = Common Sandpiper *Actitis hypoleucos*, comparing its wing and tail structure with that of Least Seedsnipe; **J**: 111; **Z**: 118.

Tringa = CD refers his reader to Macgillivray's *History of British Birds* for examples of young generally resembling the adult females of their kind; **D**: 740 footnote.

sandpiper = unspecified, in noting wing shape similar to that of **Tinochorus rumicivorus**; **Z**: 118.

Strepsilas interpres = Ruddy Turnstone *Arenaria interpres*, collected at Iquique, Peru and Galapagos Archipelago; **Z**: 132.

The pelvis of this bird, and of the ***Charadriidae*** in general, agrees "perfectly" with that of **Tinochorus rumicivorus**; **Z**: 156.

Tringa Morinellus = as a synonym of **Strepsilas interpres**; **Z**: 132.

Strepsilas = *Arenaria*, the genus of the two turnstone species, with a pelvis like that of **Tinochorus rumicivorus**; **Z**: 155.

Knot (*Tringa canutus*) = Red Knot *Calidris canutus*, eight or nine individuals in London Zoo retained winter plumage throughout the year, indicating that "the summer plumage, though common to both sexes, partakes of the nature of the exclusively masculine plumage of many other birds"; **D**: 596.

Pelidina minutilla = Least Sandpiper *Calidris minutilla*, collected on Galapagos Islands; **Z**: 131.

Tringa minutilla as a synonym of **Pelidna minutilla**; **Z**: 131.

Pelidna schinzii = White-rumped Sandpiper *Calidris fuscicollis*, flocks common on inland bay shores, southern Tierra del Fuego; **Z**: 131.

Tringa Schinzii = as a synonym of **Pelidna schinzii**; **Z**: 130.

Pelidna cinclus = as a synonym of **Pelidna schinzii**; **Z**: 130.

Tringa rufescens = Buff-breasted Sandpaper *Tryngites subruficollis*, recorded at Monte Video, Rio Plata; **Z**: 130.

Machetes = *Philomachus*, the genus for the Ruff *P. pugnax*, has a sternum structure most like that of **Tinochorus rumicivorus** among the gallinaceous birds; **Z**: 156. See figure 6.1.

CD refers his reader to Macgillivray's *History of British Birds* for an example of young generally resembling the adult females of their kind; **D**: 740 footnote.

Ruff (*Machetes pugnax*) = *Philomachus pugnax*, males erect their "collar of feathers when fighting"; **E**: 99. See figure 6.1.

Within the **Grallatores**, "extremely few species differ sexually"; but the **ruff** is an exception, and CD notes that Montagu believes it to be polygamous (which it is); **D**: 339. See chapter 6.

Males much more numerous than females; **D**: 382.

Males are "notorious" for "extreme pugnacity" in spring, and Col. Montagu stated the ruff functions to "defend the more tender parts" in battle, illustrated; **D**: 552. See chapter 6.

Males are ready to fight whenever they meet; **D**: 560.

The sexes differ in both summer and winter plumage, but males undergo a greater change at each recurrent season than do females; **D**: 595.

Males retain their ruff in the spring for barely two months; **D**: 598. See figure 6.1.

The "fowlers" discover traditional leks of these "polygamists" by "grass being trampled bare"; **D**: 617–618.

Individuals are attracted to "any bright object" out of "curiosity"; **D**: 629.

Phalaropus hyperboreus = Red Phalarope *P. fulicarius*, females are larger and in breeding plumage slightly brighter than males; **D**: 728. See chapter 7.

Phalaropus fulicarius = Red Phalarope *P. fulicarius*, females are larger and in breeding plumage slightly brighter than males, and males incubate the clutch; **D**: 728–729. See chapter 7.

waders = mentioned in passing as present on Galapagos Islands, saying difficult to know if any new species there; **J**: 461.

Believed by CD to be the first colonists of islands; **J**: 543. See chapter 2.

Unspecified, in CD noting that those on the Galapagos Islands were not endemic, in contrast to most of the land birds, and that of "waders and water-birds," he was able to get only eleven kinds; **J2**: 380.

The summer and winter plumages of many species differ very little in colour; **D**: 705.

In some few species, the females are the larger and more strongly plumaged sex; **D**: 729 footnote.

Yarrell (1837–1843) insists that young of many species wear characters intermediate between those of adult summer and winter plumages; **D**: 741. See chapter 7.

Most species can be tamed, but several are short-lived in confinement and are unproductive; **V2**: 156.

GLAREOLIDAE—Coursers, Pratincoles

Glareola = a genus of the sub-family Glareolinae for the seven of the eight pratincole species, the bill and wing structure of which "much resembles" that of **Tinochorus rumicivorus**; **Z**: 155.

swallow-plovers = unspecified pratincoles (*Glareola* and *Stiltia*), in some species, the "sexes resemble each other and do not change colour at any season"; **D**: 594.

Glarecola Australis [*sic*] = Australian Pratincole *Stiltia isabella*, tail shape is as in ***Tinochorus***; **Z**: 156.

Glareola praticola = Collared Pratincole, tail length like that of **Tinochorus rumicivorus**; **Z**: 155.

LARIDAE—Gulls, Terns, and Skimmers

noddy = most likely Brown Noddy *Anous stolidus*, possibly Black Noddy *A. tenuirostris*, on St. Paul Rocks, Atlantic; **J**: 9. See chapter 2.

Megalopterus stolidus = Brown Noddy *Anous stolidus*, specimens from Galapagos, nest, incubating in February, one flew aboard the *Beagle* in the Pacific several hundred miles from land; **Z**: 145.

Sterna stolida = as a synonym of **Megalopterus stolidus**; **Z**: 145.

snow-white tern = White Tern *Gygis alba*, nesting on (Cocos) Keeling Island; **J**: 545.

Scissor-beak (Rhyncops nigra) = Black Skimmer *Rynchops niger*, described and its foraging detailed; **J**: 161–162. See chapters 2 and 3, and figure 3.14.

Rhyncops nigra = Black Skimmer *Rynchops niger*, distribution, habitats, feeding technique, flight, common far inland, flocking and flying out to sea at sunset; **Z**: 143–144; **J**: 475. See chapters 2 and 3, and figure 3.14.

gull (*Larus tridactylus*) = Black-legged Kittiwake *Rissa tridactyla*, one fed predominantly grain for a year developed thickened muscles on the stomach wall; **V2**: 302.

Xema ridibundum = Black-headed Gull *Chroicocephalus ridibundus*, see **Xema (Chroicocephalus) cirrocephalum**.

Xema (Chroicocephalus) cirrocephalum = Grey-Headed Gull *Chroicocephalus cirrocephalus*, very similar to **X. ridibundum** and the two compared, seen far inland, breed in marshes, sometimes feeds at slaughter houses; **Z**: 142–143.

Larus cirrocephalus = as a synonym of **Xema (Chroicocephalus) cirrocephalum**; **Z**: 142.

Larus maculipennis = as a synonym of **Xema (Chroicocephalus) cirrocephalum**; **Z**: 142.

Larus glaucodes = as a synonym of **Xema (Chroicocephalus) cirrocephalum**; **Z**: 142.

Larus haematorhynchus = Dolphin Gull *Leucophaeus scoresbii*, specimen from Port St. Julian, Patagonia; **Z**: 142.

Having many characters in common with **Larus fuliginosus**; **Z**: 142. See also chapter 2.

Larus fuliginosus = Lava Gull *Leucophaeus fuliginosus*, specimen from James Island, Galapagos; described as having many characters in common with **L. haematorhynchus**, but duskier and beak longer; **Z**: 141–142. See also chapter 2 and colour figure 2.9

Larus Franklinii = Franklin's Gull *Leucophaeus pipixcan*, CD cites a record at Concepcion; **Z**: 143 footnote.

English gull = Mew Gull *Larus canus*, call of **L. dominicanus** like that of this gull; **Z**: 142.

Larus dominicanus = Kelp Gull, abounds in flocks on Pampas, to 60 miles inland, attends slaughter houses and feeds with caracaras and vultures on garbage and offal, call like **L. canus** (Mew Gull); **Z**: 142–143.

herring gull (*Larus argentatus*) = bred many times at London Zoo and Knowsley since 1848; **V2**: 157.

Larus argentatus = European Herring Gull, birds feeding on corn seed in spring develop a more muscular stomach wall; **V2**: 302.

gull = CD mentions the only gull he encountered in the Galapagos as being new to science (subsequently named *Larus fuliginosus* by Gould, in Gould and Darwin [1838–1841], and now called Lava Gull); **J**: 461.

Gulls = Darwin gained pleasure seeing them flying home in England; **L**: 354.

Unspecified, young seagulls killed and eaten by Antarctic Giant-Petrels *Macronectes giganteus*; **J**: 354.

Feeding at slaughterhouses with Black Vultures and caracaras; **Z**: 7.

Colouration of bare facial skin changes seasonally; **D**: 593.

Both sexes in several species acquired white plumage through sexual selection; **D**: 751.

In many species, the head and neck becomes pure white in summer, being grey or mottled in winter and in young birds, but the reverse is true with some smaller species; **D**: 751.

Although many kept at London Zoo and the Old Surrey Gardens, none bred there prior to 1848; **V2**: 157.

In some species, coloured parts appear as if almost washed out; **V2**: 350.

Viralva aranea = Gull-billed Tern *Gelochelidon nilotica*, specimen from Bahia Blanca, Patagonia, a species seen in open ocean—a flock likely of this species fishing 70 and 120 miles from land; **Z**: 145.

Sterna aranea = as a synonym of **Viralva aranea**; **Z**: 145.

terns of unknown species feeding on large prawn-like crustacea; **J**: 18.
Unspecified, nesting on (Cocos) Keeling Island; **J**: 544–545.
Both sexes in several species acquired white plumage through sexual selection; **D**: 751.
Sexual selection has brought about both white and black species; **D**: 752.
A group with both black and white species but that also includes "piebald" ones; **D**: 753.

terns (Sterna) = unspecified, the adult head and neck becomes grey or mottled during summer and white in winter in adults and young; **D**: 751.

ALCIDAE—Auks

guillemots *Uria lacrymans* = Common Murre *Uria aalge*, CD notes that if the bridled form of this bird, estimated to be one-fifth of those on the Faroe Islands, were "of a beneficial nature," it would supplant the more common form through "the survival of the fittest"; **O**: 113.

common Guillemot (*Uria troile*) = Common Murre *Uria aalge*, CD describes the sympatric two forms, notes that one of each may form a pair, and repeats notes made under **guillemots *Uria lacrimans***; **D**: 647.

razor-bill (*Alca torda*) = Razorbill, young, "in an early state of plumage," are coloured like adults during summer; **D**: 741. See chapter 7.

auk = unspecified, in noting flight, swimming, and diving of the Common Diving-Petrel *Pelecanoides urinatri* is auk-like; **O**: 221, 223, 646.

COLUMBIFORMES

COLUMBIDAE—Pigeons, Doves

Columbidae = many of the genera do not differ greatly from each other;**V1**: 157.
Species never have less than 12 or more than 16 tail feathers; **V1**: 158.
The various species always have nine or 10 primaries, the shape of the "first few" [? outer ones] varies far more between the species than within domestic pigeons; **V1**: 159.
The beak wattle and that about the eyes in domestic pigeons increase with age but not in the Rock Dove or other wild pigeon species; **V1**: 162. See colour figure 5.2.
Certain more terrestrial groups have larger feet; **V1**: 174 footnote.
In several members, the proportionate length of toes and of primaries are apt to differ; **V1**: 194.
CD quotes Prince C. L. Bonaparte (1855) who finds 288 pigeon species in 85 genera; **V1**: 133.

dodo = *Raphus cucullatus*, in passing CD says the Falkland Island Fox (*Dusicyon australis*) will probably soon "be classed with the dodo" (he was correct, as it became extinct about 1876); **J**: 250.

Columba = two species of this genus lack the oil gland typical of most birds; **V1**: 147 footnote.

C. livia = Rock Dove *Columba livia,* as origin of fantail and pouter is indiscernible from its external appearance; **O**: 413–414. See colour figure 5.2.

Skulls of this and of C. *oenas, palumbus*, and *turtur* show extremely slight differences; **V1**: 163.

CD found only this, ***C. leuconota*** and ***C. rupestris,*** with two black wing bars among all pigeons; **V1**: 195.

Reversion to plumage characters of the Rock Dove in pure breeds of the domestic pigeon; **V2**: 29. See colour figure 5.2.

rock pigeon = Rock Dove *Columba livia*, with respect to ancestor of domestic pigeon; **O**: 26–31, 40, 478, 649. See chapters 4, 5, and 8, and colour figure 5.2.

Variation in characters in domestic forms, including reversion to those of this bird; **O**: 194–198, 202–203, 478. See chapter 4, 5, and 8, and colour figure 5.2.

In stating that the parent form of any two or more species would not exhibit characters "directly intermediate" between them, CD goes on to state "any more than the rock-pigeon is directly intermediate in crop and tail between its decendants, the pouter and fantail pigeons"; **O**: 637. See chapter 4.

common pigeon = feral pigeon or Rock Dove *Columba livia*, has iridescent breast feathers that males display in courtship; **D**: 612. See colour figure 5.2.

Columba amaliæ = Rock Dove *Columba livia*, a no longer used name once applied to the population on the Faroe and Hebrides Islands; **V1**: 183.

Columba affinis = Rock Dove *Columba livia*, a disused name once applied to a population on English cliffs; **V1**: 183–186.

Columba turricola = Rock Dove *Columba livia*, a disused name once applied to a population of Italy; **V1**: 184.

Columba Schimperis = Rock Dove *Columba livia*, a disused name once applied to a population of Abyssinia; **V1**: 184.

Columba gymnocyclus = Rock Dove *Columba livia*, a disused name once applied to a population of Senegal; **V1**: 184.

Columba intermedia = Rock Dove *Columba livia*, a disused name once applied to a population of India; **V1**: 184–187.

Pigeon(s) = domestic pigeon or Rock Dove *Columba livia*, those "with feathered feet have skin between their outer toes; pigeons with short beaks have small feet, and those with long beaks large feet"; **O**: 14.

A major part of CD's chapter "Variation Under Domestication" is given over to "Breeds of the Domestic Pigeon, Their Differences and Origin"; in **O**: 23–34, 44–45, 47. See chapters 4, 5, and 8, and figures 5.2–5.5.

Briefly alludes to numerous domestic forms; **O**: 64. See figure 5.1.

On continental Europe, people advised not to keep white ones, as they are commonly predated; **O**: 104; **V2:** 230.

A greater number of "the best" of the short-beaked tumbler-pigeon are asserted to perish in the egg than hatch [i.e., because artificial selection in captivity has brought about an inadequately adapted embryo] unless assisted by the pigeon keeper; **O**: 106.

Pigeon fanciers select for extreme characters such as short or long beaks; **O**: 135.

CD remarks on a "prodigious amount" of variety within the same breed of domestic pigeon in beak, wattles, tail, and so forth; **O**: 187–188.

Variation in characters including reversion to those of the **rock pigeon**; **O**: 194–198, 202–203, 478; **V1**: 239. See chapters 4, 5, and 8, and colour figure 5.2.

Young tumbler pigeons tumble having never seen an adult do so; **O**: 327.

Fantail and pouter pigeon origin from **rock pigeon** indiscernible from external appearances; **O**: 413–414.

CD asks if all domestic fantail pigeons were destroyed, would fanciers be able to "make a new breed hardly distinguishable from the present breed" but considers it hardly likely, even if from any other established domestic form because successive variations would be different; **O**: 457.

If "principal living and extinct races of the domestic pigeon were arranged in serial affinity," it would not accord with the "order in time of their production, and even less with the order of their disappearance; for the parent rock-pigeon still lives"; **O**: 478.

All forms of tumbler pigeon tumble, regardless of variation in beak length; but whereas the short-faced form has almost lost tumbling, all are "kept in the same [taxonomic] group"; **O**: 580.

Embryos of the various forms of domestic pigeon, highly distinctive in shape and proportions in adults, are near identical (but the short-faced tumbler less so); **O**: 612–624. See chapter 4.

Descended from "the blue and barred rock-pigeon"; **O**: 649. See chapter 4.

Eggs of **Prion vittatus** the size of those of a pigeon; **Z**: 141.

Some breeds "have learnt to coo in a new and quite peculiar manner"; **E**: 86.

Females "prefer the more vigorous and lively males"; **D**: 330.

Whereas the ancestral **rock pigeon** does not change plumage with age, save the breast becoming more iridescent upon maturity, some domestic breeds do not acquire their characteristic colours until their second, third, or fourth moult; **D**: 354.

Whereas the sexes of the ancestral **rock pigeon** do not differ externally, they do so in some domestic breeds but not because artificially selected for; **D**: 355. See chapter 5.

Limiting a character exhibited by both sexes to one sex by artificial selection is possible but extremely difficult; **D**: 357–358.

Regarding his rule that male-specific nuptial characters are acquired relatively late in life, CD notes an unspecified domestic breed in which feather streaking in males only can be detected in nestlings but become more conspicuous with each moult; **D**: 367.

In English Carrier and Pouter pigeons, the full development of wattles and crop occurs rather late in life and are transmitted "in full perfection" to males only; **D**: 367.

Sub-breeds described by Neumeister change their colour in both sexes during two or three months, as in the Almond Tumbler; but although these changes occur rather late in life, they are common to both sexes; **D**: 367–368.

Good evidence indicates "males are produced in excess [of females] or that they live longer"; usually two eggs in a nest produce one of each sex [the female proving the weaker], but although two males may result two females, seldom do; **D**: 382.

The soft cooing of [males of] many species presumed to please females; **D**: 572.

Swell their oesophagus to act as a vocal resonator; **D**: 575.

In certain breeds, "the feathers are plumose, with some tendency in the shafts to be naked"; **D**: 589.

The tail of the fantail, hood of the Jacobin, and beak and wattle of the carrier pigeons demonstrate results of artificial selection "by man"; **D**: 592.

Individuals of same sex sometimes form pairs or small parties; **D**: 623.

Have "excellent local memories" in returning to their homes after as long as nine months away; but if a pair, which would naturally remain mated for life, are separated for a few winter weeks and are then matched with other birds, they would rarely, if ever, subsequently recognize their original mate; **D**: 626.

Under domestication and "high feeding," birds' natural instincts may be corrupted; **D**: 634.

In domestication, both sexes prefer mating with their own breed; and dovecot birds dislike all "highly improved" breeds, whereas colour has little influence on mate choice; **D**: 636–638.

A domestic variety noted to have wing bars "symmetrically zoned with three shades, instead of being simply black on a slaty-blue ground, as in the parent-species." Breeds usually retain the two wing bars, if coloured red, yellow, white, black, or blue, with the rest of the plumage of a different tint; and the tail may be pigmented the reverse of the parent-species; **D**: 651–652.

Feathers may have separated or decomposed barbs for a significant part of their length; **D**: 656.

CD discusses the difficulty of producing a domestic breed in which the sexes differ in colour; **D**: 677–678.

There are breeds in Belgium in which only males are marked with black striae; **D**: 679–680.

The inheritance of characters between the sexes under domestication briefly discussed; **D**: 680–681.

Females of numerous species are brightly coloured and thus notoriously liable to attack from birds of prey, yet attend sparse open nests as exceptions to the rule that such nesting females are colourless—in discussing Wallace's objections to sexual selection for plumage colouration with regard to open nesting; **D**: 691.

In most Australian species, the slight differences in sexual dimorphism are of the same nature as the occasional greater differences in a minority of species; **D**: 699.

In the numerous distinct domestic breeds, the sexes appear alike, with only rare exceptions; and CD predicts this would hold true of other species if so diversified under domestication; **D**: 701.

Adult males differ from adult females, and young of both sexes resemble adult females; **D**: 712.

Dovecot pigeons do not willingly associate with coloured fancy breeds; **D**: 756.

Domestic pigeons pair with ***C. oenas*** and will cross with *C. palumbus*, ***Turtur risoria***, and ***T. vulgaris***; **V1**: 193 footnote.

Having examined the British Museum collection, CD found a dark bar at the end of the tail common, white edging to outer tail feathers not rare, but a white croup extremely rare—with two black wing bars in no pigeons other than ***C. livia,*** with the exceptions of ***C. leuconata*** and **C. rupestris**; **V1**: 195. See chapter 5.

Despite their great diversity of colour, domestic pigeons never show pencilled or spangled feathers, understandable as the wild **rock pigeon** and closely allied species lack them; **V1**: 244.

The influence of a first male fertilizing a female sometimes makes itself apparent in successive broods sired by a subsequent male; **V1**: 405.

A paragraph discussing "reversion," or the appearance of ancestral traits, in individuals at various ages; **V2**: 40, 48.

pigeon(s) and/or **rock pigeon(s)** or ***C. livia*** = the domesticated species or Rock Pigeon *Columbia livia* and its various breeds; **V1**: 29, 61, 73, 131–224, 239–240, 244, 405; **V2**: 29, 33, 38, 54–56, 66–67, 69, 74, 76, 83, 86, 95, 97, 103–104, 112, 117, 125–126, 161–162, 189, 195, 197–8, 202, 204, 211, 214, 221, 226–227, 229–230, 234–235, 238–240, 242, 245–246, 257, 264, 268, 278, 286–287, 289, 298, 304, 321–325, 332, 349–350, 354, 373, 386, 392–393, 401, 407–408, 410, 412–417, 419, 420–426, 429, 431. See chapters 4, 5, and 8, and figures 5.2–5.5.

C. rupestris = Hill Pigeon *Columba rupestris,* of Central Asia, is intermediate between *C. leuconota* and *C. livia* but with nearly the same tail as *C. leuconota* and is unlikely to have parented domestic pigeons (and did not); **V1**: 182, 184.

CD found only this, ***C. livia***, and ***C. leuconata*** with two black wing bars among all pigeons; **V1**: 195. See chapter 5.

Columba leuconota = Snow Pigeon, of the Himalaya, resembles some domestic pigeon breeds but has a white band across the tail and is unlikely to have parented domestic pigeons (and did not); **V1**: 182.

CD found only this, ***C. livia***, and ***C. rupestris*** with two black wing bars among all pigeons; **V1**: 195. See chapter 5.

Columba Guinea = Speckled Pigeon *C. guinea* (formerly of ***Strictœnas***), of Africa, roosts on trees or rocks; and although coloured like some domestic pigeon breeds, it is unlikely to have parented domestic pigeons (and did not); **V1**: 182–183.

Stictoenas = see ***Columba Guinea***.

C. oenas = Stock Dove *Columba oenas*, from its external appearance, could be seen as origin of fantail and pouter pigeons as equally as **C. livia**; **O**: 414.

Skulls of this and of *C. livia, palumbus,* and *turtur* show extremely slight differences; **V1**: 163. See chapter 5.

Roosts on trees and builds nests in holes; and although it hybridizes with *C. livia,* the offspring are sterile and is unlikely to have parented domestic pigeons (and did not); **V1**: 182.

Domestic pigeons readily pair with this species; **V1**: 193.

Hybrids with ***C. gymnophthalmos*** were sterile; **V1**: 19[illegible]

Palumbus = a disused genus that included *C. palumbus*, some species showing variation that could represent speciation and/or subspeciation; **V1**: 204 footnote.

English woodpigeon = Common Wood Pigeon *Columba palumbus*; in discussing relative wildness and tameness of birds in various places, CD notes that the typically wild Common Wood-Pigeon "should very frequently rear its young in shrubberies close to houses!"; **J2**: 400 footnote.

Skulls of this and of *C. livia, oenas*, and *turtur* show extremely slight differences; **V1**: 163. See chapter 5.

Columba palumbus = Common Wood Pigeon, "display singular vagaries" in their flight; **V2**: 350.

Columba leucocephala = White-crowned Pigeon *Patagioenas leucocephala*, of the West Indies, stated to be a rock-pigeon by Temminck; but Mr. Gosse tells CD (correctly) this is an error; **V1**: 183 footnote.

Invariably lays two eggs in the wild; but in Lord Derby's [UK] menagerie, only one; **V2**: 155–156.

Columba gymnophthalmos = Bare-eyed Pigeon *Patagioenas corensis*, hybrids with ***C. oenas***, were sterile and produced a hybrid offspring with ***C. maculosa*** in captivity; **V1**: 193–194 footnotes.

Columba loricata = Picazuro Pigeon *Patagioenas picazuro*, flocking on cornfields at Maldonado; **Z**: 115.

Columba gymnophthalmus = as a synonym of **Columba loricata**; **Z**: 115.

Columba leucoptera = as a synonym of **Columba loricata**; **Z**: 115.

Columba picazuro = as a synonym of **Columba loricata**; **Z**: 115.

Columba maculosa = Spot-winged Pigeon *Patagioenas maculosa*, produced a hybrid offspring with ***C. gymnophthalmos*** in captivity; **V1**: 194 footnote.

Columba fitzroyii = Chilean Pigeon *Patagioenas araucana*, locations, habitats; **Z**: 114.

Columba denisea = as a synonym of **Columba fitzroyii**; **Z**: 114.

Columba araucana = as a synonym of **Columba fitzroyii**; **Z**: 114.

turtle-dove = unspecified, but doubtless of the genus *Streptopelia* and most likely *S. turtur*, the soft cooing [of males] is assumed to please females; **D**: 572, 580.

Skulls of this and of *C. livia, oenas*, and *palumbus* show extremely slight differences; **V1**: 163. See chapter 5.

When "white and common collared turtle-doves" are paired, the offspring are as one or the other parent; **V2**: 92.

Turtur vulgaris = European Turtle Dove *Streptopelia turtur*, will cross with domestic pigeon and ***T. risoria***; hybrids with ***T. suratensis*** and ***Ectopistes migratorius*** are sterile; **V1**: 193 footnote.

Turtur auritus = European Turtle Dove *Streptopelia turtur*, hybrids with ***T. cambayensis*** and ***T. suratensis*** have been raised; **V1**: 194 footnote.

collared turtle-dove = European Turtle Dove *Streptopelia turtur*, when "white and common collared turtle-doves" are paired, the offspring are as one or the other parent; **V2**: 92.

Columba torquatrix = Eurasian Collared Dove *Streptopelia decaocto*, "display singular vagaries" in their flight; **V2**: 350.

Turtur risoria = African Collared Dove *Streptopelia roseogrisea*, will cross with domestic pigeon and ***T. vulgaris***; **V1**: 193 footnote.

Turtur suratensis = Spotted Dove *Spilopelia chinensis*, hybrids with ***T. vulgaris*** are sterile; **V1**: 193 footnote.

Turtur cambayensis = Laughing Dove *Spilopelia senegalensis*, hybrids with ***Turtur auritus*** have been raised; **V1**: 194 footnote.

Turtur = a pigeon genus, some species showing variation that could represent speciation and/or subspeciation; **V1**: 204 footnote.

Indian pigeon (*Chalcophaps indicus*) = Common Emerald Dove *Chalcophaps indica*, the young are transversely striped below; and certain allied species or whole genera are similarly marked when adult; **D**: 709.

Phaps chalcoptera = Common Bronzewing, its wing bars are "beautifully edged with different zones of colour" (in contrast to the black ones of the **rock pigeon**); **V2**: 349.

bronze-winged pigeon (*Ocyphaps lophotes*) = Crested Pigeon, Mr. Weir describes its courtship display; **D**: 612,

Ectopistes = of the two species (now one), the Passenger Pigeon *E. migratorius* has 12 tail feathers, whereas the other (? Mourning Dove *Zenaida macroura*) has 14; **V1**: 159 footnote.

Ectopistes migratorius = Passenger Pigeon, hybrids with ***Turtur vulgaris*** are sterile; **V1**: 193 footnote.

Columba migratoria = Passenger Pigeon *Ectopistes migratorius*, invariably lays two eggs in the wild, but in Lord Derby's [UK] menagerie only one; **V2**: 155–156.

Zenaida aurita = Eared Dove *Zenaida auriculata*, locations acquired given; **Z**: 115.

Columba aurita = as a synonym of **Zenaida aurita**; **Z**: 116.

Zenaida = Galapagos Dove *Zenaida galapagoensis*, CD notes that Mr. Sclater informs him (wrongly) that this bird is possibly not endemic to the Galapagos, as it also occurs on the American continent (wrong); **J2**: vii. See colour figure 3.12.

Zenaida galapagoensis = Galapagos Dove, described, compared with **Z. aurita**; illustrated, habitats, habits, feeds with *Geospiza*, tameness; **Z**: 115–116; **J**: 475. See chapters 3 and 8, and colour figure 3.12.

Turtle-doves = Galapagos Doves *Zenaida galapagoensis*, reported by CD to have been found extremely tame in 1684 by the author Cowley until shot at, when "They were rendered more shy"; **J**: 476. See colour figure 3.12.

dove = CD merely states that a species "like, but distinct from, the Chilean species" occurs as endemic on the Galapagos (now the Galapagos Dove *Zenaida galapagoensis*); **J**: 461. See chapter 2.

Listed among birds that were tame on the Galapagos; **J**: 475.

A boy killed many doves and finches with a "switch," or stick, as they came to a Galapagos well; **J**: 476.

Columba passerina = Common Ground Dove *Columbina passerina*, CD merely directs his reader to see an Audubon publication on the young of the two sexes; **D**: 713 footnote.

Columbina talpacoti = Ruddy Ground Dove, collected at Rio de Janeiro; **Z**: 116.
Columbina Cabocolo = as a synonym of **Columbina talpacoti**; **Z**: 116.

Columbina strepitans = Picui Ground Dove *Columbina picui*, locations recorded; **Z**: 116.

Zenaida boliviana = Black-Winged Ground Dove *Metriopelia melanoptera*, one collected at Valparaiso; **Z**: 116.
Columba Boliviana = as a synonym of **Zenaida boliviana**; **Z**: 115.

Calænas = Nicobar Pigeon *Caloenas nicobarica*, being an exception to CD's view that domestic races of the **rock-pigeon** differ as much from each other as do most distinct natural pigeon genera; **V1**: 132.

sub-family of the Treronidæ = the sexes often differ in vividness of colour; **V1**: 162 footnote.

Treron = the genus of Green-Pigeons, some species showing variation that could represent speciation and/or subspeciation; **V1**: 204 footnote.

Carpophaga oceanica = Micronesian Imperial Pigeon *Ducula oceanica*, the "excrescence at the base of the beak" is said to be sexual [in function]; **V1**: 163 footnote.

The wattle or corrugated skin at the base of the beak of the English carrier-pigeon might be called a monstrosity, but on this wild species is not seen as such; **V2**: 413.

Columba litteralis = Pied Imperial Pigeon *Ducula bicolor*, builds and roosts on rocks but is unlikely to have parented domestic pigeons (and did not), and CD says is actually a ***Carpophaga*** species; **V1**: 182.

Columba luctuosa = Silver-tipped Imperial Pigeon *Ducula luctuosa*, merely mentioned as a closely allied species to ***Columba litteralis*** in same context; **V1**: 182.

Goura = the genus of the three crowned-pigeons New Guinea, being an exception to CD's view that domestic races of the **rock-pigeon** differ as much from each other as do most distinct natural pigeon genera; **V1**: 132.

Goura coronata = Western Crowned Pigeon *G. cristata*, produced a hybrid with ***Goura victoria*** in captivity; **V1**: 194 footnote; **V2**: 155.

More than 12 kept in Penang "under a perfectly well-adapted climate" never bred there; **V2**: 155.

Goura victoria Victoria Crowned Pigeon *G. victoria*, produced a hybrid with ***Goura coronata*** in captivity; **V1**: 194 footnote; **V2**: 155.

Didunculus = Tooth-billed Pigeon *D. strigirostris*, being an exception to CD's view that domestic races of the **rock-pigeon** differ as much from each other as do most distinct natural pigeon genera; **V1**: 132.

pigeons = unspecified, CD notes that the London Zoo, in contrast to poor results with parrots, bred more than 13 pigeon species; **V2**: 155.

MUSOPHAGIFORMES

MUSOPHAGIDAE—Turacos

plantain-eaters (Musophagæ) = Turacos, Musophagidae, unspecified, the females are conspicuously coloured and build concealed nest; **D**: 694.

There are few "large groups of birds" in which both sexes are brilliantly coloured, but this is an exception; **D**: 701.

CUCULIFORMES

CUCULIDAE—Cuckoos

cuckoo family = several species have the same bill structure as **Pterotochos albicollis**; **Z**: 152.

Diplopterus guira = Guira Cuckoo *Guira guira*, small noisy flocks at Buenos Ayres; **Z**: 114.

Cuculus guira = as a synonym of **Diplopterus guira**; **Z**: 114.
Crotophaga Piririgua = as a synonym of **Diplopterus guira**; **Z**: 114.
Ptiloleptus cristatus = as a synonym of **Diplopterus guira**; **Z**: 114.

Crotophaga ani = Smooth-billed Ani, recorded at Buenos Ayres, food; **Z**: 114.

Diplopterus naevius = Striped Cuckoo *Tapera naevia*, recorded at Rio de Janeiro; **Z**: 114.

Cuculus naevius = as a synonym of **Diplopterus naevius**; **Z**: 114.

cuckoos, in the genus *Pelophilus* = the genus *Coua* of Madagascar, have same form of claw as **Pterotochos albicollis**; **Z**: 152.

North American cuckoo or **American cuckoo** = unspecified supposedly non-parasitic species [of the genus *Coccyzus*], mentioned in discussing parasitic habits of *Molothrus* cowbirds; **1881** paper.

Discussed in comparing nest parasite habits with the **cuckoo**; **O**: 330–336. See chapters 3 and 4.

As host to nestling **Blue Jay** and compared with other nest parasite cuckoos; **O**: 330–332. See chapter 3.

Australian cuckoos = three unspecified nest parasite species [of the genus *Chrysococcyx*], compared with European and American nest parasite cuckoos; **O**: 332–333. See chapters 3 and 4.

Bronze cuckoo = unspecified, one of four Australian species of the genus *Chrysococcyx*, in comparing their nest parasite habits with European and American parasitic cuckoos; **O**: 332–333. See chapters 3 and 4.

Indian cuckoos (Chrysococcyx) = unspecified, "the mature species differ considerably from one another in colour, but their young cannot be distinguished"; **D**: 709.

cuckoo(s) = Common Cuckoo *Cuculus canorus*, mentioned only in passing, regarding South American birds as also being nest brood parasite; **J**: 62; **1881** paper. See chapters 2, 3, and 8.

Mentioned as sharing nest parasite habits with **Molothrus niger**, and its habits; **Z**: 108–109.

Its nest parasite habits discussed and compared with some American and Australian species; **O**: 330–336.

In the nest, young cuckoos "raise their feathers, open their mouths widely, and make themselves as frightful as possible"; **E**: 99–100.

European cuckoo = Common Cuckoo *Cuculus canorus,* nest parasite habits discussed and compared with some American and Australian species; **O**: 333; but see also pp. 330–336, 364. See chapters 3, 4, and 8.

STRIGIFORMES

TYTONIDAE—Barn Owls

Strix flammea = Western Barn Owl *Tyto alba*, specimen from Bahia Blanca—Gould tells CD the barn owls require revision; **Z**: 34.

White owl (*Strix flammea*) = Western Barn owl *Tyto alba*, if one of a pair is shot the survivor readily finds another mate; **D**: 622.

Strix punctatissima = Western (Galapagos) Barn Owl *Tyto* (*punctatissima*) *alba*; in a postscript, CD notes that Mr. Sclater informs him that this is not endemic to the Galapagos Islands, as it also occurs on the American continent; **J2**: vii.

A "FitzRoy" specimen from Galapagos Islands described and illustrated; **Z**: 34–35.

CD notes it is a third less in size and darker in plumage than *Tyto alba* of Europe; **Z**: 35.

Barn-owl = *Tyto alba* Western Barn Owl, display and vocalizations given when approached; **E**: 99.

two owls = Western (Galapagos) Barn Owl *Tyto* (*punctatissima*) *alba* and Short-eared Owl *Asio flammeus* listed as part of the Galapagos avifauna; **J**: 461. See chapter 2.

STRIGIDAE—Owls

Sub-family: Surninae, Ululinae, and **Striginae** = all now disused, and are Strigidae; **Z**: 31–34.

Eagle Owl (*Bubo maximus*) = Eurasian Eagle-Owl *Bubo bubo*, shows a special inclination to breed in captivity; **V2**: 154.

Ulula rufipes = Rufous-legged Owl *Strix rufipes*, a specimen obtained at Tierra del Fuego, possible food discussed; **Z**: 34.

  Strix rufipes = as a synonym of **Ulula rufipes**; **Z**: 34.

Strix passerina = Eurasian Pygmy Owl *Glaucidium passerinum*, Mr. Gurney reports captive breeding; **V2**: 154.

little owls of the Pampas (*Noctua cunicularia*) = Burrowing Owl *Athene cunicularia*, make their own burrows in the absence of the Bizcacha (a rodent); **J**: 82. See chapter 2.

  Noted in passing, and its biology subsequently described; **J**: 143, 145.

Athene cunicularia = Burrowing Owl *Athene cunicularia*, distribution, habits, food; **Z**: 31–32. See chapter 3.

  Strix cunicularia = as a synonym of **Athene cunicularia**; **Z**: 31.

owl (*Strix grallaria*) = Burrowing Owl *Athene cunicularia*, one captive that was long fed vegetable food developed a "leathery" stomach and larger liver; **V2**: 302.

Otus Galapagoensis = Short-eared Owl *Asio flammeus*, CD notes that Mr. Sclater informs him that this bird is not endemic to the Galapagos Islands, as it also occurs on the American continent; **J2**: vii.

  Described and illustrated; **Z**: 32–33.

Otus palustris = Short-eared Owl *Asio flammeus*, specimen locations and wider distribution noted; **Z**: 33.

  Strix brachyota as a synonym of **Otus palustris**; Z: 33.

short-eared owl of Eurpoe (*Strix brachyota*) = Short-eared owl *Asio flammeus*, has "most of the essential characters" of **Otus Galapagoensis** (and is now considered the same species); **Z**: 33.

two owls = Western (Galapagos) Barn Owl *Tyto* (*punctatissima*) *alba* and Short-eared Owl *Asio flammeus* listed as part of the Galapagos Islands avifauna; **J**: 461.

owl(s) = an unidentified owl on the Chonos Archipelago had stomach full of good-sized crabs; **J**: 145. See chapter 2.

  Unspecified, species found in more open parts of Tierra del Fuego; **J**: 301.

  CD wonders if owls ever bring live prey to the nest, because in so doing, they may introduce at least small mammals to islands; **J**: 351.

Two endemic Galapagos species briefly alluded to; **J**: 461.

In swallowing their avian prey whole, owls may disperse seeds that were within them (particularly small migrant birds); **O**: 510.

Display and vocalizations given when approached described; **E**: 99. See also **Barn-owl**.

None had laid fertile eggs in zoos (many have since); **V2**: 153–154.

CD theorizes that if a bird received advantage from seeing well in low light, those individuals with the ability would succeed best and be most likely to survive and thus lead to owl-like vision; **V2**: 222.

Their fine leg down is nothing like the primary feather-like ones on the feet of domestic fowls; **V2**: 323.

CAPRIMULGIDAE—Nightjars

Australian night-jar (Eurostopodus) = White-throated Nightjar *E. mystacalis*, females are larger and brighter than males; **D**: 731.

Virginian goat-sucker (*Caprimulgus virginianus*) = Common Nighthawk *Chordeiles minor*, Audubon described how courtship is followed by the female-selected male driving off his rivals; **D**: 561.

Sub-family Caprimulginae = of the Caprimulgidae, **Z**: 36.

Caprimulgus bifasciatus = Band-winged Nightjar *Systellura longirostris*, appearance described and discussed; **Z**: 36–37.

Caprimulgus parvulus = Little Nightjar *Setopagis parvula*, appearance and size described and habitat and habits noted; **Z**: 37–38.

Caprimulgus Europaeus = European Nightjar, similar to **Caprimulgus bifasciatus** but with shorter wing, longer tarsus, and white band across tail base; **Z**: 37.

Rises from ground like **Caprimulgus parvulus**; **Z**: 38.

small African night-jar (*Cosmetornis vexillarius*) = Pennant-winged Nightjar *Caprimulgus vexillarius*, while breeding, one primary on each wing of this 10-inch bird attains 26 inches in length; **D**: 586.

closely-allied genus of night-jars (to **small African night-jar** = Standard-winged Nightjar *Caprimulgus longipennis*), the shafts of "elongated wing-feathers are naked except at the extremity, where there is a disc"; **D**: 586.

African night-jar (Cosmetornis) = Pennant-winged Nightjar *Caprimulgus vexillarius*, its seasonal wing pennants retard its flight, which at other times is swift; **D**: 613.

Cosmetornis = Pennant-winged Nightjar *Caprimulgus vexillarius*, males acquire wing feathers when breeding that are so long as to impede their flight; **D**: 705.

another genus of nightjars = now of the genera *Hydropsalis, Uropsalis,* and *Macropsalis*, "the tail-feathers are even still more prodigiously developed" (i.e., than wing feathers of the last previous species); **D**: 586.

night-jars (Caprimulgus) = unspecified, in the courting season males of certain species make a strange booming noise with their wings; **D**: 574. See chapter 6.

night-jar(s) = unspecified, Audubon described several males performing aerial displays until the female chooses from among them, and the other males are then driven away; **D**: 634.

Are difficult to see when crouched on the ground, D: 746. Are coloured for concealment but "with extreme elegance," leading CD to conclude that both natural and sexual selection were involved [a questionable conclusion]; D: 748.

goatsucker = unspecified nightjar, calling on Bell Mountain, near Valparaiso, central Chile; given the location, at altitude, most likely Band-winged Nightjar *Systellura longirostris* or possibly Lesser Nighthawk *Chordelies acutipennis*; **J**: 313.

APODIFORMES

APODIDAE—Swifts

Swift of North America = Chimney Swift *Chaetura pelagica*, makes a nest of sticks glued together with saliva; **O**: 355. See chapter 4.

English Swift (*Hirundo Apus*, Linn.) = Common Swift *Apus apus*, flight and call like those of **Progne purpurea**; **Z**: 39.

swifts = Common Swift *Apus apus*, the migratory instinct will cause them to "desert their tender young, leaving them to perish miserably in their nests"; **D**: 165. See **swallow** and **house-martins.**

Cypselus = Common Swift *Apus apus*, an inconspicuous British species that nests in a hole, mentioned in discussing Wallace's objections to sexual selection for plumage colouration with regard to nesting; **D**: 693 footnote.

Cypselus unicolor = Plain Swift *Apus unicolor*, specimen from St. Jago, Cape Verde Islands; CD noted their flight as more swallow- than swift-like; **Z**: 41.

gigantic swift (Cypselus) = unspecified, a long-winged and long-tailed breed of domestic pigeon had a similar appearance; **V1**: 157.

TROCHILIDAE—Hummingbirds

humming-birds (*Grypus*) = Hook-billed Hermit *Glaucis dohrnii*, the beak of the male is "serrated along the margin and hooked at the extremity" unlike in the female; **D**: 550–551.

Campylopterus hemileucurus = Violet Sabrewing, on one occasion five males found for every two females, but the reverse ratio was found on another; **D**: 383.

Eupetomena macroura = Swallow-tailed Hummingbird; as females are brightly coloured, it was mentioned in discussing Wallace's objections to sexual selection for plumage colouration with regard to open nesting; **D**: 691 footnote.

Florisuga mellivora = White-necked Jacobin, courtship of a female by two males described; **D**: 675.

Lampornis porphyrurus = Jamaican Mango *Anthracothorax mango*, ditto ***Eupetomena macroura***; **D**: 691 footnote.

Eulampis jugularis = Purple-throated Carib, ditto ***Eupetomena macroura***; **D**: 691 footnote.

Lophornis ornatus = Tufted Coquette, a pair illustrated (from Brehm et al. 1876–1879); **D**: 590. See figure 6.9.

Trochilus flavifrons = Glittering-bellied Emerald *Chlorostilbon lucidus*, not abundant at Monte Video; **Z**: 110.

Cynanthus = a genus (now of three species), which near Bogata are "divided into two or three races or varieties" differing in tail colouring, mentioned regarding intraspecific variability; **D**: 646, 674.

Aïthurus polytmus = Red-billed Streamertail *Trochilus polytmus*, the highly colourful males have two immensely long tail feathers, whereas the dull females lack these; but young males do not resemble females but rather males, and they quickly acquire the long tail feathers; **D**: 743. See chapter 7.

Trochilus pella = Crimson Topaz *Topaza pella*, the skeleton of which is the same form as that of ***Trochilus gigas***; **Z**: 154.

Trochilus gigas = Giant Hummingbird *Patagona gigas*, migration in Valparaiso area, flight and hovering abilities; **J**: 331–332. See chapters 2 and 3.

Orsimya tristis as a synonym of ***Trochilus gigas***; Z: 111.

Migration, habitats, nest, habits, flight, food; **Z**: 111–112.

Anatomical notes by T. C. Eyton (skeleton is the same form as in ***Trochilus pella***); **Z**: 154.

Trochilus forficatus = Green-backed Firecrown *Sephanoides sephaniodes*, distribution, habitats, habits, food, migration, nest, egg; **Z**: 110–111; **J**: 475. See chapter 3.

Ornismya Kingii = as a synonym of **Trochilus forficatus**; **Z**: 110.

Two humming-birds belonging to the genus *Eustephanus* = *Sephanoides* on Juan Fernandez Island, long thought specifically distinct, but Gould informs CD they are the sexes of one species; **D**: 550, 743.

Urosticte benjamini = Purple-bibbed Whitetip, exhibits sexual dimorphism in tail pattern, affecting different numbers of feathers in each sex; **D**: 673–675. See chapter 6.

Spathura underwoodi = Booted Racket-tail *Ocreatus underwoodi*, a pair illustrated (from Brehm et al. 1876–1879); **D**: 590.

Metallura = this generic name applied to the eight species of metaltail for the splendour of their iridescent tail plumage; **D**: 675.

Mellisuga Kingii = Green-backed Firecrown Sephanoides sephanoides, range, migration, habitats, habits, food; **J**: 330. See chapter 2 and colour figure 2.6.

Heliothrix auriculata = Black-eared Fairy *Heliothryx auritus*, the colourful male differs conspicuously from the female, but the female and young have a much longer tail than the male [which thus gets shorter with age]; **D**: 713–714, 753. See chapter 7.

Selasphorus platycercus = Broad-tailed Hummingbird, males have the first primary abruptly excised to produce a whistling sound in flight, whereas females do not; that of both sexes illustrated; **D**: 577–578.

humming-bird(s), of unknown species at Rio de Janeiro, reminded CD of sphinx moths; **J**: 37. See chapter 2.

Unspecified, alluded to as possibly inhabiting forests as far north as today's central Denmark in earlier geological time; **J**: 291.

CD notes two species common in Chile, with a third "within the Cordillera, at an elevation of about 10,000 feet" and discusses hummingbird ecology and migration; **J**: 330–332.

Unspecified, in Chiloe and Chonos Islands noted in passing; **J**: 353.

Alpine species obviously colonized mountains as they were uplifted; **O**: 558. See chapter 2.

Migration of them on east and west coast of North America; **Z**: 111.

Their nests "are tastefully ornamented with gaily-coloured objects"; **D**: 140, 630.

Mr. Salvin believed hummingbirds to be polygamous (and was correct); **D**: 338.

Mr. Salvin convinced that in most Central American species, males are in excess of females; **D**: 383.

Cited as "smallest of birds" but as being "one of the most quarrelsome," examples given; **D**: 551–2.

Males chirp in triumph over a defeated rival; **D**: 563.

Only males of some species have their primary shafts "broadly dilated" or the webs "abruptly excised towards the extremity," **D**: 577.

A species has a tail including bare-shafted feathers terminating in a disc of vanes; **D**: 586.

They "almost vie with birds of paradise in their beauty," and this "is due to the selection by females of the more beautiful males"; **D**: 591–592. See chapter 8.

Gould, having described "some peculiarities in a male," has no doubt it displays them to greatest advantage to females; **D**: 601.

Tails differ between species in many ways, and birds "take especial pains" in displaying them; **D**: 674.

A large family in which all build open nests, and yet some have both sexes brightly plumaged—in discussing Wallace's objections to sexual selection for plumage colouration with regard to open nesting; **D**: 691.

Mr. Salvin observed birds in Guatemala were more unwilling to leave their nest in hot sunny weather "as if their eggs would be thus injured," **D**: 692 footnote.

Brief discussion of one sex exceeding the other in numbers in some species, their respective plumages, and the influence of sexual selection on these; **D**: 744. See also chapter 6.

Long- and short-billed species feeding on nectar, some having beaks adapted to specific plant genera; **F**: 371. See chapter 7.

TROGONIFORMES

TROGONIDAE—Trogons

trogon(s) = unspecified, the females are conspicuously coloured and build concealed nest; **D**: 694.

All species nest in holes, and while females are brightly coloured, the males are even brighter; **D**: 696.

CORACIIFORMES

CORACIIDAE—Rollers

roller = unspecified, they utter harsh cries (as is typical of colourful species CD observes); see **bee-eater, kingfisher, hoopoe, woodpeckers**; **D**: 568.

Coracias = European Roller *Coracias garrulus*, a British species with conspicuous females that nest in holes, mentioned in discussing Wallace's objections to sexual selection for plumage colouration with regard to nesting; **D**: 693 footnote.

ALCEDINIDAE—Kingfishers

Carcineutes = *Lacedo*, the difference between the sexes is conspicuous; **D**: 697.

Australian kingfisher (*Tanysiptera sylvia*) = Buff-breasted Paradise-Kingfisher, both sexes have greatly lengthened central tail feathers that in the hole-nesting females become "much crumpled"; **D**: 688.

kingfishers (Tanysiptera) = geographically males of some species have undergone intraspecific morphological change into distinguishable forms, but not so the females and young; **D**: 714.

Dacelo = in the three species (now four), the sexes differ only in tail pattern/colouration; **D**: 697, 713.

D. gaudichaudi = Rufous-bellied Kookaburra, *Dacelo gaudichaud*, the tail of young males is brown [but is bright blue in adults]; **D**: 713.

kingfisher (Dacelo jagoensis) = Grey-headed Kingfisher *Halcyon leucocephala*, observed on Cape Verde Islands and briefly compared with the Common Kingfisher; **J**: 2. See chapters 1 and 2.

Halcyon erythrorhyncha = Grey-headed Kingfisher *Halcyon leucocephala*, observed as common on Cape Verde Islands at start of the *Beagle* voyage but not seen upon return, so CD (wrongly) thought it a winter visitor there; describes habits; **J**: 2; **Z**: 41–42. See chapters 1 and 2.

Alcedo Senegalensis = as a synonym of **Halcyon erythrorhyncha**; **Z**: 41.

Cyanalcyon = Red-backed Kingfisher *Todirhamphus pyrrhopygius*, shows sexual dimorphism so marked that the sexes were once thought of as being different species; **D**: 697 footnote.

Adult males differ from adult females, and young of both sexes resemble dull adult females; **D**: 712.

kingfisher, European species = Common Kingfisher *Alcedo atthis*, briefly compared with the "**kingfisher (Dacelo jagoensis)**" of Cape Verde Islands; **J**: 2; **Z**: 41–42. See chapters 1 and 2.

kingfisher = Common Kingfisher *Alcedo atthis*, CD states that Saurophagus sulphuratus dashes into water in foraging like a "kingfisher"; **O**: 220. See chapter 4.

Utter harsh cries (as is typical of colourful species CD observes); see **bee-eater**, **roller**, **hoopoe**, **woodpeckers**; **D**: 568.

Both adult sexes and their young are of similar appearance; **D**: 734. See chapter 7.

Alcedo = Common Kingfisher *Alcedo atthis*, a species found in Britain with conspicuous females that nest in holes, mentioned in discussing Wallace's objections to sexual selection for plumage colouration with regard to nesting; **D**: 693 footnote.

small kingfisher (*Alcedo Americana*) = Green Kingfisher *Chloroceryle americana*, appearance, habits, and call briefly described; **J**: 162–163. See chapter 2.

Ceryle americana = Green Kingfisher *Chloroceryle americana*, habitat and habits; **Z**: 42.

Alcedo Americana = as a synonym of **Ceryle americana**; **Z**: 42.

Ceryle = Belted *Megaceryle alcyon* and Green *Chloroceryle americana* Kingfishers have males that have a breast belted with black; **D**: 697.

Ceryle torquata = Ringed Kingfisher *Megaceryle torquata*, distribution, habitat, and habits; **Z**: 42.

Alcedo torquata = as a synonym of **Ceryle torquata**; **Z**: 42.

Ispida torquata = as a synonym of **Ceryle torquata**; **Z**: 42.

Kingfisher(s) = unspecified, always beat caught fish to death in the wild and do so instinctively with meat in captivity; **E**: 50. See chapter 7.

Utter harsh cries (as is typical of colourful species CD observes); see **bee-eater, roller, hoopoe, woodpeckers**; **D**: 568.

A species has a tail including bare-shafted feathers terminating in a disc of vanes; **D**: 586.

The females are conspicuously coloured and build concealed nests; **D**: 694.

All species build nests in holes; and in most, both sexes are equally brilliantly coloured; **D**: 696.

In most species, the slight differences in sexual dimorphism are of the same nature as the occasional greater differences in a minority of species; **D**: 699.

Unspecified, the adult sexes are alike, and young of both sexes in their first plumage resemble them, as in many parrots, crows, and hedge-warbler; **D**: 711. See chapter 7.

Adult males differ from adult females, and young of both sexes resemble adult females; **D**: 712.

The young of some species are not only less vividly coloured than adults, they also have brown feather edging on their underparts; **D**: 735.

MOMOTIDAE—Motmots

motmot (*Eumomota superciliaris*) = Turquoise-browed Motmot *E. superciliosa*, its tail includes bare-shafted feathers terminating in a disc of vanes; **D**: 586.

motmots = unspecified, Mr. Salvin demonstrated that they create their racket-shaped tail feather tips by "biting off the barbs" and that this has "produced a certain amount of inherited effect"; **D**: 587.

MEROPIDAE—Bee-eaters

bee-eater = unspecified, utter harsh cries (as is typical of colourful species CD observes); see **kingfisher, roller, hoopoe, woodpeckers**; **D**: 568.

Merops = European Bee-eater *Merops apiaster*, a species found in Britain with conspicuous females that nests in holes, mentioned in discussing Wallace's objections to sexual selection for plumage colouration with regard to nesting; **D**: 693 footnote.

BUCEROTIFORMES

UPUPIDAE—Hoopoes

hoopoe = unspecified, but the two species of the genus *Upupa* utter harsh cries (as is typical of colourful species, CD observes); see kingfisher, roller, woodpeckers; **D**: 568.

Males combine "vocal and instrumental music" in the breeding season by tapping the beak on a hard substrate to force out air with the oesophagus swollen; **D**: 575. See chapter 6.

Upupa = Eurasian Hoopoe *U. epops*, as a species found in Britain with conspicuous females that nests in holes—mentioned in discussing Wallace's objections to sexual selection for plumage colouration with regard to nesting; **D**: 693 footnote.

BUCORVIDAE—Ground Hornbills

African hornbill (*Bucorax abyssinicus*) = Abyssinian Ground Hornbill *Bucorvus abyssinicus*, in display males inflate a scarlet bladder-like wattle on the neck, droop wings, and expand the tail; **D**: 584.

BUCEROTIDAE—Hornbills

hornbills = males (in fact only females in most species but also males in some) of African and Indian species have the same habit of plastering their mates into a tree hole; **O**: 364.

Hornbill (*Buceros*) . . . of India and Africa = unspecified, females (in fact only females in most species but also males in some) plaster themselves into tree hole nest—mentioned in discussing Wallace's objections to sexual selection for plumage colouration with regard to nesting; **D**: 692.

Buceros = unspecified, sexual dimorphism in iris, casque, and mouth colour briefly described and discussed; **D**: 649. See also *Buceros bicornis*.

Buceros bicornis = Great Hornbill, sexual dimorphism in iris, casque, and mouth colour briefly described and discussed; **D**: 649. See also *Buceros*.

Buceros corrugatus = Wrinkled Hornbill *Aceros corrugatus*, bill colour and structure more colourful and complex, respectively, in males than in females; **D**: 585.

PICIFORMES

BUCCONIDAE—Puffbirds

puffbirds (Capitonidae) = Bucconidae, unspecified, the females are conspicuously coloured and build concealed nests; **D**: 694.

RAMPHASTIDAE—**Toucans**

toucans, a few of unspecified species shot near Rio de Janeiro; **J**: 32. See chapter 2.

The females are conspicuously coloured and build concealed nests; **D**: 694.

May owe their enormous beaks and naked facial skin to sexual selection "for the sake of displaying the diversified and vivid stripes of colour" they are ornamented with; **D**: 749–750.

Rhamphastos carinatus = Keel-billed Toucan *Ramphastos sulfuratus*, Gould states the colour of the beak is doubtless most brilliant at breeding time; **D**: 750.

PICIDAE—**Woodpeckers**

Picidae = in some anatomical features **Pterotochos albicoillis** similar to ***Picus*** and to this family as a whole; **Z**: 151.

Position and length of hydroids in **Trochilus gigas** resemble those of the **Picidae**; **Z**: 154.

Yunx = Eurasian Wryneck *Jynx torquilla*, an inconspicuous British species that nests in a tree hole, mentioned in discussing Wallace's objections to sexual selection for plumage colouration with regard to nesting; **D**: 693 footnote.

Mexican Colaptes = Acorn Woodpecker *Melanerpes formicivorus*, acorn storing habit; **O**: 221.

English woodpecker = unspecified but probably Great Spotted Woodpecker *Dendrocopos major*, its undulatory flight and call being like that of **Chrysoptilus campestris**; **Z**: 113–114.

Picus major = Great Spotted Woodpecker *Dendrocopos major*, crimson under tail coverts are conspicuous without them being displayed by the bird; **D**: 612.

Picus kingii = Striped Woodpecker *Venilliornis lignarius*, locations and sexual difference in head plumage colour; **Z**: 113.

Picus melanocephalus = as a synonym of **Picus kingii**; **Z**: 113.

(*Picusauratus*) (*sic*) = Northern Flicker *Colaptes auratus*, according to Audubon, it never fights intraspecifically (but untrue); **D**: 554.

Colaptes chilensis = Chilean Flicker *Colaptes pitius*, habitat, habits, nest; **Z**: 114.

Picus Chilensis = as a synonym of **Colaptes chilensis**; **Z**: 114.

Chrysoptilus campestris = Campo Flicker *Colaptes campestris*, locations, habitats, habits, flight, food; **Z**: 113–114.

Picus campestris = as a synonym of **Chrysoptilus campestris**; **Z**: 113.

ground woodpeckers = Campo Flicker *Colaptes campestris*, use their tails little; **Z**: 114.

woodpecker (*Colaptes campestris*) = Campo Flicker, a structurally typical woodpecker, which does not climb trees; **O**: 221–222, 646. See chapters 3 and 4.

black woodpecker = Magellanic Woodpecker *Campephilus magellanicus*, its rare strange cry heard in woods of Tierra del Fuego; **J**: 301. See chapter 2 and colour figure 2.5.

green woodpecker = European Green Woodpecker *Picus viridis*, CD suggests its plumage colour is not solely selected for by natural selection, as being cryptic, but chiefly by sexual selection; **O**: 247.

Picus, 4 sp. = unspecified but including *P. viridis, Dendrocopos major,* and *D. minor*, as British species with conspicuous females that nests in holes, mentioned in discussing Wallace's objections to sexual selection for plumage colouration with regard to nesting; **D**: 693 footnote.

Indopicus carlotta = Buff-Spotted Flameback *Chrysocolaptes lucidus*, young females show some red of adult males but then loose it as adults; **D**: 698, 724.

See **Woodpecker(s)**.

Megapicus validus = Orange-backed Woodpecker *Reinwardtipicus validus*, sexual dimorphism described, and loss of head colour in females possibly adapted to nesting habits; **D**: 698.

Woodpecker(s) = unspecified, mentioned as an example of adaptation; **1858**: 50, 52. See chapter 4.

Unspecified, most briefly mentioned with reference to discussing how the anonymously published Vestiges of Creation cannot explain adaptations to a woodpecker's "peculiar habits of life"; **O**: xxiv, 8.

In briefly referring to "beautiful co-adaptations" between life forms, CD obscurely states that such are seen "most plainly in the woodpecker and the mistletoe" but meaning as quite separate examples; **O**: 76, 78.

Diversity of foraging and foods among the species of; **O**: 220–221, 223.

Utter harsh cries (as is typical of colourful species CD observes); see bee-eater, roller, hoopoe, kingfisher; **D**: 568.

During the breeding season, "various species" strike "a sonorous branch with their beaks very rapidly to produce a sound audible at a considerable distance as a "love call"; **D**: 574, 580. See chapter 6.

Audubon stated a female was pursued by six displaying suitors until she selected one; **D**: 634.

The females are conspicuously coloured and build concealed nests; **D**: 694.

The sexes are generally nearly alike, and in "several," the head of the male is bright crimson, whereas that of the female is plain, and reasons for this are discussed; **D**: 698.

The young of some species are transversely striped below, and certain allied species or whole genera are similarly marked when adult; **D**: 709.

The young of certain species are an exception to the rule that female birds generally are not dull, whereas their young are bright coloured because in those woodpeckers, their young have all or part of the head tinged red, which decreases with age in both sexes or entirely disappears in females; **D**: 724.

In some British species, both adult sexes and their young are of similar appearance; **D**: 734–735.

Although "eminantly arboreal," not all are green, and many are black or black-and-white, although all are apparently exposed to "nearly the same dangers"; **D**: 746.

See ***Indopicus carlotta***.

FALCONIFORMES

FALCONIDAE—Caracaras, Falcons

Falconidae = mentioned in observing it is strange that the caracaras of this family can run so well; **Z**: 18.

Sub-family: Falconina = a now disused taxon of the Falconidae; **Z**: 28.

Sub-family: Polyborinae = the now disused caracara sub-family of the Falconidae; **Z**: 9, 12, 18.

Caracaridae = a disused family taxon absorbed into the Falconidae; **Z**: 9.

Phalcobaenus = mentioned in discussing characters of the various caracaras, notably **Milvago**; **Z**: 12.

Milvago montanus = Mountain Caracara *Phalcoboenus megalopterus*, listed among *Milvago* species; **Z**: 13, 19, 20–21.

Phalcobaenus montanus = as a synonym of **Milvago montanus**; **Z**: 13.

Phalcobaenus montanus = Mountain Caracara *Phalcoboenus megalopterus*, running ability; **Z**: 18. See chapter 2.

Milvago megalopterus = Mountain Caracara *Phalcoboenus megalopterus*, location, description, habits; **Z**: 21.

Aquila megaloptera = as a synonym of **Milvago megalopterus**; **Z**: 21.

Milvago albogularis = White-throated Caracara *Phalcoboenus albogularis*, described illustrated, compared with congeners, distribution, habits; **Z**: 18–21. See chapter 2.

Polyborus (Phalcobaenus) albogularis = as a synonym of **Milvago albogularis**; **Z**: 18.

Polyborus Novae Zelandiae = Striated Caracara *Phalcoboenus australis*, feeding, tameness, and habits; **J**: 64, 66. See colour figure 3.2.

Milvago leucurus = Striated Caracara *Phalcoboenus australis*, nomenclature, description, habits; **Z**: 15–18. See colour figure 3.2.

Falco lecurus = as a synonym of **Milvago leucurus**; **Z**: 15.

Falco NovaeZelandiae = as a synonym of **Milvago leucurus**; **Z**: 15.

Falco australis = as a synonym of **Milvago leucurus**; **Z**: 15.
Circaëtus antarcticus = as a synonym of **Milvago leucurus**; **Z**: 15.

Milvago leucurus = Striated Caracara *Phalcoboenus australis*, CD found females of those he dissected the more colourful sex (but was probably misled by age differences); **D**: 731. See chapter 3 and colour figure 3.2.

Carrancha *Polyborus Braziliensis* = Southern Crested Caracara *Caracara plancus*, range, habitat, feeding, nesting, and other habits, local names; **J**: 64–68; **Z**: 9–12, 17–18, 22, 24, 26. See chapters 2 and 3.
In many respects resembles **Milvago leucurus; Z**: 17.
Edibility of its flesh; **Z**: 18.

Polyborus brasiliensis = see ***Carrancha* Polyborus Braziliensis**. See chapter 3.
Polyborus vulgaris = as a synonym of **Polyborus brasiliensis**; **Z**: 9.
Falco Brasiliensis = as a synonym of **Polyborus brasiliensis**; **Z**: 9.

Milvago ochrocephalus = Yellow-headed Caracara *Milvago chimachima*, listed among *Milvago* species; **Z**: 12–13. See chapter 2.
Polyborus chimachima = as a synonym of **Milvago ochrocephalus**; **Z**: 11–13.
Falco degener = as a synonym of **Milvago ochrocephalus**; **Z**: 13.
Haliaëtus chimachima = as a synonym of **Milvago ochrocephalus**; **Z**: 13.

Milvago pezoporus = Yellow-headed Caracara *Milvago chimachima*, specimen locations, distribution; **Z**: 13–14.
Aquila pezopora = as a synonym of **Milvago pezoporus**; **Z**: 13.

Polyborus Chimango = Chimango Caracara *Milvago chimango*, range, habitats, feeding, and other habits. One pounced on a dog, and several attacked wounded geese; **J**: 64, 66, 68.
Feed with P. plancus in great numbers at estancias and slaughterhouses near the La Plata; **Z**: 9.
Under the name **Milvago chimango** said to have "some resemblance in general habits" to **Circus cinerius**; **Z**: 31.

Milvago chimango = Chimango Caracara, distribution, habits; **Z**: 14–15. See chapter 2.
Polborus chimango = as a synonym of **Milvago chimango**; **Z**: 14.
Haliaëtus chimango = as a synonym of **Milvago chimango**; **Z**: 14.

Chimango—see ***Polyborus Chimango*.** See chapter 3.

Polybori = of the disused *Polyborinae*, which attend slaughterhouses to feed with **Cathartes** and **Larus dominicanus**; **Z**: 142.

Polyborus = a genus of **caracara**, compared with **Craxirex galapagoensis**; **Z**: 24.

Milvago = a genus of caracara, an outdated discussion of the taxonomy of the genus followed by descriptions and notes on distribution and habits of five species (**M. pezoporos** = *M. chimachima*, **chimango, leucurus** = *Phalcoboenus australis*, **albogularis illustrated** = *P. albogularis*, **megalopterus** = *P. megalopterus*); **Z**: 12–21, 27, 28. See chapter 3.

Compared with **Craxirex galapagoensis; Z**: 24.

Aquila pezopora = as a synonym of **Milvago pezoporos**; **Z**: 13.

Polyborus ocrocephalus = as a synonym of **Milvago pezoporos**; **Z**: vii, 13.

Polyborus chimango = as a synonym of **Milvago chimango**; **Z**: 13-14.

Haliaëtus chimango = as a synonym of **Milvago chimango**; **Z**: 13-14.

Falco leucurus = as a synonym of **Milvago leucurus**; **Z**: 13, 15.

Falco Novae Zealandiae = as a synonym of **Milvago leucurus**; **Z**: 13, 15.

Falco Australis or **australis** = as a synonym of **Milvago leucurus**; **Z**: 13, 15.

Circaëtus antarcticus = as a synonym of **Milvago leucurus**; **Z**: 13, 15.

Polyborus (Phalcobaenus?) albogularis = a synonym of **Milvago albogularis**; **Z**: 13, 18.

Aquila megaloptera = as a synonym of **Milvago megalopterus**; **Z**: 13, 21.

eagles = CD quotes Captain Cook as seeing eagles at Christmas Sound, Tierra del Fuego, and expressed "little doubt" they would have been **Milvago leucurus**; **Z**: 15.

CD quotes Brehm, who saw an unspecified African eagle seize a young monkey only to have the troop attack and displace it from its prey; **D**: 155, 157.

Caracara or *Polyborus/Polybori* = unspecified, feeding and other habits; **J**: 64–65. See chapter 2.

Feeding on bullock carcass on East Falklands; **J**: 251.

Compared with Galapagos Hawk; **J**: 461. See chapter 2.

A Caracara tame on the Falkland Islands; **J**: 476.

Feeding at slaughterhouses with Black Vultures and gulls; **Z**: 7. See chapter 2.

carrion feeding hawks of extratropical parts of South America = the feeding habits of caracaras and vultures, respectively; **J**: 63–69. See chapter 2.

carrion hawks = CD mentions three recorded on the arid plains of Patagonia, and Gauchos making a hot fire from a bullock's skeleton picked clean by these birds; **J**: 194.

carrion-hawk = unspecified, the call of Galapagos birds reminiscent of North American ones; **V1**: 9. See chapter 5.

kestrel = Common Kestrel *Falco tinnunculus*, the tyrant-flycatcher **Saurophagus sulphuratus** hovers like a kestrel in foraging; **O**: 220. See chapter 4.

Three males were killed one after the other attending the same nest, two in mature plumage and one "in the plumage of the previous year"; **D**: 622.

English kestrel (*Falco tinnunculus*) = Common Kestrel, hovers like **Tinnunculus sparverius**; **Z**: 29.

Mr. Morris mentions it being unique that the species bred in an aviary; **V2**: 154.

Falco tinnunculus = Common Kestrel, seen to mate in London Zoo; **V2**: 154.

Tinnunculus sparverius = American Kestrel *Falco sparverius*, specimen locations, range, and hovering to hunt noted; **Z**: 29.

Falco sparverius = as a synonym of **Tinnunculus sparverius**; **Z**: 29.

Falco femoralis = Aplomado Falcon, specimen described, distribution, and habitat; **Z**: 28.

Falco subbuteo = Eurasian Hobby, seen to mate in London Zoo; **V2**: 154.

peregrine-falcon (*Falco peregrinus*) = CD cites two people reporting that one of a nesting pair shot is replaced within a few days by a new mate; **D**: 622.

Females acquire the blue adult plumage more slowly than do males; **D**: 703.

PSITTACIFORMES

Order: Scansores = a disused taxon that included the old family groupings of Psittacini (parrots), Serrati (toucans, trogons, turacos, etc.), Amphiboli (cuckoos), Sagittilingues (woodpeckers), and Syndactyli (jacamars); **Z**: 112.

Mentioned in passing; **Z**: 152.

CACATUIDAE—Cockatoos

black Cockatoos = *Calyptorhynchus* species, the adult sexes differ, but the young of both sexes resemble females; **D**: 713 footnote.

It is probable that their complete or partial blackness is the result of sexual selection "accompanied by equal transmission to both sexes"; **D**: 749.

white cockatoo = Sulphur-crested Cockatoo *Cacatua galerita*, large flocks seen by CD in the Blue Mountains, Australia, in January 1836; **J**: 526. See chapter 2.

Unspecified but likely this species—as a pair made a nest in a tree [hole], conspecifics took great interest and were observed to evince "unbounded curiosity" and had "the idea of property and possession"; and the species to have good memories, as captives clearly recognize their former masters after an absence of months; **D**: 627.

white cockatoos = unspecified, both sexes acquired white plumage through sexual selection; **D**: 751.

Cockatoos = unspecified, sexual selection has brought about both white and black species; **D**: 752.

PSITTACIDAE—Parrots

lories, brush-tongued = unspecified, of the Moluccas seen with flower pollen on them; **F**: 371. See chapter 7.

Lorius garrulus = Chattering Lory, natives of Gilolo change its feather colouring with a food and "thus produce the *Lori rajah* or King-Lory"; **V2**: 280, 290. See chapter 5.

parrots abound = Red-crowned Parakeet *Cyanoramphus novaezelandiae*, Dr. Dieffenbach quoted as stating unspecified species abound in the "Macquarrie Islands" [*sic*]; **J2**: 244.

Australian parrakeets (*Platycercus*) = rosellas, the young of some species closely resemble, whereas those of other differ considerably, from their parents; **D**: 735.

Euphema splendida = Scarlet-chested Parrot *Neophema splendida*, sexual dimorphism described; **D**: 697.

grass-parrakeet = Budgerigar *Melopsittacus undulatus*, adpress, or sleek, their feathers in fright; **E**: 100.

King Lory or **King Lory (*Aprosmictus scapulatus*)** = Australian King Parrot *Alisterus scapularis*, a captive male was killed and its mate "fretted and moped, refused her food, and died of a broken heart"; **D**: 626 footnote.

Sexual dimorphism described; **D**: 697.

The adult sexes differ, but the young of both sexes resemble females; **D**: 713 footnote.

Palæornia rosa = Plum-headed Parakeet *Psittacula cyanocephala*, the young are more like adult females than males; **D**: 713 footnote.

Indian parrakeet (*Palaeornis javanicus*) = Red-breasted Parakeet *Psittacula alexandri*, in males, the upper mandible is coral-red "from earliest youth"; but in females, it is black for a year, after which the sexes are alike; **D**: 703.

Psittacus = several species of this genus lack the oil gland typical of most birds; **V1**: 147 footnote.

African parrot = unspecified, but very probably Grey Parrot *Psittacus erithacus*, captive birds saying the names of specific people as they entered or left a room; **D**: 130.

African *Psittacus erithacus* = Grey Parrot, breeds more often (in captivity?) than any other [parrot] species; **V2**: 155.

Psittacus macoa (*sic*) = Scarlet Macaw *Ara macao*, captives occasionally lay fertile eggs but rarely hatch them, but are used to incubate fowl or pigeon eggs; **V2**: 155.

macaws = CD quotes Mr. Buxton who observed his free living macaws in Norfolk, England, appeared to "honour" a nesting female by loudly calling whenever she left her nest; **D**: 156.

CD appears to make a rare, tongue-in-cheek, humorous remark about their harsh screams indicating as much bad taste for musical sounds in them as for gaudy plumage; **D**: 573.

Conurus patachonicus = Burrowing Parrot *Cyanoliseus patagonus*, locations, habitat, nest (together in same cliffs with ***Hirundo cyanoleuca***), eggs, call; **Z**: 113.

Psittacus Patagonus = as a synonym of **Conurus patachonicus**; **Z**: 113.
Psittacara Patagonica = as a synonym of **Conurus patachonicus**; **Z**: 113.
Psittacara Patachonica = as a synonym of **Conurus patachonicus**; **Z**: 113.

ground parrot of Patagonia (*Psittacara Patagonica*) = Burrowing Parrot *Cyanoliseus patagonus*, nesting in holes in cliffs together with **Progne purpurea**; **Z**: 39.

Conurus murinus = Monk Parakeet *Myiopsitta monachus*, flocking, habitat, destructive to corn, nesting colonially; **Z**: 112. See chapter 2.

Psittacus murinus = as a synonym of **Conurus murinus**; **Z**: 112.

small green parrot = Monk Parakeet *Myiopsitta monachus*, flocking, communal nest building, 2,500 killed in a year as a crop pest; **J**: 163.

Also 3,500 "killed in one field of corn near Colonia" in CD's "Banda oriental notebook" (in Chancellor and van Wyhe 2009: 266).

common green parrot (*Chrysotis festiva*) = Festive Amazon *Amazona festiva*, Amazonian natives feed them fish fat, causing them to become variegated with red and yellow feathers; **V2**: 280. See chapter 5.

small green parrots, of unknown species, shot at Botofogo Bay, Rio de Janeiro; **J**: 32.

Parrot(s) unspecified, briefly alluded to as possibly inhabiting forests as far north as today's central Denmark in earlier geological time; **J**: 291.

Natives of Lemuy Island, Chile, said to each other upon the arrival of the *Beagle*, "This is the reason we have seen so many parrots lately"; **J**: 339.

In the Blue Mountains, Australia, CD found "a few most beautiful parrots"; **J**: 526. See chapter 2.

Erect their feathers [in fear]; **E**: 99.

Are "notorious imitators of any sound which they often hear"; **D**: 110.

CD cites examples of parrots saying appropriate words in response to seeing specific people or events; **D**: 130.

A species (in fact, the 10 species of the genus *Prioniturus*) has a tail including bare-shafted feathers terminating in a disc of vanes; **D**: 586.

Sometimes "live in triplets"; **D**: 624.

Pairs become so deeply attached that upon the death of one, the other pines for a long time; **D**: 626.

Mr. Buxton reports of a parrot "which took care of a frost-bitten and crippled bird of a distinct species, cleansed her feathers, and defended her"; **D**: 627.

Females are conspicuously coloured and build concealed nest; **D**: 694.

In most species of these hole nesters, the colourful sexes are indistinguishable; but in some species, males are brighter than females and sometimes very different from them; **D**: 697.

The Australian species include every gradation of sexual dimorphism; **D**: 698–699.

Adult males differ from adult females, and young of both sexes resemble adult females; **D**: 712. See chapter 7.

Are predominantly green and are difficult to distinguish in tree foliage, but most do also have patches of brilliant colours, "which can hardly be protective"; **D**: 746.

Their typically limited sexual dimorphism briefly discussed; **D**: 753.

Are singularly long-lived in captivity, up to 100 years, but rarely breed; one spoke an extinct South American Indian tribe language; **V2**: 155.

In UK aviaries, "some few species" have coupled; but, save for three parakeet species, none bred; **V2**: 155.

Two Guianan species often kept tame in Jamaica but do not breed; **V2**: 155

Their reproduction is much more easily affected by changed conditions than that of pigeons; **V2**: 268.

See **Kingfisher(s)**.

parrakeets = unspecified but most likely Crimson Rosella *Platycercus elegans*, their blue tail feathers used as bower decorations by bowerbirds; **D**: 630.

In males of one Australian species, the thighs may be scarlet or grass-green (Superb Parrot *Polytelis swainsonii*), whereas in another of the same country (varied Lorikeet *Psitteuteles versicolor*), the band across the wing coverts may be yellow or tinged with red, in discussing intraspecific variability; **D**: 646.

See **parrot(s)**.

PASSERIFORMES

Order of Insessores ("**common songsters**") = a disused taxon, established in 1823, that contained nearly all of the then Picæ and Passeres; **D**: 567.

As the **house-sparrow** and **linnet** (see both) are members of this order, the progenitor of the sparrow may have been a songster; **D**: 869.

PITTIDAE—Pittas

Pittidae = erroneously mentioned (as pittas build domed and not open nests as CD suggests) in discussing Wallace's objections to sexual selection for plumage colouration with regard to open nesting; **D**: 691. See chapters 7 and 8.

FURNARIDAE—Ovenbirds

Furnarius = anatomically compared to ***Uppucerthis*** and **Synallaxis maluroides**; **Z**: 152.

species of Furnarius, called Casarita by Spaniards = Common Miner *Geositta cunicularia*, nest burrowing described; **J**: 112–113. See chapter 2.

Furnarius cunicularius = Common Miner *Geositta cunicularia*, distribution, habitat, habits [compared to European Robin], nesting, food, call, plus anatomical notes; **Z**: 65–66, 149. See chapter 3.

Alauda cunicularia = as a synonym of **Furnarius cunicularius**; **Z**: 65.

Alauda fissirostra = as a synonym of **Furnarius cunicularius**; **Z**: 65.

Certhilauda cunicularia = as a synonym of **Furnarius cunicularius**; **Z**: 65.

Uppucerthia = *Upucerthia*, tongue, trachea, and oesophagus are as in **Furnarius cunicularius**; **Z**: 148.

Anatomically compared to ***Furnarius*** and **Synallaxis maluroides**; **Z**: 152.

Uppucerthia dumetoria = Scaly-throated Earthcreeper *Upucerthia dumetaria*, illustrated, locations, habitat, habits, variation in bill length as also in Dark-Bellied Cinclodes, anatomical notes; **Z**: 66, 148–149.

Furnarius dumetorum = as a synonym of **Uppucerthia dumetoria**; **Z**: 66.

Uppucerthia dumetorum = as a synonym of **Uppucerthia dumetoria**; **Z**: 66.

Upercerthia dumeteria = as a synonym of **Uppucerthia dumetoria**; **Z**: 66.

Opetiorhynchus antarcticus = Blackish Cinclodes *Cinclodes antarcticus*, CD amends this name to **Cinclodes antarcticus**; **Z**: vii.

Extremely tame on Falkland Islands, where confined to, plus anatomical notes (compared with **O. patagonicus** and **O. vulgaris**); **Z**: 67–68, 149–150. See chapter 3.

Certhia antarctica = as a synonym of **Opetiorhynchus antarcticus**; **Z**: 67.

Cinclodes fuliginosus = as a synonym of **Cinclodes antarcticus**; **Z**: viii.

Furnarius fuliginosus = as a synonym of **Opetiorhynchus antarcticus**; **Z**: 67.

Furnarius fuliginosus = Blackish Cinclodes *Cinclodes antarcticus*, mentioned as "a *Furnarius* allied to *fuliginosus*" was a "small dusky-coloured bird" CD saw on the coast of the Chonos Archipelago, Chile; **J**: 353. See chapter 2.

Opetiorhynchus vulgaris = Buff-winged Cinclodes *Cinclodes fuscus*, locations; CD amends this name, applied by Gould, to **Cinclodes vulgaris**; **Z**: vii.

Habits (some similar to Common Miner), habitats, food, call (like Common Miner), anatomical notes (very similar to **Furnarius** and **Uppucerthis**); **Z**: 66–67, 149. See chapter 3.

Anatomically compared with ***O. patagonicus*** and ***O. antarcticus***; **Z**: 150.

Uppucerthia vulgaris = as a synonym of **Opetiorhynchus vulgaris**; **Z**: 66.

Cinclodes vulgaris = see **Opetiorhynchus vulgaris**.

Opetiorhynchus Patagonicus = Dark-bellied Cinclodes *Cinclodes patagonicus*, CD amends this name, applied by Gould, to **Cincloides**; **Z**: viii.

Distribution, noted as geographically replaced by Seaside Cinclodes *C. nigrofumosus* (colour figure 3.6) and Blackish Cinclodes *C. antarcticus*, variation in bill length noted as like in Scaly-Throated Earthcreeper, anatomical notes (compared with ***O. vulgaris*** and ***C. antarcticus***, skeleton like that of ***Furnarius cunicularius***); **Z**: 67, 150. See chapter 3.

Habits, nest, and egg described under **O. nigrofumosus**; **Z**: 69.

Motacilla Patagonica = as a synonym of **Opetiorhynchus patagonicus**; **Z**: 67.

Motacilla Gracula = as a synonym of **Opetiorhynchus patagonicus**; **Z**: 67.

Sylvia Patagonica = as a synonym of **Opetiorhynchus patagonicus**; **Z**: 67.

Furnarius Lessonii = as a synonym of **Opetiorhynchus patagonicus**; **Z**: 67.

Furnarius Chilensis = as a synonym of **Opetiorhynchus patagonicus**; **Z**: 67.

Opetiorhynchus rupestris = as a synonym of **Opetiorhynchus patagonicus**; **Z**: 67.

Uppucerthia rupestris = as a synonym of **Opetiorhynchus patagonicus**; **Z**: 67.

Opetiorhynchus nigofumosus = Chilean Seaside Cinclodes *Cinclodes nigrofumosus*, CD amends this name to **Cinclodes nigofumosus**; **Z**: vii. See colour figure 3.6.

Described, illustrated, and compared to **O. patagonicus, vulgaris, antarcticus**; see also under **Opetiorhynchus patagonicus**; **Z**: 68–69. See chapter 3 and colour figure 3.6.

Uppucerthia nigrofumosa = as a synonym of **Opetiorhynchus nigofumosus**; **Z**: 68.

Opetiorhynchus lanceolatus = as a synonym of **Opetiorhynchus nigofumosus**; **Z**: 68.

Opetiorhynchi = *Cincloides* species, two unspecified species mentioned as frequenting more open parts of Tierra del Fuego; **J2**: 237. See chapter 3.

furnarii = two species, found in more open parts of Tierra del Fuego; **J**: 301.

Furnarius = of this genus of "several species" (now consisting of eight species of hornero), CD briefly describes their habits and mud nests with particular reference to ***F. rufus***; **J**: 112–113.

Mention of an unidentified species "allied to *fuliginosus*" as very common on the Chiloe and Chonos Islands; **J**: 353.

An unspecified "dark-coloured Furnarius" was tame on the Falkland Islands; **J**: 476.

CD cites "Pernety (1763)" as having reported an unspecified "Furnarius" on the Falkland Islands that "would almost perch on his finger; and that with a wand he killed ten in half an hour"; **J**: 477.

hornero = see **Furnarius**.

Furnarius rufus = Rufous Hornero, habits and nesting (very different to that of **Furnarius cunicularius**); **J**: 112–113. See chapter 2.

Distribution, nesting, habitat, habits, food, call; **Z**: 64. See chapter 2.

Merops rufus = as a synonym of **Furnarius rufus**; **Z**: 64.

Opetiorhynchus rufus = as a synonym of **Furnarius rufus**; **Z**: 64.

Turdus vadius = as a synonym of **Furnarius rufus**; **Z**: 64.

Figulus albogularis = as a synonym of **Furnarius rufus**; **Z**: 64.

creeper (*Synallaxis Tupinieri*) = Thorn-tailed Rayadito *Aphrastura spinicauda*, abundance, habitats, habits, foraging, call; **J**: 301, 353. See chapter 2.

CD notes it as foraging like the **willow wren** (*Phylloscopus trochilus*); **J**: 301.

creeper (Oxyurus tupiniero) = Thorn-tailed Rayadito *Aphrastura spinicauda*, commonest bird about Tierra del Fuego; **J2**: 237.

Oxyurus tupinieri = Thorn-tailed Rayadito *Aphrastura spinicauda*, range, abundance, habits, foraging; **Z**: 81.

Synallaxis tupinieri = as a synonym of **Oxyurus tupinieri**; **Z**: 81.

Oxyurus ornatus = as a synonym of **Oxyurus tupinieri**; **Z**: 81.

Synallaxis aegithaloides = Plain-mantled Tit-Spinetail *Leptasthenura aegithaloides*, a nest CD thought to be of Chilean Mockingbird was apparently of this bird; **Z**: 61.

Synallaxis aegithaloides = Plain-mantled Tit-Spinetail *Leptasthenura aegithaloides*, habitats, habits [some like titmouse (*Parus*)], call; nest led CD to see it as similar to **Synallaxis major**; **Z**: 79.

Synallaxis ruficapilla = Rufous-capped or Sooty-fronted Spinetail *Synallaxis ruficapilla* or *S. frontalis*, locations, tongue structure, habits like **Synallaxis maluroides** but morphologically approaches *Anumbius ruber*; **Z**: 79.

Parulus ruficeps = as a synonym of **Synallaxis ruficapilla**; **Z**: 79.

Sphenura ruficeps = as a synonym of **Synallaxis ruficapilla**; **Z**: 79.

Synallaxis = unspecified, occurs rarely in central Chile; **J**: 353–354.

Likened to the Vermillion Flycatcher; **Z**: 45.

Synallaxis flavogularis = Sharp-billed Canastero *Asthenes pyrrholeuca*, described, illustrated, thought closest to **S. flavogularis**, habitats, habits; **Z**: 78.

Synallaxis rufogularis = Austral Canastero *Asthenes anthoides*, described, illustrated, habitat, habits; **Z**: 77.

Synallaxis humicola = Dusky-tailed Canastero *Pseudashenes humicola*, CD says shows some affinity with **Pterophtochos** but much closer to **Eremobius** and "impossible to represent by a linear arrangement . . . relations between Furnarius, Uppucerthia, Opetiorhynchus, Eremobius, Anumbius, Synallaxis, Limnornis, Oxyurus; and again, Rhynomya, Pteroptochos, Scytalopus, and Troglodytes"; habits, tongue structure; **Z**: 75.

Synallaxis brunnea = Dusky-tailed Canastero *Pseudashenes humicola*, described, habitat, habits; **Z**: 78–79.

Anumbius ruber = Freckle-breasted Thornbird *Phacellodomus striaticollis*, habitats, habits like those of **Synallaxis maluroides** and **Limnornis** and thought most close to ***L. curvirostris***; **Z**: 80.

Furnarius ruber = as a synonym of **Anumbius ruber**; **Z**: 81.

Synallaxis maluroides = Bay-capped Wren-Spinetail *Spartonoica maluroides*, habitats, habits, call, tongue structure, food, seen as ecologically Old World warbler-like, anatomical notes; **Z**: 77–78, 152–153.

Habits very similar to those of **Anumbius ruber** and **Limnornis rectirostris**; **Z**: 80.

Oxyurus? dorso-maculatus = Wren-like Rushbird *Phleocryptes melanops*, frequents same locations as **Synallaxis maluroides** and two **Limnornis** species and has similar habits; **Z**: 82.

Synallaxis dorso-maculata = as a synonym of **Oxyurus? dorso-maculatus**; **Z**: 82.

Limnornis curvirostris = Curve-billed Reedhaunter, described, most common land bird of Tierra del Fuego, habitats, call; **Z**: 81.

Limnornis rectirostris = Straight-billed Reedhaunter *Limnoctites rectirostris*, described, illustrated, habitats, habits found like those of **Synallaxis maluroides**; **Z**: 80.

Synallaxis major = Firewood-gatherer *Anumbius annumbi*, CD notes this name should read **Anumbius acuticaudatus**; **Z**: viii.

Described, illustrated, habitat, habits (some like *Furnarius cunicularius*), tongue structure, nest; **Z**: 76.

Furnarius annumbi = as a synonym of **Anumbius acuticaudatus**; **Z**: viii.

Anthus acuticaudatus = as a synonym of **Anumbius acuticaudatus**; **Z**: viii.

Anumbius anthoides = as a synonym of **Anumbius acuticaudatus**; **Z**: viii.

Anumbius acuticaudatus = see **Synallaxis major**.

Eremobius phoenicurus = Band-tailed Earthcreeper *Ochetorhynchus phoenicurus*, CD notes the name **Eremobius** as preoccupied and thus changes it to **Enicornis**; **Z**: viii.

Described, illustrated, locations, allied to *Furnarius* and *Opetiorhynchus*, habits, call, food; **Z**: 69–70.

Distantly related to **Rhinomya lanceolata**; **Z**: 70.

Enicornis = see **Eremobius phoenicurus**.

Dendrodramus leucosternus = White-throated Treerunner *Pygarrhichas albogularis*, described, illustrated, range, habits (unlike those of *Oxyurus tupinieri*) like those of *Certhia familiaris*, food, considered closely allied to **Dendroplex**; **Z**: 82–83.

THAMNOPHILIDAE—**Antbirds**

Sub-family Thamnophilinae = Thamnophilidae; **Z**: 58.

Thamnophilus doliatus = Barred Antshrike, a specimen from Moldonado, markings, habitat, call; **Z**: 58.

Lanius doliatus = as a synonym of **Thamnophilus doliatus**; **Z**: 58.

RHINOCRYPTIDAE—**Tapaculos**

Guid-guid (*Hylactes Tarnii*) = Black-throated Huet-huet *Pteroptochos tarnii*, habitats, habits, diet, dog-like calls; **J**: 352–353.

Pterophtochos tarnii = Black-Thoated Huet-huet, distribution, call, habits, anatomical notes; **Z**: 70–71, 150–151. See chapter 3.

Hylactes Tarnii = as a synonym of **Pterophtochos tarnii**; **Z**: 70.

Megalonyx ruficeps = as a synonym of **Pterophtochos tarnii**; **Z**: 70.

Leptonyx Tarnii = as a synonym of **Pterophtochos tarnii**; **Z**: 70.

Pterophtochos megapodius = Moustached Turca, appearance, behaviour, ecology, and diet; **J**: 329–330, 351–352. See chapter 2.

Geographically replaces *P. tarnii*, habitat, habits, calls; **Z**: 71–72. See chapter 3.

Megalonyx rufus = as a synonym of **Pterophtochos megapodius**; **Z**: 71.

Leptonyx macropus = as a synonym of **Pterophtochos megapodius**; **Z**: 71.

Pterophtochos albicollis = White-throated Tapaculo *Scelorchilus albicollis*, appearance, behaviour, ecology, and diet; **J**: 329–330, 351–352. See chapter 2.

Geographically replaces **P. rubecula**, habitat, habits and nest like **P. megapodius** but unlike **P. rupecula** and **P. tarnii**; food, anatomical notes (skeleton as that of ***P. Tarnii*** save in measurements); **Z**: 72, 151–152. See chapter 3.

Megalonyx medius = as a synonym of **Pterophtochos albicollis**; **Z**: 72.

Megalonyx albicollis = as a synonym of **Pterophtochos albicollis**; **Z**: 72.

Leptonyx albicollis = as a synonym of **Pterophtochos albicollis**; **Z**: 72.

Cheucau (*Pteroptochos rubecula*) = Chucao Tapaculo *Scelorchilus rubecula*, habitats, habits, diet, calls, and relationship to people's superstitions; **J**: 339, 352. See chapter 2.

Pterophtochos rupecula = Chucao Tapaculo *Scelorchilus rubecula*, distribution, habitat, habits, food, nest, calls; **Z**: 73. See chapter 3.

Megalonyx rubecula = as a synonym of **Pterophtochos rupecula**; **Z**: 73.

Megalonyx rufogularis = as a synonym of **Pterophtochos rupecula**; **Z**: 73.

Leptonyx rubecula = as a synonym of **Pterophtochos rupecula**; **Z**: 73.

Rhinomya lanceolata = Crested Gallito *Rhinocrypta lanceolata*, CD notes the name **Rhinomya** as preoccupied and changes it to **Rhinocrypta**; **Z**: viii.

Geographically replaces several **Pterophtochos** species, "distantly allied to *Eremobius phoenicurus*," habits, call; **Z**: 70.

Rhinocrypta = see **Rhinomya lanceolata**.

Pterophtochos paradoxus = Ochre-flanked Tapaculo *Eugralla paradoxa*, close to **Scytalopus**. **P. tarnii**, and **P. megapodius**; form a close pair as do ***P. albicollis*** and **P. rubecula**. Habitat, food, call; **Z**: 73–74.

Troglodytes paradoxus = as a synonym of **Pterophtochos paradoxus**; **Z**: 73.

Malacorhynchus Chilensis = as a synonym of **Pterophtochos paradoxus**; **Z**: 73.

Leptonyx paradoxus = as a synonym of **Pterophtochos paradoxus**; **Z**: 73.

wren and **black wren (*Scytalopus fuscus*)** = Magellanic Tapaculo *Scytalopus magellanicus*, occurs rarely in central Chile; **J**: 301, 353–354. See chapter 2.

Distribution, habitat, habits; **Z**: 74.

Scytalopus megallanicus (*sic*) = Magellanic Tapaculo *S. magellanicus*, in appearance could be mistaken for a *Troglodytes*, but in habits closely allied to **Pterophtochos**. Locations, habitats, habits; **Z**: 74.

Sylvia Magellanica = as a synonym of **Scytalopus magallanicus**; **Z**: 74.

Scytalopus fuscus = as a synonym of **Scytalopus magallanicus**; **Z**: 74.

Platyurus niger = as a synonym of **Scytalopus magallanicus**; **Z**: 74.

TYRANNIDAE—Tyrant Flycatchers

Sub-family Fluvicolinae = a disused taxon of the Tyrannidae; **Z**: 51.

Sub-family Tyranninae = of the **Tyrannidae**; **Z**: 43.

Pachyrhynchus = a generic name CD noted as changed to **Pachyramphus** (now *Suiriri*) because previously used in entomology; **Z**: 50.

Pachyramphus albescens = Suiriri Flycatcher *Suiriri suiriri*, described and illustrated; **Z**: 50.

Myiobus albiceps = White-crested Elaenia *Elaenia albiceps*, distribution, habitat, and habits described; **Z**: 47.

Muscipeta albiceps = as a synonym of **Myiobus albiceps**; **Z**: 47.

Myiobius auriceps = White-crested Elaenia *Elaenia albiceps*, described; **Z**: 47.

Tyrannula auriceps = as a synonym of **Myiobus auriceps**; **Z**: 47.

white-tufted tyrant-flycatcher (Myiobius albiceps) = White-crested Elaenia *Elaenia albiceps*, CD records it singing; **J**: 301; **J2**: 237. See chapter 2.

Present on Chiloe and Chonos Islands; **J**: 353.

Myiobius parvirostris = Small-billed Elaenia *Elaenia parvirostris*, described, Galapagos; **Z**: 48.

Tyrannula parvirostris = as a synonym of **Myiobus parvirostris**; **Z**: 48.

Myiobus [*sic*] or **Myiobius** = a genus (now *Elaenia*) of the **Tyrannidae**; **Z**: 46.

Tyrannula = as a synonym of the genus **Myiobus**; **Z**: 46.

Serpophaga = a genus of Tyrannidae, "probably synonymous with **Euscarthmus**"; **Z**: vii.

Serpophaga nigricans = Sooty Tyrannulet, habitats and habits; **Z**: 50.

Sylvia nigricans = as a synonym of **Serpophaga nigricans**; **Z**: 50.

Tachuris nigricans = as a synonym of **Serpophaga nigricans**; **Z**: 50.

Serpophaga albo-coronata = White-crested Tyrannulet *Serpophaga subcristata*, like **Serpophaga parulus** in appearance and habits, described and habits noted as like tit-mice, plus anatomical notes ["precisely that of the smaller and weaker species of Laniadae"]; **Z**: 49–50, 147.

Serpophaga parulus = Tufted Tit-Tyrant *Anairetes parulus*. Distribution, habitat, nest, and habits; **Z**: 49.

Muscicapa parulus = as a synonym of **Serpophaga parulus**; **Z**: 49.
Sylvia Bloxami = as a synonym of **Serpophaga parulus**; **Z**: 49.
Culicivora parulus = as a synonym of **Serpophaga parulus**; **Z**: 49.

Cyanotis omnicolor = Many-colored Rush Tyrant *Tachuris rubrigastra*, locations, habits; **Z**: 86.

Regulus omnicolor = as a synonym of **Cyanotis omnicolor**; **Z**: 86.
Sylvia rubrigastra = as a synonym of **Cyanotis omnicolor**; **Z**: 86.
Regulus Byronensis = as a synonym of **Cyanotis omnicolor**; **Z**: 86.
Tachuris omnicolor = as a synonym of **Cyanotis omnicolor**; **Z**: 86.

Pachyramphus minimus = Bearded Tachuri *Polystictus pectoralis*, described and illustrated; **Z**: 51.

Pachyrhynchus minimus = as a synonym of **Pachyramphus minimus**; **Z**: 51.

Sub-genus Pyrocephalus = the genus ***Pyrocephalus***; **Z**: 44.

Muscicapa, Muscipeta, Tyrannula = as synonyms of the **Sub-genus *Pyrocephalus***; **Z**: 44.

Euscarthmus = a genus of Tyrannidae, for the pygmy-tyrants, which CD states is probably synonymous with **Serpophaga**; **Z**: vii.

Tyrannula ferruginea = Bran-coloured Flycatcher *Myiophobus fasciatus*, "nearly allied to" **Myiobius auriceps**; **Z**: 47.

Pyrocephalus = CD states this may be considered a sub-genus of **Myiobius** or a distinct genus; **Z**: 44.

Pyrocephalus nanus = Vermillion Flycatcher *Pyrocephalus rubinus*, CD notes that Mr. Sclater informs him that this bird is not endemic to the Galapagos, as it also occurs on the American continent; **J2**: vii.

Pyrocephalus parvirostris = Vermillion Flycatcher *Pyrocephalus rubinus*, described, illustrated, and months seen noted; habits likened to some European birds; **Z**: 44–45.

Differs from *Pyr. Coronatus* or *Muscicapa coronata* chiefly in size; Z: 45.

Pyrocephalus obscurus = Vermillion Flycatcher *Pyrocephalus rubinus*, one specimen described; **Z**: 45.

P. lividus rufotinctus = as a synonym of **Pyrocephalus obscurus**, **Z**: 45.

Pyrocephalus nanus = Vermillion Flycatcher *Pyrocephalus rubinus*, described, illustrated, habitat on Galapagos Islands; **Z**: 45–46.

Pyrocephalus dubius = Vermillion Flycatcher *Pyrocephalus rubinus,* most briefly described from Galapagos Islands, closely resembles **P. nanus**; **Z**: 46.

small tyrant-flycatchers = Vermillion Flycatcher *Pyrocephalus rubinus,* on Galapagos Islands, these live in most favourable part for them, "only land bird with bright colours," and "seems to be a wanderer from the continent"; **J**: 473–474.

three tyrant-flycatchers (two of them species of Pyrocephalus, one or both of which would be ranked by some ornithologists as only varieties) = Vermillion Flycatcher *Pyrocephalus rubinus* and Galapagos Flycatcher *Myiarchus magnirostris* on Galapagos Islands; **J2**: 378.

Tyrant-flycatchers = extremely abundant on grassy plains around Maldonado; **J**: 60.

Three new species briefly alluded to as endemic on Galapagos, these now being only two species—the Vermillion Flycatcher *Pyrocephalus rubinus*, and Eastern Kingbird *Tyrannus tyrannus*; **J**: 461. See chapter 2.

Tyrant-Flycatchers listed among birds tame on the Galapagos Islands; **J**: 475. See chapter 2.

Myiobius parvirostris = Patagonian Tyrant *Colorhamphus parvirostris,* described, distribution, and habitat; **Z**: 48.

Xolmis = it is suggested that perhaps ***Muscicapa thamnophiloides*** and ***M. cinerea*** (= Cinnamon Attila *Attila cinnamomeus* and Grey Monjita *Xolmis cinereus,* respectively, cf. Gill and Donsker 2015) belong to this genus; **Z**: 56 footnote.

Xolmis pyrope = Fire-eyed Diucon, distribution, habitat, habits; **Z**: 55.

Xolmis nengeta = Grey Monjita *Xolmis cinereus*, specimen from Maldonado, La Plata, habits like Black-and-White Monjita; **Z**: 54.

Habits like those of **Fluvicola Azarae**; **Z**: 54.

Plumage much like that of **Mimus**; **Z**: 60.

Lanius nengeta = as a synonym of **Xolmis nengeta**; **Z**: 54.

Tyrannus nengeta = as a synonym of **Xolmis nengeta**; **Z**: 54.

Fluvicola nengeta = as a synonym of **Xolmis nengeta**; **Z**: 54.

Tyrannus pepoaza = as a synonym of **Xolmis nengeta**; **Z**: 54.

Muscicapa polyglotta = as a synonym of **Xolmis nengeta**; **Z**: 54.

Tyrannus polyglottus = as a synonym of **Xolmis nengeta**; **Z**: 54.

Muscicapa coronata? = Black-crowned Monjita *Xolmis coronatus*—a widespread South American species, which a tyrant flycatcher found on the Galapagos, is puzzlingly stated by CD to be identical to; **J**: 461. See chapter 2.

Pyr. coronatus or ***Muscicapa coronata*** = Black-crowned Monjita *Xolmis coronatus*, location, differs from **Pyrocephalus parvirostris** chiefly in size; **Z**: 45.

Xolmis coronata = Black-crowned Monjita *X. coronatus*, one specimen from wooded banks of Parana River, near Santa Fé; **Z**: 53–54.

Tyrannus coronatus = as a synonym of **Xolmis coronata**; **Z**: 54.

Muscicapa vittiger = as a synonym of **Xolmis coronata**; **Z**: 54

Fluvicola irupero = White Monjita *Xolmis irupero*, locations observed, and habits; **Z**: 53.

Tytannus Irupero = as a synonym of **Fluvicola irupero**; **Z**: 53.

Muscicapa moesta = as a synonym of **Fluvicola irupero**; **Z**: 53.

Muscicapa nivea = as a synonym of **Fluvicola irupero**; **Z**: 53.

Pepoaza nivea = as a synonym of **Fluvicola irupero**; **Z**: 53.

Fluvicola azarae = Black-and-white Monjita *Heteroxolmis dominicana*, described, said closely related to *Taenioptera* and *Pepoaza*, particularly *P. Dominicana*; illustrated, habitat, habits; **Z**: 53–54.

Habits like those of **Xolmis nengeta**; **Z**: 54.

Pepoaza Dominicana = Black-and-white Monjita *Heteroxolmis dominicana*, "closely allied to, if not identical with" **Fluvicola azarae**; **Z**: 54.

Xolmis variegata = Chocolate-vented Tyrant *Neoxolmis rufiventris*, illustrated, compared with the Fieldfare *Turdus pilaris* in appearance and habits, food; **Z**: 55.

Feeding in "small flocks, mingled with the icteri" and **Oreophilus totanirostris**; **Z**: 125.

Pepoaza variegata = as a synonym of **Xolmis variegata**; **Z**: 55.

Taenioptera variegata = as a synonym of **Xolmis variegata**; **Z**: 55.

Taenioptera = a genus most close to **Fluvicola**; **Z**: 54.

Agriornis = a genus of the Tyrannidae; **Z**: 56.

Dasycephala = as a synonym of **Agriornis**; **Z**: vii.

Tamnolanius = as a synonym of **Agriornis**; **Z**: vii.

Tyrannus = as a synonym of **Agriornis**; **Z**: 56.

Pepoaza = as a synonym of **Agriornis**; **Z**: 56.

Agriornis leucurus = Black-billed Shrike-Tyrant *Agriornis montanus*, the illustration of CD's *A. maritimus* is captioned with this name; **Z**: 58–59.

Agriornis maritimus = Black-billed Shrike-Tyrant *Agriornis montanus*, illustrated, distribution, habitat, habits; **Z**: 57.

Pepoaza maritima = as a synonym of **Agriornis maritimus**; **Z**: 57.

Agriornis leucurus = as a synonym of **Agriornis maritimus**; **Z**: 57.

Agriornis gutturalis = Great Shrike-Tyrant *Agriornis lividus*, distribution, habits; **Z**: 56.

Tyrannus gutturalis = as a synonym of **Agriornis gutturalis**; **Z**: 56.

Pepoaza gutturalis = as a synonym of **Agriornis gutturalis**; **Z**: 56.

Agriornis striatus = Grey-bellied Shrike-Tyrant *Agriornis micropterus*, described, habits, habitat, and thought as an immature of *A. micropterus*; **Z**: 56–57.

Agriornis micropterus = Grey-bellied Shrike-Tyrant *Agriornis micropterus*, described, illustrated, habitat, habits; **Z**: 57.
Habits as those of **Agriornis striatus**; **Z**: 57.

Muscisaxicola brunnea = Spot-billed Ground Tyrant *Muscisaxicola maculirostris*, one immature specimen from Port St. Julian, Patagonia, briefly described; **Z**: 84.

Muscisaxicola mentalis = Dark-faced Ground Tyrant *Muscisaxicola maclovianus*, distribution on South America, habitats, habits [some of which are like the "whinchat (*Motacilla rubetra*)"], *Muscisaxicola* "probably synonymous with Lessonia"; **Z**: 83.

Muscisaxicola macloviana = Dark-faced Ground Tyrant *Muscisaxicola maclovianus*, one specimen from Falklands Islands. Gould thought it a distinct species, but CD could not see differences other than in size and disagreed; **Z**: 83–84. See chapter 3.
Sylvia macloviana = as a synonym of **Muscisaxicola macloviana**; **Z**: 83.
Curruca macloviana = as a synonym of **Muscisaxicola macloviana**; **Z**: 83.

Muscisaxicola nigra = Austral Negrito *Lessonia rufa*, distribution, habitats (some shared with **Opetiorhynchus**), habits; **Z**: 84.
Alaunda nigra = as a synonym of **Muscisaxicola nigra**; **Z**: 84.
Alauda rufa = as a synonym of **Muscisaxicola nigra**; **Z**: 84.
Alauda fulva = as a synonym of **Muscisaxicola nigra**; **Z**: 84.
Anthus fulvus = as a synonym of **Muscisaxicola nigra**; **Z**: 84.
Anthus variegatus = as a synonym of **Muscisaxicola nigra**; **Z**: 84.
Sylvia dorsalis = as a synonym of **Muscisaxicola nigra**; **Z**: 84.
Lessonia erythronotus = as a synonym of **Muscisaxicola nigra**; **Z**: 84.

Lichenops perspicillatus = Spectacled Tyrant *Hymenops perspicillatus*, said to be of the sub-genus *Perspicilla*, distribution, habits, food; **Z**: 51–52.
Sylvia perspicillata = as a synonym of **Lichenops perspicillatus**; **Z**: 51.
Oenanthe perspicillata = as a synonym of **Lichenops perspicillatus**; **Z**: 51.
Ada Commersoni = as a synonym of **Lichenops perspicillatus**; **Z**: 51.
Perspicilla = a sub-genus that **Lichenops perspicillatus** belongs to; **Z**: 51.
Perspicilla leucoptera = as a synonym of **Lichenops perspicillatus**; **Z**: 51.
Fluviola perspicillata = as a synonym of **Lichenops perspicillatus**; **Z**: 51.

Lichenops erythropterus = Spectacled Tyrant *Hymenops perspicillatus*, erroneously described as the female of **Lichenops perspicillatus** and so the two compared; described, illustrated, habitats, habits; **Z**: 52–53.

Alecturus guirayetupa = Strange-tailed Tyrant *Alectrurus risora*, brief note of habitat, habits, food; **Z**: 51.

Muscicapa psalura = as a synonym of ***Alecturus guirayetupa***; **Z**: 51.
Muscicapa risoria = as a synonym of ***Alecturus guirayetupa***; **Z**: 51.
Yetapa psalura = as a synonym of ***Alecturus guirayetupa***; **Z**: 51.

Fluvicola icterophrys = Yellow-browed Tyrant *Satrapa icterophys*, specimen locations, beetles eaten; **Z**: 53.

Muscicapa icterophrys = as a synonym of **Fluvicola icterophrys**; **Z**: 53.

Muscicapa thamnophiloides = Cinnamon Attila *Attila cinnamomeus*, CD suggests perhaps this belongs to the genus ***Xolmis***; **Z**: 56.

Muscicapa cinerea = Grey-hooded Attila *Attila rufus*, CD suggests perhaps this belongs to the genus ***Xolmis***; **Z**: 56.

Myiobius magnirostris = Galapagos Flycatcher *Myiarchus magnirostris*, described, illustrated, "Not very uncommon" on Chatham Island, Galapagos; **Z**: 48.

Tyrannula magnirostris = as a synonym of **Myiobus magnirostris**; **Z**: 48.

Milvulus or "more properly" **Milvilus** = a genus CD states **Muscivora tyrannus** was placed in by Mr. Swainson, but G. R. Gray pointed out was already proposed for *M. forficate* (and now *Tyrannus forficatus*); **Z**: 43.

Musc. forficata = see **Milvulus.**

Milvulus forficatus = Swallow-tailed Flycatcher *Tyrannus forficatus*, common near Buenos Ayres, habits, food, flight; **J** 163. See chapter 2.

Muscivora tyrannus = Fork-tailed Flycatcher *Tyrannus savana*, foraging and flight described; **Z**: 43–44.

Muscicapa Tyrannus = as a synonym of **Muscivora tyrannus**; **Z**: 43.
Tyrannus Savana = as a synonym of **Muscivora tyrannus**; **Z**: 43.

Saurophagus sulphureus and **Saurophagus sulphuratus** = Great Kiskadee *Pitangus sulphuratus*, foraging and other habits, also held in captivity; **J**: 62; **Z**: 43. See chapter 2.

Diversified habits of; **O**: 220. See chapter 2.

Lanius sulphuratus = as a synonym of **Saurophagus sulphuratus**; **Z**: 43.

Tyrannus maganimus = as a synonym of **Saurophagus sulphuratus**; **Z**: 43.

Tyrannus sulphuratus = as a synonym of **Saurophagus sulphuratus**; **Z**: 43.

Sub-family Tityranae. (Psarianae, *Sw.*) = Tityrinae of the Tyrannidae; **Z**: 50.

COTINGIDAE—Cotingas

Chatterers or Cotingidae = monogamous species show well-marked sexual differences (see **blackbird** and **bullfinch)**; **D**: 338.

Sexual dimorphism within the family, as then perceived, briefly discussed; **D**: 701.

Females are generally similar to those of allied species, whereas the males differ dramatically; **D**: 717.

Phytotoma rara = Rufous-tailed Plantcutter, habits, destructive feeding on tree buds, "manners" like those of **bullfinch (Loxia Pyrrhula**), plus anatomical notes; **Z**: 106, 153–154. See also chapter 3.

P. Bloxami = as a synonym of **Phytotoma rara**; **Z**: 106.

P. rutila = as a synonym of **Phytotoma rara**; **Z**: 106.

P. silens = as a synonym of **Phytotoma rara**; **Z**: 106.

crow-like bird or Cephalopterus ornatus = Amazonian Umbrellabird, umbrella-like crest and extensible throat wattle described, and the latter stated to enhance vocalizations, illustrated; **D**: 570–571. See colour figure 6.4.

Umbrella-bird (*Cephalopterus ornatus*) = Amazonian Umbrellabird, umbrella-like crest and extensible throat wattle described and the latter stated to enhance vocalizations, illustrated; **D**: 570–571. See chapter 6 and colour figure 6.4.

The genus [***Chasmorhynchus***] or **bell-birds** = the bellbirds *Procnias*, discussed as a genus within which the males of the species "differ much more from each other than do the females"; **D**: 593. See chapter 6 and colour figure 6.7.

In several species, the males are wholly or partially white, doubtless through sexual selection; **D**: 750. See colour figure 6.7.

(C. *tricarunculatus*) = Three-wattled Bellbird *Procnias tricarunculatus*, male described; **D**: 593. See colour figure 6.7.

bell-bird (*Chasmorhynchus niveus*) = White Bellbird *Procnias albus*, "remarkable from the extreme contrast in colour between the sexes"; the call is audible at nearly three miles away; **D**: 592. See colour figure 6.7.

The white males "appreciated" by females [and thus are the result of sexual selection]; **D**: 674.

(C. *nudicollis*) = Bare-throated Bellbird *Procnias nudicollis*, male described and the seasonal nature of their bare facial skin noted; **D**: 593. See colour figure 6.7.

rock-thrush = Guianan Cock-of-the-rock *Rupicola rupicola*, male lekking and display; **O**: 109. See chapters 4, 6, and figure 4.2.

Rupicola crocea = Guianan Cock-of-the-rock *Rupicola rupicola*, the sexes, lek, display, and killing of them by local people described, and an adult male illustrated; **D**: 601–602. See figure 4.2.

cocks of the Rock = Guianan Cock-of-the-rock *Rupicola rupicola*, Indians of Guiana are "well acquainted with the cleared arenas" where they expect to find them; **D**: 617–618. See figure 4.2.

PIPRIDAE—Manakins

a sub-genus of Pipra or Manakin = the manakin genus *Pipra* (no sub-genera now acknowledged), males of some species have secondary feathers modified for sound production, illustrated; **D**: 578–579. See chapter 6.

Pipra deliciosa = Club-winged Manakin *Machaeropterus deliciosus*, primaries modified for sound production in display described and illustrated; **D**: 578–579. See chapter 6 and figure 6.5.

M. cinnamonea = Cinnamon Neopipo *Neopipo cinnamomea*, "nearly allied to" **Myiobius auriceps**; **Z**: 47.

MENURIDAE—Lyrebirds

Menura = this genus probably allied to ***Pterochos*** (true today); **Z**: 152.

An exception to CD's statement that "only small birds properly sing"; **D**: 567.

Menura Alberti = Albert's Lyrebird *Menura alberti*, "not only mocks other birds" but produces "beautiful and varied" whistles of its own; CD states (erroneously) that males congregate at leks; **D**: 567–568. See chapter 6.

Scratches "for itself shallow holes," called "*corroborying places*" by the natives, where both sexes assemble; **D**: 619. See chapter 6.

Menura lyra = Superb Lyrebird *Menura novaehollandiae*, T. C. Eyton claims an examination of a specimen supports his conclusion that ***Menura*** is allied to ***Pterochos*** (true today); **Z**, 152 footnote.

lyre-bird (*Menura superba*) = Superb Lyrebird *Menura novaehollandiae*, forms "small round hillocks" as display site; and CD cites T. W. Wood who (inexplicably) claims to have seen ca. 150 adult males "fighting with indescribable fury"; **D**: 619. See chapter 6.

Menura superba = Superb Lyrebird *Menura novaehollandiae*, females have a long tail yet build a domed nest, "which is a great anomaly," but he explains how it is accommodated; **D**: 687–688.

PTILONORHYNCHIDAE—Bowerbirds

Regent bird = Regent Bowerbird *Sericulus chrysocephalus*, bower decorations described and the bird's "taste for the beautiful" noted; **D**: 630.

Satin Bower-birds = Satin Bowerbirds *Ptilonorhynchus violaceus*, brief description of male courtship display to a female at a bower in captivity; **D**: 583. See colour figures 6.6 and A.1.

Their bower decorations detailed and their "taste for the beautiful" noted (see **parrakeets**), **D**: 630. See chapter 6.

Great Bower-bird = Great Bowerbird *Chlamydera nuchalis*, brief description of a bird decorating its "play-house" or bower by Captain Stokes cited; **D**: 583.

Bower-bird *Chlamydera maculata* = Spotted Bowerbird, two birds at a bower illustrated; **D**: 582. See chapter 6.

Calodra maculata = Spotted Bowerbird *Chlamydera maculata*, CD compares the habit the bizcacha (*Calomys bizcacha*, a rabbit-like vesper mouse) has of accumulating objects at its burrow entrance to that of the bowerbird doing so about its bower; **J2**: 125. See chapter 2 and colour figure A.2, appendix 4.

Spotted bower-bird = *Chlamydera maculata*, bower and decorations described and the bird's "taste for the beautiful" noted; **D**: 630. See colour figure 6.6.

Fawn-breasted Bower-bird = Fawn-breasted Bowerbird *Chlamydera cervini-ventris*, its bower briefly described; **D**: 583.

bower-birds or **Bower-birds** = unspecified, build bowers "tastefully ornamented with gaily-coloured objects"; **D**: 140. See colour figure 6.6.

A description and illustration of their avenue bower structures, which CD thought both sexes build [but only males do so]; **D**: 583. See chapter 6 and colour figure 6.6.

Their bowers "are the resort of both sexes during the breeding season" and "with two of the genera the same bower is attended for many years" according (correctly) to John Gould; **D**: 619. See colour figure 6.6.

They appreciate bright and beautiful objects, and this is involved in female sexual selection; **D**: 941–942. See colour figure 6.6.

CLIMACTERIDAE—**Australasian Treecreepers**

Australian tree-creeper (*Climacteris erythrops*) = Red-browed Treecreeper, sexual dimorphism described, in which females are more colourful than males; **D**: 731.

MALURIDAE—**Australasian Wrens**

Superb Warblers (Maluridæ) = fairywrens of the genus *Malurus*, males are colourful and the dull females incubate in a domed nest—mentioned in discussing Wallace's objections to sexual selection for plumage colouration with regard to nesting; **D**: 692.

Maluri = Australasian fairywrens of the genus *Malurus*, males differ from females during summer, and the young generally resemble females; **D**: 741. See chapter 7.

Splendid maluri = Splendid Fairywren *Malurus splendens*, mentioned as an example of a species in which young generally resembling the adult females of their kind; **D**: 740–741. See chapter 7.

MELIPHAGIDAE—**Honeyeaters**

Anthornis melanura = New Zealand Bellbird, with their heads covered in native Fushia pollen; **F**: 371. See chapter 7.

honey-sucking birds = doubtless largely, if not exclusively, refers to this family—CD does not doubt they fertilize many Australian plants; **F**: 371. See chapter 7.

CALLAEIDAE—**New Zealand Wattlebirds**

Huia = *Heterolocha acutirostris* (an extinct New Zealand bird), exhibit a "wonderfully great" sexual dimorphism in bill size and structure, reflected in different foraging between the sexes; **D**: 321.

Neomorpha of New Zealand = Huia *Heterolocha acutirostris*, wide difference in beak between the sexes relates to their feeding ecology; **D**: 551.

TEPHRODORNITHIDAE—**Woodshrikes and allies**

shrikes (Tephrodornis) = the genus for the (now four) Asian woodshrike species—males, but not the females and young, of some species have geographically undergone intraspecific morphological change into distinguishable forms; **D**: 714.

CAMPEPHAGIDAE—**Cuckooshrikes**

Oxynotus (shrikes) = the cuckoo-shrike genus *Coracina*, in the two different species on Mauritius and Bourbon Islands, the males differ little, but the females differ much; **D**: 717.

LANIIDAE—**Shrikes**

true shrikes = species of the Laniidae; **Saurophagus sulphureus** seen as like them structurally; **J**: 62. See chapter 2.

shrike = unspecified, but a species of *Lanius*, probably Great Grey Shrike *L. exubitor*, in noting that Great Tits *Parus major* kill small birds "like a shrike" by blows to the head; **O**: 220.

Lanius = unspecified, young fertile females may acquire the characters of males; **D**: 704.

shrikes (Lanius) = the young of many species are transversely striped below, and certain allied species or whole genera are similarly marked when adult; **D**: 709.

Lanius rufus = Woodchat Shrike *Lanius senator*, Mr. Blyth reports specimens that had anomalously assumed adult plumage while young; **D**: 736.

VIREONIDAE—**Vireos, Greenlets**

Family Laniadae, Sub-family Lanianae = Vireonidae; **Z**: 58.

Skeleton of **Serpophaga albocoronata** "precisely that of the smaller and weaker species of Laniadae"; **Z**: 147.

Cyclarhis guianensis = Rufous-browed Peppershrike *Cyclarhis gujanensis*, specimen from Maldonado with Coleoptera (beetles) in stomach; **Z**: 58.

Tanagra Guianensis = as a synonym of **Cyclarhis guianensis**; **Z**: 58.

Laniagra Guyanensis = as a synonym of **Cyclarhis guianensis**; **Z**: 58.
Falcunculus Guianensis = as a synonym of **Cyclarhis guianensis**; **Z**: 58.

ORIOLIDAE—**Figbirds, Orioles**
Oriolus = unspecified, immature males may rarely breed; **D**: 739 and footnote.

In confinement, a male assumed the female plumage of his species; **V2**: 158.

Orioles = unspecified, members of the tribe Oriolini, mentioned in discussing Wallace's objections to sexual selection for plumage colouration with regard to open nesting; **D**: 691.

Oriolus melanocephalus = Black-hooded Oriole *Oriolus xanthornus*, females of this, as some closely related, species differ from males in plumage at initial breeding maturity, but after several subsequent moults do so only in beak colour; **D**: 702.

DICRURIDAE—**Drongos**
King-crows (Dicrurus) = mentioned in discussing Wallace's objections to sexual selection for plumage colouration with regard to open nesting; **D**: 691.

Indian drongos (Dicrurus and Edolius) = both now *Dicrurus*, have a tail including bare shafted feathers terminating in a vertical disc of vanes; **D**: 586–587.

Drongo shrikes (*Dicrurus macrocercus*) = Black Drongo, the male "while almost a nestling" moults soft brown plumage into a uniform glossy greenish-black one, but females retain white markings on the "axillary feathers" until becoming uniform greenish-black after three years; **D**: 703.

drongos (*Bhringa*) or **Drongo-shrikes in India** = *Dicrurus* species, highly modified tail feathers are acquired by them only during spring; **D**: 597.

CORVIDAE—**Crows, jays**
Blue jay (Garrulus cristatus) = Blue Jay *Cyanocitta cristata*, as a host to American cuckoos' eggs and nestlings; **O**: 330–332. See chapter 3.

jays (*Garrulus glandarius*) = Eurasian Jay, a gamekeeper repeatedly shot one of a pair to find the survivor re-matched with another mate shortly afterward; **D**: 621.

Always seen during spring in pairs and not alone; **D**: 623.

Captives distinguish different individual people and exhibit affection to them, **D**: 627.

Both adult sexes and their young are of similar appearance; **D**: 734–735. See chapter 7.

common jay = Eurasian Jay *Garrulus glandarius*, both sexes and young are closely similar; **D**: 735. See chapter 7.

Canada jay (*Perisoreus canadensis*) = Grey Jay, the young differ so much from their parents, they were once thought to be of a different species; **D**: 735. See chapter 7.

Magpie(s) = Eurasian Magpie *Pica pica,* most briefly alluded to with regard to Galapagos birds ignoring people just as these magpies ignore grazing cows and horses in England; **J**: 476; **V1**: 21.

A "bird something like the magpie" seen by CD in the Blue Mountains, Australia (would have been an Australasian Magpie *Cracticus tibicen*); **J**: 526.

Magpies wary in England but tame in Norway; **O**: 325.

Learn to pronounce words, even short sentences, in captivity, but do not imitate in the wild; **D**: 137.

Have the complex vocal organs of the **Insessores** but "never sing"; **D**: 567.

CD reports that the rapid replacement of a lost mate is more frequently observed in this species than in any other bird, possibly due to its conspicuousness; **D**: 620. See chapter 6.

Always seen during spring in pairs and not alone; **D**: 623.

CD asks, "Is it admiration or curiosity" that makes this bird, the **raven,** and some other species "steal and secrete bright objects" such as jewels?; **D**: 630.

Both adult sexes and their young are of similar appearance; **D**: 734.

common magpie (*Corvus pica*) = Eurasian Magpie *Pica pica,* sometimes forming large aggregations, particularly in early spring prior to pairing off; **D**: 619–620. See chapter 6.

A "piebald" species within a group largely of black ones; **D**: 753.

chough (*Corvus gracilis*) = Red-billed Chough *Pyrrhocorax pyrrhocorax*, both sexes are black throughout the plumage, but this is contrasted with a bright red beak; **D**: 749.

jackdaws = Western Jackdaw *Corvus monedula*, mentioned because "crows" (*Corvus* spp.), seen by CD in the Blue Mountains, Australia, struck him as being "like our jackdaws"; **J**: 526. See **Magpie(s)**.

Unite in flocks with **rooks** and **starlings**; **D**: 153.

Indian crows = House Crow *Corvus splendens*, Mr. Blyth informs CD that birds fed their blind companions (see **blind pelican**); **D**: 157.

English Rook = Rook *Corvus frugilegus*, calls likened to those of the Falkland Hawk; **J**: 67. See chapter 3.

Unite in flocks with **jackdaws** and **starlings**; **D**: 153.

common rook = Rook *Corvus frugilegus*, its call alters during the breeding season "and is therefore in some way sexual"; **D**: 537.

rook = *Corvus frugilegus*, offspring of a family of pied birds were pied only until their first moult—but their subsequent offspring were also pied; **V2**: 77.

carrion-crows = Carrion Crows *Corvus corone*, individuals killed by game-keepers in Delamere Forest were all males; **D**: 621.

People have shot one of a nesting pair only to find the nest "soon again tenanted by a pair"; **D**: 621.

Always seen during spring in pairs and not alone; **D**: 623.

Sometimes "live in triplets"; **D**: 624.

Both adult sexes and their young are of similar appearance; **D**: 734. See chapter 7.

Hybrid offspring with **hooded crows** show no sign of [morphological] "fusion"; **V2**: 94.

common crow = Carrion Crow *Corvus corone*, hybrid offspring with **hooded crows** show no sign of [morphological] "fusion"; **V2**: 94.

hooded crow = Carrion Crow *Corvus corone*, tame in Egypt [unlike in England]; **O**: 325.

Hybrid offspring with **carrion-crows** show no sign of [morphological] "fusion"; **V2**: 94.

crows = unspecified, species of *Corvus* "like our jackdaws" (Eurasian Jackdaw *Corvus monedula*) seen by CD in the Blue Mountains, Australia, and most likely Australian Raven *Corvus coronoides*; **J**: 526. See **magpie(s)** and chapter 2.

Have the complex vocal organs of the **Insessores** but "never sing"; **D**: 567.

It is probable that their complete or partial blackness is the result of sexual selection "accompanied by equal transmission to both sexes"; **D**: 749.

See **Kingfisher(s)**.

raven(s) = unspecified, but larger members of the genus *Corvus*, and most likely Northern Raven *C. corax*, have the complex vocal organs of the **Insessores** but "never sing"; **D**: 567.

CD asks "Is it admiration or curiosity" that makes the **raven** and some other species "steal and secrete bright objects" such as jewels?; **D**: 630.

A pied Feroe Islands population, persecuted by normally plumaged ravens, was wrongly thought a distinct species, and CD likens the behaviour of typical ravens toward them to that suffered by **albinos** in other species; **D**: 646–648, 756.

One individual long fed vegetable food developed a great change in its stomach;**V2**: 302.

CORCORACIDAE—Australian Mudnesters

Grallinae = a sub-family name for what was the, now disused, family Grallinidae consisting of three groups but now treated as two unrelated groups—the two mudlarks *Gralllina* being one (in the family Monarchidae), the White-Winged Chough *Corcorax melanorhamphos* and Apostlebird *Struthidea cinerea* (now constituting the family Corcoracidae) the other. Unless he misspelt this sub-family name, CD erroneously states the males are more colourful than the females and females incubate in a domed nest. Mentioned in discussing Wallace's objections to sexual selection for plumage colouration with regard to nesting; **D**: 692. See chapter 8.

PARADISAEIDAE—Birds-of-paradise

birds of paradise = of the family Paradisaeidae, male lekking and display; **O**: 109.

CD writes that Lesson says that birds of paradise are polygamous, but "Mr. Wallace doubts whether he had sufficient evidence" (perhaps not, but Lesson was correct); **D**: 338.

They [males] "rattle their quills together"; **D**: 537. See chapter 6.

Some species have a tail including bare-shafted feathers terminating in a disc of vanes, as do some head feathers; and in some, barbs of feathers are filamentous or plumose; **D**: 587.

Females "are obscurely coloured and destitute of all ornaments, whilst males are probably the most highly decorated of all birds, and in so many different ways"; **D**: 589.

Variation in annual moult of males of different species described and briefly discussed; **D**: 597.

Unspecified, but undoubtedly *Paradisaea* species, 12 or more adult males congregate in a tree [lek] "to hold a dancing-party, as it is called by the natives" to display and become so absorbed that a "skilful archer may shoot nearly the while party." Captive males take great care to clean their plumage, and their display is doubtless "intended to please the female" according to Mr. Wallace; **D**: 603.

The plumes of males "trouble them during a high wind"; **D**: 613.

Natives of New Guinea know lek trees where 10 to 20 males congregate; **D**: 618.

Some species do not cast their plumes during the winter; **D**: 705.

The plumage of [males of] some species increases in beauty for many years after they mature; **D**: 740. See chapter 7.

It could not be supposed that males display their plumage to females for no purpose; **D**: 942.

(*Lophorina atra*) = Superb Bird-of-paradise *Lophorina superba*, males are black and the females brown or mottled, doubtless through sexual selection; **D**: 749.

another most beautiful species = Wilson's Bird-of-paradise *Diphyllodes respublica*, CD quotes Wallace who described the head as bald "and of a rich cobalt blue, crossed by several lines of black velvety feathers"; **D**: 589–590.

Paradisaea apoda = Greater Bird-of-paradise, barbless tail plumes attain 34 inches in length; **D**: 587.

Brief description of adult male courtship display; **D**: 589.

Females of this and of ***P. papuana*** differ more from each other than do their respective [adult] males; **D**: 717. See chapter 7.

P. Papuana = Lesser Bird of paradise *Paradisaea minor*, barbless tail plumes are "much shorter and thin" compared to those of ***Paradisaea apoda***, illustrated; **D**: 587–588. See figure 6.8.

Females of this and of ***P. apoda*** differ more from each other than do their respective [adult] males; **D**: 717. See chapter 7.

BOMBYCILLIDAE—Waxwings

Bombycilla carolinensis = Cedar Waxwing *B. cedrorum*; females differ very little from males, but the red wing "wax" is not attained as early in life as it is by males; **D**: 703.

PARIDAE—Tits, Chickadees

Parinæ = the sub-family of the Paridae, which build concealed nests; **D**: 698.

Sultan yellow tit = Sultan Tit *Melanochlora sultanea*, shows sexual dimorphism greater than in the **common blue tomtit**; **D**: 698.

tit-mice = of the genus *Parus*, which CD merely compares with the Tufted Tit-Tyrant in habits; **Z**: 49.

Serpophaga albo-coronata said to be like tit-mice in habits and appearance; **Z**: 50.

titmouse (*Parus*) = compared in habits with **Synallaxis aegithaloides**; **Z**: 79.

titmouse (*Parus major*) = Great Tit, foraging behaviour and associated morphology; **O**: 220, 354–355. See chapter 4.

tomtits (*Parus major*) = Great Tit, hard-shelled peas escaped attack by this bird better than thin-shelled peas (and walnuts); **V2**: 231.

Parus major = Great Tit, as a British species with conspicuous females that nests in holes, mentioned in discussing Wallace's objections to sexual selection for plumage colouration with regard to nesting; **D**: 693 footnote.

P. caeruleus = Eurasian Blue Tit *Cyanistes caeruleus*, as a British species with conspicuous females that nests in holes, mentioned in discussing Wallace's objections to sexual selection for plumage colouration with regard to nesting; **D**: 693 footnote.

common blue tomtit (*Parus cæruleus*) = Eurasian Blue Tit *Cyanistes caeruleus*, females are much duller than males; **D**: 698.

Parus, 3 sp. = unspecified, as inconspicuous British species that nests in holes, mentioned in discussing Wallace's objections to sexual selection for plumage colouration with regard to nesting; **D**: 693 footnote.

ALAUDIDAE—Larks

Melanocorypha cinctura = Bar-tailed Lark *Ammomanes cinctura,* described, habitats, habits; **Z**: 87.

Pyrrhalauda nigriceps = Black-crowned Sparrow-Lark *Eremopterix nigriceps,* described, habitats, habits; **Z**: 87–88.

Wood Lark = Woodlark *Lullula arborea*, see **Anthus furcatus**; **Z**: 85.

skylark = Eurasian Skylark *Alauda arvensis*, **common cuckoo** eggs said to not exceed those of this species in size; **O**: 332.

skylark (*Alauda arvensis*) = Eurasian Skylark, have lived seven years in captivity (in UK) without breeding, but they did so once elsewhere; **V2**: 154.

Lark(s) or **common lark** = unspecified, but likely Eurasian Skylark *Alauda arvensis*; CD informed that the sexes are about equal (probably in UK) population; **D**: 383.

Females may sing, especially widowed individuals (see **robin** and **bullfinch**); **D**: 566.

Individuals are drawn down from the sky and caught by a small mirror being moved on the ground; **D**: 629–630.

Unspecified, some are difficult to see when crouched on the ground; **D**: 746.

PYCNONOTIDAE—Bulbuls

bulbuls = Red-vented Bulbul *Pycnonotus cafer*, males kept captive in India as fighting birds; **D**: 552. See chapter 6.

Indian bulbul (*Pycnonotus hæmorrhous*) = Red-Vented Bulbul *Pycnonotus cafer*, presents crimson under tail coverts in display; **D**: 612.

HIRUNDINIDAE—Swallows, Martins

Hirundo frontalis = White-rumped Swallow *Tachycineta leucorrhoa*, size, appearance described and discussed and distribution noted; **Z**: 40.

Hirundo leucopygia = Chilean Swallow *Tachycineta leucopyga*, specimen locations and nesting briefly noted; **Z**: 40.

Progne purpurea = Galapagos Martin *Progne modesta*, CD notes a "swallow" he saw in the Galapagos Islands was similar to *Progne purpurea* but that Gould found it "specifically distinct"; **J2**: 379.

Specimens from Monte Video and Bahia Blanca, range, abundance, nesting, flight, and screaming call [like **English Swift**]; **Z**: 38–40.

Hirundo purpurea = as a synonym of **Progne purpurea**; **Z**: 38.

Progne modesta = Galapagos Martin, specimen locations, migration, range and habits discussed in **Z**: 39–40.

Described, illustrated, and discussed briefly, observed about coastal lava cliffs; **Z**: 39–40.

Hirundo concolor = as a synonym of **Progne modesta**; **Z**: 38.

Hirundo cyanoleuca = Blue-and-white Swallow *Notiochelidon cyanoleuca*, specimen locations and bank-hole nesting together with **P. purpurea** noted; **Z**: 41, 113.

Common house swallow *Hirundo rustica* = Barn Swallow, flight of **Muscivora tyrannus** like a "caricature likeness of that of this swallow"; **Z**: 44.

Swallow[s] = unspecified, but likely Barn Swallow *Hirundo rustica* and mentioned only to note flight similar to that of Black Skimmer of the non-passerine family Laridae; **J**: 162.

CD states that a species "belonging to the American division of that genus [i.e., of swallows]" occurs as endemic on the Galapagos Islands, which is now known to be the Southern Martin *Progne modesta*; **J**: 461. See chapter 2.

One species extending into range of another in America to cause decline of the latter; **O**: 93.

The migratory instinct will cause them to "desert their tender young, leaving them to perish miserably in their nests" (see **house martin** and **swifts)**; **D**: 165, 173.

Hirundo, 3 sp. = Sand martin *Riparia riparia*, Barn Swallow *Hirundo rustica*, and Northern House Martin *Delichon urbica*, as inconspicuous British species that nests in holes, mentioned in discussing Wallace's objections to sexual selection for plumage colouration with regard to nesting; **D**: 693 footnote.

H. urbica = Common House Martin *Delichon urbicum*, plumage compared to that of **H. frontalis** and **H. leucopygia**; **Z**: 40.

house-martins = Common House Martin *Delichon urbicum*, the migratory instinct will cause them to "desert their tender young, leaving them to perish miserably in their nests" (see swallows and swifts); **D**: 165.

Sparrows deprive these birds of their nests; **D**: 622.

AEGITHALIDAE—**Bushtits**

long-tailed titmouse (*Parus caudatus*) = Long-tailed Tit *Aegithalos caudatus*, with hopping gait like **Synallaxis aegithaloides**; **Z**: 79.

Mecistura = Long-tailed Tit *Aegithalos caudatus*, as an inconspicuous British species that nests in a domed nest, mentioned in discussing Wallace's objections to sexual selection for plumage colouration with regard to nesting; **D**: 693 footnote.

PHYLLOSCOPIDAE—Leaf Warblers and allies

willow wren = Willow Warbler *Phylloscopus trochilus*, as foraging like **Synallaxis Tupinieri**; **J**: 301. See chapter 2.

ACROCEPHALIDAE—Reed Warblers and allies

Sedge Warbler = *Acrocephalus schoenobaenus*, its song compared with that of a mockingbird; **J**: 63. See chapters 1 and 3.

sedge-bird (*Motacilla salicaria*) = Sedge Warbler *Acrocephalus schoenobaenus,* its song compared to that of **Mimus patagonicus**; **Z**: 60. See chapter 1.

LOCUSTELLIDAE—Grassbirds and allies

Cincloramphus cruralis = Brown Songlark *Megalurus cruralis*, males are twice as large as females; **D**: 554–555.

SYLVIIDAE—Sylviid Babblers

Family Sylviadae, Sub-family Motacillinae = disused taxa, alluding to the Sylviidae and unrelated Motacillidae; **Z**: 83, 105. See chapter 3.

black-cap *Sylvia atricapilla* = Eurasian Blackcap, likened to Vermillion Flycatcher; **Z**: 45.

Migrant males invariably arrive on breeding grounds before females (see **nightingale** and **Ray's wagtail)**; **D**: 327.

Blackcap (*Sylvia atricapilla*) = Eurasian Blackcap, adult male has a black head and the female a reddish-brown one, and nestling sex distinguishable by this difference in them; **D**: 742. See chapter 7.

Sylvia, 3 sp. = unspecified, probably Garden Warbler *S. borin*, Eurasian Blackcap *S. atricapilla*, and Common Whitethroat *Sylvia communis*, as inconspicuous British species that "lay their eggs in holes or in domed nests" (which they do not, all making well-concealed open cup nests), mentioned in discussing Wallace's objections to sexual selection for plumage colouration with regard to nesting; **D**: 693 footnote.

white-throat (*Sylvia cinerea*) = Common Whitethroat *Sylvia communis*, brief description of male courtship flight display; **D**: 581.

REGULIDAE—Goldcrests, Kinglets

golden-crested wren = Goldcrest *Regulus regulus*, CD argues it is improbable that the duller crest of the female is for protection [at the nest]; **D**: 700.

TROGLODYTIDAE—Wrens

Troglodytes platensis = Sedge Wren *Cistothorus platensis*, locations, habitat, habits; **Z**: 75.

Common Troglodytes = Eurasian Wren *Troglodytes troglodytes,* nest site like that of **Troglodytes magellanicus**; **Z**: 74.

The obscure tints of both sexes have likely been "acquired and preserved" for protection; **D**: 722.

Both adult sexes and their young are of similar appearance; **D**: 735.

Anorthura = Eurasian Wren *Troglodytes troglodytes,* as an inconspicuous British species that nests in a domed nest, mentioned in discussing Wallace's objections to sexual selection for plumage colouration with regard to nesting; **D**: 693 footnote.

Kitty-wrens = Winter Wren *Troglodytes hiemalis*, build "cock-nests" to roost in (see **wrens**); **O**: 364.

A little, dusky-coloured wren (Scytalopus Magellanicus) = House Wren *Troglodytes aedon*, brief note on habits; **J2**: 237.

Troglodytes magellanicus = House Wren *Troglodytes aedon*, distribution, habitats, habits like *T. troglodytes*, nest building; **Z**: 74.

Troglodytes = unspecified, some anatomical characters agree with **Synallaxis maluroides**; **Z**: 152.

wrens (Troglodytes) = of North America build "cock-nests" to roost in (see **Kitty-wrens**); **O**: 364.

SITTIDAE—Nuthatches

nuthatch = unspecified, but doubtless the Eurasian Nuthatch *Sitta europaea*, in noting that the Great Tit *Parus major* breaks yew seeds like a nuthatch; **O**: 220. See chapter 4.

Climbing and power of; **O**: 355. See chapter 4.

Sitta = Eurasian Nuthatch *Sitta europaea*, as an inconspicuous British species that nests in a tree hole, mentioned in discussing Wallace's objections to sexual selection for plumage colouration with regard to nesting; **D**: 693 footnote.

Japanese nut-hatches = Eurasian Nuthatch *Sitta europaea*, a captive placing hazelnuts, too hard for it to crack, into water and this thought done in an attempt to soften them; **D**: 625.

blue nuthatch *Dendrophila frontalis* = Velvet-fronted Nuthatch *Sitta frontalis*, nestling males are distinguishable from females; **D**: 743 footnote.

CERTHIIDAE—Treecreepers

true creeper (Certhia familiaris) = Eurasian Treecreeper, nest and foraging compared with that of the **creeper (*Synallaxis Tupinieri*)**; **J**: 301. See chapters 2 and 4.

Merely briefly compared with **creeper (Oxyurus tupiniero)** in habits; **J2**: 237.

creeper = Eurasian Treecreeper *Certhia familiaris*, alluded to in noting that insects are typically found in its stomach (indicating CD had dissected one) in a form much like that found in hummingbirds; **J**: 331.

creeper (*Certhia familiaris*) = Eurasian Treecreeper, nest and foraging compared with that of the **creeper (Oxyurus tupiniero)**; **Z**: 81. See chapter 2.

Certhia = Eurasian Tree-creeper *Certhia familiaris*, as an inconspicuous British species that nests in a crevice, mentioned in discussing Wallace's objections to sexual selection for plumage colouration with regard to nesting; **D**: 693 footnote.

MIMIDAE—Mockingbirds, Thrashers

mocking-thrushes (*Mimus polyglottus*) = Northern Mockingbird, John Audubon observed that some remain all year at Louisiana, whereas others migrate, but the former always attack the latter upon their return; **D**: 627.

mocking bird (*Turdus polyglottus*) = Northern Mockingbird *Mimus polyglottus*, the sexes differ little, but males can be distinguished at a very early age because they show more pure white plumage; **D**: 742. See chapter 7.

Mimus thenca = Chilean Mockingbird, distribution, habits (as those of *M. patagonicus*), feeding on flowers, possibly for insects, nesting; **Z**: 61. See chapter 3.

  Turdus Thenca as a synonym of **Mimus thenca**; **Z**: 61.
  Orpheus Thenca as a synonym of **Mimus thenca**; **Z**: 61.

A "Calandria" mockingbird, *Opheus modulator* = Chalk-browed Mockingbird *Mimus saturninus modulator*, song described and compared to that of **Sedge Warbler** *Acrocephalus schoenobaenus* of the Sylviidae; habitat, habits, tameness; **J**: 62. See chapters 1 and 3.

Mimus = mockingbirds, seen with flower pollen on their heads; **F**: 371.

Mimus orpheus = Chalk-browed Mockingbird *Mimus saturninus*, distribution, habitat, habits (like those of *Xolmis*); **Z**: 60. See chapter 3.

  Orpheus Calandria = as a synonym of **Mimus orpheus**; **Z**: 60.
  Turdus Orpheus = as a synonym of **Mimus orpheus**; **Z**: 60.
  Mimus saturninus = as a synonym of **Mimus orpheus**; **Z**: 60.
  Orpheus modulator = as a synonym of **Mimus orpheus**; **Z**: 60.

Opheus Patagonica = Patagonian Mockingbird *Mimus patagonicus*, voice, habitat, habits (slightly different to **Mimus orpheus**); **J**: 63. See chapter 3.

  As a synonym of **Mimus orpheus**; **Z**: 60.

Mimus patagonicus = Patagonian Mockingbird, distribution, habits (slightly different to *M. Orpheus*), song; **Z**: 60–61. See chapter 3.

Orpheus parvulus = Galapagos Mockingbird *Mimus parvulus*, only found on Albemarle Island; **J**: 475. See chapter 2.

Mimus parvulus = Galapagos Mockingbird *Mimus parvulus*, described, illustrated, distribution, habits, all compared with other *Mimus* species; **Z**: 63–64.

Orpheus parvulus = as a synonym of **Mimus parvulus**; **Z**: 63.

Mimus trifasciatus = Floreana Mockingbird *Mimus trifasciatus*, described and illustrated; **Z**: 62. See chapter 3 and colour figure 3.4.

Only found on Charles Island; **J**: 475. See chapter 2 and colour figure 3.4.

Orpheus trifasciatus as a synonym of **Mimus trifasciatus**; **Z**: 62.

Orpheus trifasciatus = Floreana Mockingbird *Mimus trifasciatus*, exclusive to Charles Island, Galapagos; **J**: 475. See chapter 2 and colour figure 3.4.

Orpheus melanotis = Floreana Mockingbird *Mimus trifasciatus*, common to James and Chatham Islands; **J**: 475. See chapter 2 and colour figure 3.4.

Mimus melanotis = Floreana Mockingbird *Mimus trifasciatus*, described and illustrated; **Z**: 62. See chapter 3 and colour figure 3.4.

Orpheus melanotis = as a synonym of **Mimus melanotis**; **Z**: 62.

mockingbird[s] = unspecified, probably a *Mimus* species, abundant on grass plains around Maldonado; **J**: 60.

CD merely states that three new species, of "a genus common to both Americas," occur as endemic on the Galapagos Islands—these being now known as Floreana *Mimus trifasciatus*, San Cristobel *M. melanotis*, and Galapagos *M. parvulus* Mockingbirds (but a fourth species, the Hood Mockingbird *M. macdonaldi*, now acknowledged); **J**: 461, 474–475. See chapter 5 and colour figures 3.4 and 3.5.

See also **J2**: 394–395. See chapter 2.

mocking thrush(es) or mocking-thrush(es) = mockingbirds of the genus *Mimus*, three endemic species noted on the Galapagos Islands (now four); **J2**: 379, 380, 394–395.

Species on the Galapagos Islands well adapted to flying, yet differ from island to island [due to competitive exclusion]; **O**: 557. See chapters 3, 4, and 8.

CD found the song of birds on the Galapagos reminiscent of North American birds; **V1**: 9. See chapter 5.

Observed with pollen on their heads; **F**: 371. See chapter 7.

STURNIDAE—Starlings, Rhabdornis

Pastor = Rosy Starling *Pastor roseus*, as a British species with conspicuous females that nests in holes, mentioned in discussing Wallace's objections to sexual selection for plumage colouration with regard to nesting; **D**: 693 footnote.

Starling[s] = Common Starling *Sturnus vulgaris*, mentioned in passing, regarding South American birds thought by CD to be related; **J**: 60–61.

Seen at Angra, Terceira, Azores; **J**: 595.

Habits like those of **Leistes anticus**, **Agelaius chopi**, and **Molothrus niger**; **Z**: 107–109.

Allied to the nest parasites of **Molothrus**; **O**: 334. See chapter 2.

Unite in flocks with **rooks** and **jackdaws**; **D**: 153.

CD is assured that "it is somewhat common for three [adult] starlings to frequent the same nest; but whether this is a case of polygamy or polyandry has not been ascertained" (it is polygamous); **D**: 338.

One of a wild pair was shot twice on the same day and both were replaced that day, and this was the experience of another observer who shot 35 birds at the same nest in one season; **D**: 623, 626.

Sometimes "live in triplets"; **D**: 624.

Sturnus = Common Starling *Sturnus vulgaris*, an inconspicuous British species that nests in holes, mentioned in discussing Wallace's objections to sexual selection for plumage colouration with regard to nesting; **D**: 693 footnote.

TURDIDAE—Thrushes

ring-ouzel (*T. torquatus*) = Ring Ouzel *Turdus torquatus*, the female differs less from the male than is the case in the Blackbird; mentioned in discussing Wallace's objections to sexual selection for plumage colouration with regard to nesting; **D**: 693 footnote, 694. See also **blackbird** and **common thrush.**

blackbird or **common blackbird** = Common Blackbird *Turdus merula*, seen at Angra, Terceira, Azores; **J**: 595.

A monogamous species showing well-marked sexual differences (see **bullfinch**); **D**: 338.

Males "by far" more numerous than females; **D**: 383.

The beak is more brightly coloured in males than in females; **D**: 585.

Is far more pugnacious than the less colourful **thrush**; **D**: 609.

Macgillivray relates that a wild male with a female **thrush** produced hybrid offspring; **D**: 631.

The female plumage used as a "standard of degree of conspicuousness" in discussing Wallace's objections to sexual selection for plumage colouration with regard to nesting; **D**: 693.

In the young, the wing feathers [in males] become black after the others; **D**: 743.

Males are black with a bright yellow beak, and the females brown or mottled, doubtless through sexual selection; **D**: 749.

Blackbird (*Turdus merula*) = Common Blackbird *Turdus merula*, the female differs much from the male, mentioned in discussing Wallace's objections to sexual selection for plumage colouration with regard to nesting; **D**: 693 footnote. See also **ring-ouzel** and **common thrush.**

The sexes are identifiable in nestlings by characters found in the adults; **D**: 742. See chapter 7.

Species within a genus may include those with identical sexes or very different sexes; **D**: 702.

In "almost the whole large group," the young have their breasts spotted, a character retained throughout life by many species but lost in others, as in ***Turdus migratorius***; in many species, back feathers are mottled before lost with the first postnatal moult but are retained by "certain eastern species"; **D**: 709.

In the thrush family, there are an unusual number of cases in which nestlings can be sexed by one or more characters of their dimorphic parents' plumage; **D**: 742. See chapter 7.

MUSCICAPIDAE—Chats, Old World Flycatchers

Indian chats (Thamnobia) = *Copsychus* and/or related genera, males have geographically undergone intraspecific morphological change into distinguishable forms, but females and young have not; **D**: 714.

Fly-catchers (*Muscicapa grisola*) = Spotted Flycatcher *Muscicapa striata*, said to perch atop plants like **Lichenops perspicillatus**; **Z**: 52.

common fly-catcher (*Muscicapa grisola*) = Spotted Flycatcher *Muscicapa striata*, the sexes can hardly be distinguished contrary to those of the **pied fly-catcher**, but both build nests concealed in holes—mentioned in discussing Wallace's objections to sexual selection for plumage colouration with regard to nesting; **D**: 692.

Muscicapa, 2 sp. = Spotted Flycatcher *Muscicapa striata* and European Pied Flycatcher *Ficedula hypoleuca*, as inconspicuous British species that nest in a tree hole or crevice, mentioned in discussing Wallace's objections to sexual selection for plumage colouration with regard to nesting; **D**: 693 footnote.

Robin *Sylvia rubecula* = European Robin *Erithacus rubecula*, likened to Vermillion Flycatcher; **Z**: 45.

Its tame nature compared with that of the **Common Miner**; **Z**: 65.

Males are pugnacious each spring; **D**: 551.

Females may sing, especially widowed individuals (see **lark** and **bullfinch**); **D**: 566.

Males singing in autumn; **D**: 566. See chapters 6 and 8.

Captives fiercely attack all birds in the same cage that have any red plumage, one killing a **red-breasted crossbill** and nearly killing a **goldfinch**; **D**: 629.

Robin = European Robin *Erithacus rubecula*; in this species the adult sexes are alike, but the young of both sexes wear a peculiar first plumage, as is widespread within birds; **D**: 711. See chapter 7.

The adult sexes can hardly be distinguished, but the young are very differently plumaged; **D**: 733–734. See chapter 7.

Erithacus(?) = European Robin *Erithacus rubecula*, an inconspicuous British species that nests in holes, mentioned in discussing Wallace's objections to sexual selection for plumage colouration with regard to nesting; **D**: 693 footnote.

red-throated blue-breast (*Cyanecula suecica*) = Bluethroat *Luscinia svecica*, sexual dimorphism in breast markings described; **D**: 720.

nightingale = Common Nightingale *Luscinia megarhnchos*, in migrating the males invariably arrive on the breeding grounds before the females (see **blackcap** and **Ray's wagtail**); **D**: 327.

Males certainly attract females with song from a conspicuous perch; **D**: 564. See chapter 6.

Male song is "pleasing to the females"; **D**: 580.

A bird deprived of its nesting partner very quickly obtains another; **D**: 622.

pied fly-catcher (*M. luctuosa*) = European Pied Flycatcher *Ficedula hypoleuca*, the sexes are readily distinguished contrary to those of the **common fly-catcher**, but both build nests concealed in holes—mentioned in discussing Wallace's objections to sexual selection for plumage colouration with regard to nesting; **D**: 692.

fly-catchers = unspecified, species within a genus may include those with identical or very different sexes; **D**: 702.

redstart = Common Redstart *Phoenicurus phoenicurus*, a bird deprived of its nesting partner very quickly obtains another; **D**: 622.

Ruticilla, 2 sp. = Common Redstart *Phoenicurus phoenicurus* and Black Redstart *P. ochruros*, inconspicuous British species that nest in holes—mentioned in discussing Wallace's objections to sexual selection for plumage colouration with regard to nesting; **D**: 693 footnote.

Ruticilla = unspecified, young fertile females may acquire the characters of the males; **D**: 704.

Monticola cyanea = Blue Rock Thrush *M. solitarius*, the male of this desert-dweller is conspicuous bright blue and the female a conspicuous mottled brown and white—contrary to the typically cryptic plumages of birds in this habitat; **D**: 695.

Rock thrush (*Petrocincla cyanea*) = Blue Rock Thrush *Monticola solitarius*, males have much plumage blue, whereas females are brown; and nestling males have their main wing and tail feathers edged blue, whereas those of females are edged brown; **D**: 742–743.

forest-thrush (*Orocetes erythrogastra*) = Chestnut-bellied Rock Thrush *Monticola rufiventris*, males have much plumage blue, whereas females are brown; and nestling males have their main wing and tail feathers edged blue, whereas those of females are edged brown; **D**: 742–743.

whinchat (*Motacilla rubetra*) = Whinchat *Saxicola rubetra*, in certain habits compared to **Muscisaxicola mentalis**; **Z**: 83.

whinchats (Saxicolae) = *Saxicola rubetra*, newly arrived migrants to south English coast had seeds in mud attached to their legs and were thus potential plant dispersers; **O**: 512. See chapter 3.

Saxicola = unspecified, but Whinchat *S. rubetra* and/or European Stonechat *S. rubicola*, as inconspicuous British species that nests in holes—mentioned in discussing Wallace's objections to sexual selection for plumage colouration with regard to nesting; **D**: 693 footnote.

stonechat *Saxicola rubicola* = European Stonechat *Saxicola rubicola*, the sexes are distinguishable at a very early age; **D**: 743 footnote.

Fruticola, 2 sp. = European Stonechat *Saxicola rubicola*, cited as an inconspicuous British species that nests in holes—mentioned in discussing Wallace's objections to sexual selection for plumage colouration with regard to nesting; **D**: 693 footnote.

Dromolæa = *Oenanthe*, two unspecified desert-dwelling wheaters *Oenanthe* are conspicuously black, contrary to the typically cryptic plumages of birds in this habitat; **D**: 695.

wheatears = Northern Wheatear *Oenanthe oenanthe*, newly arrived migrants to south English coast had seeds in mud attached to their legs and were thus potential plant dispersers; **O**: 512. See chapter 3.

CINCLIDAE—**Dippers**

water–ouzel = White-throated Dipper *Cinclus cinclus*, foraging habits; **O**: 222–223, 646.

The sexes differ about as much as do open nesting **ring-ouzel**; but dippers build a domed nest—mentioned in discussing Wallace's objections to sexual selection for plumage colouration with regard to nesting; **D**: 694.

Cinclus = White-throated Dipper *Cinclus cinclus*, an inconspicuous British species that makes a domed nest—mentioned in discussing Wallace's objections to sexual selection for plumage colouration with regard to nesting; **D**: 693 footnote.

NECTARINIIDAE—**Sunbirds**

Nectarinidae [*sic*] = members of this family seen fertilizing *Strelitzia* plants; **F**: 371. See chapter 7.

honey-suckers (Nectariniæ) of India = unspecified Nectariniidae, some "have a double, whilst others have only a single annual moult"; **D**: 597.

honey-suckers = unspecified, males of some species have geographically undergone intraspecific morphological change into distinguishable forms but not females and young; **D**: 714.

Sun-birds (Nectariniæ) = unspecified, males are colourful, and the dull females incubate in a domed nest—mentioned in discussing Wallace's objections to sexual selection for plumage colouration with regard to nesting; **D**: 692.

PASSERIDAE—**Old World Sparrows, Snowfinches, and allies**
English sparrow = presumably the House Sparrow *Passer domesticus*, **Zonotrchia matutina** being mentioned as similarly domesticated about human dwellings; **Z**: 91.

CD says it occupies in European ornithology the place **Zonotrichia ruficollis** does in South America; **Z**: 108.

common sparrow = presumably the House Sparrow *Passer domesticus*, whereas males are dull, they differ much from females through sexual selection, CD concludes; **D**: 748.

Sparrow(s) = unspecified, but doubtless House (*Passer domesticus*) and/or Eurasian Tree (*Passer montanus*) Sparrow, in stating males are pugnacious each spring; **D**: 551.

Have learned to sing like a **linnet**; **D**: 567.

When one of a nesting pair was shot, the survivor would procure another mate; **D**: 622.

Species within a genus may include those with identical sexes or very different sexes; **D**: 702.

The females compared with those of other **British finches** differ in the characters in which they partially resemble their respective males (which "may be safely attributed to sexual selection"); **D**: 718.

house sparrow (*Passer domesticus*) = males differ much from females, whereas in the **tree-sparrow** hardly at all, yet both build well-concealed nests—in discussing Wallace's objections to sexual selection for plumage colouration with regard to nesting; **D**: 693.

Young of both sexes closely resemble adult females, whereas adult males differ conspicuously; **D**: 712.

Although both sexes are dull, males differ much from females and are probably attractive to them; **D**: 723.

Males differ much from the similar [adult] females and young, which are largely like both sexes and young of ***P. brachydactylus*** and some allied species; and CD assumed that the female and young of the house-sparrow exhibit the approximate appearance of the *Passer* progenitor; **D**: 737.

One learned the song of a **linnet,** and it is possible the progenitor of the sparrow may have been a songster; **D**: 869.

Passer, 2 species = ***Passer domesticus*** and ***P. montanus*** (House and Eurasian Tree Sparrow), inconspicuous British birds that nest in holes or domed nests, mentioned in discussing Wallace's objections to sexual selection for plumage colouration with regard to nesting; **D**: 693 footnote.

Passer hispaniolensis = Spanish Sparrow, one specimen from St. Jago, Cape Verde Islands; **Z**: 95.

Fringilla Hispaniolensis = as a synonym of **Passer hispaniolensis**; **Z**: 95.

Passer jagoensis = Iago Sparrow *Passer iagoensis*, described, illustrated, habitat, habits, nesting; **Z**: 95.

Pyrgita Jagoensis = as a synonym of **Passer jagoensis**; **Z**: 95.

tree-sparrow (*P. montanus*) = Eurasian Tree Sparrow, males hardly differ from females, whereas in the **house-sparrow,** the sexes differ much; yet both build well-concealed nests—in discussing Wallace's objections to sexual selection for plumage colouration with regard to nesting; **D**: 693.

Both sexes and young closely resemble male house sparrows, giving CD cause to think that they have all been modified in the same manner and all departing from the typical colouring of their early progenitor; **D**: 737. See ***P[asser]. Brachydactylus***

P[asser]. Brachydactylus = Pale Rockfinch *Carpospiza brachydactyla*, males of the **house-sparrow** differ much from the similar [adult] females and young, which are largely like both sexes and young of the Pale Rock-Finch and some allied species; and CD assumed that the female and young of the House Sparrow exhibit the approximate appearance of the *Passer* progenitor; **D**: 737.

PLOCEIDAE—Weavers, Widowbirds

weaver-bird (Ploceus) = weavers of the large genus *Ploceus*, in captivity "amuse" themselves "by neatly weaving blades of grass between the wires" of their cage; **D**: 56.

African weaver (*Ploceus*) = unspecified, small parties congregate in the breeding season and display for hours; **D**: 618.

Little black-weavers (Ploceus?) = possibly Vieillot's Black Weaver *P. nigerrimus*, on the west coast of Africa small parties congregate on bushes round a small open space to sing and glide through the air with quivering wings that produce a rapid "whirring sound like a child's rattle"; **D**: 574, 581. See chapter 6.

Vidua axillaris = Fan-tailed Widowbird *Euplectes axillaris*, mention as a "see also" citation to a publication dealing with the species regarding **widow-bird**; **D**: 338.

widow-bird = Long-tailed Widowbird *Euplectes progne*, "seems to be a polygamist" (and it is); **D**: 338.

widow-bird (*Chera progne*) = Long-tailed Widowbird *Euplectes progne*, in Natal males acquire their fine plumage and long tail in December or January and lose them in March; **D**: 598.

Females disown males lacking their long tail feathers; **D**: 639. See chapter 6.

widow-birds (Vidua) = unspecified *Euplectes*, the extremely long tail feathers of Southern Africa males "render their flight heavy"; but as soon as they moult them, they fly as well as females; **D**: 613.

Vidua = unspecified, *Euplectes* or *Vidua* males acquire such long tail feathers when breeding as to impede their flight; **D**: 705.

ESTRILDIDAE—**Waxbills, Munias and allies**

amadavat (*Estrelda amandava*) = Red Avadavat *Amandava amandava*, Bengali people cause captive males to fight in the presence of a female; **D**: 560.

Amadina Lathami = Diamond Firetail *Stagonopleura guttata*, appearance and courtship display described; **D**: 611.

Amadina castanotis = Zebra Finch *Teaniopygia guttata*, appearance and courtship display described; **D**: 611–612.

PRUNELLIDAE—**Accentors**

Hedge-warbler(s) = Dunnock *Prunella modularis*, cites the **Common Cuckoo** laying in the nest of this species; **O**: 333.

Both adult sexes and their young are of similar appearance; **D**: 734. See chapter 7.

See **Kingfisher(s)**.

hedge-sparrow = Dunnock *Prunella modularis*, likened to the Vermillion Flycatcher; **Z**: 45.

hedge-warbler (*Accentor modularis*) = Dunnock *Prunella modularis*, the obscure tints of both sexes have likely been "acquired and preserved" for protection; **D**: 722.

MOTACILLIDAE—**Wagtails, Pipits**

Ray's wagtail (*Budytes Raii*) = Western Yellow Wagtail *Motacilla flava*, migrant males arrive on breeding grounds before females (see **nightingale** and **blackcap**); **D**: 327.

water-wagtail = Grey Wagtail *Motacilla cinerea*, seen at Angra, Terceira, Azores; **J**: 595.

Motallica boarula(?) = Grey Wagtail *Motacilla cinerea*, an inconspicuous British species that nests in holes—mentioned in discussing Wallace's objections to sexual selection for plumage colouration with regard to nesting; **D**: 693 footnote.

Motacilla alba = White Wagtail, a British species with conspicuous females that nest in holes—mentioned in discussing Wallace's objections to sexual selection for plumage colouration with regard to nesting; **D**: 693 footnote.

wagtails (Motacillae) = unspecified, newly arrived migrants to south English coast had seeds in mud attached to their legs and were thus potential plant dispersers; **O**: 512. See chapter 3.

Indian wag-tails or Motacillæ = species of *Motacilla*, the sexes are "nearly alike"; and some closely allied species are easily distinguished in summer but not so in winter and immature plumage; **D**: 714–715.

Anthus chii = Yellowish Pipit *Anthus lutescens*, a specimen from Rio de Janeiro; **Z**: 85.

Anthus furcatus = Short-billed Pipit, said to be most closely allied in habits and structure to the Wood Lark *Lullula arborea*, but eggs very different; habits, nest, egg; **Z**: 85.

Titlark, *Anthus* = Correndera Pipit *Anthus correndera*, CD collected this bird in the Falkland Islands and stated that in "Georgia, lat. 54° to 55°," it was the only land bird and is a species that "inhabits a more inhospitable region than any other terrestrial animal" (in fact, the bird on South Georgia is the closely related South Georgia Pipit *Anthus antarcticus*). CD wrote "This bird, if undescribed, certainly well deserves the name of antarcticus"; **J** 273.

Anthus correndera = Correndera Pipit *Anthus correndera*, distribution, habits, said very closely allied to the similar Meadow Pipit *Anthus pratensis*; **Z**: 85.

Pipit (*aenthus chu*) = Correndera Pipit *Anthus correndera*, as a host of the parasitic cowbird; **1881** paper.

pipits (Anthus) = some sub-genera have a double, whereas others have only a single, annual moult; **D**: 597.

FRINGILLIDAE—**Finches**

Family Fringillidae, Sub-family Alaudinae = the Alaudinae alludes to the true larks, which now constitute the family Alaudidae and are removed from the Fringillidae; **Z**: 87.

Fringillidae = a large number of British species kept by a Mr. Weir, who provides notes to CD; **D**: 610.

Females are generally similar to those of allied species, whereas the males differ dramatically; **D**: 717.

The Canary has been hybridized with nine or ten other species of Fringillidae; **V1**: 295.

Sub-family Pyrrhulinae = a disused taxon, now absorbed into the family Fringillidae; **Z**: 88.

Sub-family Fringillinae = a taxon indicating a sub-family of the Fringillidae; **Z**: 90.

Chaffinch = Common Chaffinch *Fringilla coelebs*, seen at Angra, Terceira, Azores; **J**: 595.

Males in "large excess" to females; **D**: 382.

The sexes kept apart, and females were far more numerous than males on Corfu and Epirus; **D**: 383.

Using a captive "call bird," 50 and 70 wild males were caught at different times; **D**: 565.

Male courtship display described; **D**: 610.

A bird deprived of its nesting mate very quickly obtains another; **D**: 622.

CD argues it is improbable that the duller head of the female is for protection [at the nest]; **D**: 700.

Females compared with those of other **British finches** differ in the characters in which they partially resemble their respective males (which "may be safely attributed to sexual selection"); **D**: 718.

gros-beak = unspecified, most likely Pine Grosbeak *Pinicola enucleator,* merely mentioned with respect to a species of finch on the Galapagos Islands with a similar beak; **J**: 462.

bullfinch (Loxia Pyrrhula) = Eurasian Bullfinch *Pyrrhula pyrrhula*, "Manners" like those of **Phytotomara rara**; **Z**: 106.

A monogamous species showing well-marked sexual differences (see **blackbird**); **D**: 338.

One taught to "pipe a German waltz" (see **linnet** and **canary**); **D**: 564.

Females may sing, especially widowed individuals (see **robin** and **lark**); **D**: 566.

This and the **Goldfinch** are the only colourful British "best songsters," the rest being "plain-coloured"; **D**: 568.

Male courtship display described; **D**: 610.

At Blackheath, a Mr. Jenner Weir never sees or hears wild individuals; and yet when one of his captive females is widowed, a wild male approaches closely in response to her soft calls within days; **D**: 623.

Captives distinguish different individual people and exhibit affection to them; **D**: 627.

A black-headed **reed-bunting** introduced into an aviary of mixed birds was attacked by a bullfinch (which is black headed) that had previously ignored a reed-bunting lacking the black head; **D**: 628–629.

In captivity, a newly introduced female attacked a pair-bonded one and replaced her only to subsequently be replaced by the original "wife"; **D**: 640.

Has bred with the **Canary** in captivity, as it has with its own kind; **V2**: 154.

bullfinch(s) = Eurasian Bullfinch *Pyrrhula pyrrhula*, the female plumage used as a "standard of degree of conspicuousness" in discussing Wallace's objections to sexual selection for plumage colouration with regard to nesting; **D**: 693.

CD argues it is improbable that the duller breast of the female is for protection [at the nest]; **D**: 700.

The females compared with those of other **British finches** differ in the characters in which they partially resemble their respective males (which "may be safely attributed to sexual selection"); **D**: 718.

Bird-fanciers pull breast feathers from nestlings to sex them by the underfeather colour; **D**: 738.

Pigeon fanciers compare the head and beak of this bird to that of the Barb pigeon breed; **V1**: 146.

Feeding them hemp-seed causes this and some other birds to become black; **V2**: 280. See chapter 5.

Bullfinches (*Pyrrhula vulgaris*) = Eurasian Bullfinch *Pyrrhula pyrrhula*, injure fruit trees by eating flower-buds; **V2**: 232.

Pyrrhula = unspecified, in confinement, a male assumed the female plumage of his species; **V2**: 158.

greenfinch = European Greenfinch *Chloris chloris*; see **canary**.

CD argues it is improbable that the duller green of the female is for protection [at the nest]; **D**: 700.

The females compared with those of other **British finches** differ in the characters in which they partially resemble their respective males (which "may be safely attributed to sexual selection"); **D**: 718.

See *Loxia*.

twite (*Linaria montana*) = Twite *Linaria flavirostris*, CD informed that the sexes about equal in this species population; **D**: 383.

Linaria = unspecified, young fertile females may acquire the characters of the males; **D**: 704.

common linnet = Common Linnet *Linaria cannabina*, CD informed that the females predominate greatly in numbers over males, but migration may influence this impression; **D**: 383.

Birds listened intently to a **Bullfinch** taught to "pipe a German waltz" (see **canary**); **D**: 564.

Sparrows have learned to sing like them; **D**: 567.

In England, the crimson forehead and breast are worn only in summer, whereas in Madeira, they are worn year round; **D**: 600.

Is far less pugnacious than the more colourful **goldfinch**; **D**: 609.

Male courtship display described; **D**: 610.

See **canary** and ***Loxia***.

common linnet (*Linota cannabina*) = Common Linnet *Linaria cannabina*, when caged, males do not acquire the fine crimson breast colour; **V2**: 158.

linnet = Common Linnet *Linaria cannabinai*, a **house-sparrow** learnt the song of this species; **D**: 869.

redpole = Common Redpoll *Acanthis flammea*, see ***Loxia***.

Loxia = the genus for the five crossbill species, anatomically like **Phytotoma rara**; **Z**: 154.

red-breasted crossbill = Scottish *Loxia scotica* or Red *L. curvirostra* Crossbill, one killed by a fellow captive **robin** because of its red plumage; **D**: 629.

crossbill = unspecified, the females compared with those of other **British finches** differ in the characters in which they partially resemble their respective males (which "may be safely attributed to sexual selection"); **D**: 718.

Unspecified, but Red Crossbill *L. curvirostra* young initially have straight bills and striated plumage in which they resemble the mature **redpole**, female **siskin**, young **goldfinch**, **greenfinch**, and some allied species; **D**: 708–709. See chapter 7.

goldfinch = European Goldfinch *Carduelis carduelis*, CD informed that the sexes about equal in this species population; **D**: 383.

Exhibits bright plumage year round, unlike ***Fringilla tristis***; **D**: 600.

Is far more pugnacious than the less colourful **linnet**; **D**: 609.

Male courtship display described; **D**: 610. See chapter 6.

Females may attend to each detail of the male's plumage during his courtship; **D**: 643. See chapter 6.

The female plumage used as a "standard of degree of conspicuousness" in discussing Wallace's objections to sexual selection for plumage colouration with regard to nesting; **D**: 693.

The females compared with those of other **British finches** differ in the characters in which they partially resemble their respective males (which "may be safely attributed to sexual selection"); **D**: 718.

See *Loxia*.

Goldfinch (*Carduelis elegans*) = European Goldfinch *Carduelis carduelis*, sexes identifiable by difference in beak length; **D**: 551.

This and the **Bullfinch** are the only colourful British "best songsters," the rest being "plain-coloured"; **D**: 568.

One nearly killed by a fellow captive **robin** because of its red plumage; **D**: 629.

See canary.

a hybrid goldfinch = European Goldfinch *Carduelis carduelis* hybrid (i.e., crossed with a canary or some other species) would become "a ball of ruffled feathers" if approached too closely; **E**: 100.

canary-bird(s), **canary** or **canaries** = Atlantic Canary *Serinus canaria*, noted as crossed with 9 or 10 different finch species; **O**: 373. See chapter 5.

Wild birds pair monogamously, but breeders "successfully put one male to four or five females"; **D**: 339.

A captive variety exhibits a change of colour rather late in life, but in both sexes; **D:** 368.

Mr. Bechstein asserts that the female "always chooses the best singer"; **D**: 564.

Birds listened intently to a **Bullfinch** taught to "pipe a German waltz" (see **linnet**); **D**: 564.

A captive sterile hybrid sang to its mirrored reflection and attacked it and a female placed with it; **D**: 565.

Captives distinguish different individual people and exhibit affection to them; **D**: 627.

To best hybridize this with a **Siskin**, it is best to place birds of the same colour tint together; **D**: 633.

A female placed into an aviary containing male linnets, goldfinches, siskins, greenfinches, chaffinches, and other birds paired and produced offspring with the [most similarly coloured] greenfinch; **D: 633.**

Has been hybridized with nine or ten other species of Fringillidae; **V1**: 295; or with "more than a dozen species"; **V2**: 154. See chapter 5.

Offspring of some paired birds with a particular colour or crest do not inherit the character; **V2**: 21–22.

Prize birds have black wings and tail, but these last only until their first moult; **V2**: 77.

Breed freely in confinement; **V2**: 154.

It "was long before the canary-bird was fully fertile [as captives in the UK], and even now first-rate breeding birds are not common"; **V2**: 161.

As long ago as 1780–1790, rules were established for breeding captives to a standard of perfection; **V2**: 195. See also chapter 5.

Of the sub-varieties of the Belgian breed fanciers always select for extremes; **V2**: 240.

Feather-footed forms of the domesticated canary alluded to; **V2**: 349. See chapter 8.

Few would maintain the species was created with a tendency to vary only after ages so that contemporary fanciers could then select for curious breeds; **V2**: 406–407.

See chapter 8.

Fringilla tristis = American Goldfinch *Spinus tristis*, exhibits bright plumage only after winter, unlike the **goldfinch**; **D**: 600.

goldfinch of North America *Fringilla tristis* = American Goldfinch *Spinus tristis*, the young "generally resemble the females"; **D**: 740, 741 footnote. See chapter 7.

siskin = Eurasian Siskin *Spinus spinus*, retains bright plumage year round, unlike ***Fringilla tristis***; **D**: 600.

Male courtship display described; **D**: 610.

See canary and *Loxia*.

Siskin (*Fringilla spinus*) = Pine Siskin *Spinus pinus*, to best hybridize this with a **canary**, it is best to place birds of the same colour tint together; **D**: 622.

Has raised offspring in captivity; **V2**: 154.

Chrysometris campestris = Black-chinned Siskin *Spinus barbata*, described, habitat; **Z**: 89.

Fringilla campestris = as a synonym of **Chrysometris campestris**; **Z**: 89.

Chrysomitris magellanica = Hooded Siskin *Spinus magellanica*, abundant in large flocks in May at Maldonado, and also seen a Rio Negro; **Z**: 97.

Fringilla Magellanica = as a synonym of **Chrysomitris magellanica**; **Z**: 97.

Fringilla icterica = as a synonym of **Chrysomitris magellanica**; **Z**: 97.

British finches = of the **Fringillidae** plus the "sparrow" in which the sexes differ either very slightly or considerably; **D**: 718.

Fringilla cyanea = Blue-naped Chlorophonia *Chlorophonia cyanea*, males take three years to acquire full adult plumage; **D**: 738 footnote.

PARULIDAE—New World Warblers

Trichas velata = Masked Yellowthroat *Geothlypis aequinoctialis*, a specimen from Maldonado; **Z**: 87.

Sylvia velata = as a synonym of **Trichas velata**; **Z**: 87.

Tanagra canicapilla = as a synonym of **Trichas velata**; **Z**: 87.

Trichas canicapilla = as a synonym of **Trichas velata**; **Z**: 87.

sylvicola = Mangrove Warbler *Setophaga petechia,* present on Galapagos Islands but "an American form, and especially common in the northern division of the Continent"; **J**: 461. See chapter 2.

Listed among birds tame on the Galapagos Islands; **J**: 475.

Sylvicola aureola = Mangrove Warbler *Setophaga petechia*, described, illustrated, habits on Galapagos; **Z**: 86.

wren = CD mentions a wren "with a fine yellow breast" as one of only two "brilliantly coloured" birds on the Galapagos Islands; and although it is unclear what he refers to, it is probably the Mangrove Warbler *Setophaga petechia*; **J2**: 381.

redstart of America (***Muscapica ruticilla***) = American Redstart *Setophaga ruticilla*, occasionally breed in immature plumage; **D**: 739 footnote.

ICTERIDAE—**Oropendolas, Orioles, and Blackbirds**

Dolichonyx oryzivorus = Bobolink, a specimen taken on James Island, Galapagos; **Z**: 106.

Emberiza oryzivorus = as a synonym of **Dolichonyx oryzivorus**; **Z**: 106.

Lark-like finch from North America (*Dolichonyx oryzivorus*) = Bobolink, the only one of 26 land birds CD recorded on the Galapagos Islands as not endemic to them; **J2**: 378 (but see **J2:** vii). See chapter 2.

Sturnella ludoviciana = Eastern Meadowlark *S. magna*, males fight fiercely, but at the sight of a female, all pursue her; **D**: 562.

Sturnella militaris = Red-Breasted Blackbird, locations found; **Z**: 110.

Sturnus militaris = as a synonym of **Sturnella militaris**; **Z**: 110.

Starling (*Sturnella bellicosa*) = Peruvian Meadowlark, as host of parasitic cowbirds; **1881** paper.

Amblyramphus ruber = Yellow-billed Cacique *Amblycercus holosericeus*, habitat, habits, call, flight; **Z**: 109–110.

Oriolus ruber = as a synonym of **Amblyramphus ruber**; **Z**: 109.

Amblyramphus bicolour = as a synonym of **Amblyramphus ruber**; **Z**: 109.

Sturnus pyrrhocephalus = as a synonym of **Amblyramphus ruber**, **Z**: 109.

Sturnella rubra = as a synonym of **Amblyramphus ruber**, **Z**: 109.

Leistes erythrocephala = as a synonym of **Amblyramphus ruber**, **Z**: 109.

Cassicus = a genus, now spelt *Cacicus*, of New World Blackbirds; **J**: 60.

Icterus = unspecified American oriole, mentioned as laying six eggs a clutch, in discussing cowbirds as nest parasites; **1881** paper.

Starling (or *Icterus*) = one New World Blackbird species found in more open parts of Tierra del Fuego; **J**: 301. See chapter 2.

Agelaeus phoeniceus = Red-winged Blackbird *Agelaius phoeniceus*, the sexes are differently affected by the north–south variation in plumage colour that typically affects both sexes in the same way; **D**: 347.

red-winged starling (*Agelœus phœniceus*) = Red-winged Blackbird *Agelaius phoeniceus*, the female is pursued by several courting males from which she "makes a choice"; **D**: 634.

Molothrus niger = Shiny Cowbird *Molothrus bonariensis,* extensive notes on flocking [often with **Leistes anticus**], habits, calls, cuckoo-like nest parasitism (possibly of Rufous-collared Sparrow), closely allied to ***M. pecoris***, eggs, association with cattle; **Z**: 107–109; **J**: 475. See chapters 2, 3, and 8.

Tanagra bonariensis = as a synonym of **Molothrus niger**; **Z**: 107.
Icterus niger = as a synonym of **Molothrus niger**; **Z**: 107.
Passrina discolor = as a synonym of **Molothrus niger**; **Z**: 107.
Icterus maxillaris = as a synonym of **Molothrus niger**; **Z**: 107.
Icterus sericeus = as a synonym of **Molothrus niger**; **Z**: 107.
Psarocolius = as a synonym of **Molothrus niger**; **Z**: 107.

Molothrus bonariensis = Shiny Cowbird, its nest parasite behaviour and nidification discussed and compared to that of congeners and Common Cuckoo; **O**: 335–336. See chapters 2, 3, 4, and 8.

Molothrus perpurascens = Shiny Cowbird *Molothrus bonariensis*, egg laying intervals and their significance and compared to those of cuckoos; **1881** paper.

Molothrus pecoris = Brown-headed Cowbird *Molothrus ater,* a North American nest parasite similar in some respects to **Le Troupiale commum**; **J**: 61. See chapters 2 and 4.

Molothrus pecoris = Brown-headed Cowbird *Molothrus ater*, noted as closely allied to **Molothrus niger**; **Z**: 107–109.

Molothrus pecoris = Brown-headed Cowbird *Molothrus ater*, its nest parasite behaviour and nidification; **O**: 335.
Noted as closely allied to **Molothrus niger**; **Z**: 107–109. See chapter 3.

Molothrus = the genus of cowbirds, of the New World Blackbirds; **J**: 60–62. See chapter 2.

Molothrus = cowbirds, their parasitic habits discussed and compared with cuckoos; **1881** paper. See chapter 7.

Comparative notes on the nest parasite behaviour and nidification of the Common Cuckoo; **O**: 334–336. See chapter 3.

Le Troupiale commum = unspecified cowbird = Brown-headed Cowbird *Molothrus ater*, habitat, flocking, habits, call, a nest parasite; **J**: 60–61. See chapter 2.

Quiscalus major = Boat-tailed Grackle, the sexes are differently affected by the north–south variation in plumage colour that typically affects both sexes in the same way; **D**: 347.

Mr. G. Taylor found "very few females in proportion to the males" in Florida but the opposite in Honduras, where they behaved as "polygamists"; **D**: 383.

Agelaius chopi = Austral Blackbird *Curaeus curaeus*, common in flocks on pastures of Chile, noisy and runs like "our starling" (Common Starling *Sturnus vulgaris*), can be taught to speak and sometimes caged, nests in bushes; **Z**: 107.

Turdus curaeus = as a synonym of **Agelaius chopi**; **Z**: 107.
Icterus unicolour = as a synonym of **Agelaius chopi**; **Z**: 107.
Icterus sulcirostris = as a synonym of **Agelaius chopi**; **Z**: 107.

Oriolus ruber = Scarlet-headed Blackbird *Amblyramphus holosericeus*, CD mentions this as representing an exception in colouration for its group "in having its head, shoulders, and thighs of the most splendid scarlet. This bird differs from its congeners in being solitary. It frequents marshes . . ." (from description, I think he can only have meant this species, although its shoulders are actually black); **J**: 60. See chapter 2.

Agelaius fringillarius = Baywing *Agelaioides badius,* rare at Maldonado, more common on Parana River; **Z**: 107.

Icterus fringillarius = as a synonym of **Agelaius fringillarius**; **Z**: 107.
Psarocolius sericeus = as a synonym of **Agelaius fringillarius**; **Z**: 107.

Molothrus badius = Baywing *Agelaioides badius,* an occasional flocking, and also nest parasite, species, its nidification discussed and compared to that of Common Cuckoo; **O**: 334–336. See chapter 3.

Xanthornus chrysopterus = Yellow-winged Blackbird *Agelasticus thilius,* range, locations, habitat, nest site; **Z**: 106.

Oriolus cayennensis = as a synonym of **Xanthornus chrysopterus**; **Z**: 106.

Agelaius chrysopterus = as a synonym of **Xanthornus chrysopterus**; **Z**: 106.

Psarocolius chrysopterus = as a synonym of **Xanthornus chrysopterus**; **Z**: 106.

Xanthornus flavus = Saffron-cowled Blackbird *Xanthopsar flavus*, illustrated, common in large flocks at Maldonado; **Z**: 107.
Oriolus flavus = as a synonym of **Xanthornus flavus**; **Z**: 107.
Psarocolius flaviceps = as a synonym of **Xanthornus flavus**; **Z**: 107.

Leistes anticus = Brown-and-yellow Marshbird *Pseudoleistes virescens*, very abundant in large flocks on grassy plains of La Plata, sometimes with **Molothrus niger**, habits like **starling** (= Common Starling *Sturnus vulgaris*); **Z**: 107; **J**: 475. See chapter 3.
Icterus anticus = as a synonym of **Leistes anticus**; **Z**: 107.
Agelaius virescens = as a synonym of **Leistes anticus**; **Z**: 107.
Psarocolius anticus = as a synonym of **Leistes anticus**; **Z**: 107.

EMBERIZIDAE—**Buntings, New World Sparrows, and allies**
Emberizidae = a large number of British species were kept by a Mr. Weir, who provides notes on them to CD; **D**: 610.

Sub-family Emberizinae = of the Emberizidae; **Z**: 89.

Sub-family tanagrinae = a disused taxon, once placed within the Thraupinae of the Emberizidae (Bock 1994: 155) and subsequently in the Emberizinae of the Fringillidae (Sibley and Monroe 1990: xv; Gill and Donsker 2015); **Z**: 97.

Emberiza = unspecified, noted to have been tame on the Falkland Islands; **J**: 477.

buntings (Emberiza) = the young of many species resemble one another as do adult *E. miliaria*; **D**: 709.

British Buntings = most "are plain coloured birds"; **D**: 611.

common bunting, *E. miliaria* = Corn Bunting *Emberiza calandra*, see **buntings (Emberiza)**.

reed-bunting (Emberiza schœniculus) = Common Reed Bunting, the head feathers of males become black by the abrasion of the dusky tips, and these are erected in display; **D**: 611.

A black-headed individual introduced into an aviary of mixed birds was attacked by a **bullfinch** (which is black headed) that had previously ignored a reed bunting lacking the black head; **D**: 628–629.

Bunting (*Emberiza passerina*) = Common Reed Bunting *Emberiza schœniculus*, loses the black head in captivity; **V2**: 158.

Zonotrichia matutina = Rufous-collared Sparrow *Zonotrichia capensis*, domesticated about humans like the **English sparrow**, locations, habitats, habits, nest, eggs of a parasitic **Molothrus** species found in a clutch; **Z**: 91.

Habits similar to those of **Fringilla diuca**; **Z**: 93.

Fringilla matutina = as a synonym of **Zonotrichia matutina**; **Z**: 91.

Tanagra ruficallis = as a synonym of **Zonotrichia matutina**; **Z**: 91.

Zonotrichia canicapilla = Rufous-collared Sparrow *Zonotrichia capensis*, described, locations, habitat, nest, egg; **Z**: 91–92.

Zonotrichia ruficollis = Rufous-collared Sparrow *Zonotrichia capensis*, brief mention as possibly having eggs laid in its nest by **Molothrus niger**; **Z**: 108.

white-crowned sparrow (*Fringilla leucophrys*) = *Zonotrichia leucophrys*, young have white head stripes as soon as fledged, which are lost by both young and old in winter; **D**: 741.

sparrow, a *Zonotrichia* = one of the "fox-sparrows," also treated as of the genus *Passerella*; **J**: 61.

As host to a parasitic cowbird, **1881** paper. See chapter 2.

Ammodramus manimbe = Grassland Sparrow *Ammodramus humeralis*, one specimen, illustrated, from Maldonado; **Z**: 90.

Ammodramus xanthornus = as a synonym of **Ammodramus manimbe**; **Z**: 90.

Fringilla Manimbè = as a synonym of **Ammodramus manimbe**; **Z**: 90.

Emberiza Manimbè = as a synonym of **Ammodramus manimbe**; **Z**: 90.

Ammodramus = a genus of American sparrow, which CD briefly and ambiguously mentions in an Addenda to note that, contrary to what he stated on page 461, one species was not peculiar to the Galapagos Islands but is also known from North America; **J**: 628.

Zonotrichia strigiceps = Stripe-capped Sparrow *Rhynchospiza strigiceps*, described, CD notes it geographically replaces **Z. matutina**, which in turn replaces **Z. canicapilla**; **Z**: 92.

Emberiza gubernatrix = Yellow Cardinal *Gubernatrix cristata*, one specimen from banks of Parana River, near Santa Fe; **Z**: 89.

Emberiza cristata = as a synonym of **Emberiza gubernatrix**; **Z**: 89.

Emberiza cristatella = as a synonym of **Emberiza gubernatrix**; **Z**: 89.

THRAUPIDAE—**Tangers and allies**

Aglaia striata = Blue-and-yellow Tanager *Thraupis bonariensis*, one specimen, illustrated, from Maldonado, considered same species as **Aglaia striata**, food; **Z**: 97.

Tanagra striata = as a synonym of **Aglaia striata**; **Z**: 97.
Tanagra Darwinii = as a synonym of **Aglaia striata**; **Z**: 97.

Tanagra Darwinii = Blue-and-yellow Tanager *Thraupis bonariensis*, mentioned as being same species as **Aglaia striata**; **Z**: 97.

Aglaia vittata = Fawn-breasted Tanager *Pipraeidea melanonota*, at Maldonado, not common; **Z**: 98.

Tanagra vittata = as a synonym of **Aglaia vittata**; **Z**: 98.

Spiza cyanea = Masked Flowerpiecer *Diglossa cyanea*, is bright blue in breeding season and "though generally peaceable," a captive attacked a blue-headed ***Spiza ciris*** and scalped it; **D**: 629.

Fringilla gayi = Grey-hooded Sierra Finch *Phrygilus gayi*, abundant in southern Patagonia; **Z**: 93.

Emberiza Gayi = as a synonym of **Fringilla gayi**; **Z**: 93.

Fringilla formosa = Patagonian Sierra Finch *Phrgilus patagonicus*, described and considered "nearly allied" to ***F. gayi***; **Z**: 93.

Fringilla fruticeti = Mourning Sierra Finch *Phrgilus frutceti*, CD notes that this name "gives place to" (a synonym of) **Fingilla erythrorhyncha**; **Z**: viii.

Distribution, habitats, habits; **Z**: 94.
Emberiza luctuosa = as a synonym of **Fringilla fruticeti**; **Z**: 94.

Fringilla erythrorhyncha = see **Fringilla fruticeti**.

Fringilla carbonaria = Carbonated Sierra Finch *Phrgilus carbonarius*, seen once, on desert plains between Negro and Colorado Rivers; **Z**: 94.

Emberiza carbonaria = as a synonym of **Fringilla carbonaria**; **Z**: 94.

Fringilla alaudina = Band-tailed Sierra Finch *Phrgilus alaudinus*, specimens from neighbourhood of Valparaiso; **Z**: 94.

Emberiza guttata = as a synonym of **Fringilla alaudina**; **Z**: 94.
Passerina guttata = as a synonym of **Fringilla alaudina**; **Z**: 94.

Chlorospiza? melanodera = White-Bridled Finch *Melanodera melanodera*, described, illustrated, locations, flocking; **Z**: 96.

Emberiza melanodera = as a synonym of **Chlorospiza? melanodera**; **Z**: 95.

Chlorospiza? xanthogramma = Yellow-Bridled Finch *Melanodera xanthogramma*, described, illustrated, locations, sometimes flocks with *M. melanodera*, morphology compared with *M. melanodera*; **Z**: 97.

Ammodramus longicaudatus = Long-tailed Reed Finch *Donacospiza albifrons*, described, illustrated, habitats, habits (similar to **Synallaxis** and **Limnornis**), food, morphology like that of **Synallaxis**; **Z**: 90.

bunting = Inaccessible Island Finch *Neospiza acunhae*, CD notes a Captain D. Carmichael reports this as being very tame on Tristan da Cunha or Inaccesible Island; **J**: 629.

Fringilla diuca = Common Diuca Finch *Diuca diuca*, locations, habitats, habits (similar to those of **Zonotrichia matutina**), calls, nest, eggs; **Z**: 93.
  Fringilla Diuca = as a synonym of **Fringilla diuca**; **Z**: 93.
  Emberiza Douca = as a synonym of **Fringilla diuca**; **Z**: 93.

Pipillo personata = Black-and-rufous Warbling Finch *Poospiza nigrorufa*, at Maldonado, not common, eats seeds, illustrated; **Z**: 98.

Crithagra? brasiliensis = Saffron Finch *Sicalis flaveola*, obtained on the La Plata; **Z**: 88.
  Fringilla Brasiliensis = as a synonym of **Crithagra? brasiliensis**; **Z**: 88.

Crithagra? brevirostris = Grassland Yellow Finch *Sicalis luteola*, described, habits, call; **Z**: 88–89.

Emberiza luteoventris = Grassland Yellow Finch *Sicalis luteola*, collected at Santa Cruz, where rare; **Z**: 89.
  Fringilla luteoventris = as a synonym of **Emberiza luteoventris**; **Z**: 89.

Emberizoides poliocephalus = Pampa Finch *Embernagra platensis*, described, locations, food, call, flight; **Z**: 98.

Passerina jacarina = Blue-black Grassquit *Volatinia jacarina*, a specimen from Rio de Janeiro; **Z**: 92.
  Tanagra jacarina = as a synonym of **Passerina jacarina**; **Z**: 92.
  Emberiza jacarina = as a synonym of **Passerina jacarina**; **Z**: 92.
  Euphone jacarina = as a synonym of **Passerina jacarina**; **Z**: 92.
  Frigilla splendens = as a synonym of **Passerina jacarina**; **Z**: 92.

Spermophila nigrogularis = Double-collared Seedeater *Sporophila caerulescens*, described, from Monte Video; Z: 88. Family Coccothraustinae [*sic*] = a disused taxon for the Galapagos Finches, once within the Carduelinae of the Fringillidae (Bock 1994: 156) and subsequently within the Emberizinae of the Fringillidae (Sibley and Monroe 1990: xv) and the Thraupidae (Gill and Donsker 2015); Z: 98. See chapter 3.

Geospiza magnirostris = Large Ground Finch, described, illustrated, from Charles and Chatham Islands, Galapagos; **Z**: 100. See figures 2.8, 3.8, 8.5.

Gopiza strenua = Large Ground Finch *Geospiza magnirostris*, described, illustrated, from James and Chatham Islands, Galapagos; **Z**: 100–101. See figures 2.8, 3.8, 8.5.

Geospiza fortis = Medium Ground Finch, described, illustrated, from Charles and Chatham Islands, Galapagos; **Z**: 101. See figures 2.8, 3.7.

Geospiza dentirostris = Medium Ground Finch *Geospiza fortis*, described, from "Galapagos Archipelago"; **Z**: 102. See figures 2.8, 3.7.

Geospiza dubia = Medium Ground Finch *Geospiza fortis*, described, from Chatham Island; **Z**: 103. See figures 2.8, 3.7.

Geospiza fuliginosa = Small Ground Finch, described, from Chatham and James' Island; **Z**: 101–102.

Geospiza nebulosa = Sharp-beaked (Floreana) Ground Finch *Geospiza (nebulosa) difficilis*, described, from Charles Island; **Z**: 101.

Cactornis scandens = Sharp-beaked (Floreana) Ground Finch *Geospiza (nebulosa) difficilis*, described, illustrated; CD thought this distinguishable in habits from several of his **Geospiza** and **Camarhynchus** species; cactus feeding and other habits noted; **Z**: 104–105.

Cactornis assimilis = Common Cactus Finch *Geospiza scandens*, described, illustrated, from "Galapagos Archipelago"; mention made of a **Cactornis** species discovered on Cocos **Island** (= Cocos Finch *Pinaroloxias inornata*), **Z**. 103. See also chapter 2 and colour figures 3.10, 4.3.

Cactornis = Common Cactus Finch *Geospiza scandens* and Large Cactus Finch *G. conirostris*, CD notes all of the Galapagos Finches are endemic to the islands but that "the sub-group Cactornis, lately brought from Bow Island, in the Low Archipelago" is a member of the larger group; **J2**: 379. See colour figures 3.10, 4.3.

CD states "Two species of the sub-group Cactornis, and two of the Camarhynchus, were procured in the archipelago"; **J2**: 395. See chapter 2.

Given sub-genus status but no longer used; **Z**: 104.

Camarhynchus crassirostris = Vegetarian Finch *Platyspiza crassirostris*, described, illustrated, from "Charles Island?" where CD thought it replaces **C. psittaculus**; **Z**: 103–104.

Camarhynchus psittaculus = Large Tree Finch *Camarhynchus psittacula*, described, illustrated, from James Island; flocks with *Geospiza* species; **Z**: 103.

Geospiza parvula = Small Tree Finch *Camarhynchus parvulus*, described, illustrated, from James' Island; **Z**: 102. See figures 2.8, 3.9.

Geospiza = a genus of Galapagos Finches, CD noting the "perfect gradation in the size of beaks in the different species"; **J2**: 379. See figure 2.8.

CD states he has "strong reasons to suspect that some of the species of the sub-group Geospiza are confined to separate islands"; **J2**: 395. See figure 2.8.

Camarhynchus = a genus of Galapagos Finches, CD noting the beak of this sub-group of finches being "slightly parrot-shaped"; **J2**: 380. See figure 2.8.

CD stated "Two species of the sub-group Cactornis, and two of the Camarhynchus, were procured in the archipelago"; **J2**: 395. See chapter 2.

Certhidea = Green Warbler-Finch *Certhida olivacea*, CD noting that the "perfect gradation in the size of beaks in the different species" of Galapagos Finches even includes a beak like that "of a warbler" if Gould was correct in including the Warbler Finch within the group (which he was); **J2**: 379–380. See figures 2.8, 3.11.

Given sub-genus status, but no longer used; **Z**: 105.

Certhidea olivacea = Green Warbler-Finch, described, illustrated, from Chatham and James Islands; **Z**: 106. See figures 2.8, 3.11.

warbler = Green Warbler-Finch *Certhidea olivacea*, CD merely states that a species of finch on the Galapagos Islands has a beak "differing but little from that of a warbler"; **J**: 462. See figures 2.8, 3.11.

Thick-billed finches = unspecified Galapagos finch, feeding on same cactus as a land iguana *Conolophus subcristatus*; **J**: 472.

group of finches = 13 species of Galapagos finches included in list of land birds of the archipelago; **J**: 461; **J2**: 379–380. See chapters 2, 3, 4, and 8.

ground-finches = 13 species of Galapagos finches, CD coming to believe that each species is confined to one or more islands; **J**: 475. See chapters 3 and 8.

Plumages discussed; **Z**: 98–106. See chapter 3.

CARDINALIDAE—**Cardinals, Grosbeaks, and allies**

Scarlet Tanager (*Tanagra rubra*) = Summer Tanager *Piranga rubra*, a few males have a band of glowing red on smaller wing coverts, mentioned in discussing intraspecific variability; **D**: 646.

Tanagra rubra = Summer Tanager *Piranga rubra*, young males distinguishable from young females; **D**: 743 footnote.

Pyranga æstiva = Summer Tanager *Piranga rubra*, briefest discussion of significance of and selection pressure on marked sexual dimorphism in colour; **D**: 690.

Tanagra aestiva = Summer Tanager *Piranga rubra*, males take three years to acquire full adult plumage; **D**: 738 footnote.

Pyranga = *Piranga*, unspecified, young fertile females may acquire the characters of the males; **D**: 704.

Cardinalis virginianus = Northern Cardinal *Cardinalis cardinalis*, the sexes are differently affected by the north–south variation in plumage colour that typically affects both sexes in the same way; **D**: 347.

Pitylus superciliaris = Greyish Saltator *Saltator coerulescens*, a specimen from Santa Fé; **Z**: 97.

Tanagra superciliaris = a synonym of **Pitylus superciliaris**; **Z**: 97.

Spiza ciris = Painted Bunting *Passerina ciris*, is blue-headed, and an all blue "though generally peaceable" captive ***Spiza cyanea*** attacked a Painted Bunting and scalped it; **D**: 629.

Fringilla ciris = Painted Bunting *Passerina ciris*, males take four years to acquire adult plumage; **D**: 738 footnote.

Breeds "as perfectly as the canary" in confinement; **V2**: 154.

Finch[es] = unspecified, three or four species found in more open parts of Tierra del Fuego; **J**: 301. See chapter 2.

Mentioned in passing as being the commonest birds in some South American districts; **J**: 353.

CD states that a species "with a stiff tail and a long claw to its hinder toe, closely allied to a North American genus" occurs as endemic on the Galapagos (= Bobolink *Dolichonyx oryzivorus*) and that a group of 13 finches are considered by Gould to constitute four new sub-genera (now of the Thraupidae); **J**: 461–462. See chapter 2.

Finches listed among birds that were tame on the Galapagos Islands; **J**: 475.

CD watched a boy killing many doves and finches with a "switch," or stick, as they came to drink at a well on Charles Island, Galapagos; **J**: 476.

Nine finch species noted as crossed with the **canary-bird**; **O**: 373.

A species has a tail including bare-shafted feathers terminating in a disc of vanes; **D**: 586.

Certain species "shed the margins of their feathers in the spring, and then become brighter feathered," whereas other species do not; **D**: 599–600.

See **small birds**; **E**: 100.

Finches and insect eaters, recorded on the arid plains of Patagonia; **J**: 194.

bird(s) = a "nearly perfect" collection of them obtained at Maldonado; **J**: 46.

CD collected 80 species within a morning walk of Maldonado; **J**: 55. See chapter 2.

An account of various species of the Maldonado area; **J**: 60–69.

Many more birds noted at Santa Fe than Buenos Aires; **J**: 147–148.

"Very great numbers of birds" died at northern Buenos Aires Province and southern Santa Fe in drought, with "partridges" having "hardly the strength to fly away when pursued"; **J**: 155–156.

Same birds seen throughout an arid Patagonian trek; **J**: 215.

Other than geese and a snipe, few seen on East Falklands; **J**: 246; including "hawks, owls, and a few small land-birds"; **J**: 256.

Birds "scanty' at high altitude in the Andes; **J**: 387, 394.

Different bird species on east and west sides of the Andes briefly mentioned; **J**: 399.

"Birds are strangers to man" on the Galapagos Islands; **J**: 455.

Birds on Galapagos "nearly all" live in the "lower country" owing to scarcity of insects higher up; **J**: 473.

Tameness and wildness of birds on isolated islands; **J**: 455, 475–478; **V1**: 20–1. See chapters 2 and 5.

CD saw "very few birds" in New Zealand woods; **J**: 511.

In the woods of the Blue Mountains, Australia, CD found "not many birds"; **J**: 526.

Birds reported by CD as "very few in number" on St. Helena Island, Atlantic Ocean, where he believed them all "introduced within late years" (by which he would have meant land birds); **J**: 584.

Birds merely mentioned in the context of their "wild flight" during storms over land; **J**: 603.

Unspecified birds dispersing mistletoe seeds by having them stick to their feathers; **O**: 76.

CD notes that people do not see or forget that birds about them "mostly live on insects or seeds, and are thus constantly destroying life" or that their eggs and nestlings are destroyed by birds and other predators; **O**: 77–78, 91.

Unspecified species disseminate mistletoe seeds; **O**: 78.

CD estimated that the winter of 1854–1855 destroyed four-fifths of birds in the grounds of his home; **O**: 84.

In the presence of a superabundance of corn and rape-seed, and so forth, in CD's time, birds left adequate crops nor increased in numbers greatly as their numbers were "checked during winter"; **O**: 86.

Among birds, competition through sexual selection is often "more peaceful" than in mammals; **O**: 109.

Sexual selection by females and its influence on the plumage of adult and young; **O**: 109.

Birds used as an example of natural selection favouring individuals with a novel useful trait, for example, a curved beak; **O**: 112.

CD cites Gould's belief that "birds of the same species are more brightly coloured under a clear atmosphere, than when living near the coast or on islands" (in this, Gould may have observed a phenomenon akin to Gloger's Rule, which states that races of a species in warm and humid areas are more heavily pigmented than those in cooler and drier ones); **O**: 165.

Richard Owen remarked "there is no greater anomaly in nature than a bird that cannot fly"; **O**: 167.

Some authors believe that the shape of the pelvis causes the observed diversity in the shape of kidneys within the birds; **O**: 178.

Individuals "of the same species, inhabiting the same country, vary extremely little"; **O**: 186.

Wing use other than for flight noted; **O**: 218.

Wings of birds differently constructed to those of bats; **O**: 240.

Attracted by colourful fruits to eat and then disperse their seed; **O**: 240, 250.

Sexual selection for beauty in male appearance and song, and sometimes "transmitted to both sexes"; **O**: 252–253.

Bird excrement mimicked by insects in their body form; **O**: 283.

Insectivorous birds place selective pressure on insects for cryptic appearance; **O**: 284.

That birds' wings originated by a "comparatively sudden modification," according to Mr. Milvart, is discussed; **O**: 313, 317.

Greater wildness in larger birds in England than in small ones is because larger ones are more persecuted by people; **O**: 325.

Some build nests of "mud, believed to be moistened with saliva"; **O**: 355.

Using birds as an example, a critic of CD failed to see how successive modifications of the anterior limb could have been advantageous; **O**: 442.

A bird lived during the deposition of the tertiary "upper greensand"; **O**: 444.

Good authority states that the rudiments of teeth can be detected in the beaks of certain embryonic birds; **O**: 450–451.

CD notes that the "wide interval between birds and reptiles" to be shown by T. H. Huxley to be "partially bridged" on one hand "by the ostrich and extinct Archeopteryx"; **O**: 472.

Birds of the caves of Brazil (presumably fossil of subfossil) as an example of "this wonderful relationship in the same continent between the dead and the living"; **O**: 485.

The geographical replacement of closely related species over continents, reflected in morphology, song, nests, and eggs; **O**: 496.

The dispersal of seeds by birds, in their carcasses, crop, intestines, excreta, via the stomach of predatory birds and fish, on their beaks, legs and feet; **O**: 509–512. See chapters 3 and 4.

Wind-blown and migrant birds carrying seeds internally or in dirt on feet and beak; **O**: 512, 514, 538–539, 541. See chapter 3.

North American ones occasionally to frequently visit Bermuda, while European and African ones are blown to Madeira—both islands being stocked from neighbouring continents and with few endemics; **O**: 544.

Species on the Galapagos Islands are well adapted to flying yet differ from island to island [due to competitive exclusion], **O**. 557. See chapter 3.

Those on the southeast and southwest of Australia "have nearly the same physical conditions" but are inhabited by distinctly different species; **O**: 557.

Within genera ranging widely globally, many species have "very wide ranges"; **O**: 559–560.

The use of geographical distributions in their classification, particularly in large groups of closely related forms; **O**: 575.

Extinction has played a part in isolating birds from other vertebrates; **O**: 592–593. See chapter 4.

Limb structure of early vertebrate progenitor supposed to have been of existing general pattern; **O**: 598.

E. Ray Lankester finds the hearts of birds and mammals "are as a whole homogenous"; **O**: 601.

Von Baer observes that early stage embryos of mammals, birds, and reptiles are "exceedingly like one another" as is the development of their parts; **O**: 605, 655.

Birds of a genus and of closely related genera have similar immature appearance; **O**: 605.

The young of a species in which adults have a "much-curved" beak do not suffer the consequences of having a short straight one while its parents continue to feed it; **O**: 610.

CD observes that birds and other vertebrate groups are modified descendants of an ancient progenitor in which the adults had "branchiae, a swim-bladder, four fin-like limbs, and a long tail, all fitted for an aquatic life"; **O**: 617.

CD states that the "bustard-wing" of birds is a rudimentary digit; **O**: 619.

Birds on islands lacking predators loose the power of flight, their wings becoming rudimentary; **O**: 625.

Sexual selection has "rendered the voice of the male musical to the female"; **O**: 647.

Mentioned in the last paragraph of *The Origin of Species*, as CD alludes to the ecological complexity of a tangled, vegetated, bank with birds singing; **O**: 669.

Many "during their courtship produce diversified sounds by means of specially adapted feathers"; **E**: 95.

Birds "belonging to all the chief Orders ruffle their feathers when angry or frightened," with domestic cocks detailed; **E**: 98–99.

Fuegians skilfully kill birds with a thrown stone; **D**: 73.

Birds show individual variation in "mental characteristic"; **D**: 100.

They "imitate the songs of their parents, and sometimes of other birds"; **D**: 110.

Even "birds have vivid dreams"; **D**: 113.

They doubtless "both acquire and lose caution in relation to man and other enemies"; **D**: 122.

Sounds uttered by them offer in several respects nearest analogy to language; **D**: 131.

Require time to perform the song of their kind well, which may show local dialects; **D**: 132.

It "might be urged" that the "aesthetic faculty" of "most savages . . . was not so highly developed . . . as in birds"; **D**: 142.

Many species "post sentinels" (or "look-outs," to warn of danger); **D**: 153.

Even birds "certainly sympathise with each other's distress or danger"; **D**: 156.

A strong feeling of satisfaction must impel a bird to brood her eggs day after day; **D**: 160.

Migratory birds are "miserable" if stopped from migrating, and a pinioned goose started to do so on foot; **D**: 160.

The seasonal migratory instinct will cause a confined bird to "beat her breast against the wires of her cage, until it is bare and bloody"; **D**: 164.

The maternal instinct drives "even timid birds to face great danger, though with hesitation, and in opposition to the instinct of self-preservation"; **D**: 165.

Many birds "not yet old enough for a prolonged flight" are deserted by their migrating adults (see **swallows**, **house-martins**, and **swifts**); **D**: 165, 173.

The transmission of traits fully developed in males, such as spurs, plumes, and colours, to females in rudimentary form; **D**: 250.

With "some few birds the beak of the male differs from that of the female (see **Huia**); **D**: 321. See chapter 6.

In observing "several male birds displaying" to "an assembled body of females, we cannot doubt that, though led by instinct, they know what they are about, and consciously exert their mental and bodily powers"; **D**: 325.

Males of migratory species "generally arrive at their places of breeding before the females" (see **nightingale**, **blackcap**, and **Ray's wagtail**); **D**: 327.

In the majority of migratory birds of the United States, males arrive on the breeding grounds before females; **D**: 327.

As the first of both sexes of migratory species to arrive on the breeding grounds are likely the "most vigorous, best-nourished," they form pairs to be more successful breeders than later arrivals, and their male offspring will thus, over generations, improve in fitness; **D**: 329–330.

Females of some species appear to die earlier than males and are liable to be killed at the nest or while caring for fledged offspring; **D**: 333.

Males in many species are considerably more numerous than females; **D**: 333.

CD states, "some few birds are polygamous"; **D**: 334.

Many monogamous birds "display strongly-marked secondary sexual characters"; **D**: 334, 337.

Among them, "there often exists a close relationship between polygamy and the development of strongly marked sexual differences"; **D**: 339

While males do pursue females, many do more displaying of their plumage, perform courtship postures, or sing to females; **D**: 341.

CD quotes Mr. Allen who noted that in many species of the United States, those of the south are darker than those of the north, and that this relates to differences in temperature, light, and so forth; **D**: 347.

Brief mention of species in which the females are the brighter and/or more "eager in courtship" sex, whereas males are more passive but select the more attractive females; **D**: 347–348.

Many species acquire bright colours and decoration only during breeding seasons; **D**: 354.

CD quotes Dr. W. Marshall who found that in species with protuberances on the head only in males, they are developed late in life; whereas in birds with such in both sexes, they are developed "at a very early period"; **D**: 363.

In wild birds, "Mr. Gould and others" were convinced that males are the more numerous sex; and as young males of many species resemble the female, the latter would naturally appear the more numerous; **D**: 382.

Secondary sexual characters are "more diversified and conspicuous in birds"; males rarely "possess special weapons for fighting with each other," charm females with vocal or other sounds, are ornamented by various body part characters, perform courtship dances and postures, in at least one species emits a musky odour, and are "the most aesthetic of all animals, excepting of course man"; **D**: 549–550.

"Almost all male birds are extremely pugnacious, using their beaks, wings, and legs for fighting together"; **D**: 551.

Birds "which seem ill-adapted for fighting engage in fierce conflicts"; **D**: 554.

The males of many birds "are larger than the females, and this no doubt is the result of the advantage gained by the larger and stronger males over their rivals during many generations"; **D**: 554.

In many species, females are larger than males; and the explanation that this is because females have most of the work in feeding young is inadequate because in some cases, females have become the larger sex to conquer other females and obtain males; **D**: 555.

When many males congregate at an "appointed spot" and fight, they are generally attended by females (e.g., **grouse**); **D**: 560

The voice serves to express emotions, such as distress, fear, anger, triumph, or happiness; and nestlings of some species hiss "to excite terror," with vocal and instrumental "music" discussed; **D**: 562–580.

Gliding and sailing through the air "obviously for pleasure"; **D**: 566.

Males continue singing "for their own amusement" after the courtship season; **D**: 567.

CD states "brilliant birds of the tropics are hardly ever songsters," **D**: 568.

Mr. Genner Weir finds that "all male birds with rich or strongly-characterised plumage are more quarrelsome than the dull-coloured species belonging to the same group," and CD gives a few English examples; **D**: 609.

There are few "large groups of birds" in which both sexes are brilliantly coloured, but the **plantain-eaters** (turacos) are an exception; **D**: 701.

In many species elegant plumes, long pendant feathers, crests, and the like worn during summer serve ornamental and nuptial purposes although common to both sexes; **D**: 704.

Nearly all desert- or ground-dwelling species depend on their cryptic colours for concealment; **D**: 746–748.

Worldwide, both sexes of many soft-billed species, "especially those which frequent reeds or sedges," are obscurely coloured as protection from predation; **D**: 748.

CD thinks that males, although dull coloured, often differ much from females as a result of sexual selection; **D**: 748.

Some species that naturally never sing can be taught to do so by people; **D**: 869.

Distinct evidence demonstrates that "the individuals of one sex are capable of feeling a strong antipathy or preference for certain individuals of the other sex;" **D**: 941.

Birds appreciate bright and beautiful objects, as with **bower-birds**, and the power of song; and in the case of females, sexual selection plays a role; **D**: 941–942.

There "is no a priori difficulty in the belief" that several species were involved in the origin of some domesticated birds; **V1**: 20.

Five digits not normally exceeded; **V2**: 13—and have no power of regrowth; **V2**: 15.

Widely kept tame in South America, Africa, and South Pacific Islands but rarely breed; **V2**: 150.

CD theorizes that if a bird received advantage from seeing well in low light, those individuals with the ability would succeed best and be most likely to survive and lead to owl-like vision; **V2**: 222.

That fruits eaten and not eaten by birds is relative to their colour briefly discussed; **V2**: 230.

South American Indians change the feather colour of "many birds" by plucking an area and then "inoculate the flesh wound" with a milky toad skin secretion; **V2**: 280. See chapter 5.

Species with long wing-feathers usually have long tail-feathers; **V2**: 354.

Facts and traditional statements by local inhabitants regarding the first arrival of species onto islands of coral-reefs should be collected; **1849**: 183.

Darwin learnt to skin birds from a Negro who travelled with Waterton; **A**: 22. See chapter 1.

Darwin kept an exact record of every British game bird he shot; **A**: 24. See chapter 1.

bird-like animals = mention made of at least 30 from the tertiary deposits; **O**: 443.

nests of birds = intraspecific variation in; **O**: 324.

nestling birds = instinctive and learned fear in; **O**: 324.

native birds = CD found none on Ascension Island, Atlantic Ocean; **J**: 587.
Slowly acquire and inherit a dread of man on islands; **V1**: 20.

insectivorous birds = six species found inhabiting a botanically diversely planted heath, whereas only two or three found in an original and less diverse adjacent heath; **O**: 87–88.

Should such birds decrease in Paraguay, "parasitic insects would probably increase"; **O**: 89.

predaceous birds = are deceived by butterflies mimicking distasteful species, **O**; 587–588.

carnivorous birds = CD cites reference to as many as 18 species used in Europe for hawking, and several others in Persia and India, but with no record of any producing offspring and no hawk, vulture, or owl having laid fertile eggs in zoos; **V2**: 153–154.

When confined to captivity are extremely sterile; **V2**: 159.

Their reproduction is much more easily affected by changed conditions than that of carnivorous mammals; **V2**: 268.

graminivorous birds = many kinds kept tame and lived long but "uncommonly difficult" to breed; **V2**: 154.

Any such species caught far out at sea should have their intestine content dried for possible seed content; **1849**: 183.

insessorial birds = insessorial birds, a disused taxon that contained nearly all the Picae and Passeres of Linnæus; **V2**: 154.

In a nine-year report of the Zoological Society (of London), none of 24 species bred, and only four were observed to mate; **V2**: 154.

aquatic birds = seen in the Galapagos as appearing to be the same as well-known species of "American birds"; **J**: 461; and "some are peculiar to these islands, and some common to North and South America"; **J**: 462.

land birds = CD writes "In my collections from these islands [Galapagos], Mr. Gould considers that there are twenty-six different species of land bird. With the exception of one, all probably are undescribed kinds, which inhabit this archipelago, and no other part of the world"; **J**: 461. See chapters 2 and 4.

CD writes of having not met with any account of land birds being so tame as those of the Galapagos and Falkland Islands; **J**: 477.

CD noted that there were "no true land-birds" on the Keeling or Cocos Islands (but in fact, there are several regular visiting ones and a good number of vagrant land bird species that visit there); **J**: 543.

Brief mention of the tameness of land birds on Tristan da Acunha [*sic*]; **J**: 629.

Proportion of these on Galapagos and their affinity with birds of the Americas; **O**: 543, 552. See chapter 4.

marine birds = proportion of these endemic on the Galapagos, and on islands in general, is small compared to endemic land birds; **O**: 543. See chapter 4.

It is probable that the complete or partial blackness of many is the result of sexual selection "accompanied by equal transmission to both sexes"; **D**: 749.

endemic birds = proportion of these on the Galapagos Islands, and islands in general, large compared to mainlands; **O**: 543. See chapter 4.

domesticated birds = with "anciently domesticated birds" wings have little use and are slightly reduced as are anatomical adaptations associated with flight; **V2**: 353.

hybrids = a brief discussion of wild hybrid birds; **D**: 631–633.

albino(s) = CD cites an observer with much experience of albinos who never saw one paired to another bird, and he attributes this to their being rejected by normally coloured conspecifics; **D**: 639–640, 756.

dull-coloured birds = unspecified, on Chatham Island, Galapagos, CD noted a few of these "cared no more for me than they did for the great tortoises"; **J2**: 375.

little birds = unspecified, on Galapagos Islands, noted to be unafraid of iguanas; **J**: 471.

Several "beautiful little birds" seen by CD ca. 30–40 miles NE of Cape Town, South Africa; **J**: 577.

small birds = "such as various finches, buntings, and warblers, when angry, ruffle all their feathers, or only those round the neck; or they spread out their wings and tail-feathers"; **E**: 100.

smaller birds brilliantly coloured = CD was struck by these, presumably passerines, at Maldonado, La Plata, where he made a "nearly perfect collection of the animals, birds, and reptiles" after 10 weeks; **J**: 46.

smaller birds, as ducks, hawks, and partridges, were killed = by "hail as large as small apples, and extremely hard"; **J**: 134.

flocks of migratory birds . . . and **the multitude of birds of prey** = in late August 1833, merely comparing the former with herds of antelope and the latter as indicative of the "abundance of smaller quadrupeds" in South Africa; **J**: 100.

soft-billed birds = many are songsters, and the best songsters are rarely brightly ornamented, suggesting to CD that females select for song or bright colours but not both combined; **D**: 748.

web-footed birds = unspecified, in CD noting that these on the Galapagos Islands were not endemic, in contrast to most of the land birds there; **J2**: 380.

Their vocal organs are "extraordinarily complex" and differ slightly between the sexes; **D**: 571.

unhatched birds = unspecified, CD noting that the egg tooth is used only the once by birds, in hatching (see **short-beaked tumbler-pigeon**); **O**: 106.

nearly wingless = a term used by CD for birds on islands lacking predators, and caused "by disuse"; **O**: 167.

larger ground-feeding birds = as these "seldom take flight except to escape danger, it is probable that the nearly wingless condition" of them on islands lacking predators is caused by disuse (see **nearly wingless**); **O**: 167.

ground-birds = young will run from under their mother to hide, allowing her to fly away; **O**: 329.

Wading birds = dispersing sea-shell eggs and larvae on their feet; **O**: 544.

Their vocal organs are "extraordinarily complex" and differ slightly between the sexes; **D**: 571.

bird(s) of prey = unspecified, retain seeds for long periods in stomach and thus may disperse plants widely; **O**: 510. See chapter 3.

Notoriously attack female pigeons attending their sparse open nests, in discussing Wallace's objections to sexual selection for plumage colouration with regard to open nesting; **D**: 691.

Cause **ptarmigans** to suffer greatly; **D**: 723.

A breed of domestic pigeon said to fly like one; **V1**: 156.

White varieties of birds are most likely to be attacked by birds of prey; **V2**: 229.

gigantic wingless birds = take, or recently took, the place of mammals in New Zealand; **O**: 545.

domestic(ated) birds = in several species, certain coloured marks are "either strongly inherited or tend to reappear after having been long lost"; **V1**: 55.

feather = structure discussed in context of CD's pangenesis theory; **V2**: 382.

Law of Battle = a broad discussion of this (i.e., fighting) in birds; **D**: 551–562. See chapter 6.

Vocal and instrumental music = a broad discussion of these in birds; **D**: 562–580.

Love antics and Dances = a broad discussion of these in birds; **D**: 580–583.

Decoration = a broad discussion of this in birds; **D**: 583–600.

Display by Male Birds of their Plumage = a broad discussion of this; **D**: 600–515. See chapter 6.

Length of Courtship = a broad discussion of this in birds; **D**: 617–620. See chapter 6.

Unpaired Birds = a broad discussion of these; **D**: 620–625.

Mental Qualities of Birds, and their Taste for the Beautiful = a broad discussion of this; **D**: 625–631. See chapter 6.

Preference for particular Males by the Females = a broad discussion of this in birds; **D**: 631–643. See chapter 6.

Variability of Birds, and especially of their Secondary Sexual Characters = a broad discussion of this; **D**: 643–652. See chapter 6.

Formation and Variability of the Ocelli or eye-like Spots on the Plumage of Birds = a broad discussion of this; **D**: 652–655. See chapter 6.

Gradation of Secondary Sexual Characters = a broad discussion of this in birds (mostly of plumage ocelli); **D**: 655–675. See chapter 6.

Sexually Limited Inheritance = a broad discussion of this in birds; **D**: 676–685. See chapter 7.

Development of Spurs = a broad discussion of this in birds; **D**: 685–686.

Length of female's tail = a broad discussion of this in birds; **D**: 687–689.

Colour and Nidification = a broad discussion of the colour of males and females in species relative to their nesting habits; **D**: 689–706. See chapter 7.

Inheritance Limited by Age = a broad discussion of this in birds; **D**: 707–745. See chapter 7.

On the Colour of the Plumage in relation to Protection = a broad discussion of this in birds; **D**: 745–749. See chapter 7.

Conspicuous Colours = a broad discussion of this in birds; **D**: 749–753. See chapter 7.

Novelty Admired = a broad discussion of this in courting birds; **D**: 753–756.

APPENDIX 2

Birds Named after Charles Darwin

This list includes all bird forms named after Charles Darwin by application to their vernacular and/or scientific name(s), be they in use today or no longer in use. Those names given after the equals (=) sign are those in the International Ornithological Committee's (IOC) "IOC master list" of Gill and Donsker (2015; http://www.worldbirdnames.org) where different from the first name given. The citations to the title, author, and year of publication of new bird species descriptions are not necessarily included in the References of this book, but the journal or book involved is indicated for each.

Darwin's Nothura [*Nothura darwinii*] G. R. Gray 1867, *List of Birds of the British Museum, Gallinae*: 104; of the non-passerine family Tinamidae.

Darwin's Rhea [*Pterocnemia pennata*] d'Orbigny 1834, *Voyage Amérique Méridionale* **2**, 67. = Lesser Rhea [*Rhea pennata*]; of the non-passerine family Rheidae.

Darwin's [Pucras] Pheasant [*Pucrasia macrolopha darwini*] Swinhoe 1872, *Proceedings of the Zoological Society of London*: 552. = Koklass Pheasant [*Pucrasia macrolopha darwini*]; of the non-passerine family Phasianidae.

Darwin's Rail [*Coturnicops notatus*] (Gould 1841) in Darwin, *The Zoology of the Voyage of H. M. S. Beagle, Part III: Birds*: 132, plate 48. = Speckled Rail; of the non-passerine family Rallidae.

Darwin's Caracara [*Phalcoboenus albogularis*] (Gould 1837) *Proceedings of the Zoological Society of London* **5**, 9. = White-throated Caracara; of the non-passerine family Falconidae.

Upucerthia darwini (Scott 1900) *Bulletin of the British Ornithologists' Club* **10**, xliii. = Scaly-throated Earthcreeper; of the non-passerine family Furnariidae.

Lanius darwini (Severtzov 1879) *Tashkent-Zapiski I.* **1**, 51; a name applied to what is now perceived to be a hybrid between two subspecies of the Red-Backed Shrike [*Lanius collurio*] (Mayr and Greenway 1960: 345) or between the Red-tailed Shrike [*Lanius phoenicuroides*] and Long-tailed Shrike [*L. Schach*] (McCarthy 2006: 218); of the passerine family Laniidae.

Thraupis darwinii (Bonaparte 1838) *Proceedings of the Zoological Society of London* **5** (1837), 121. = Darwin's Tanager *Thraupis bonariensis darwinii* = Blue-and-yellow Tanager; of the passerine family Fringillidae.

Darwin's Finches, the collective name for members of the endemic Galapagos finches apparently first applied by the ornithologist Percy R. Lowe, in giving a 1935 lecture at the British Association, but given wider currency by a book on the birds by David Lack (see chapter 3); of the passerine family Thraupidae, **as are all species following**.

Darwin's Large Ground Finch [*Geospiza magnirostris*] (Gould 1837) *Proceedings of the Zoological Society of London* **5**, 5. = Large Ground Finch.

Darwin's Medium Ground Finch [*Geospiza fortis*] (Gould 1837) *Proceedings of the Zoological Society of London* **5**, 5. = Medium Ground Finch.

Darwin's Small Ground Finch [*Geospiza fuliginosa*] (Gould 1837) *Proceedings of the Zoological Society of London* **5**, 5. = Small Ground Finch.

Darwin's Sharp-beaked Ground Finch [*Geospiza difficilis*] (Sharpe 1888) *Catalogue of Birds in the British Museum* **12**, 12. = Sharp-beaked Ground Finch.

Darwin's Cactus Ground Finch [*Geospiza scandens*] (Gould 1837) *Proceedings of the Zoological Society of London* **5**, 7. = Common Cactus Finch.

Darwin's Large Cactus Ground Finch [*Geospiza conirostris*] (Ridgway 1890) *Proceedings of the United States National Museum* **12** (1899), 106. = Large Cactus Finch.

Darwin's Vegetarian Finch [*Camarhynchus crassirostris*] (Gould 1837) *Proceedings of the Zoological Society of London* **5**, 6. = Vegetarian Finch [*Platyspiza crassirostris*].

Darwin's Large Tree Finch [*Camarhynchus psittacula*] (Gould 1837) *Proceedings of the Zoological Society of London* **5**, 6. = Large Tree-Finch.

Darwin's Medium Tree Finch [*Geospiza pauper*] (Ridgway 1890) *Proceedings of the United States National Museum* 12 (1899). 111. = Medium Tree Finch [*Camarhynchus pauper*].

Darwin's Small Tree Finch [*Camarhynchus parvulus*] (Gould 1837) *Proceedings of the Zoological Society of London* **5**, 6. = Small Tree Finch.

Darwin's Woodpecker Finch [*Camarhynchus pallidus*] (Sclater and Salvin 1870) *Proceedings of the Zoological Society of London*: 327. = W oodpecker Finch.

Darwin's Mangrove Finch [*Camarhynchus heliobates*] (Snodgrass and Heller 1901) *Condor* **3**, 96. = Mangrove Finch.

Darwin's Warbler Finch [*Certhidea olivaea*] (Gould 1837) *Proceedings of the Zoological Society of London* **5**, 7. = Green Warbler-Finch.

Darwin's Cocos Island Finch [*Pinaroloxias inornata*] (Gould 1843) *Proceedings of the Zoological Society of London* **11**, 104. = Cocos Finch.

It has been suggested that "Charles's Mockingbird [*Nesomimus trifasciatus*]" (now the Floreana Mockingbird) and "Charles's Tree Finch [*Camarhynchus pauper*]" (now the Medium Tree Finch) were named after Charles Darwin (Beolens and Watkins 2003: 99). However, no such eponymic acknowledgement to Charles Darwin appears in John Gould's original description and naming of these birds (Gould 1837b,c). Perhaps, if they were ever so named in the vernacular, it should read "Charles" as it would probably have been for Charles (now Floreana) Island of the Galapagos. Both birds were collected on Charles Island, to which the Medium Tree Finch is endemic (and is now critically endangered there) and on which the Floreana Mockingbird is now extinct, although surviving on two small adjacent islands as an endangered species (Jaramillo 2011: 682; Cody 2005: 488, respectively).

APPENDIX 3

Birds Collected by Charles Darwin During the Voyage of the *Beagle*

This introductory text and subsequent listing is closely based on that of Frank D. Steinheimer (2004), with three additions kindly provided by him (personal communication, email, F. D. Steinheimer–C. B. Frith, February 2015) and with additions and modifications by myself. It includes all actual specimens that Steinheimer could locate. Similar lists of Darwin's specimens have been published for other animals than birds, for example, invertebrates at Oxford University Museum (G. Chancellor et al. 1988) and fishes by Daniel Pauly (2004). The following list does not follow the general procedure in giving the entries in Darwin's number sequence, as I have revised the avian orders, families, genera, and species to the systematic order in which they appear in the IOC World Bird List (Gill and Donsker 2015) to comply with the taxonomy and nomenclature of the rest of this book. Vernacular names in normal UPPER CASE typeface are those applied by that IOC list where they differ from those presented by Steinheimer (2004). The first scientific name for each species (not subspecies) listed following, in **bold typeface**, is that given by Steinheimer (2004, from Peters et al. 1934–1987); the second, in underlined *UPPERCASE ITALIC* typeface, is that of the IOC World Bird List where it differs from the previous one; and the third, in *lowercase italics*, is the name applied in *The Zoology of the Voyage of H. M. S. Beagle, under the Command of Captain Fitzroy, R. N., During the Years 1832 to 1836. Part III: Birds* (Gould and Darwin 1838–1841).

Squared parentheses contents indicate comments and references added by Steinheimer (2004). Sex and age, when given, are as on the specimen label, in Darwin's notes, or in the *Catalogue of Birds* (Sharpe et al. 1874–1895). The determination of subspecies by Steinheimer (2004) are mainly based on collecting locality, cited from Peters et al. (1934–1987), except for the three genera of Darwin's Finches that follow Sulloway (1982a). The total number of specimens slightly differs to Sulloway (1982a) and Steinheimer (2003).

The authorship of new names has mainly been attributed to John Gould by Steinheimer (2004) and to George Robert Gray only in the few cases when the accompanying text refers to him. However, a small number of corrections to authorship of scientific names have subsequently been made, which are not included here (but see Steinheimer et al. 2006). The majority of the numerous citations to the title, author, and year of publication of new bird species descriptions are not included in the References of this book, as indeed they were not included in the bibliography of Steinheimer's paper, as this is standard practice in the ornithological literature. Also the abbreviation of names of widely known ornithologists (e.g., Wagl. [Johann Wagler 1800–1832] and Lath. [John Latham 1740–1837]) are not clarified below, as is also standard practice in the ornithological literature.

The names and authors cited from *The Zoology of the Voyage of H. M. S. Beagle* (Gould and Darwin 1838–1841; indicated by "Z:") may sometimes differ from what modern nomenclature would regard as correct. If a name on a colour plate has been published earlier than its accompanying text according to Sherborn (1897), then the name on the plate is given when referring to type status followed by the plate number in squared parentheses. Differences in the plate sequence between a facsimile copy and two original sets of *The Zoology of the Voyage of H. M. S. Beagle*—now housed at The Natural History Museum, Tring, and the University College, London—are addressed for each case (cf. Steinheimer 2003, 2004; Steinheimer et al. 2006). The word "status" refers to the known status/locality of the specimen(s).

It is important to note that the citations to Gould 1839 refer to the publication treated in the body of this book and in the References as Gould and Darwin (1838–1841).

Abbreviations used:

BMNH = British Museum (Natural History), now The Natural History Museum, Tring.
Cat. = *Catalogue of Birds in the British Museum* (Sharpe et al. 1874–1895).
CD = Darwin's Ornithological Notes (Barlow 1963), Specimen Lists (R. Keynes 2000).
CUMZ = Cambridge University Museum of Zoology, Cambridge.
ex. coll. = a specimen previously in the collection of the named institution or person.
fide = according to.

imm. = immature.
juv. = juvenile.
Leg. = Lat. legit; collected by.
LIVCM = Liverpool Museum National Museums & Galleries on Merseyside, Liverpool.
Loc. = collecting locality as on label and/or in CD.
MANCH = Manchester Museum University of Manchester, Manchester, England.
MNHN = Muséum National d'Histoire Naturelle, Paris.
nec. = not being, or not.
NMSE = National Museums of Scotland, Edinburgh.
RMNH = Rijks Museum van Natuurlijke Historie, now Naturalis Nationaal Natuurhistorisch Museum, Leiden.
ssp. = subspecies.
USNM = United States National Museum, The Smithsonian Institution, Washington.
VMM = Victoria Museum, Melbourne.
Z: = *The Zoology of the Voyage of H. M. S. Beagle, Part III: Birds* (Gould and Darwin 1838–1841), followed by the pagination.
ZMD = The Zoology Museum, University of Dundee, Dundee (cf. Steinheimer 2004).
ZSL = Zoological Society of London.

NON-PASSERINES

TINAMIDAE:

Rhynchotus rufescens **(Temminck, 1815).** RED-WINGED TINAMOU [CD 1382]. Z: 120: *Rhynchotus rufescens* Wagl. Loc.: Uruguay: Maldonado. July 1833. Remark: perhaps a second bird of this species had been collected; cf. Z: 120: "my specimens were procured at Maldonado." Material: skin/mount. Status: missing.

Nothoprocta perdicaria perdicaria **(Kittlitz, 1830).** CHILEAN TINAMOU [CD 2159]. Z: 119–120: *Nothura perdicaria* G. R. Gray. Loc.: Chile: Valparaso. August 1834. Material: skin/mount. Status: missing.

Nothoprocta perdicaria perdicaria **(Kittlitz, 1830)**. CHILEAN TINAMOU [CD 2427]. Z: 119–120: *Nothura perdicaria* G. R. Gray. Loc.: Chile: Valparaso. August–November 1834. Material: single egg. Status: missing.

Nothura darwinii darwinii **(Gray, 1867).** DARWIN'S NOTHURA [CD 1447]. Z: 119: *Northura* [*sic*] *minor* Wagl. Holotype *Nothura darwinii* Gray, 1867.

Loc.: Argentina: Bahia Blanca. 1833. Ex. coll. ZSL. Material: ex mount, ad. Status: BMNH 1855.12.19.107. Cat. XXVII: 563: a.

***Nothura maculosa maculosa* (Temminck, 1815).** SPOTTED NOTHURA [CD 1223]. Z: 119: *Northura* [*sic*] *major* Wagl. Loc.: Uruguay: Maldonado: northern shores of the Plata. 1833. Ex. coll. ZSL. Material: ex mount, ad. Status: BMNH 1855.12.18.34. Cat. XXVII: 560: e.

***Nothura maculosa maculosa* (Temminck, 1815).** SPOTTED NOTHURA [CD 1378]. Z: 119: *Northura* [*sic*] *major* Wagl. Loc.: Uruguay: Maldonado: northern shores of the Plata. 1833. Material: single egg. Status: Cambridge University Museum of Zoology and Laboratory for Development and Evolution, Department of Zoology, Cambridge, UK. Ex. coll. Alfred Newton.

RHEIDAE:

***Rhea americana americana* (Linnaeus, 1758).** GREATER RHEA [CD 814]. Z: 120–123: *Rhea americana* Lath. Loc.: Argentina: Bahia Blanca. October 1832. Material: single egg. Status: missing.

***Pterocnemia pennata pennata* (d'Orbigny, 1834).** LESSER RHEA *RHEA PENNATA* [CD 1832–1838, 2004, 2147, 2148]. Z: 123–125: *Rhea darwinii* Gould. Holotype *Rhea Darwinii* Gould, 1837 [based on a composite specimen]. Loc.: Argentina: Port Desire, Port Famine, Port St. Julian, Santa Cruz: February–July 1834. Ex. coll. ZSL. Material: mount. Status: missing.

***Pterocnemia pennata pennata* (d'Orbigny, 1834).** LESSER RHEA *RHEA PENNATA* [CD 1814]. Z: 123–125: *Rhea darwinii* Gould. Paratype *Rhea darwinii* Gould, 1837. Loc.: Argentina: Port Desire. January–February 1834. Remark: BMNH possesses one data-less egg, BMNH 1859.3.25.31, from "Rhea darwinii" = *Pterocnemia pennata*, which is registered as from John Gould's collection; see also Keynes 2000: 189 for detailed description of the egg.

ANATIDAE:

***Tachyeres patachonicus* (King, 1828).** FLYING STEAMER-DUCK [CD 1661]. Z: 136: *Micropterus brachypterus* Eyton. Loc.: Argentina: Port Desire, 20 miles up the creek. January 1834. Remark: no authentic locality data on label, was always believed to come from the Falklands, but Darwin did not collect any duck there. If locality is correct, then the duck was shot by Fuller or Covington. Material: mount. Status: probably CUMZ 12/Ana/60/a/5.

***Chloephaga melanoptera* (Eyton, 1838).** ANDEAN GOOSE [CD n/a]. Z: 134: *Anser melanopterus* Eyton. Loc.: Chile: Valparaso. Remark: bird mentioned in Z: was bought by FitzRoy on a local market at Valparaso. Material & Status: not collected by Darwin.

***Chloephaga picta leucoptera* (Gmelin, 1789).** UPLAND GOOSE [CD 576 specimens in spirit]. Z: 134: *Chloephaga magellanica* Eyton. Loc.: Falkland Islands/Islas Malvinas: East Falkland Island. March 1833. Ex. coll. Eyton. Material: trachea in spirit. Status: missing.

***Chloephaga hybrida malvinarum* (Phillips, 1916).** KELP GOOSE [CD 577 specimens in spirit]. Z: 134–135: *Bernicla antarctica* Steph. Loc.: Falkland Islands/Islas Malvinas: East Falkland Island. March 1833. Ex. coll. Eyton. Material: trachea in spirit. Status: missing.

***Amazonetta brasiliensis ipecutiri* (Vieillot, 1816).** BRAZILIAN TEAL [? CD 1419/1421/1436, see also following]. Z: 135: *Querquedula erythrorhyncha* Eyton. Loc.: Argentina: Buenos Aires. October 1833. Material: skin/mount. Status: missing.

?*Amazonetta brasiliensis* ss (Gmelin, 1789). BRAZILIAN TEAL [? CD 1778/1779, see following]. Z: 135: *Querquedula erythrorhyncha* Eyton Loc.: Chile: Straits of Magellan: Cape Negro (fresh water). February 1834. Remark: Z: claims this species to occur near Straits of Magellan–either very rare vagrant or different species. Material: skin/mount. Status: missing.

***Anas platalea* (Vieillot, 1816).** RED SHOVELER [? CD 1419/1421/1436, see previously]. Z: 135: *Rhynchaspis maculatus* Gould. Loc.: Argentina: Rio Plata near Buenos Aires. October 1833. Material: skin/mount. Status: missing.

***Anas bahamensis galapagensis* (Ridgway, 1889).** WHITE-CHEEKED PINTAIL [CD 3299]. Z: 135: *Paecilonitta bahamensis* Eyton. Loc.: Ecuador: Galápagos Archipelago. October 1835. Material: skin/mount, male. Status: missing.

***Anas flavirostris flavirostris* (Vieillot, 1816).** YELLOW-BILLED TEAL [? CD 1419/1421/1436, see previous and following]. Z: 135–136: *Querquedula creccodes* Eyton. Loc.: Argentina: Rio Plata near Buenos Aires. 1833. Material: skin/mount. Status: missing.

***Anas flavirostris flavirostris* (Vieillot, 1816).** YELLOW-BILLED TEAL [? CD 1778/1779, see previously]. Z: 135–136: *Querquedula creccoïdes* Eyton. Loc.: Chile: Straits of Magellan: Cape Negro (fresh water). February 1834. Material: skin/mount. Status: missing.

***Anas georgica spinicauda* (Vieillot, 1816).** YELLOW-BILLED PINTAIL [CD 1454]. Z: 135: *Dafila urophasianus* Eyton. Loc.: Argentina: Bahia Blanca. 1833. Remark: Cat. (Salvadori 1895: 282) listed *D. urophasianus* under *P. bahamensis*, which does not fit with the locality of the specimen. Therefore, it is more likely that the taxon of Z: is *Anas spinicauda.* Material: skin/mount. Status: missing.

SPHENISCIDAE:

***Spheniscus humboldti* (Meyen, 1834).** HUMBOLDT PENGUIN [CD 2321]. Z: 137: *Spheniscus humboldtii* [*sic*] Meyen. Loc.: Chile: coast near Valparaso. September–November 1834. Material: skin/mount. Status: missing.

HYDROBATIDAE:

***Oceanites oceanicus* ssp. (Kuhl, 1820).** WILSON'S STORM PETREL [CD 1349]. Z: 141: *Thalassidroma oceanica* Bona Loc.: Uruguay: Maldonado. July 1833. Material: skin/mount. Status: missing.

PROCELLARIIDAE:

***Macronectes giganteus* (Gmelin, 1789).** SOUTHERN GIANT PETREL [CD 2080]. Z: 139–140: *Procellaria gigantea* Gmel. Loc.: Chile: Port Famine. June 1834. Material: skin/ex mount, imm. Status: missing.

***Fulmarus glacialoides* (Smith, 1840).** SOUTHERN FULMAR [CD 1335]. Z: 140: *Procellaria glacialodes* A. Smith. Loc.: Argentina: Bay of St Mathias [Golf S. Matias]. [? April] 1833. Material: skin/ex mount. Status: missing.

***Daption capense australe* (Mathews, 1913).** CAPE PETREL [? CD 3413]. Z: 140–141: *Daption capensis* Steph. Loc.: New Zealand: Bay of Islands. December 1835. Remark: probably listed as "bird" in CD. Keynes 2000 accidentally referred to CD No. 3189, which is *Puffinus griseus*. Material: skin/mount. Status: missing.

***Pachyptila desolata? banksi* (Smith, 1840).** ANTARCTIC PRION [CD n/a]. Z: 141: *Prion vittatus* Cuv. Loc.: South America and off shore islands. Remark: Ranges suggests *P. desolata* versus *vittata*. Material & Status: not collected.

***Procellaria cinerea* (Gmelin, 1789).** GREY PETREL [CD 1624]. Z: 137–138: *Puffinus cinereus* Steph. Loc.: Argentina: little south of the mouth of the Plata. November 1833. Remark: species attribution uncertain, could also be *P. griseus*. Material: skin/mount. Status: missing.

***Procellaria cinerea* (Gmelin, 1789).** GREY PETREL [CD 1816]. Z: 137–138: *Puffinus cinereus* Steph. Loc.: Chile: Tierra del Fuego: Port Famine in the Straits of Magellan. January–February1834. Remark: species attribution uncertain, could also be *P. griseus*. Material: skin/mount. Status: missing.

***Puffinus griseus* (Gmelin, 1789).** SOOTY SHEARWATER CD 3189. Z: 137–138: *Puffinus cinereus* Steph. Loc.: Peru: Lima: Callao Bay. August 1835. Ex. coll. ZSL. Remark: accidentally referred to as *Daption capensis* in Keynes 2000. Material: ex mount, ad. Status: BMNH 1855.12.19.133. Cat. XXV: 388: o.

PELECANOIDIDAE:

***Pelecanoides garnotii* (Lesson, 1828).** PERUVIAN DIVING PETREL [CD 3190]. Z: 139: *Pelecanoides garnotii* G. R. Gray. Loc.: Peru: Callao Bay [on label], Iquique [Z:]. July 12–15, 1835. Ex. coll. Salvin & Godman, ex. Eyton. Remark: On label as Calao [Callao Bay], which is near Iquique. Very likely one of Darwin's birds. Material: skin, ad. Status: probably BMNH 1888.5.18.167. Cat. XXV: 439: g.

***Pelecanoides urinator berard* (Gaimard, 1823).** COMMON DIVING PETREL *PELECANOIDES URINATRIX* [CD 1782]. Z: 138–139: *Pelecanoides berardi* G. R. Gray. "Holotype" [fide Cat.] *Haladroma tenuirostris* Eyton MS [not valid]. Loc.: Chile: Straits of Magellan. January–February 1834. Ex. coll. Gould, ex. ZSL. Remark: no indication of Darwin on label. Material: skin, imm. Status: probably BMNH 1881.5.1.6015. Cat. XXV: 438: r.

PODICIPEDIDAE:

***Rollandia rolland chilensis* (Lesson, 1828).** WHITE-TUFTED GREBE [CD 1429]. Z: 137: *Podiceps chilensis* Garnot. Loc.: Argentina: near Buenos Aires. October 1833. Ex. coll. Gould, ex. ZSL. Remark: Preparation style and stuffing would speak for a Darwin bird, though locality "Chile" on label. Material: skin, imm. Status: possible BMNH 1867.3.16.78. Cat. XXVI: 526: h.

***Rollandia rolland chilensis* (Lesson, 1828).** WHITE-TUFTED GREBE CD 2435 [original field label]. Z: 137: *Podiceps rollandii* Quoy et Gaim. Loc.: Chile: eastern coast of Isla de Chiloé. December 1834. Ex. coll. Gould, ex. ZSL. Material: skin, imm. Status: BMNH 1881.5.1.6001. Cat. XXVI: 526: a'.

***Rollandia rolland chilensis* (Lesson, 1828).** WHITE-TUFTED GREBE CD 1780 [original field label]. Z: 137: *Podiceps rollandii* Quoy et Gaim. Loc.: Chile: eastern coast of Isla de Chiloé [on label], Straits of Magellan [CD]. February 1834. Ex. coll. Gould, ex. ZSL. Material: skin, ad. Status: BMNH 1881.5.1.5999. Cat. XXVI: 526: z' [in error: z].

***Rollandia rolland rolland* (Quoy & Gaimard, 1824).** WHITE-TUFTED GREBE [CD 1917]. Z: 137: *Podiceps rollandii* Quoy et Gaim. Loc.: Falkland Islands/Islas Malvinas: East Falkland Island. March 1834. Material: skin/mount. Status: missing.

***Podiceps occipitalis occipitalis* (Garnot, 1826).** SILVERY GREBE [CD 1918]. Z: 136: *Podiceps kalipareus* Quoy & Gaim. Loc.: Falkland Islands/Islas Malvinas: East Falkland Island. March 1834. Ex. coll. Gould, ex. ZSL. Remark: the specimen listed would fit to Darwin's specimen style. Material: skin, ad. Status: perhaps 1860.1.16.89. Cat. XXVI: 537: p/q.

***Podiceps occipitalis occipitalis* (Garnot, 1826).** SILVERY GREBE [CD 713]. Z: 136: *Podiceps kalipareus* Quoy & Gaim. Loc.: Argentina: Bahia Blanca. September 1832. Material: skin/mount. Status: missing.

PHOENICOPTERIDAE:

***Phoenicopterus ruber ruber* (Linnaeus, 1758).** GREATER FLAMINGO *PHOENICOPTERUS ROSEUS* [CD 3362]. Z: n/a. Loc.: Ecuador: Galápagos Archipelago. October 1835. Material: stomach contents in spirit. Status: missing.

THRESKIONITHIDAE:

***Theristicus melanopis melanopis* (Gmelin, 1789).** BLACK-FACED IBIS [CD 1773]. Z: 128–129: *Theristicus melanops* Wagl. Loc.: Argentina: Port Desire: desert gravel plains. 23 December 1833–4 January 1834. Remark: Z: mentions at least two specimens, but only one in CD. Material: skin/mount. Status: missing.

***Plegadis chihi* (Vieillot, 1817).** WHITE-FACED IBIS [CD 1458]. Z: 129: *Ibis (falcinellus) ordi* Bona Loc.: Argentina: Rio Negro. August 1833. Material: skin/mount. Status: missing.

ARDEIDAE:

***Nycticorax nycticorax obscurus* (Bonaparte, 1855).** BLACK-CROWNED NIGHT HERON [CD 2184]. Z: 128: *Nycticorax americanus* Bona Loc.: Chile: Valparaso. August–September 1834. Material: skin/mount, female/imm. Status: missing.

***Nyctanassa violacea pauper* (Sclater & Salvin, 1870).** YELLOW-CROWNED NIGHT HERON [CD 3300]. Z: 128: *Nycticorax violaceus* Bona Loc.: Ecuador: Galápagos Archipelago. October 1835. Material: skin/mount, female/imm. Status: missing.

***Nyctanassa violacea pauper* (Sclater & Salvin, 1870).** YELLOW-CROWNED NIGHT HERON [CD 3301]. Z: 128: *Nycticorax violaceus* Bona Loc.: Ecuador: Galápagos Archipelago. October 1835. Remark: Z: speaks only of one specimen, although two are entered in CD. Material: skin/mount, female/imm. Status: missing.

***Ardea herodias cognata* (Bangs, 1903).** GREAT BLUE HERON [CD 3296]. Z: 128: *Ardea herodias* Linn. Loc.: Ecuador: Galápagos Archipelago. October 1835. Material: skin/mount, fem. Status: missing.

***Ardea alba egretta* (Gmelin, 1789).** GREAT EGRET [CD 1269]. Z: 128: *Egretta leuce* Bona Loc.: Uruguay: Maldonado. 1833. Material: skin/mount. Status: missing.

FREGATIDAE:

***Fregata aquila* (Linnaeus, 1758).** ASCENSION FRIGATEBIRD [CD n/a]. Z: 146: *Fregata aquila* Cuv. Loc.: United Kingdom: Atlantic Ocean: Ascension Island. July 1836. Material & Status: not collected.

***Fregata magnificens/minor ridgwayi* (Mathews, 1914).** MAGNIFICENT/ GREAT FRIGATEBIRD [CD n/a]. Z: 146: *Fregata aquila* Cuv. Loc.: Ecuador: Galápagos Archipelago. September–October 1835. Material & Status: not collected.

SULIDAE:

***Sula leucogaster leucogaster* (Boddaert, 1783).** BROWN BOOBY [CD 413, shared with an egg of *Anous*]. Z: n/a. Loc.: Brazil: St. Paul Rocks [= Sao Paulo Island, Atlantic Ocean]. April 1832. Material: several eggs. Status: missing.

PHALACROCORACIDAE:

***Phalacrocorax atriceps atriceps* (King, 1828).** IMPERIAL SHAG *LEUCOCARBO ATRICEPS* [CD 1756]. Z: 145: *Phalacrocorax carunculatus* Stephens. Loc.: Argentina: Port St. Julian. January 1834. Material: skin/ mount. Status: missing.

CARTHARTIDAE:

***Cathartes aura falklandicus* (Sharpe, 1873).** TURKEY VULTURE [CD 1915]. Z: 8–9: *Cathartes aura* Illi. Loc.: Falkland Islands/Islas Malvinas: East Falkland Island. March 1834. Material: skin/mount, female. Status: Not yet looked for by Steinheimer.

***Coragyps atratus* (Bechstein, 1793).** BLACK VULTURE [CD n/a]. Z: 7–8: *Cathartes atratus* Rich. & Swain. Loc.: Argentina: Rio Negro: Colorado. Material & Status: not collected.

***Vultur gryphus* (Linnaeus, 1758).** ANDEAN CONDOR [CD n/a]. Z: 3–6: *Sarcoramphus gryphus* Bona Loc.: Chile, Argentina. Material & Status: not collected.

ACCIPITRIDAE:

***Circus buffoni* (Gmelin, 1788).** LONG-WINGED HARRIER [CD 1396]. Z: 29–30: *Circus megaspilus* Gould. Holotype *Circus megaspilus* Gould, 1837. Loc.: Uruguay: Maldonado: La Plata. July 1833. Ex. coll. ZSL. Material: ex mount, imm. Status: BMNH 1855.12.19.258. Cat. I: 63: a.

***Circus cinereus* (Vieillot, 1816).** CINEREOUS HARRIER [CD 2822]. Z: 30–31: *Circus cinerius* [*sic*] Vieill. Loc.: Chile: Concépcion or Coquimbo. 1835. Ex. coll. Norwich Castle Museum/Gurney Collection, ex. S. G. Buxton, ex. ZSL. Material: ex mount, male [CD], ad. [label]. Status: Probably BMNH 1955.6.N20.3488.

***Circus cinereus* (Vieillot, 1816).** CINEREOUS HARRIER CD 1881. Z: 30–31: *Circus cinerius* [*sic*] Vieill. Loc.: Falkland Islands/Islas Malvinas: East Falkland Island. March 1834. Ex. coll. Norwich Castle Museum/Gurney Collection, ex. S. G. Buxton, ex. ZSL. Material: skin, female/imm. Status: BMNH 1955.6.N20.3497.

***Circus cinereus* (Vieillot, 1816).** CINEREOUS HARRIER [CD 1160]. Z: 30–31: *Circus cinerius* [*sic*] Vieill. Loc.: Falkland Islands/Islas Malvinas: East Falkland

Island. March 1833. Material: skin/mount, male. Status: RMNH, Cat. No. 7, Ex. coll. Gustav Adolph Frank, acquisition 1860, John Gould/ZSL. cf. van Grouw & Steinheimer 2008.

***Circus cinereus* (Vieillot, 1816).** CINEREOUS HARRIER [CD 1054]. Z: 30–31: *Circus cinerius* [*sic*] Vieill. Loc.: Falkland Islands/Islas Malvinas: East Falkland Island. March 1833. Ex. coll. Norwich Castle Museum/Gurney Collection, ex. ZSL. Material: ex mount, female/imm. Status: BMNH 1955. 6.N20.3487.

***Buteo polyosoma polyosoma* (Quoy & Gaimard, 1824).** VARIABLE HAWK *GERANOAETUS POLYSOMA* [CD 1781]. Z: 26–27: *Buteo varius* Gould. Holotype [fide Warren 1966] *Buteo varius* Gould, 1837. Loc.: Chile: Straits of Magellan: Cape Negro, 20 km NNE of Punta Arenas. February 1834. Ex. coll. Norwich Castle Museum/Gurney Collection, ex. ZSL. Remark: Type designation uncertain. Material: ex mount, imm. Status: BMNH 1955.6.N20.2410.

***Buteo polyosoma polyosoma* (Quoy & Gaimard, 1824).** VARIABLE HAWK *GERANOAETUS POLYSOMA* [CD 1758]. Z: 26–27: *Buteo varius* Gould. Loc.: Argentina: Port St. Julian. January 1834. Ex. coll. ZSL. Remark: In Gould's first description, only one bird is mentioned; but two birds are referred to in Z. The bird listed here is not the BMNH specimen collected by King. Material: ex mount, female/imm. Status: BMNH 1855.12.19.208. Cat. I: 173: m.

***Buteo polyosoma polyosoma* (Quoy & Gaimard, 1824).** VARIABLE HAWK *GERANOAETUS POLYSOMA* [CD 2136]. Z: 26: *Buteo erythronotus* [King]. Loc.: Chile: Isla de Chiloé. July 1834. Ex. coll. Norwich Castle –Museum/ Gurney Collection, ex. ZSL. Material: ex mount, female, ad. Status: probably BMNH 1955.6.N20.2412.

***Buteo polyosoma polyosoma* (Quoy & Gaimard, 1824).** VARIABLE HAWK *GERANOAETUS POLYSOMA* [CD 1916]. Z: 26: *Buteo erythronotus* [King]. Loc.: Falkland Islands/Islas Malvinas: East Falkland Island. March 1834. Ex. coll. "old collection." Remark: This specimen can only derive from three collections: FitzRoy, King, or Darwin. Material: ex mount, male, ad. Status: perhaps BMNH unregistered specimen. Cat. I: 163: a.

***Buteo galapagoensis* (Gould, 1837).** GALAPAGOS HAWK [CD 3297]. Z: 23–25: *Craxirex galapagoensis* Gould. Syntype *Polyborus galapagoensis* Gould, 1837. Loc.: Ecuador: Galápagos Archipelago. October 1835. Ex. coll. ZSL. Material: ex mount, male, ad. Status: BMNH 1855.12.19.202. Cat. I: 171: a.

***Buteo galapagoensis* (Gould, 1837).** GALAPAGOS HAWK [CD 3298]. Z: 23–25: *Craxirex galapagoensis* Gould. Syntype *Polyborus galapagoensis* Gould, 1837. Loc.: Ecuador: Galápagos Archipelago. October 1835. Ex. coll. ZSL. Material: ex mount, female/imm. Status: BMNH 1855.12.19.203. Cat. I: 171: b.

***Buteo ventralis* (Gould, 1837).** RUFOUS-TAILED HAWK [CD 2030]. Z: 27–28: *Buteo ventralis* Gould. Holotype *Buteo ventralis* Gould, 1837. Loc.: Argentina: Santa Cruz: April 1834. Ex. coll. ZSL. Material: ex mount, imm. Status: BMNH 1855.12.19.204. Cat. I: 190:

RALLIDAE:

***Coturnicops notata notata* (Gould, 1841).** SPECKLED RAIL COTURNICOPS NOTATUS [? CD 1453/1424, see following]. Z: 132: *Zapornia notata* Gould. Holotype *Zapornia notata* Gould, 1841. Loc.: Argentina: Rio Plata: shot on board of the Beagle near Buenos Aires. 1833. Material: skin, ad. Status: BMNH 1964.44.1. Cat. XXIII: 129: a.

***Laterallus melanophaius melanophaius* (Vieillot, 1819).** RUFOUS-SIDED CRAKE [? CD 1235/1295, see following]. Z: 132: *Crex lateralis* Licht. Loc.: Uruguay: Maldonado near Rio Plata. May–June 1833. Material: skin/mount. Status: missing.

***Laterallus melanophaius melanophaius* (Vieillot, 1819).** RUFOUS-SIDED CRAKE [? CD 1295/1235, see previously]. Z: 132: *Crex lateralis* Licht. Loc.: Uruguay: Maldonado near Rio Plata. May–June 1833. Material: skin/mount. Status: missing.

***Laterallus spilonotus* (Gould, 1841).** GALAPAGOS CRAKE\[? CD 3353/3351, see following]. Z: 132–133: *Zapornia spilonota* Gould. Syntype *Zapornia spilonota* Gould, 1841. Loc.: Ecuador: Galápagos Archipelago [James Isl. fide Rothschild/Hartert Nov. Zool. VI, 1899: 185]. October 1835. Material: skin/mount. Status: missing [unless the specimen of FitzRoy is the bird listed in CD: BMNH 1837.2.21.404. Cat. XXIII: 113: b. Vell. Cat. 40: 180 a, cf. Warren 1966].

***Laterallus spilonotus* (Gould, 1841).** GALAPAGOS CRAKE [CD 3352]. Z: 132–133: *Zapornia spilonota* Gould. Syntype *Zapornia spilonota* Gould, 1841 & Syntype *Porzana galapagoensis* Sharpe, 1894. Loc.: Ecuador: Galápagos Archipelago [James Isl. fide Rothschild/Hartert Nov. Zool.VI, 1899: 185]. October 1835. Material: skin, female, ad. Status: BMNH 1964.45.1. Cat. XXIII: 113: a.

***Laterallus spilonotus* (Gould, 1841).** GALAPAGOS CRAKE [? CD 3351/3353, see previously]. Z: 132–133: *Zapornia spilonota* Gould. Syntype *Zapornia spilonota* Gould, 1841. Loc.: Ecuador: Galápagos Archipelago [James Isl. fide Rothschild/Hartert Nov. Zool. VI, 1899: 185]. October 1835. Remark: This specimen is not a syntype of *Porzana galapagoensis*. The second syntype of this name is a FitzRoy specimen. Material: skin, female, imm. Status: BMNH 1964.46.1. Cat. XXIII: 138: e.

***Rallus philippensis andrewsi* (Mathews, 1911).** BUFF-BANDED RAIL *GALLIRALLUS PHILLIPENSIS* [CD 3591]. Z: 133: *Rallus phillipensis* [*sic*] Linn. Loc.: Australia: Cocos/Keeling Islands. April 1836. Material: skin/mount. Status: missing.

***Aramides ypecaha* (Vieillot, 1819).** GIANT WOOD-RAIL CD 1435 [original field label]. Z: 133: *Rallus ypecaha* Vieill. Loc.: Argentina: Buenos Aires. 1833. Remark: Darwin added the species name on his label. Material: skin. Status: BMNH unregistered specimen.

***Ortygonax rytirhynchos landbecki* (Hellmayr, 1932).** PLUMBEOUS RAIL *PARDIRALLUS SANGUINOLENTUS* [CD 2183]. Z: 133: *Rallus sanguinolentus* Swains. Loc.: Chile: Valparaso. August–September 1834. Material: skin/mount, male. Status: missing.

***Porphyrula alleni* (Thomson, 1842).** ALLEN'S GALLINULE *PORPHYRIO ALLENI* [CD 3900]. Z: 133–134: *Porphyrio simplex* Gould. Holotype *Porphyrio simplex* Gould, 1841. Loc.: United Kingdom: Atlantic Ocean: Ascension Island. July 1836. Remark: Barlow 1933: 413–415 gives no further details. Taylor 1998: 478 added Darwin's record with question mark. One other record than Darwin's is known for Ascension Island. The nomenclature and taxonomy of this taxon would need further studies. Material: skin/mount, female/imm. Status: missing.

***Gallinula chloropus garmani* (Allen, 1876).** COMMON MOORHEN [CD 2821]. Z: 133: *Fulica galeata* G. R. Gray. Loc.: Chile: Concépcion. 1835. Material: skin/mount. Status: missing.

***Porphyriops melanops melanops* (Vieillot, 1819).** SPOT-FLANKED GALLINULE *GALLINULA MELANOPS* [? CD 1424/1453, see previously]. Z: 133: *Gallinula crassirostris* J. E. Gray. Loc.: Argentina: Buenos Aires: banks of the Plata. 1833. Material: skin/mount. Status: missing.

***Porphyriops melanops crassirostris* (J. E. Gray, 1829).** SPOT-FLANKED GALLINULE *GALLINULA MELANOPS* [CD 2164]. Z: 133: *Gallinula crassirostris* J. E. Gray. Loc.: Chile: Valparaso. August–September 1834. Material: skin/mount, male. Status: missing.

***Porphyriops melanops crassirostris* (J. E. Gray, 1829).** SPOT-FLANKED GALLINULE *GALLINULA MELANOPS* [CD 2165]. Z: 133: *Gallinula crassirostris* J. E. Gray. Loc.: Chile: Valparaso. August–September 1834. Material: skin/mount, female. Status: missing.

CHIONIDIDAE:

***Chionis alba* (Gmelin, 1789).** SNOWY SHEATHBILL [CD n/a]. Z: 118–119: *Chionis alba* Forst. Loc.: Falkland Islands/Islas Malvinas: East Falkland Island. March 1833 or 1834. Material & Status: shot, but not preserved.

HAEMATOPODIDAE:

***Haematopus ostralegus durnfordi* (Sharpe, 1896).** EURASIAN OYSTERCATCHER [CD 1383]. Z: 128: *Haematopus palliatus* Temm. Loc.: Argentina/Uruguay: Guritti Island: Rio de la Plata or Maldonado. July 1833. Material: skin/mount. Status: missing.

***Haematopus ostralegus durnfordi* (Sharpe, 1896).** EURASIAN OYSTER-CATCHER [? CD 1420]. Z: 128: *Haematopus palliatus* Temm. Loc.: Argentina: Rio de la Plata near Buenos Aires. July 1833. Material: skin/mount. Status: missing.

RECURVIROSTRIDAE:

***Himantopus himantopus melanurus* (Vieillot, 1817).** BLACK-WINGED STILT [CD 1221]. Z: 130: *Himantopus nigricollis* Vieill. Loc.: Uruguay: Maldonado. 1833. Material: skin/mount. Status: missing.

***Himantopus himantopus melanurus* (Vieillot, 1817).** BLACK-WINGED STILT [?? CD 1422/1423]. Z: 130: *Himantopus nigricollis* Vieill. Loc.: Argentina: provinces bordering the Plata. 1833. Material: skin/mount. Status: missing.

CHARADRIIDAE:

***Belonopterus chilensis lampronotus* (Wagler, 1827).** SOUTHERN LAPWING *VANELLUS CHILENSIS* [CD 1602]. Z: 127: *Philomachus cayanus* G. R. Gray. Loc.: Uruguay: Rio de la Plata near Montevideo. November 1833. Material: skin/mount. Status: missing.

***Pluvialis dominica dominica* (L. S. Mller, 1776).** AMERICAN GOLDEN PLOVER [CD 1606]. Z: 126: *Charadrius virgininus* Borkh. Loc.: Uruguay: banks of the Plata. November 1833. Material: skin/mount. Status: missing.

***Charadrius hiaticula semipalmatus* (Bonaparte, 1825).** SEMIPALMATED PLOVER *CHARADRIUS SEMIPALMATUS* [CD 3357]. Z: 128: *Hiaticula semipalmata* G. R. Gray. Loc.: Ecuador: Galápagos Archipelago. September–October 1835. Material: skin/mount, female. Status: missing.

***Charadrius alexandrinus occidentalis* (Cabanis, 1872).** KENTISH PLOVER CD 2188. Z: 127: *Hiaticula azarae* G. R. Gray. Loc.: Chile: Valparaso [not on label]. August–September 1834. Ex. coll. Salvin & Godman, ex. Gould, ex. ZSL. Material: skin, ad. Status: BMNH 1891.10.20.337. Cat. XXIV: 292: w'.

***Charadrius collaris* (Vieillot, 1818).** COLLARED PLOVER [CD 1208]. Z: 127: *Hiaticula azarae* G. R. Gray. Loc.: Uruguay: banks of the Plata. 1833. Material: skin/mount. Status: missing.

***Charadrius collaris* (Vieillot, 1818).** COLLARED PLOVER [? CD 1435]. Z: 127: *Hiaticula azarae* G. R. Gray. Loc.: Argentina: banks of the Plata near Buenos Aires. October 1833. Material: skin/mount. Status: missing.

***Charadrius collaris* (Vieillot, 1818).** COLLARED PLOVER [CD 712]. Z: 127: *Hiaticula azarae* G. R. Gray. Loc.: Argentina: Bahia Blanca. September 1832. Material: skin/mount. Status: missing.

***Charadrius falklandicus* (Latham, 1790).** TWO-BANDED PLOVER [CD 1449; versus Keynes 2000]. Z: 127: *Hiaticula trifasciatus* G. R. Gray. Loc.: Argentina: Bahia Blanca. 1833. Material: skin/mount. Status: missing.

***Charadrius falklandicus* Latham, 1790.** TWO-BANDED PLOVER [CD 1433]. Z: 127: *Hiaticula trifasciatus* G. R. Gray. Loc.: Argentina: Bahia Blanca. 1833. Material: skin/mount. Status: missing.

***Zonibyx modestus* (Lichtenstein, 1823).** RUFOUS-CHESTED PLOVER *CHARADRIUS MODESTUS* CD 901. Z: 126: *Squatarola cincta* Jard. & Selby. Loc.: Chile: Tierra del Fuego: summits of highest mountains of Good Success Bay. 20 December 1832. Ex. coll. Gould, ex. ZSL. Material: skin, ad. Status: BMNH 1857.10.16.69. Cat. XXIV: 239: i.

***Zonibyx modestus* (Lichtenstein, 1823).** RUFOUS-CHESTED PLOVER *CHARADRIUS MODESTUS* CD 1817 [original field label]. Z: 126: *Squatarola cincta* Jard. & Selby. Loc.: Chile: Tierra del Fuego: Port Famine. Early February 1834. Ex. coll. Gould, ex. ZSL. Material: skin, ad. Status: BMNH 1857.10.16.68. Cat. XXIV: 239: k.

***Zonibyx modestus* (Lichtenstein, 1823).** RUFOUS-CHESTED PLOVER *CHARADRIUS MODESTUS* [CD 1145]. Z: 126: *Squatarola cincta* Jard. & Selby. Loc.: Falkland Islands/Islas Malvinas: East Falkland Island. March 1833. Ex. coll. Gould, ex. ZSL. Material: skin, ad. Status: probably BMNH 1859.3.25.84.

***Zonibyx modestus* (Lichtenstein, 1823).** RUFOUS-CHESTED PLOVER *CHARADRIUS MODESTUS* [? CD 1403]. Z: 126: *Squatarola cincta* Jard. & Selby. Loc.: Falkland Islands/Islas Malvinas: East Falkland Island. March 1833. Material: skin/mount. Status: missing.

***Zonibyx modestus* (Lichtenstein, 1823).** RUFOUS-CHESTED PLOVER *CHARADRIUS MODESTUS* [CD 2123]. Z: 126: *Squatarola cincta* Jard. & Selby. Loc.: Chile: Isla de Chiloé. July 1834. Material: skin/mount. Status: missing.

***Zonibyx modestus* (Lichtenstein, 1823).** RUFOUS-CHESTED PLOVER *CHARADRIUS MODESTUS* [CD 1236]. Z: 126–127: *Squatarola fusca* Gould. Holotype *Squatarola fusca* Gould, 1841. Loc.: Uruguay: Maldonado. 1833. Material: skin/mount. Status: missing.

***Oreopholus ruficollis* (Wagler, 1829).** TAWNY-THROATED DOTTEREL [CD 1263]. Z: 125–126: *Oreophilus totanirostris* Jard. & Selby. Loc.: Uruguay: Maldonado. 1833. Material: skin/mount. Status: missing.

***Oreopholus ruficollis* (Wagler, 1829).** TAWNY-THROATED DOTTEREL *OREOPHOLUS RUFICOLLIS* [CD 2166]. Z: 125–126: *Oreophilus totanirostris* Jard. & Selby. Loc.: Chile: Valparaso. August–September 1834. Material: skin/mount. Status: missing.

ROSTRATULIDAE:

***Nycticryphes semicollaris* (Vieillot, 1816).** SOUTH AMERICAN PAINTED SNIPE [? CD 1214]. Z: 131: *Rhynchaea semicollaris* G. R. Gray. Loc.: Uruguay:

Montevideo or Maldonado. November 1833. Material: skin/mount. Status: missing.

THINOCORIDAE:

***Attagis gayi gayi* (Geoffroy Saint-Hilaire & Lesson, 1830).** RUFOUS-BELLIED SEEDSNIPE [CD 2823]. Z: 117: *Attagis gayii* Less. Loc.: Chile: Cordillera of Coquimbo or Copiapò. 1835. Material: skin/mount. Status: missing.

***Attagis malouinus malouinus* (Boddaert, 1783).** WHITE-BELLIED SEED-SNIPE [CD 1402]. Z: 117: *Attagis falklandica* G. R. Gray. Loc.: Chile: extreme Southern part of Tierra del Fuego: summit of Katers peak (1,700 feet high) on Hermit Island. 1833. Material: skin/mount. Status: missing.

***Thinocorus rumicivorus rumicivorus* (Eschscholtz, 1829).** LEAST SEED-SNIPE [CD 710]. Z: 117–118: *Tinochorus* [*sic*] *rumicivorus* Eschsch. Loc.: Argentina: Bahia Blanca: sterile plains. February–March 1832. Material: skin/mount. Status: missing.

***Thinocorus rumicivorus rumicivorus* (Eschscholtz, 1829).** LEAST SEED-SNIPE [CD 711]. Z: 117–118: *Tinochorus* [*sic*] *rumicivorus* Eschsch. Loc.: Argentina: Bahia Blanca: sterile plains. February–March 1832. Material: tail, added to No. CD 710. Status: missing.

***Thinocorus rumicivorus rumicivorus* (Eschscholtz, 1829).** LEAST SEED-SNIPE [CD 1224]. Z: 117–118: *Tinochorus* [*sic*] *rumicivorus* Eschsch. Loc.: Uruguay: Maldonado. May–June 1833. Material: skin/mount, female. Status: missing.

***Thinocorus rumicivorus rumicivorus* (Eschscholtz, 1829).** LEAST SEEDSNIPE [CD 1273]. Z: 117–118: *Tinochorus* [*sic*] *rumicivorus* Eschsch. Loc.: Uruguay: Maldonado. May–June 1833. Material: skin/mount, male. Status: missing.

***Thinocorus rumicivorus rumicivorus* (Eschscholtz, 1829).** LEAST SEED-SNIPE [CD 707, specimens in spirit]. Z: 117–118 & 155–156: *Tinochorus* [*sic*] *rumicivorus* Eschsch. Loc.: Uruguay: Maldonado. June 1833. Material: specimen in alcohol. Status: missing.

***Thinocorus rumicivorus rumicivorus* (Eschscholtz, 1829).** LEAST SEED-SNIPE [CD 388, specimens in spirit]. Z: 117–118 & 155–156: *Tinochorus* [*sic*] *rumicivorus* Eschsch. Loc.: Argentina: Bahia Blanca. September 1832. Material: specimen in alcohol. Status: missing.

SCOLOPACIDAE:

***Capella paraguaiae paraguaiae* (Vieillot, 1816).** SOUTH AMERICAN SNIPE *GALLINAGO PARAGUAIAE* CD 1203 [original field label]. Z: 131: *Scolopax* (*Telmatias*) *paraguaiae* Vieill. Loc.: Uruguay: Rio Plata near Maldonado.

May–June 1833. Ex. coll. Gould, ex. ZSL. Material: skin, ad. Status: BMNH 1860.1.16.74. Cat. XXIV: 651: h.

***Capella paraguaiae paraguaiae* (Vieillot, 1816).** SOUTH AMERICAN SNIPE *GALLINAGO PARAGUAIAE* [CD 1243]. Z: 131: *Scolopax (Telmatias) magellanicus* King. Loc.: Uruguay: Maldonado. May–June 1833. Ex. coll. ZSL via Darwin. Material: skin, ad. Status: BMNH unregistered specimen. Cat. XXIV: 652: n'.

***Capella paraguaiae magellanica* (King, 1828).** SOUTH AMERICAN SNIPE *GALLINAGO PARAGUAIAE* CD 1048. Z: 131: *Scolopax (Telmatias) magellanicus* King. Loc.: Falkland Islands/Islas Malvinas: East Falkland Island. March 1833. Ex. coll. Salvin & Godman, ex. Gould, ex. ZSL. Material: skin, ad. Status: BMNH 1891.10.20.563. Cat. XXIV: 652: i.

***Capella paraguaiae magellanica* (King, 1828).** SOUTH AMERICAN SNIPE *GALLINAGO PARAGUAIAE* [CD 2168]. Z: 131: *Scolopax (Telmatias) paraguaiae* Vieill. Loc.: Chile: Valparaso. August–September 1834. Material: skin/mount, female Status: missing.

***Limosa haemastica* (Linnaeus, 1758).** HUDSONIAN GODWIT CD 1147 [original field label]. Z: 129. *Limosa hudsonica* Swains. Loc.: Falkland Islands/Islas Malvinas: East Falkland Island: Port Louis. March 1833. Ex. coll. Gould, ex. ZSL. Material: skin. Status: USNM No. 8074.

***Limosa haemastica* (Linnaeus, 1758).** HUDSONIAN GODWIT [CD 1148]. Z: 129: *Limosa hudsonica* Swains. Loc.: Falkland Islands/Islas Malvinas: East Falkland Island: Port Louis. March 1833. Material: skin/mount. Status: missing.

***Limosa haemastica* (Linnaeus, 1758).** HUDSONIAN GODWIT [? CD 1880]. Z: 129: *Limosa hudsonica* Swains. Loc.: Falkland Islands/Islas Malvinas: East Falkland Island. March 1834. Material: skin/mount. Status: missing.

***Limosa haemastica* (Linnaeus, 1758).** HUDSONIAN GODWIT [CD 2434]. Z: 129: *Limosa hudsonica* Swains. Loc.: Chile: mud-banks of East coast of Isla de Chiloé. December 1834. Material: skin/mount. Status: missing.

***Numenius borealis* (J. R. Forster, 1772).** ESKIMO CURLEW [? CD 684]. Z: 129: *Numenius brevirostris* Licht. Loc.: Argentina: Buenos Aires or Brazil: Rio de Janeiro. Summer 1832 or September 1833. Remark: Darwin in CD reported to have collected a "Numenius" at Rio, and it is assumed that it is indeed this species. Material: skin/mount. Status: missing.

***Numenius phaeopus hudsonicus* (Latham, 1790).** WHIMBREL [CD 2501]. Z: 129: *Numenius hudsonicus* Lath. Loc.: Chile: mud-banks of Isla de Chiloé or Chonos Archipel. January 1835. Material: skin/mount. Status: missing.

***Tringa melanoleuca* (Gmelin, 1789).** GREATER YELLOWLEGS [?? CD 1430/1431]. Z: 130: *Totanus melanoleucos* Licht. et Vieill. Loc.: Uruguay: Rio Plata near Maldonado. October 1833. Material: skin/mount. Status: missing.

***Tringa flavipes* (Gmelin, 1789).** LESSER YELLOWLEGS [? CD 1603/1607/1608, see following]. Z: 129: *Totanus flavipes* Vieill. Loc.: Uruguay: Rio Plata near Montevideo. November 1833. Material: skin/mount. Status: missing.

***Tringa solitaria cinnamomea* (Brewster, 1890).** SOLITARY SANDPIPER [? CD 1603/1607/1608, see previously & following]. Z: 129: *Totanus macropterus* G. R. Gray. Loc.: Uruguay: Rio Plata near Montevideo. November 1833. Material: skin/mount. Status: missing.

***Heteroscelus incanus* (Gmelin, 1789).** WANDERING TATTLER *TRINGA INCANA* [CD 3355]. Z: 130: *Totanus fuliginosus* Gould. Holotype *Totanus fuliginosus* Gould, 1841. Loc.: Ecuador: Galápagos Archipelago. October 1835. Remark: The BMNH holds a specimen from FitzRoy's collection (BMNH 1837.2.21.263), considered by Warren (1966) to be the type. Material: skin/mount. Status: missing.

***Arenaria interpres interpres* (Linnaeus, 1758).** RUDDY TURNSTONE [CD 3191]. Z: 132: *Strepsilas interpres* Ill. Loc.: Peru: coast near Iquique. August 1835. Material: skin/mount. Status: missing.

***Arenaria interpres interpres* (Linnaeus, 1758).** RUDDY TURNSTONE [CD 3354]. Z: 132: *Strepsilas interpres* Ill. Loc.: Ecuador: Galápagos Archipelago. October 1835. Material: skin/mount, female. Status: missing.

***Erolia minutilla* (Vieillot, 1819).** LEAST SANDPIPER [CD 3358]. *CALIDRIS MINUTILLA* Z: 131: *Pelidna minutilla* Gould. Loc.: Ecuador: Galápagos Archipelago. October 1835. Material: skin/mount. Status: missing.

***Erolia minutilla* (Vieillot, 1819).** LEAST SANDPIPER [CD 3359]. *CALIDRIS MINUTILLA* Z: 131: *Pelidna minutilla* Gould. Loc.: Ecuador: Galápagos Archipelago. October 1835. Material: skin/mount. Status: missing.

***Erolia fuscicollis* (Vieillot, 1819).** WHITE-RUMPED SANDPIPER *CALIDRIS FUSCICOLLIS* [CD 970]. Z: 131: *Pelidna schinzii* Bona Loc.: Chile: Tierra del Fuego: Goree Sound. January 1833. Material: skin/mount. Status: missing.

***Tryngites subruficollis* (Vieillot, 1819).** BUFF-BREASTED SANDPIPER [? CD 1603/1607/1608, see previously]. Z: 130: *Tringa rufescens* Vieill. Loc.: Uruguay: Rio Plata near Montevideo. November 1833. Material: skin/mount. Status: missing.

LARIDAE:

***Anous stolidus galapagensis* (Sharpe, 1879).** BROWN NODDY [CD 3302]. Z: 145: *Megalopterus stolidus* Boie. Loc.: Ecuador: Galápagos Archipelago. October 1835. Material: skin/mount, female. Status: missing.

***Anous stolidus galapagensis* (Sharpe, 1879).** BROWN NODDY [CD 3375]. Z: 145: *Megalopterus stolidus* Boie. Loc.: Pacific Ocean: many 100 miles east from the Galápagos Archipelago. Night of 3 November 1835. Material: skin/mount. Status: missing.

***Anous minutus atlanticus* (Mathews, 1912).** BLACK NODDY [CD 413, shared with eggs of *Sula*]. Z: 145: *Megalopterus stolidus* Boie. Loc.: Brazil: St. Paul Rocks [=Sao Paulo Island, Atlantic Ocean]. April 1832. Material: single egg. Status: missing.

***Rynchops nigra intercedens* (Saunders, 1895).** BLACK SKIMMER *R. NIGER* [CD 1264]. Z: 143–144: *Rhynchops* [*sic*] *nigra* Linn. Loc.: Uruguay: Maldonado. May 1833. Ex. coll. ZSL [fide Barlow 1963: 221]. Material: skin/mount. Status: missing.

***Larus maculipennis* (Lichtenstein, 1823).** BROWN-HOODED GULL *CHROICOCEPHALUS MACULIPENNIS* [CD 1268]. Z: 142–143: *Xema (chroicocephalus) cirrocephalum* G. R. Gray. Loc.: Uruguay: Maldonado. June 1833. Material: skin/mount. Status: missing.

***Larus maculipennis* (Lichtenstein, 1823).** BROWN-HOODED GULL *CHROICOCEPHALUS MACULIPENNIS* [CD 1783]. Z: 142–143: *Xema (chroicocephalus) cirrocephalum* G. R. Gray. Loc.: Chile: Straits of Magellan. January–February 1834. Ex. coll. ZSL. Material: ex mount, imm. Status: BMNH 1855.12.19.140. Cat. XXV: 206: d'.

***Larus maculipennis* (Lichtenstein, 1823).** BROWN-HOODED GULL *CHROICOCEPHALUS MACULIPENNIS* [CD 1390]. Z: 142–143: *Xema (chroicocephalus) cirrocephalum* G. R. Gray. Loc.: Chile: Straits of Magellan. July 1833. Ex. coll. ZSL. Material: skin/mount, imm. Status: former BMNH specimen, missing. Cat. XXV: 206: e'.

***Larus maculipennis* (Lichtenstein, 1823).** BROWN-HOODED GULL *CHROICOCEPHALUS MACULIPENNIS* [CD 748]. Z: 142–143: *Xema (chroicocephalus) cirrocephalum* G. R. Gray. Loc.: Argentina: Bahia Blanca. September 1832. Ex. coll. ZSL via Darwin. Material: ex mount, ad. Status: BMNH unregistered specimen. Cat. XXV: 203: h'.

***Gabianus scoresbii* (Traill, 1822).** DOLPHIN GULL *LEUCOPHAEUS SCORESBII* [CD 1757]. Z: 142: *Larus haematorhynchus* King. Loc.: Argentina: Port St. Julian. January 1834. Material: skin/mount. Status: missing.

***Larus fuliginosus* (Gould, 1841).** LAVA GULL *LEUCOPHAEUS FULIGINOSUS* [CD 3304]. Z: 141–142: *Larus fuliginosus* Gould. Holotype *Larus fuliginosus* Gould, 1841. Loc.: Ecuador: Galápagos Archipelago: James Island. October 1835. Ex. coll. ZSL. Material: ex mount, male, ad. Status: BMNH 1855.12.19.218. Cat. XXV: 223: b.

***Larus dominicanus* (Lichtenstein, 1823).** KELP GULL [CD 1455]. Z: 142: *Larus dominicanus* Licht. Loc.: Argentina: Pampas near Rio Plata. 1833. Material: skin/mount. Status: missing.

"Gull." [CD 185]. Loc.: Portugal: Cape Verde Islands: So Tiago: Porto Praia (1455'N 23?31W'). 9 January–7 February 1832. Remark: This specimen is mentioned as "Gull" from Porto Praya and would be the first record of any gull sighting on the Cape Verde Islands [cf. Hazevoet 1995: 124]. Material: skin/mount. Status: missing.

***Gelochelidon nilotica aranea* (Wilson, 1814).** GULL-BILLED TERN [CD 745]. Z: 145: *Viralva aranea* G. R. Gray. Loc.: Argentina: Bahia Blanca. September 1832. Ex. coll. ZSL. Material: ex mount, ad. Status: BMNH 1855.12.19.134. Cat. XXV: 31: n.

"Sterna." [CD 1384]. Loc.: Argentina: Guritti Island in Rio de la Plata. May–June 1833. Remark: Might be an additional record of *Rynchops nigra* (see following). Locality not traced in Paynter 1994, 1995. Material: skin/mount. Status: missing.

COLUMBIDAE:

***Columba picazuro picazuro* (Temminck, 1813).** PICAZURO PIGEON *PATAGIOENAS PICAZURO* [CD 1340]. Z: 115: *Columba loricata* Licht. Loc.: Uruguay: Maldonado. June 1833. Ex. coll. ZSL via Darwin. Material: skin/mount, ad. Status: former BMNH specimen, missing. Cat. XXI: 272: c.

***Columba araucana* (Lesson, 1827).** CHILEAN PIGEON *PATAGIOENAS ARAUCANA* [CD 2481]. Z: 114: *Columba fitzroyii* King. Loc.: Chile: Peninsula Tres Montes. January 1835. Material: skin/mount. Status: missing.

***Columba araucana* (Lesson, 1827).** CHILEAN PIGEON *PATAGIOENAS ARAUCANA* [CD 2160]. Z: 114: *Columba fitzroyii* King. Loc.: Chile: Valparaso. August–September 1834. Material: skin/mount. Status: missing.

***Zenaidura auriculata auriculata* (Des Murs, 1847).** EARED DOVE *ZENAIDA AURICULATA* [CD 2220]. Z: 115: *Zenaida aurita* G. R. Gray. Loc.: Chile: Valparaso. August–September 1834. Material: skin/mount. Status: missing.

***Zenaidura auriculata virgata* (Bertoni, 1823).** EARED DOVE *ZENAIDA AURICULATA* [CD 1385]. Z: 115: *Zenaida aurita* G. R. Gray. Loc.: Uruguay: La Plata near Maldonado. July 1833. Material: skin/mount. Status: missing.

***Nesopelia galapagoensis galapagoensis* (Gould, 1841).** GALAPAGOS DOVE *ZENAIDA GALAPAGOENSIS* [CD 3305]. Z: 115–116: *Zenaida galapagoensis* Gould. Syntype *Zenaida galapagoensis* Gould, 1841. Loc.: Ecuador: Galápagos Archipelago. September–October 1835. Ex. coll. Eyton, ex. ZSL. Remark: Darwin collected a single dove on the Galápagos Islands. RMNH and BMNH (coll. Eyton/ZSL) claim to possess this specimen, cf. Sulloway 1982; however, it is now believed that the two specimens at Leiden (Cat. No. 1, 2) probably derive from coll. Fuller/Covington and that the BMNH specimen, formerly from the ZSL, is Darwin's; nevertheless, the RMNH specimens have probably type status of Gould's name. Material: skin, ad. Status: BMNH 1881.2.18.84. Cat. XXI: 391: c.

***Columbigallina talpacoti talpacoti* (Temminck, 1811).** RUDDY GROUND-DOVE *COLUMBINA TALPACOTI* [? CD No. between 412–446, number not in Barlow 1963 nor in Keynes 2000]. Z: 116: *Columbina talpacoti* G. R. Gray. Loc.: Brazil: Rio de Janeiro. April–May 1832. Remark: Z: cites at least two specimens. Material: skin/mount. Status: missing.

***Columbigallina talpacoti talpacoti* (Temminck, 1811).** RUDDY GROUND-DOVE *COLUMBINA TALPACOTI* [? CD No. between 412–446, number not in Barlow 1963 nor in Keynes 2000]. Z: 116: *Columbina talpacoti* G. R. Gray. Loc.: Brazil: Rio de Janeiro. April–May 1832. Remark: Z: cites at least two specimens. Material: skin/mount. Status: missing.

***Columbina picui picui* (Temminck, 1813).** PICUI GROUND DOVE CD 1272 [original field label]. Z: 116: *Columbina strepitans* Spix. Loc.: Uruguay: Maldonado: on the banks of the Plata. 1833. Ex. coll. ZSL via Darwin. Material: skin, ad. Status: BMNH unregistered specimen. Cat. XXI: 472: u.

***Columbina picui picui* (Temminck, 1813).** PICUI GROUND-DOVE [CD 1463]. Z: 116: *Columbina strepitans* Spix. Loc.: Argentina: Rio Negro. August 1833. Material: skin/mount. Status: missing.

***Metriopelia melanoptera melanoptera* (Molina, 1782).** BLACK-WINGED GROUND-DOVE [CD 2163]. Z: 116: *Zenaida boliviana* G. R. Gray. Loc.: Chile: Valparaso. Late August 1834. Remark: BMNH possess an old data-less bird Cat. XXI: 499: p/q. Material: skin/mount, female. Status: probably missing.

CUCULIDAE:

***Guira guira* (Gmelin, 1788).** GUIRA CUCKOO [CD 1427]. Z: 114: *Diplopterus guira* G. R. Gray. Loc.: Argentina: Buenos Aires. October 1833. Ex. coll. ZSL. Material: skin. Status: perhaps BMNH 1858.4.3.143 [although "?Chile" on label].

***Crotophaga ani* (Linnaeus, 1758).** SMOOTH-BILLED ANI [CD 455]. Z: 114: *Crotophaga ani* Linn. Loc.: Brazil: Rio de Janeiro. May 1832. Material: skin/mount. Status: Not yet looked for by author.

***Tapera naevia chochi* (Vieillot, 1817).** STRIPED CUCKOO [? CD No. between 412–446, number not in Barlow 1963 and Keynes 2000]. Z: 114: *Diplopterus naevius* Boie. Loc.: Brazil: Rio de Janeiro. April 1832. Material: skin/mount. Status: Not yet looked for by author.

TYTONIDAE:

***Tyto alba tuidara* (J. E. Gray, 1829).** WESTERN BARN OWL CD 1446. Z: 34: *Strix flammea* Linn. Loc.: Argentina: Bahia Blanca. 1833. Ex. coll. ZSL via Darwin. Material: ex mount, ad. Status: BMNH 1841.1.18.16. Cat. II: 302: l."

***Tyto alba punctatissima* (G. R. Gray, 1838).** WESTERN BARN OWL [CD n/a]. Z: 34–35: *Strix punctatissima* G. R. Gray. Loc.: Ecuador: Galápagos Islands: James Island. October 1835. Remark: Holotype *Strix punctatissima* Gray, 1838 [pl. IV] = BMNH 1837.2.21.244 is from FitzRoy's collection. The main text of this paper erroneously cites the year of original description from the textual account (1839), not from the plate (1838). The authorship is attributed to Gray on the basis that his name appears after the new name; however, plate IV was probably sketched by John Gould, and executed on stone by Elizabeth Gould. This matter would need further investigations. Material/Status: not collected by Darwin.

STRIGIDAE:

***Strix rufipes rufipes* (King, 1828).** RUFOUS-LEGGED OWL [CD 1875]. Z: 34: *Ulula rufipes* [King]. Loc.: Chile: Tierra del Fuego: extreme southern Islands: Ponsonby Sound. February 1834. Ex. coll. ZSL. Material: ex mount, ad. Status: probably BMNH 1855.12.19.65. Cat. II: 261: b.

***Speotyto cunicularia cunicularia* (Molina, 1782).** *ATHENE CUNICULARIA* BURROWING OWL CD 1293. Z: 31–32: *Athene cunicularia* Bona Loc.: Uruguay: Maldonado. June 1833. Ex. coll. ZSL via Darwin. Material: ex mount, ad. Status: BMNH 1841.1.18.17 [probably this registration number, although not on label]. [Cat. II: 144: t = 1855.12.19.144 nec. Darwin].

***Speotyto cunicularia cunicularia* (Molina, 1782).** BURROWING OWL *ATHENE CUNICULARIA* [CD 2162]. Z: 31–32: *Athene cunicularia* Bona Loc.: Chile: Valparaso. August–September 1834. Ex. coll. Norwich Castle Museum/ Gurney Collection, ex. ZSL. Material: ex mount. Status: perhaps BMNH 1955.6.N20.4499.

***Asio flammeus suinda* (Vieillot, 1817).** SHORT-EARED OWL CD 1270. Z: 33: *Otus palustris* Gould. Loc.: Uruguay: Maldonado. 1833. Ex. coll. ZSL via Darwin. Material: skin/mount. Status: BMNH 1841.1.18.15.

***Asio flammeus suinda* (Vieillot, 1817).** SHORT-EARED OWL [CD 2031]. Z: 33: *Otus palustris* Gould. Loc.: Argentina: Santa Cruz: April 1834. Material: skin/mount. Status: missing.

***Asio flammeus sanfordi* (Bangs, 1919)**. SHORT-EARED OWL CD 1901 [original field label]. Z: 33: *Otus palustris* Gould. Loc.: Falkland Islands/ Islas Malvinas: East Falkland Island. March 1834. Ex. coll. Norwich Castle Museum/Gurney Collection, ex. ZSL. Material: ex mount. Status: BMNH 1955.6.N20.3684.

***Asio flammeus galapagoensis* (Gould, 1837).** SHORT-EARED OWL [CD 3303]. Z: 32–33: *Otus galapagoensis* Gould. Holotype *Otus galapagoensis* Gould, 1837. Loc.: Ecuador: Galápagos Archipelago: James Island. October 1835. Ex. coll. ZSL. Material: ex mount, male, imm. Status: BMNH 1855.12.19.153. Cat. II: 238: x.

CAPRIMULGIDAE:

***Caprimulgus parvulus parvulus* (Gould, 1837).** LITTLE NIGHTJAR *SETOPAGIS PARVULUS* CD 1623. Z: 37–38: *Caprimulgus parvulus* Gould. Holotype *Caprimulgus parvulus* Gould, 1837. Loc.: Argentina: Santa Fé de Bajada. 27–30 September 1833. Ex. coll. ZSL. Material: ex mount, female, ad. Status: BMNH 1855.12.19.158. Cat. XVI: 575: m.

***Caprimulgus longirostris bifasciatus* (Gould, 1837).** BAND-WINGED NIGHTJAR *SYSTELLURA LONGIROSTRIS* [CD 2171]. Z: 36–37: *Caprimulgus bifasciatus* Gould. Holotype *Caprimulgus bifasciatus* Gould, 1837. Loc.: Chile: Valparaso. August 1834. Ex. coll. ZSL. Material: ex mount, male, ad. Status: BMNH 1855.12.19.241. Cat. XVI: 586: u.

APODIDAE:

***Apus unicolor alexandri* (Hartert, 1901).** PLAIN SWIFT [CD 3907]. Z: 41: *Cypselus unicolor* Jard. Loc.: Portugal: Cape Verde Islands: So Tiago. September 1836. Material: skin/mount. Status: missing.

TROCHILIDAE:

***Chlorostilbon aureoventris berlepschi* (Pinto, 1938).** GLITTERING-BELLIED EMERALD *CHLOROSTILBON LUCIDUS* [CD 1610]. Z: 110: *Trochilus flavifrons* [n.n.]. [Nomen nudum *Trochilus flavifrons* Gould, 1841]. Loc.: Uruguay: Montevideo. November 1833. Ex. coll. ZSL via Darwin. Material: ex mount, male, ad. Status: BMNH unregistered specimen. Cat. XVI: 50: x.

***Patagona gigas gigas* (Vieillot, 1824).** GIANT HUMMINGBIRD [? CD 2179/ 2180, see following]. Z: 111–112: *Trochilus gigas* Vieill. Loc.: Chile: Valparaso. September 1834. Ex. coll. Balston, ex. Eyton,? ex. ZSL. Material: skin, female, ad. Status: perhaps BMNH 1913.3.20.292.

***Patagona gigas gigas* (Vieillot, 1824).** GIANT HUMMINGBIRD [? CD 2179/ 2180, see previously]. Z: 111–112: *Trochilus gigas* Vieill. Loc.: Chile: Valparaso. September 1834. Material: skin/mount. Status: missing.

***Patagona gigas gigas* (Vieillot, 1824).** GIANT HUMMINGBIRD [CD 2319]. Z: 111–112: *Trochilus gigas* Vieill. Loc.: Chile: Valparaso. September–October 1834. Material: nest. Status: missing.

***Patagona gigas gigas* (Vieillot, 1824).** GIANT HUMMINGBIRD [CD 1050, specimens in spirit]. Z: 111–112 & 154: *Trochilus gigas* Vieill. Loc.: Chile: Valparaso. August 1834. Material: specimen in alcohol. Status: missing.

***Sephanoides sephanoides* (Lesson, 1826).** GREEN-BACKED FIRECROWN [CD 2134]. Z: 110–111: *Trochilus forficatus* Lath. Loc.: Chile: Isla de Chiloé. July 1834. Ex. coll. Salvin & Godman, ex. ZSL. Remark: on younger label as leg. Leybold, but on older ZSL. Material: skin, male, ad. Status: perhaps BMNH 1887.3.22.924. Cat. XVI: 157: b/c.

***Sephanoides sephanoides* (Lesson, 1826).** GREEN-BACKED FIRECROWN [CD 2135]. Z: 110–111: *Trochilus forficatus* Lath. Loc.: Chile: Isla de Chiloé. July 1834. Material: skin/mount. Status: missing.

***Sephanoides sephanoides* (Lesson, 1826).** GREEN-BACKED FIRECROWN [CD 2503]. Z: 110–111: *Trochilus forficatus* Lath. Loc.: Chile: Chonos Archipel. January 1835. Material: skin/mount. Status: missing.

***Sephanoides sephanoides* (Lesson, 1826).** GREEN-BACKED FIRECROWN [CD 2425]. Z: 110–111: *Trochilus forficatus* Lath. Loc.: Chile: near South end of Isla de Chiloé: Island S. Pedro. December 1834. Material: egg clutch & nest. Status: missing.

ALCEDINIDAE:

***Halcyon leucocephala acteon* (Lesson, 1830).** GREY-HEADED KINGFISHER CD 192. Z: 41–42: *Halcyon erythrohyncha* Gould. Holotype *Halcyon erythrorhynchus* [*sic*] Gould, 1837. Loc.: Portugal: Cape Verde Islands: So Tiago: Porto Praia. January 1832. Ex. coll. Gould, ex. ZSL. Material: ex mount, ad. Status: BMNH 1881.5.1.3018. Cat. XVII: 235: a.

***Chloroceryle americana mathewsii* (Laubmann, 1927).** GREEN KINGFISHER [? CD not traced]. Z: 42: *Ceryle americana* Boie. Loc.: Argentina: banks of the Parana, Buenos Aires. October 1833. Material: skin/mount. Status: shot by Darwin, perhaps not preserved or missing.

***Ceryle torquata stellata* (Meyen, 1834).** RINGED KINGFISHER *MEGACERYLE TORQUATA* [CD 2122]. Z: 42: *Ceryle torquata* Bona Loc.: Chile: Isla de Chiloé. July 1834. Material: skin/mount, female. Status: missing.

***Ceryle torquata stellata* (Meyen, 1834).** RINGED KINGFISHER *MEGACERYLE TORQUATA* [CD 1210]. Z: 42: *Ceryle torquata* Bona Loc.: Uruguay: Maldonado. 1833. Material: skin/mount. Status: missing.

PICIDAE:

***Dendrocopos lignarius* (Molina, 1782).** STRIPED WOODPECKER *VENILIORNIS LIGNARIUS* CD 2480 [original field label]. Z: 113: *Picus kingii* G. R. Gray. Syntype Picus kingii Gray, 1841. Loc.: Chile: Peninsular Tres Montes: high mountains. January 1835. Ex. coll. Gould, ex. ZSL. Material: skin, female, ad. Status: BMNH 1881.5.1.3597. Cat. XVIII: 258: b/c [see following].

***Dendrocopos lignarius* (Molina, 1782).** STRIPED WOODPECKER *VENILIORNIS LIGNARIUS* CD 2185 [original field label]. Z: 113: *Picus kingii* G. R. Gray. Syntype *Picus kingii* Gray, 1841. Loc.: Chile: Valparaso. August–September 1834. Ex. coll. Gould, ex. ZSL. Material: skin, female, ad. Status: BMNH 1881.5.1.3600. Cat. XVIII: 258: b/c [see previously].

***Dendrocopos lignarius* (Molina, 1782).** STRIPED WOODPECKER *VENILIORNIS LIGNARIUS* [CD 2479]. Z: 113: *Picus kingii* G. R. Gray. Syntype *Picus kingii* Gray, 1841. Loc.: Chile: Peninsular Tres Montes: High mountains. January 1835. Ex. coll. ZMD, ex. BMNH, ex. ZSL. Material: skin, male, ad. Status: NMSE 1958.71. [batch number], ex. BMNH 1855.12.19.101. Cat. XVIII: 258: n.

***Colaptes pitius pitius* (Molina, 1782).** CHILEAN FLICKER [CD 2161]. Z: 114: *Colaptes chilensis* Vigors. Loc.: Chile: Valparaso: stony hills. August 1834. Material: skin/mount, male. Status: missing.

***Colaptes campestris campestroides* (Malherbe, 1849).** CAMPO FLICKER [CD 1238]. Z: 113–114: *Chrysoptilus campestris* Swains. Loc.: Uruguay: Maldonado. 1833. Remark: tongue of the specimen has been preserved in alcohol, cf. CD 620. Material: skin/mount. Status: missing.

***Colaptes campestris campestroides* (Malherbe, 1849).** CAMPO FLICKER [CD 620, specimens in spirit]. Z: 113–114: *Chrysoptilus campestris* Swains. Loc.: Uruguay: Maldonado. 1833. Material: tongue of CD 1238 in alcohol. Status: missing.

***Colaptes campestris campestroides* (Malherbe, 1849).** CAMPO FLICKER [CD 1428]. Z: 113–114: *Chrysoptilus campestris* Swains. Loc.: Argentina: Buenos Aires. October 1833. Material: skin/mount. Status: missing.

FALCONIDAE:

***Phalcoboenus megalopterus megalopterus* (Meyen, 1834).** MOUNTAIN CARACARA [CD n/a]. Z: 21: *Milvago megalopterus* [Meyen]. Loc.: Chile: Copiapó. 1835. Material & Status: not collected.

***Phalcoboenus megalopterus albogularis* (Gould, 1837).** WHITE-THROATED CARACARA *PHALCOBOENUS ALBOGULARIS* [CD 2029]. Z: 18–21: *Milvago albogularis* [Gould]. Holotype *Polyborus albogularis* Gould, 1837. Loc.: Argentina: Santa Cruz: April 1834. Ex. coll. ZSL. Material: ex mount, female, ad. Status: BMNH 1855.12.19.405. Cat. I: 38: a.

***Phalcoboenus australis* (Gmelin, 1788).** STRIATED CARACARA CD 1882. Z: 15–18: *Milvago leucurus* [Forster]. Syntype *Milvago leucurus* Darwin (ex Forster MS), in Darwin & Gray, 1838–1841 [Authorship here given as Darwin & Gray, on preliminary conclusion. Authorship of the species accounts of *The Zoology of HMS Beagle* need further studies]. 1 Loc.: Falkland Islands/Islas Malvinas: East Falkland Island. March 1834. Ex. coll. Norwich Castle Museum/Gurney Collection, ex. ZSL. Material: ex mount, male, imm. Status: BMNH 1955.6.N20.46.

***Phalcoboenus australis* (Gmelin, 1788).** STRIATED CARACARA [CD 1926]. Z: 15–18: *Milvago leucurus* [Forster]. Syntype *Milvago leucurus* Darwin & Gray, 1838.1 Loc.: Falkland Islands/Islas Malvinas: East Falkland Island. March 1834. Ex. coll. Norwich Castle Museum/Gurney Collection, ex. ZSL. Material: ex mount, female. Status: BMNH 1955.6.N20.47.

***Phalcoboenus australis* (Gmelin, 1788).** STRIATED CARACARA [CD 1932]. Z: 15–18: *Milvago leucurus* [Forster]. Loc.: Falkland Islands/Islas Malvinas: East Falkland Island. March 1834. Material & Status: lost [cf. Keynes 2000: 399].

***Phalcoboenus australis* (Gmelin, 1788).** STRIATED CARACARA [CD 1933]. Z: 15–18: *Milvago leucurus* [Forster]. Loc.: Falkland Islands/Islas Malvinas: East Falkland Island. March 1834. Material & Status: lost [cf. Keynes 2000: 399].

***Polyborus plancus plancus* (Miller, 1777).** SOUTHERN CRESTED CARACARA *CARACARA PLANCTUS* [? CD 2028]. Z: 9–12: *Polyborus brasiliensis* Swains. Loc.: Argentina: plains of Santa Cruz: April–May 1834. Remark: The BMNH possesses a specimen from the Norwich Castle Museum (ex. ZSL), which has not the “pale rusty brown” on head [cf. Z:: 11]. Material: skin/mount, male, imm. Status: missing.

***Polyborus plancus plancus* (Miller, 1777).** SOUTHERN CRESTED CARACARA *CARACARA PLANCTUS* [? CD 1456]. Z: 9–12: *Polyborus brasiliensis* Swains. Loc.: Argentina: [? Bahia Blanca]: plains of Santa Cruz: July–August 1833. Material: skin/mount. Status: missing.

***Milvago chimango chimango* (Vieillot, 1816).** CHIMANGO CARACARA CD 1294 [on label accidentally 1204D]. Z: 14–15: *Milvago chimango* [Vieill.]. Loc.: Uruguay: Maldonado. May 1833. Material: ex mount, ad. Status: BMNH unregistered specimen. Cat. I: 42: c.

***Milvago chimango temucoensis* (Sclater, 1918).** CHIMANGO CARACARA [CD 1772]. Z: 13–14: *Milvago pezoporos* [Meyen]. “Holotype” *Polyborus hyperstictus* Giebel MS [not valid]. Loc.: Argentina: Port Desire. January 1834. Material: ex mount, ad. Status: perhaps BMNH unregistered specimen. Cat. I: 42: d.

***Milvago chimango temucoensis* (Sclater, 1918).** CHIMANGO CARACARA [CD 1028]. Z: 13–14: *Milvago pezoporos* [Meyen]. Loc.: Chile: extreme Southern

Tierra del Fuego: Hardy Peninsula. February 1833. Material: skin/mount. Status: missing.

***Falco sparverius cinnamominus* (Swaison, 1837).** AMERICAN KESTREL [CD 2014]. Z: 29: *Tinnunculus sparverius* Vieill. Loc.: Argentina: Santa Cruz: April 1834. Material: skin/mount. Status: missing.

***Falco sparverius cinnamominus* (Swaison, 1837).** AMERICAN KESTREL [CD 1464]. Z: 29: *Tinnunculus sparverius* Vieill. Loc.: Argentina: Rio Negro. August 1833. Material: skin/mount. Status: missing.

***Falco femoralis femoralis* (Temminck, 1817).** APLOMADO FALCON [CD 1706]. Z: 28: *Falco femoralis* Temm. Loc.: Argentina: small valley on the plains at Port Desire. January 1834. Ex. coll. Norwich Castle Museum/Gurney Collection, ex. ZSL. Material: ex mount, female, ad. Status: probably BMNH 1955.6.N20.1796.

***Falco femoralis femoralis* (Temminck, 1817).** APLOMADO FALCON [CD 1710, number shared with egg(s) of *Zonotrichia capensis*, see following]. Z: 28: *Falco femoralis* Temm. Loc.: Argentina: Port Desire. January 1834. Material: egg clutch. Status: missing.

PSITTACIDAE:

***Cyanoliseus patagonus patagonus* (Vieillot, 1817).** BURROWING PARROT [CD 747]. Z: 113: *Conurus patachonicus* [Lear]. Loc.: Argentina: Bahia Blanca. September 1832. Material: skin/mount. Status: missing.

***Myiopsitta monachus monachus* (Boddaert, 1783).** MONK PARAKEET [CD 1219]. Z: 112: *Conurus murinus* Kuhl. Loc.: Uruguay: Maldonado: grassy plains. May–June 1833. Ex. coll. ZSL via Darwin. Material: ex mount, ad. Status: BMNH unregistered specimen. Cat. XX: 233: i.

PASSERIFORMES

FURNARIIDAE:

***Geositta cunicularia fissirostris* (Kittlitz, 1835).** COMMON MINER [CD 2297]. Z: 65–66: *Furnarius cunicularius* G. R. Gray. Loc.: Chile: Valparaso. August–September 1834. Material: skin/mount. Status: missing.

***Geositta cunicularia fissirostris* (Kittlitz, 1835).** COMMON MINER [CD 721, specimens in spirit]. Z: 65–66 & 148: *Furnarius cunicularius* G. R. Gray. Loc.: Uruguay: Maldonado. June 1833. Material: specimen in alcohol. Status: missing.

***Geositta cunicularia cunicularia* (Vieillot, 1816).** COMMON MINER CD 1222. Z: 65–66: *Furnarius cunicularius* G. R. Gray. Loc.: Uruguay: Maldonado.

May 1833. Ex. coll. ZSL. Material: ex mount, ad. Status: BMNH 1855.12.19.57. Cat. XV: 6: a.

***Eremobius phoenicurus* (Gould, 1839).** BAND-TAILED EARTHCREEPER *OCHETORHYNCHUS PHOENICURUS* CD 1702. Z: 69–70: *Eremobius phoenicurus* Gould. Syntype *Eremobius phoenicurus* Gould, 1839 [pl. XXI]. Loc.: Argentina: Port Desire. January 1834. Ex. coll. ZSL. Material: ex mount, ad. Status: BMNH 1855.12.19.117. Cat. XV: 27: b.

***Eremobius phoenicurus* (Gould, 1839**). BAND-TAILED EARTHCREEPER *OCHETORHYNCHUS PHOENICURUS* CD 2025 [on label 2052]. Z: 69–70: *Eremobius phoenicurus* Gould. Syntype *Eremobius phoenicurus* Gould, 1839 [pl. XXI]. Loc.: Argentina: Santa Cruz: April 1834. Ex. coll. ZSL. Material: ex mount, female, ad. Status: BMNH 1855.12.19.73. Cat. XV: 27: a.

***Eremobius phoenicurus* (Gould, 1839).** BAND-TAILED EARTHCREEPER *OCHETORHYNCHUS PHOENICURUS* CD 1754 [original field label]. Z: 69–70: *Eremobius phoenicurus* Gould. Syntype *Eremobius phoenicurus* Gould, 1839 [pl. XXI]. Loc.: Argentina: Port St. Julian. January 1834. Ex. coll. Salvin & Godman, ex. ZSL. Material: skin, ad. Status: BMNH 1889.5.14.65 [on label, this no. is a *Cinclodes fuscus* in register]. Cat. XV: 27: c.

***Upucerthia dumetaria hallinani* (Chapman, 1919).** SCALY-THROATED EARTHCREEPER [CD 2827]. Z: 66: *Uppucerthia* [*sic*] *dumetoria* J. [*sic*] Geoffr. & d'Orb. Loc.: Chile: Coquimbo. 1835. Ex. coll. ZMD, ex. BMNH, ex. ZSL. Material: ex mount, ad. Status: NMSE 1931.76.10, ex. BMNH 1855.12.19.75. Cat. XV: 17: a.

***Upucerthia dumetaria dumetaria* (Geoffroy Saint-Hilaire, 1832).** SCALY-THROATED EARTHCREEPER [CD 1467]. Z: 66: *Uppucerthia* [*sic*] *dumetoria* J. [*sic*] Geoffr. & d'Orb. Loc.: Argentina: Patagonia: Port Desire. August 1833. Ex. coll. ZSL via Darwin. Material: ex mount, ad. Status: BMNH 1839.8.4.1. Cat. XV: 17:

***Upucerthia dumetaria dumetaria* (Geoffroy Saint-Hilaire, 1832).** SCALY-THROATED EARTHCREEPER [728, specimens in spirit]. Z: 66 & 148–149: *Uppucerthia* [*sic*] *dumetoria* J. [*sic*] Geoffr. & d'Orb. Loc.: Argentina: Rio Negro. 1833. Material: specimen in alcohol. Status: missing.

***Cinclodes antarcticus antarcticus* (Garnot, 1826).** BLACKISH CINCLODES [CD 1931]. Z: 67–68: *Opetiorhynchus antarcticus* G. R. Gray. Loc.: Falkland Islands/Islas Malvinas: East Falkland Island. March 1834. Material: skin/mount. Status: missing.

***Cinclodes antarcticus antarcticus* (Garnot, 1826).** BLACKISH CINCLODES [CD not traced, specimens in spirit]. Z: 67–68 & 149–150: *Opetiorhynchus antarcticus* G. R. Gray. Loc.: Falkland Islands/Islas Malvinas:

East Falkland Island. March 1833/1834. Material: specimen in alcohol. Status: missing.

***Cinclodes fuscus fuscus* (Vieillot, 1818).** BUFF-WINGED CINCLODES [CD 1822]. Z: 66–67: *Opetiorhynchus vulgaris* G. R. Gray. Loc.: Chile: Port Famine. February 1834. Ex. coll. ZSL. Material: ex mount, ad. Status: BMNH 1855.12.19.89. Cat. XV: 23: b.

***Cinclodes fuscus fuscus* (Vieillot, 1818).** BUFF-WINGED CINCLODES [CD 1260]. Z: 66–67: *Opetiorhynchus vulgaris* G. R. Gray. Loc.: Uruguay: Maldonado. April–July 1833. Material: skin/mount. Status: missing.

***Cinclodes fuscus fuscus* (Vieillot, 1818).** BUFF-WINGED CINCLODES [CD 722, specimens in spirit]. Z: 66–67 & 149: *Opetiorhynchus vulgaris* G. R. Gray. Loc.: Uruguay: Maldonado. June 1833. Material: specimen in alcohol. Status: missing.

***Cinclodes patagonicus chilensis* (Lesson, 1828).** DARK-BELLIED CINCLODES [CD 2126]. Z: 67: *Opetiorhynchus patagonicus* G. R. Gray. Loc.: Chile: Isla de Chiloé. July 1834. Material: skin/mount,? male [CD]. Status: missing.

***Cinclodes patagonicus patagonicus* (Gmelin, 1789).** DARK-BELLIED CINCLODES [CD 1823]. Z: 67: *Opetiorhynchus patagonicus* G. R. Gray. Loc.: Chile: Tierra del Fuego: Port Famine. February 1834. Material: skin/mount. Status: missing.

***Cinclodes patagonicus patagonicus* (Gmelin, 1789).** DARK-BELLIED CINCLODES [CD 972]. Z: 67: *Opetiorhynchus patagonicus* G. R. Gray. Loc.: Chile: Tierra del Fuego: Wolsey Island. January 1833. Ex. coll. Tristram, ex. ZSL. Material: skin, female. Status: LIVCM 4280.

***Cinclodes patagonicus patagonicus* (Gmelin, 1789).** DARK-BELLIED CINCLODES [CD not traced, specimens in spirit]. Z: 67 & 150: *Opetiorhynchus patagonicus* G. R. Gray. Loc.: Chile. Material: specimen in alcohol. Status: missing.

***Cinclodes nigrofumosus* (d'Orbigny & Lafresnaye, 1838).** CHILEAN SEASIDE CINCLODES CD 2826. Z: 68–69: *Opetiorhynchus nigrofumosus* G. R. Gray. Holotype *Opetioryhnchus lanceolatus* Gould, 1839 [pl. XX]. Loc.: Chile: Coquimbo. 1835. Ex. coll. ZSL. Material: ex mount, ad. Status: BMNH 1855.12.19.244. Cat. XV: 22: c.

***Cinclodes nigrofumosus* (d'Orbigny & Lafresnaye, 1838).** CHILEAN SEASIDE CINCLODES [CD 2426]. Z: 68–69: *Opetiorhynchus nigrofumosus* G. R. Gray. Loc.: Chile: Chonos Archipelago (Lat. 45 degree): Midship Bay. December 1834. Material: single egg. Status: missing.

***Furnarius rufus rufus* (Gmelin, 1788).** RUFOUS HORNERO [CD 1619]. Z: 64: *Furnarius rufus* Vieill. Loc.: Uruguay: Montevideo. November 1833. Material: skin/mount. Status: missing.

***Furnarius rufus rufus* (Gmelin, 1788).** RUFOUS HORNERO [CD 1200]. Z: 64: *Furnarius rufus* Vieill. Loc.: Uruguay: Maldonado. 1833. Material: skin/ mount. Status: missing.

***Aphrastura spinicauda spinicauda* (Gmelin, 1789).** THORN-TAILED RAYADITO [CD 2084]. Z: 81: *Oxyurus tupinieri* Gould. Loc.: Chile: Tierra del Fuego: Port Famine. June 1834. Material: skin/mount. Status: missing.

***Aphrastura spinicauda fulva* (Angelini, 1905).** THORN-TAILED RAYADITO [CD 2130]. Z: 81: *Oxyurus tupinieri* Gould. Loc.: Chile: Isla de Chiloé. July 1834. Material: skin/mount. Status: missing.

***Leptasthenura aegithaloides pallida* (Dabbene, 1920).** PLAIN-MANTLED TIT-SPINETAIL CD 2022. Z: 79: *Synallaxis aegithaloides* Kittl. Loc.: Argentina: Santa Cruz: April 1834. Ex. coll. ZSL via Darwin. Material: ex. mount, male, ad. Status: BMNH 1856.3.15.11. Cat. XV: 35: k.

***Leptasthenura aegithaloides pallida* (Dabbene, 1920).** PLAIN-MANTLED TIT-SPINETAIL [CD 2023]. Z: 79: *Synallaxis aegithaloides* Kittl. Loc.: Argentina: Santa Cruz: April 1834. Material: skin/mount. Status: missing.

***Leptasthenura aegithaloides aegithaloides* (Kittlitz, 1830).** PLAIN-MANTLED TIT-SPINETAIL [CD 2298]. Z: 79: *Synallaxis aegithaloides* Kittl. Loc.: Chile: Valparaso. August–September 1834. Material: skin/mount. Status: missing.

***Asthenes pyrrholeuca flavogularis* (Gould, 1839).** SHARP-BILLED CANASTERO [CD 1705]. Z: 78–79: *Synallaxis brunnea* Gould. Holotype *Synallaxis brunnea* Gould, 1839. Loc.: Argentina: Port Desire. January 1834. Ex. coll. ZSL. Material: ex mount, ad. Status: BMNH 1855.12.19.99. Cat. XV: 68: m.

***Asthenes pyrrholeuca flavogularis* (Gould, 1839).** SHARP-BILLED CANASTERO CD 2024. Z: 78: *Synallaxis flavogularis* Gould. Syntype *Synallaxis flavogularis* Gould, 1839 [pl. XXIV]. Loc.: Argentina: Santa Cruz: April 1834. Ex. coll. ZSL. Material: ex mount, male, imm. Status: BMNH 1855.12.19.406. Cat. XV: 68: q.

***Asthenes pyrrholeuca flavogularis* (Gould, 1839).** SHARP-BILLED CANASTERO [? CD 751/828]. Z: 78: *Synallaxis flavogularis* Gould. Syntype *Synallaxis flavogularis* Gould, 1839 [pl. XXIV]. Loc.: Argentina: Bahia Blanca. September–October 1832. Material: skin/mount. Status: missing.

***Asthenes anthoides* (King, 1831).** AUSTRAL CANASTERO CD 2021. Z: 77: *Synallaxis rufogularis* Gould. Syntype *Synallaxis rufogularis* Gould, 1839 [pl. XXIII]. Loc.: Argentina: Santa Cruz: April 1834. Ex. coll. ZSL. Material: ex mount,? male [CD], ad. Status: BMNH 1855.12.19.170. Cat. XV: 70: h.

***Asthenes anthoides* (King, 1831).** AUSTRAL CANASTERO [CD 2190]. Z: 77: *Synallaxis rufogularis* Gould. Syntype *Synallaxis rufogularis* Gould, 1839 [pl. XXIII]. Loc.: Chile: Valparaso. September 1834. Ex. coll. ZSL. Material: ex mount,? male [CD], ad. Status: BMNH 1855.12.19.171. Cat. XV: 70:?e.

***Asthenes anthoides* (King, 1831).** AUSTRAL CANASTERO [CD 2020]. Z: 77: *Synallaxis rufogularis* Gould. Syntype *Synallaxis rufogularis* Gould, 1839 [pl. XXIII]. Loc.: Argentina: Santa Cruz: April 1834. Ex. coll. ZSL. Material: ex mount,? male [CD], ad. Status: BMNH 1855.12.19.104. Cat. XV: 70: g.

***Asthenes humicola humicola* (Kittlitz, 1830).** DUSKY-TAILED CANASTERO *PSEUDASTHENES HUMICOLA* [? CD 2191]. Z: 75: *Synallaxis humicola* Kittl. Loc.: Chile: neighbourhood of Valparaso. August–September 1834. Material: skin/mount, female. Status: missing.

***Asthenes humicola humicola* (Kittlitz, 1830).** DUSKY-TAILED CANASTERO *PSEUDASTHENES HUMICOLA* [? CD 2192]. Z: 75: *Synallaxis humicola* Kittl. Loc.: Chile: neighbourhood of Valparaso. August–September 1834. Material: skin/mount, male. Status: missing.

***Synallaxis frontalis/ruficapilla* (Pelzeln, 1859/Vieillot, 1819).** SOOTY-FRONTED/RUFOUS-CAPPED SPINETAIL [CD 1256]. Z: 79: *Synallaxis ruficapilla* Vieill. Loc.: Uruguay: Maldonado. June 1833. Material: skin/mount. Status: missing.

***Synallaxis frontalis/ruficapilla* (Pelzeln, 1859/Vieillot, 1819).** SOOTY-FRONTED/RUFOUS-CAPPED SPINETAIL [? CD 1432]. Z: 79: *Synallaxis ruficapilla* Vieill. Loc.: Argentina: Buenos Aires. October 1833. Material: skin/mount. Status: missing.

***Phacellodomus striaticollis striaticollis* (d'Orbigny & Lafresnaye, 1838).** FRECKLE-BREASTED THORNBIRD [CD 1249]. Z: 80: *Anumbius ruber* D'Orb. & Lafr. Loc.: Uruguay: Maldonado: reeds on the borders of lakes near Maldonado. 1833. Ex. coll. ZSL. Material: ex mount, ad. Status: BMNH 1855.12.19.52. Cat. XV: 82: a.

***Spartonoica maluroides* (d'Orbigny & Lafresnaye, 1837).** BAY-CAPPED WREN-SPINETAIL [CD 1250]. Z: 77–78: *Synallaxis maluroides* [D'Orb. & Lafr.]. Loc.: Uruguay: Maldonado. 1833. Material: skin/mount. Status: missing.

***Spartonoica maluroides* (d'Orbigny & Lafresnaye, 1837).** BAY-CAPPED WREN-SPINETAIL [CD 1228]. Z: 77–78: *Synallaxis maluroides* [D'Orb. & Lafr.]. Loc.: Uruguay: Maldonado. 1833. Material: skin/mount. Status: missing.

***Spartonoica maluroides* (d'Orbigny & Lafresnaye, 1837).** BAY-CAPPED WREN-SPINETAIL [CD 630, specimens in spirit]. Z: 77–78 & 152–153: *Synallaxis maluroides* [D'Orb. & Lafr.]. Loc.: Uruguay: Maldonado. May 1833. Material: specimen in alcohol. Status: missing.

***Phleocryptes melanops melanops* (Vieillot, 1817).** WREN-LIKE RUSHBIRD [? CD 1227]. Z: 82: *Oxyurus? dorsomaculatus* Gould. Loc.: Uruguay: Maldonado. June 1833. Ex. coll. ZSL. Material: skin/mount. Status: former BMNH 1855.12.19.177, missing since the 1880s.

***Limnornis curvirostris* (Gould, 1839).** CURVE-BILLED REEDHAUNTER CD 1248. Z: 81: *Limnornis curvirostris* Gould. Syntype *Limnornis curvirostris* Gould, 1839 [pl. XXVI]. Loc.: Uruguay: Maldonado. June 1833. Ex. coll. ZSL. Material: ex mount, ad. Remark: In two original copies of the "Zoology" seen, the plate to this species was issued as number XXV (1839). Status: BMNH 1855.12.19.74. Cat. XV: 77: a/b [see following].

***Limnornis curvirostris* (Gould, 1839).** CURVE-BILLED REEDHAUNTER [CD 1255]. Z: 81: *Limnornis curvirostris* Gould. Syntype *Limnornis curvirostris* Gould, 1839 [pl. XXVI]. Loc.: Uruguay: Maldonado. June 1833. Ex. coll. ZSL. Material: ex mount, ad. Remark: In two original copies of the "Zoology" seen, the plate to this species was issued as number XXV (1839). Status: BMNH 1855.12.19.56. Cat. XV: 77: a/b [see previously].

***Limnoctites rectirostris* (Gould, 1839).** STRAIGHT-BILLED REEDHUNTER [? CD 1226/1252, see following]. Z: 80: *Limnornis rectirostris* Gould. Syntype *Limnornis rectirostris* Gould, 1839 [pl. XXV]. Loc.: Uruguay: Maldonado. June 1833. Ex. coll. ZSL. Material: ex mount, ad. Remark: In two original copies of the "Zoology" seen, the plate to this species was issued as number XXVI (1839). Status: BMNH 1855.12.19.77. Cat. XV: 77: a.

***Limnoctites rectirostris* (Gould, 1839).** STRAIGHT-BILLED REEDHUNTER [? CD 1226/1252, see previously]. Z: 80: *Limnornis rectirostris* Gould. Syntype *Limnornis rectirostris* Gould, 1839 [pl. XXV]. Loc.: Uruguay: Maldonado. June 1833. Ex. coll. ZSL. Material: ex mount, ad. Remark: In two original copies of the "Zoology" seen, the plate to this species was issued as number XXVI (1839). Status: probably BMNH unregistered specimen. Cat. XV: 77: b.

***Anumbius annumbi* (Vieillot, 1817).** FIREWOOD-GATHERER [CD 1251]. Z: 76: *Synallaxis major* Gould. Holotype *Synallaxis major* Gould, 1839 [pl. XXII]. Loc.: Uruguay: Maldonado: north bank of La Plata. June 1833. Ex. coll. ZSL. Material: ex mount, ad. Status: BMNH 1855.12.19.166. Cat. XV: 76: a.

***Pygarrhichas albogularis* (King, 1831).** WHITE-THROATED TREERUNNER [CD 2129]. Z: 82–83: *Dendrodramus leucosternus* Gould. Holotype *Dendrodramus* *leucosternus* Gould, 1839 [pl. XXVII]. Loc.: Chile: Isla de Chiloé. July 1834. Material: skin/mount. Status: missing.

THAMNOPHILIDAE:

***Thamnophilus doliatus* ss (Linnaeus, 1764).** BARRED ANTSHRIKE [CD 1239]. Z: 58: *Thamnophilus doliatus* Vieill. Loc.: Uruguay: Maldonado. 1833. Remark: this species does not normally occur in Uruguay. Material: skin/mount. Status: missing.

RHINOCRYPTIDAE:

***Pteroptochos tarnii* (King, 1831).** BLACK-THROATED HUET-HUET CD 2531. Z: 70–71: *Pteroptochos tarnii* G. R. Gray. Loc.: Chile: Isla de Chiloé. January 1835. Ex. coll. ZSL via Darwin. Material: ex mount, male. Status: BMNH 1841.1.18.18. Cat. XV: 349: a.

***Pteroptochos tarnii* (King, 1831).** BLACK-THROATED HUET-HUET [CD 1157, specimens in spirit]. Z: 70–71 & 150–151: *Pteroptochos tarnii* G. R. Gray. Loc.: Chile: Isla de Chiloé. January 1835. Material: specimen in alcohol. Status: missing.

***Pteroptochos megapodius megapodius* (Kittlitz, 1830).** MOUSTACHED TURCA [CD 2172]. Z: 71–72: *Pteroptochos megapodius* Kittl. Loc.: Chile: Valparaso: dry country. August–September 1834. Material: skin/mount. Status: missing.

***Pteroptochos megapodius megapodius* (Kittlitz, 1830).** MOUSTACHED TURCA [CD 2296]. Z: 71–72: *Pteroptochos megapodius* Kittl. Loc.: Chile: Valparaso: dry country. August–September 1834. Material: skin/mount. Status: missing.

***Pteroptochos megapodius megapodius* (Kittlitz, 1830).** MOUSTACHED TURCA [CD 2824]. Z: 71–72: *Pteroptochos megapodius* Kittl. Loc.: Chile: Coquimbo: dry country. 1835. Material: skin/mount. Status: missing.

***Scelorchilus albicollis albicollis* (Kittlitz, 1830).** WHITE-THROATED TAPACULO [CD 2173]. Z: 72: *Pteroptochos albicollis* Kittl. Loc.: Chile: Valparaso. August–September 1834. Material: skin/mount. Status: missing.

***Scelorchilus albicollis albicollis* (Kittlitz, 1830).** WHITE-THROATED TAPACULO [CD 2174]. Z: 72: *Pteroptochos albicollis* Kittl. Loc.: Chile: Valparaso. August–September 1834. Material: skin/mount. Status: missing.

***Scelorchilus albicollis albicollis* (Kittlitz, 1830).** WHITE-THROATED TAPACULO [CD 2825]. Z: 72: *Pteroptochos albicollis* Kittl. Loc.: Chile: Illapel. 1835. Material: skin/mount. Status: missing.

***Scelorchilus albicollis albicollis* (Kittlitz, 1830).** WHITE-THROATED TAPACULO [CD 1037, specimens in spirit]. Z: 72 & 151–152: *Pteroptochos albicollis* Kittl. Loc.: Chile: Valparaso. August 1834. Material: specimen in alcohol. Status: missing.

***Scelorchilus rubecula rubecula* (Kittlitz, 1830).** CHUCAO TAPACULO [CD 2127]. Z: 73: *Pteroptochos rubecula* Kittl. Loc.: Chile: Isla de Chiloé. July 1834. Ex. coll. ZSL via Darwin. Material: ex mount, ad. Status: BMNH 1841.1.18.19. Cat. XV: 346: g.

***Rhinocrypta lanceolata lanceolata* (Geoffroy Saint-Hilaire, 1832).** CRESTED GALLITO [CD 1459]. Z: 70: *Rhinomya lanceolata* Is. Geoffr. & d'Orb. Loc.: Argentina: Rio Negro. August 1833. Ex. coll. ZSL. Material: ex mount, ad. Status: BMNH 1855.12.19.169. Cat. XV: 347: c.

***Eugralla paradoxa* (Kittlitz, 1830).** OCHRE-FLANKED TAPACULO [? CD 2555/2556, see following]. Z: 73–74: *Pteroptochos paradoxus* G. R. Gray. Loc.: Chile: Valdivia. January 1835. Ex. coll. ZSL. via Darwin. Material: ex mount, imm. Status: BMNH 1841.1.18.21 [wrong on label]. Cat. XV: 352: c.

***Eugralla paradoxa* (Kittlitz, 1830).** OCHRE-FLANKED TAPACULO [? CD 2555/2556, see previously]. Z: 73–74: *Pteroptochos paradoxus* G. R. Gray. Loc.: Chile: Valdivia. January 1835. Material: ex mount, imm. Status: BMNH 1855.12.19.159. Cat. XV: 352: a.

***Eugralla paradoxa* (Kittlitz, 1830).** OCHRE-FLANKED TAPACULO [CD 2436]. Z: 73–74: *Pteroptochos paradoxus* G. R. Gray. Loc.: Chile: Isla de Chiloé: East Coast. December 1834. Material: skin/mount. Status: missing.

***Scytalopus magellanicus magellanicus* (Gmelin, 1789).** MAGELLANIC TAPACULO [CD 1828]. Z: 74: *Scytalopus megallanicus* [*sic*] G. R. Gray. Loc.: Chile: Tierra del Fuego: Port Famine. February 1834. Ex. coll. ZSL. Material: ex mount, imm. Status: BMNH 1855.12.19.195. Cat. XV: 339: l.

***Scytalopus magellanicus magellanicus* (Gmelin, 1789).** MAGELLANIC TAPACULO [CD 1144]. Z: 74: *Scytalopus megallanicus* [*sic*] G. R. Gray. Loc.: Falkland Islands/Islas Malvinas: East Falkland Island. March 1833. Ex. coll. ZSL. Material: ex mount, imm. Status: BMNH 1855.12.19.180. Cat. XV: 339: m.

***Scytalopus magellanicus magellanicus* (Gmelin, 1789).** MAGELLANIC TAPACULO [CD 2502]. Z: 74: *Scytalopus megallanicus* [*sic*] G. R. Gray. Loc.: Chile: Isla de Chiloé or Chonos Archipel. January 1835. Material: skin/mount. Status: missing.

TYRANNIDAE:

***Elaenia albiceps chilensis* (Hellmayr, 1927).** WHITE-CRESTED ELAENIA [CD 2829]. Z: 47: *Myiobus* [*sic*] *albiceps* G.R. Gray. Loc.: Chile: Coquimbo. 1835. Ex. coll. ZSL. Material: ex mount, ad. Status: BMNH 1855.12.19.118. Cat. XIV: 143: g'.

***Elaenia albiceps chilensis* (Hellmayr, 1927).** WHITE-CRESTED ELAENIA [? CD 1825]. Z: 47: *Myiobus* [*sic*] *albiceps* G. R. Gray. Loc.: Chile: Tierra del Fuego:

Port Famine. February 1834. Ex. coll. Salvin & Godman, ex. Eyton, ex. ZSL, Material: skin, ad. Status: BMNH 1888.1.1.728. Cat. XIV: 143: h'.

***Elaenia albiceps chilensis* (Hellmayr, 1927).** WHITE-CRESTED ELAENIA [CD not traced]. Z: 47: *Myiobus* [*sic*] *albiceps* G. R. Gray. Loc.: Chile: Chonos Archipelago. December 1834–January 1835. Material: skin/mount. Status: not collected or missing.

***Elaenia albiceps chilensis* (Hellmayr, 1927).** WHITE-CRESTED ELAENIA [?? CD 1258]. Z: 47: *Myiobus* [*sic*] *albiceps* G. R. Gray. Loc.: Uruguay: Maldonado: banks of the Plata. 1833. Material: skin/mount. Status: missing.

***Suiriri suiriri suiriri* (Vieillot, 1818).** SUIRIRI FLYCATCHER CD 1452. Z: 50: *Pachyramphus albescens* Gould. Holotype *Pachyramphus albescens* Gould, 1838 [pl. IX]. Loc.: Argentina: Buenos Aires. 1833. Ex. coll. ZSL. Material: ex mount, ad. Remark: In two original copies of the "Zoology" seen, the plate to this species was issued as number XIV (1839). Status: BMNH unregistered specimen. Cat. XIV: 155: c.

***Anairetes parulus patagonicus* (Hellmayr, 1920).** TUFTED TIT-TYRANT [CD 2027, nec. Keynes 2000]. Z: 49: *Serpophaga parulus* Gould. Loc.: Argentina: Santa Cruz: April 1834. Ex. coll. ZSL. Material: ex mount, ad. Status: BMNH 1855.12.19.161. Cat. XIV: 107: c.

***Anairetes parulus patagonicus* (Hellmayr, 1920).** TUFTED TIT-TYRANT [CD 1469]. Z: 49: *Serpophaga parulus* Gould. Loc.: Argentina: Rio Negro. August 1833. Material: skin/mount. Status: missing.

***Anairetes parulus parulus* (Kittlitz, 1830).** TUFTED TIT-TYRANT [CD 2193]. Z: 49: *Serpophaga parulus* Gould. Loc.: Chile: Valparaso. August-September 1834. Ex. coll. ZSL. Material: ex mount, male, ad. Status: BMNH 1855.12.19.98. Cat. XIV: 107: j.

***Serpophaga nigricans* (Vieillot, 1817).** SOOTY TYRANNULET [CD 1296]. Z: 50: *Serpophaga nigricans* Gould. Loc.: Uruguay: Maldonado: on the banks of the Plata. June 1833. Material: skin/mount. Status: not yet looked for by the author.

***Serpophaga subcristata straminea* (Temminck, 1822).** WHITE-CRESTED TYRANNULET [CD 1257]. Z: 49–50: *Serpophaga albo-coronata* Gould. Holotype *Serpophaga albo-coronata* Gould, 1839. Loc.: Uruguay: Maldonado. June 1833. Remark: Warren & Harrison 1971 listed the wrong (FitzRoy) specimen as type. Material: skin/mount. Status: not yet looked for by the author.

***Serpophaga subcristata straminea* (Temminck, 1822).** WHITE-CRESTED TYRANNULET [CD 650, specimens in spirit]. Z: 49–50 & 147: *Serpophaga albo-coronata* Gould. Loc.: Uruguay: Maldonado. May–June 1833. Material: specimen in alcohol. Status: missing.

***Polystictus pectoralis pectoralis* (Vieillot, 1817).** BEARDED TACHURI [? CD 1604/1613]. Z: 51: *Pachyramphus minimus* [Gould]. Holotype *Pachyramphus minimus* Gould, 1838 [pl. X]. Loc.: Uruguay: Montevideo. November 1832. Material: skin/mount. Remark: In two original copies of the "Zoology" seen, the plate to this species was issued as number XV (1839). Status: missing.

***Myiophobus fasciatus auriceps* (Gould, 1839).** BRAN-COLORED FLYCATCHER [?? CD 847]. Z: 47: *Myiobius auriceps* [Gould]. Holotype *Tyrannula auriceps* Gould, 1839. Loc.: Argentina: Buenos Aires. August 1833. Ex. coll. ZSL. Material: ex mount, male, ad. Status: BMNH 1855.12.19.172. Cat. XIV: 210: a'.

***Tachuris rubrigastra rubrigastra* (Vieillot, 1817).** MANY-COLORED RUSH TYRANT [CD 1259]. Z: 86: *Cyanotis omnicolor* Swains. Loc.: Uruguay: Maldonado. June 1833. Material: skin/mount. Status: missing.

***Tachuris rubrigastra rubrigastra* (Vieillot, 1817).** MANY-COLORED RUSH TYRANT [CD 1277]. Z: 86: *Cyanotis omnicolor* Swains. Loc.: Uruguay: Maldonado. June 1833. Material: skin/mount. Status: missing.

***Platyrinchus mystaceus mystaceus* (Vieillot, 1818).** WHITE-THROATED SPADEBILL [CD not traced]. Z: [not traced]. Loc.: South America. Ex. coll. ZSL via Darwin. Material: ex mount, ad. Status: BMNH 1856.3.15.18. Cat. XIV: 68: t.

***Pyrocephalus rubinus nanus* (Gould, 1838).** VERMILLION FLYCATCHER CD 3309. Z: 45–46: *Pyrocephalus nanus* Gould. Syntype *Pyrocephalus nanus* Gould, 1838 [pl. VII]. Loc.: Ecuador: Galápagos Islands: Chatham Island. October [*sic*] 1835. Ex. coll. ZSL. Material: ex mount, male, ad. Status: BMNH unregistered specimen. Cat. XIV: 215: a'.

***Pyrocephalus rubinus nanus* (Gould, 1838).** VERMILLION FLYCATCHER [CD 3342]. Z: 45–46: *Pyrocephalus nanus* Gould. Syntype *Pyrocephalus nanus* Gould, 1838 [pl. VII]. Loc.: Ecuador: Galápagos Islands. October 1835. Ex. coll. ZSL. Material: ex mount, female [male, imm. in CD], ad. Status: BMNH 1855.12.19.198. Cat. XIV: 215: c'.

***Pyrocephalus rubinus nanus* (Gould, 1838).** VERMILLION FLYCATCHER CD 3344. Z: 45–46: *Pyrocephalus nanus* Gould. Syntype *Pyrocephalus nanus* Gould, 1838 [pl. VII]. Loc.: Ecuador: Galápagos Islands. October 1835. Ex. coll. ZSL. Material: ex mount, female, ad. Status: BMNH unregistered specimen. Cat. XIV: 215: b'.

***Pyrocephalus rubinus nanus* (Gould, 1838).** VERMILLION FLYCATCHER [CD 3343]. Z: 45–46: *Pyrocephalus nanus* Gould. Syntype *Pyrocephalus nanus* Gould, 1838 [pl. VII]. Loc.: Ecuador: Galápagos Islands. October 1835. Remark: Gould and Darwin 1839b: 46 probably studied also FitzRoy's specimen BMNH 1837.5.13.210/Cat. XIV: 215: e'. Material: skin/mount, male. Status: missing.

***Pyrocephalus rubinus dubius* (Gould, 1839).** VERMILLION FLYCATCHER [CD 3345]. Z: 46: *Pyrocephalus dubius* Gould. Holotype *Pyrocephalus dubius* Gould, 1839. Loc.: Ecuador: Galápagos Islands: Chatham Island. September 1835. Ex. coll. ZSL. Material: ex mount, female [male in CD], ad. Status: BMNH 1855.12.19.184. Cat. XIV: 215: d'.

***Pyrocephalus rubinus obscurus* (Gould, 1839).** VERMILLION FLYCATCHER [CD 3204]. Z: 45: *Pyrocephalus obscurus* Gould. Holotype *Pyrocephalus obscurus* Gould, 1839. Loc.: Peru: Callao. August 1835. Ex. coll. ZSL. Material: ex mount, female/imm. Status: BMNH 1855.12.19.389. Cat. XIV: 215: a.

***Pyrocephalus rubinus rubinus* (Boddaert, 1783).** VERMILLION FLYCATCHER [CD 1439]. Z: 44–45: *Pyrocephalus parvirostris* Gould. Syntype *Pyrocephalus parvirostris* Gould, 1838 [pl. VI]. Loc.: Argentina: Buenos Aires: near La Plata. October 1833. Ex. coll. ZSL via Darwin. Material: ex mount, ad. Status: BMNH 1856.3.16.17. Cat. XIV: 213: n'.

***Pyrocephalus rubinus rubinus* (Boddaert, 1783).** VERMILLION FLYCATCHER CD 1437. Z: 44–45: *Pyrocephalus parvirostris* Gould. Syntype *Pyrocephalus parvirostris* Gould, 1838 [pl. VI]. Loc.: Argentina: Buenos Aires: La Plata. October 1833. Ex. coll. ZSL via Darwin. Material: ex mount, female, ad. Status: BMNH 1856.3.15.17a. Cat. XIV: 213: o'.

***Lessonia rufa* (Gmelin, 1789).** AUSTRAL NEGRITO [? CD 749/780, see following]. Z: 84: *Muscisaxicola nigra* G. R. Gray. Loc.: Argentina: Bahia Blanca: M. Hermoso. September 1832. Material: skin/mount. Status: missing.

***Lessonia rufa* (Gmelin, 1789).** AUSTRAL NEGRITO [? CD 749/780, see previously]. Z: 84: *Muscisaxicola nigra* G. R. Gray. Loc.: Argentina: Bahia Blanca. September 1832. Material: skin/mount. Status: missing.

***Lessonia rufa* (Gmelin, 1789).** AUSTRAL NEGRITO [CD 903]. Z: 84: *Muscisaxicola nigra* G. R. Gray. Loc.: Chile: Tierra del Fuego: Good Success Bay. December 1832. Ex. coll. ZSL. Material: skin, male, ad. Status: perhaps BMNH 1858.4.3.65. Cat. XIV: 62: g.

***Hymenops perspicillata perspicillata* (Gmelin, 1789).** SPECTACLED TYRANT [CD 1231]. Z: 52–53: *Lichenops erythropterus* Gould. Holotype *Lichenops erythropterus* Gould, 1839 [pl. XI]. Loc.: Uruguay: Maldonado: banks of the Plata. 1833. Ex. coll. Eyton, ex. ZSL. Material: ex mount, female, ad. Remark: In two original copies of the "Zoology" seen, the plate to this species was issued as number IX (1838). Status: BMNH 1881.2.18.157. Cat. XIV: 49: f.

***Hymenops perspicillata perspicillata* (Gmelin, 1789).** SPECTACLED TYRANT CD 1206. Z: 51–52: *Lichenops perspicillatus* G. R. Gray. [no type; only the female has type status]. Loc.: Uruguay: Maldonado: neighbourhood of the Plata. 1833. Ex. coll. ZSL. Material: ex mount, male, ad. Status: BMNH 1855.12.19.123. Cat. XIV: 49: a.

***Satrapa icterophrys* (Vieillot, 1818).** YELLOW-BROWED TYRANT [?? CD 1601]. Z: 53: *Fluvicola icterophrys* D'Orb. & Lafr. Loc.: Uruguay: Montevideo. 1833. Ex. coll. ZSL. Material: ex mount, ad. Status: BMNH 1855.12.19.392. Cat. XIV: 42: 1.

***Satrapa icterophrys* (Vieillot, 1818).** YELLOW-BROWED TYRANT [?? CD 1271]. Z: 53: *Fluvicola icterophrys* D'Orb. & Lafr. Loc.: Uruguay: Maldonado: on the banks of the Plata. 1833. Material: skin/mount. Status: missing.

***Muscisaxicola macloviana mentalis* (d'Orbigny & Lafresnaye, 1837).** DARK-FACED GROUND TYRANT *MUSCISAXICOLA MACLOVIANUS* [? CD 2828]. Z: 83: *Muscisaxicola mentalis* D'Orb. & Lafr. Loc.: Chile: Coquimbo. 1835. Ex. coll. ZSL. Material: ex mount, ad. Status: BMNH 1855.12.19.186. Cat. XIV: 57: x.

***Muscisaxicola macloviana mentalis* (d'Orbigny & Lafresnaye, 1837).** DARK-FACED GROUND TYRANT *MUSCISAXICOLA MACLOVIANUS* [?? CD 1448]. Z: 83: *Muscisaxicola mentalis* D'Orb. & Lafr. Loc.: Argentina: Bahia Blanca. 1833. Ex. coll. Gould, ex.? ZSL, Material: skin, ad. Status: perhaps BMNH 1860.1.16.56. Cat. XIV: 57: v.

***Muscisaxicola macloviana mentalis* (d'Orbigny & Lafresnaye, 1837).** DARK-FACED GROUND TYRANT *MUSCISAXICOLA MACLOVIANUS* [?? CD 971]. Z: 83: *Muscisaxicola mentalis* D'Orb. & Lafr. Loc.: Chile: Tierra del Fuego.? January 1833. Material: skin/mount. Status: missing.

***Muscisaxicola macloviana mentalis* (d'Orbigny & Lafresnaye, 1837).** DARK-FACED GROUND TYRANT *MUSCISAXICOLA MACLOVIANUS* [CD 2128]. Z: 83: *Muscisaxicola mentalis* D'Orb. & Lafr. Loc.: Chile: Isla de Chiloé. July 1834. Material: skin/mount. Status: missing.

***Muscisaxicola macloviana mentalis* (d'Orbigny & Lafresnaye, 1837).** DARK-FACED GROUND TYRANT *MUSCISAXICOLA MACLOVIANUS* [?? CD 2208]. Z: 83: *Muscisaxicola mentalis* D'Orb. & Lafr. Loc.: Chile: Valparaso. August–September 1834. Ex. coll. ZSL. Material: skin, ad. Status: perhaps BMNH 1858.4.3.45. Cat. XIV: 57: i.

***Muscisaxicola macloviana macloviana* (Garnot, 1829).** DARK-FACED GROUND TYRANT *MUSCISAXICOLA MACLOVIANUS* [CD 1899]. Z: 83–84: *Muscisaxicola macloviana* G. R. Gray. Loc.: Falkland Islands/Islas Malvinas: East Falkland Island. March 1834. Material: skin/mount. Status: missing.

***Muscisaxicola* sp.** [?? CD 1753]. Z: 84: *Muscisaxicola brunnea* Gould. Holotype *Muscisaxicola brunnea* Gould, 1839. Loc.: Argentina: Port St. Julian. January

1834. Remark: Cat. XIV: 53 already stated that the identity of this species remains uncertain. Material: skin/mount, imm. Status: missing.

***Agriornis montana leucura* (Gould, 1839).** BLACK-BILLED SHRIKE-TYRANT [? CD 2012/2013, see following]. Z: 57: *Agriornis maritimus* G. R. Gray. Syntype *Agriornis leucurus* Gould, 1839 [pl. XV]. Loc.: Argentina: Santa Cruz: April 1834. Ex. coll. ZSL. Material: ex mount, ad. Remark: In two original copies of the "Zoology" seen, the plate to this species was issued as number XIII (1839). Status: BMNH 1855.12.19.252. Cat. XIV: 6: j/k [see following].

***Agriornis montana leucura* (Gould, 1839).** BLACK-BILLED SHRIKE-TYRANT [? CD 2012/2013, see previously]. Z: 57: *Agriornis maritimus* G. R. Gray. Syntype *Agriornis leucurus* Gould, 1839 [pl. XV]. Loc.: Argentina: Santa Cruz: April 1834. Ex. coll. ZSL. Material: ex mount, ad. Remark: In two original copies of the "Zoology" seen, the plate to this species was issued as number XIII (1839). Status: BMNH 1855.12.19.251. Cat. XIV: 6: j/k [see previously].

***Agriornis livida livida* (Kittlitz, 1835).** GREAT SHRIKE-TYRANT [CD 2167]. Z: 56: *Agriornis gutturalis* Gould. Loc.: Chile: Valparaso. August–September 1834. Ex. coll. ZSL. Material: ex mount, female, ad. Status: BMNH 1855.12.19.344. Cat. XIV: 5: g.

***Agriornis livida livida* (Kittlitz, 1835).** GREAT SHRIKE-TYRANT [? CD 2199]. Z: 56: *Agriornis gutturalis* Gould. Loc.: Chile: Valparaso. August–September 1834. Material: skin/mount. Status: missing.

***Agriornis microptera microptera* (Gould, 1839).** GREY-BELLIED SHRIKE-TYRANT *AGRIORNIS MICROPTERUS* [CD 1752]. Z: 57: *Agriornis micropterus* Gould. Syntype *Agriornis micropterus* Gould, 1839 [pl. XIV]. Loc.: Argentina: Port St. Julian. January 1834. Ex. coll. ZSL. Material: ex mount, ad. Remark: In two original copies of the "Zoology" seen, the plate to this species was issued as number XII (1839). Status: BMNH 1855.12.19.298. Cat. XIV: 5: e.

***Agriornis microptera microptera* (Gould, 1839).** GREY-BELLIED SHRIKE-TYRANT *AGRIORNIS MICROPTERUS* [CD 1699]. Z: 57: *Agriornis icropterus* Gould. Syntype *Agriornis micropterus* Gould, 1839 [pl. XIV]. Loc.: Argentina: Port Desire. January 1834. Ex. coll. ZSL. Material: ex mount, ad. Remark: In two original copies of the "Zoology" seen, the plate to this species was issued as number XII (1839). Status: BMNH 1855.12.19.253. Cat. XIV: 5: f.

***Agriornis microptera microptera* (Gould, 1839).** GREY-BELLIED SHRIKE-TYRANT *AGRIORNIS MICROPTERUS* [CD 1700]. Z: 57: *Agriornis micropterus* Gould. Syntype *Agriornis micropterus* Gould, 1839 [pl. XIV]. Loc.: Argentina: Port Desire. January 1834. Material: skin/mount, imm. Remark: In two original copies of the "Zoology" seen, the plate to this species was issued as number XII (1839). Status: missing.

***Agriornis micropteramicroptera* (Gould, 1839).** GREY-BELLIED SHRIKE-TYRANT *AGRIORNIS MICROPTERUS* [CD 2013]. Z: 56–57: *Agriornis triatus* Gould. Holotype *Agriornis striatus* Gould, 1839. Loc.: Argentina: Santa Cruz: April 1834. Ex. coll. Eyton, ex. ZSL. Material: ex mount, female, ad. Status: BMNH 1881.2.18.128. Cat. XIV: 5: d.

***Xolmis pyrope pyrope* (Kittlitz, 1830).** FIRE-EYED DIUCON [CD 1819]. Z: 55: *Xolmis pyrope* G. R. Gray. Loc.: Chile: Tierra del Fuego: Port Famine. Early February 1834. Material: skin/mount. Status: missing.

***Xolmis pyrope pyrope* (Kittlitz, 1830).** FIRE-EYED DIUCON [CD 1820]. Z: 55: *Xolmis pyrope* G. R. Gray. Loc.: Chile: Tierra del Fuego: Port Famine. Early February 1834. Material: skin/mount. Status: missing.

***Xolmis pyrope pyrope* (Kittlitz, 1830).** FIRE-EYED DIUCON [CD 2081]. Z: 55: *Xolmis pyrope* G. R. Gray. Loc.: Chile: Tierra del Fuego: Port Famine. June 1834. Material: skin/mount. Status: missing.

***Xolmis pyrope fortis* (Philippi & Johnson, 1946).** FIRE-EYED DIUCON [CD 2198]. Z: 55: *Xolmis pyrope* G. R. Gray. Loc.: Chile: Copiapó or Valparaso. August–September 1834. Material: skin/mount. Status: missing.

***Xolmis pyrope fortis* (Philippi & Johnson, 1946).** FIRE-EYED DIUCON [CD 2124]. Z: 55: *Xolmis pyrope* G. R. Gray. Loc.: Chile: Isla de Chiloé. July 1834. Material: skin/mount. Status: missing.

***Xolmis pyrope fortis* (Philippi & Johnson, 1946).** FIRE-EYED DIUCON [CD 2375]. Z: 55: *Xolmis pyrope* G. R. Gray. Loc.: Chile: Isla de Chiloé. November–December 1834. Material: single egg. Status: missing.

***Xolmis cinerea cinerea* (Vieillot, 1816).** GREY MONJITA [CD 1204]. Z: 54: *Xolmis nengeta* G. R. Gray. Loc.: Uruguay: Maldonado: banks of La Plata. 1833. Ex. coll. ZSL. Remark: This taxon is not *Fluvicola nengeta* Linnaeus, 1766. Material: ex mount, ad. Status: BMNH 1855.12.19.307. Cat. XIV: 11: e.

***Xolmis coronata* (Vieillot, 1823).** BLACK-CROWNED MONJITA *XOLMIS CORONATUS* [? CD 1414/1415/1416, see following]. Z: 54: *Xolmis coronata* G. R. Gray. Loc.: Argentina: banks of Rio Parana near Santa Fé. 1833. Material: skin/mount. Status: missing.

***Xolmis irupero irupero* (Vieillot, 1823).** [CD 1600]. WHITE MONJITA Z: 53: *Fluvicola irupero* G. R. Gray. Loc.: Argentina: Santa Fé. November 1833. Material: skin/mount. Status: missing.

***Xolmis dominicana* (Vieillot, 1823).** BLACK-AND-WHITE MONJITA *HETEROXOLMIS DOMINICANA* [CD 1205]. Z: 53–54: *Fluvicola azarae* Gould. Holotype *Fluvicola Azarae* Gould, 1839 [pl. XII]. Loc.: Uruguay: Maldonado: banks of La Plata. 1833. Ex. coll. ZSL. Material: ex mount, ad. Remark: In two original copies of the "Zoology" seen, the plate to this species was issued as number X (1838). Status: BMNH 1855.12.19.245. Cat. XIV: 13: k.

***Neoxolmis rufiventris* (Vieillot, 1823).** CHOCOLATE-VENTED TYRANT [CD 1220]. Z: 55: *Xolmis variegata* G. R. Gray. Loc.: Uruguay: Maldonado. 1833. Ex. coll. ZSL. Material: ex mount, ad. Status: BMNH 1855.12.19.276. Cat. XIV: 9: i.

***Neoxolmis rufiventris* (Vieillot, 1823).** CHOCOLATE-VENTED TYRANT [CD 1240]. Z: 55: *Xolmis variegata* G. R. Gray. Loc.: Uruguay: Maldonado. 1833. Material: skin/mount. Status: missing.

***Alectrurus risora* (Vieillot, 1824).** STRANGE-TAILED TYRANT [CD 1275]. Z: 51: *Alect[r]urus guirayetupa* Vieill. Loc.: Uruguay: Maldonado: on the banks of the Plata. 1833. Material: skin/mount. Status: missing [very unlikely BMNH 1888.1.13.125, Cat. XIV: 40: a].

***Alectrurus risora* (Vieillot, 1824).** STRANGE-TAILED TYRANT [CD 1276]. Z: 51: *Alect[r]urus guirayetupa* Vieill. Loc.: Uruguay: Maldonado: on the banks of the Plata. 1833. Material: skin/mount. Status: missing.

***Ochthoeca parvirostris* (Gould, 1839).** PATAGONIAN TYRANT *COLORHAMPHUS PARVIROSTRIS* [? CD 2197]. Z: 48: *Myiobius parvirostris* [Gould]. Syntype *Tyrannula parvirostris* Gould, 1839. Loc.: Chile: Valparaso. August–September 1834. Ex. coll. ZSL via Darwin. Material: ex mount, ad. Status: BMNH unregistered specimen. Cat. XIV: 105: d.

***Ochthoeca parvirostris* (Gould, 1839).** PATAGONIAN TYRANT *COLORHAMPHUS PARVIROSTRIS* CD 2083. Z: 48: *Myiobius parvirostris* [Gould]. Syntype *Tyrannula parvirostris* Gould, 1839. Loc.: Chile: Tierra del Fuego: Port Famine. June 1834. Ex. coll. ZSL via Darwin. Material: ex mount, ad. Status: BMNH 1856.3.15.16. Cat. XIV: 105: f.

***Ochthoeca parvirostris* (Gould, 1839).** PATAGONIAN TYRANT *COLORHAMPHUS PARVIROSTRIS* CD 1824. Z: 48: *Myiobius parvirostris* [Gould]. Syntype *Tyrannula parvirostris* Gould, 1839. Loc.: Chile: Tierra del Fuego: Port Famine. February 1834. Ex. coll. ZSL via Darwin. Material: skin/mount. Status: former BMNH 1841.1.18.25, missing since 1880s.

***Pitangus sulphuratus argentinus* (Todd, 1952).** GREAT KISKADEE [CD 1216]. Z: 43: *Saurophagus sulphuratus* Swains. Loc.: Uruguay: Maldonado: northern banks of the Plata. 1833. Material: skin/mount. Status: missing.

***Tyrannus savana savana* (Vieillot, 1808).** FORK-TAILED FLYCATCHER [CD 1022]. Z: 40–44: *Muscivora tyrannus* G. R. Gray. Loc.: Argentina: Buenos Aires. November 1833. Material: skin/mount, female. Status: missing.

***Tyrannus savana savana* (Vieillot, 1808).** FORK-TAILED FLYCATCHER [CD 1621]. Z: 43–44: *Muscivora tyrannus* G. R. Gray. Loc.: Argentina: Buenos Aires. November 1833. Material: skin/mount, male. Status: missing.

***Myiarchus magnirostris* (Gould, 1838).** GALAPAGO FLYCATCHER CD 3308. Z: 48: *Myiobius magnirostris* [Gould]. Holotype *Tyrannula magnirostris* Gould, 1838 [pl. VIII]. Loc.: Ecuador: Galápagos Islands: Chatham Island. October 1835. Ex. coll. ZSL via Darwin. Material: ex mount, female, ad. Status: BMNH 1856.3.15.10. Cat. XIV: 263: b.

COTINGIDAE:

***Phytotoma rara* (Molina, 1782).** RUFOUS-TAILED PLANTCUTTER [CD 2175]. Z: 106 + 153–154: *Phytotoma rara* Mol. Loc.: Chile: Valparaso. August–September 1834. Ex. coll. Frank, ex. Gould, ex. ZSL. Remark: On label as "Darwin's reis 1837 Chiloe" [in error]. Material: ex mount, male, ad. Status: RMNH unregistered specimen.

***Phytotoma rara* (Molina, 1782).** RUFOUS-TAILED PLANTCUTTER [CD 2176]. Z: 106: *Phytotoma rara* Mol. Loc.: Chile: Valparaso. August–September 1834. Material: skin/mount, female. Status: missing.

***Phytotoma rara* (Molina, 1782).** RUFOUS-TAILED PLANTCUTTER [CD 1043, specimens in spirit]. Z: 106 & 153–154: *Phytotoma rara* Mol. Loc.: Chile: Valparaso. August 1834. Material: specimen in alcohol. Status: missing.

VIREONIDAE:

***Cyclarhis gujanensis ochrocephala* (Tschudi, 1845).** RUFOUS-BROWED PEPPERSHRIKE [CD 1261]. Z: 58: *Cyclarhis guianensis* Swains. Loc.: Uruguay: Maldonado. 1833. Material: skin/mount. Status: missing.

ALAUDIDAE:

***Ammomanes cincturus cincturus* (Gould, 1839)**. BAR-TAILED LARK *AMMOMANES CINCTURA* CD 3905. Z: 87: *Melanocorypha cinctura* Gould. Holotype *Melanocorypha cinctura* Gould, 1839. Loc.: Portugal: Cape Verde Islands: So Tiago: Porto Praia. September 1836. Ex. coll. ZSL. Material: ex mount, ad. Status: BMNH 1855.12.19.379. Cat. XIII: 645: a.

***Eremopterix nigriceps nigriceps* (Gould, 1839).** BLACK-CROWNED SPARROW-LARK [CD 3906]. Z: 87–88: *Pyrrhalauda nigriceps* Gould. Syntype *Pyrrhalauda nigriceps* Gould, 1839. Loc.: Portugal: Cape Verde Islands: So Tiago: Porto Praia. September 1836. Material: skin/mount. Status: missing.

***Eremopterix nigriceps nigriceps* (Gould, 1839).** BLACK-CROWNED SPARROW-LARK [CD 188]. Z: 87–88: *Pyrrhalauda nigriceps* Gould. Syntype *Pyrrhalauda nigriceps* Gould, 1839. Loc.: Portugal: Cape Verde Islands: So Tiago: Porto Praia. January 1832. Material: skin/mount. Status: missing.

HIRUNDINIDAE:

***Tachycineta leucorrhoa* (Vieillot, 1817).** WHITE-RUMPED SWALLOW [? CD 1609/1618, see following]. Z: 40: *Hirundo frontalis* Gould. Holotype *Hirundo frontalis* Gould, 1839. Loc.: Uruguay: Montevideo. November 1833. Material: skin/mount. Status: missing.

***Tachycineta leucopyga* (Meyen, 1834).** CHILEAN SWALLOW [? CD 2200/2201, see following]. Z: 40: *Hirundo leucopygia* [*sic*] Licht. Loc.: Chile: Valparaso. August–September 1834. Material: skin/mount, male. Status: missing.

***Tachycineta leucopyga* (Meyen, 1834).** CHILEAN SWALLOW [CD 1827]. Z: 40: *Hirundo leucopygia* [*sic*] Licht. Loc.: Chile: Tierra del Fuego: Port Famine. February 1834. Material: skin/mount. Status: missing.

***Progne modesta modesta* (Gould, 1838).** GALAPAGO MARTIN [CD 3356]. Z: 39–40: *Progne modesta* Gould. Holotype *Hirundo concolor* Gould, 1837 & Holotype *Progne modesta* Gould, 1838 [pl. V]. Loc.: Ecuador: Galápagos Islands: James Island. October 1835. Ex. coll. Gould, ex. ZSL. Material: ex mount, male, ad. Remark: The species name differs on the plate of a facsimile copy compared to two original copies of the "Zoology"; alternative spelling is *modestus*. Status: BMNH 1860.1.16.54. Cat. X: 176: a.

***Progne tapera fusca* (Vieillot, 1817).** BROWN-CHESTED MARTIN [CD 746]. Z: 38–39: *Progne purpurea* Boie. Loc.: Argentina: Bahia Blanca. September 1832. Material: skin/mount. Status: missing.

***Progne tapera fusca* (Vieillot, 1817).** BROWN-CHESTED MARTIN [? CD 1609/1618, see previously]. Z: 38–39: *Progne purpurea* Boie. Loc.: Uruguay: Montevideo. November 1833. Material: skin/mount. Status: missing.

***Notiochelidon cyanoleuca patagonica* (d'Orbigny & Lafresnaye, 1837).** BLUE-AND-WHITE SWALLOW CD 1445. Z: 41: *Hirundo cyanoleuca* Vieill. Loc.: Argentina: Bahia Blanca. 1833. Ex. coll. ZSL via Darwin. Material: ex mount, ad. Status: BMNH 1841.1.18.20. Cat. X: 188: x.

***Notiochelidon cyanoleuca patagonica* (d'Orbigny & Lafresnaye, 1837).** BLUE-AND-WHITE SWALLOW [? CD 2200/2201, see previously]. Z: 41: *Hirundo cyanoleuca* Vieill. Loc.: Chile: Valparaso. September 1834. Material: skin/mount, male. Status: missing.

TROGLODYTIDAE:

***Cistothorus platensis platensis* (Latham, 1790).** SEDGE WREN CD 1444 [as 1443 on label]. Z: 75: *Troglodytes platensis* Gmel. Loc.: Argentina: Bahia Blanca. October 1833. Ex. coll. ZSL via Darwin. Material: ex mount, ad. Status: BMNH 1856.3.15.20. Cat. VI: 247: b.

***Cistothorus platensis falklandicus* (Chapman, 1934).** SEDGE WREN [CD 1053]. Z: 75: *Troglodytes platensis* Gmel. Loc.: Falkland Islands/Islas

Malvinas: East Falkland Island. March 1833. Ex. coll. Salvin & Godman or Gould, ex.? ZSL. Remark: not quite sure if one of these two specimens is the missing Darwin bird. Material: skin, ad. Status: perhaps BMNH 1885.3.6.480 or BMNH 1859.3.25.83 [Keynes 2000 accidentally listed the wrong specimen].

***Troglodytes aedon chilensis* (Lesson, 1830).** HOUSE WREN CD 2194. Z: 74: *Troglodytes magellanicus* Gould. Loc.: Chile: Valparaso. August–September 1834. Ex. coll. Salvin & Godman, ex. Gould, ex. ZSL. Remark: Darwin's specimens are not the types of *Troglodytes magellanicus* of Gould, 1837, which was probably based on specimens collected by Captain King (nec. Warren & Harrison 1971). Material: skin, female, ad. Status: BMNH 1885.3.6.408. Cat. VI: 257: l [annotation by Sharpe in BMNH copy].

***Troglodytes aedon musculus* (Naumann, 1823).** HOUSE WREN [CD not traced]. Z: 74: *Troglodytes magellanicus* Gould. Loc.: Brazil: Rio de Janeiro. April?–July 1832. Ex. coll. Salvin & Godman, ex. Gould, ex. ZSL. Remark: Darwin's specimens are not the types of *Troglodytes magellanicus* of Gould, 1837, which was probably based on specimens collected by Captain King (*pace* Warren & Harrison 1971). Material: skin, ad. Status: BMNH 1885.3.6.409. Cat. VI: 257: i [annotation by Sharpe in BMNH copy].

***Troglodytes aedon bonariae* (Hellmayr, 1919).** HOUSE WREN [?? CD 1425/1434/1450]. Z: 74: *Troglodytes magellanicus* Gould. Loc.: Argentina: banks of the Plata. 1833. Remark: Darwin's specimens are not the types of *Troglodytes magellanicus* of Gould, 1837, which was probably based on specimens collected by Captain King (*pace* Warren & Harrison 1971). Material: skin/mount. Status: missing.

***Troglodytes aedon chilensis* (Lesson, 1830).** HOUSE WREN [CD 2026]. Z: 74: *Troglodytes magellanicus* Gould. Loc.: Argentina: Santa Cruz: April 1834. Remark: Darwin's specimens are not the types of *Troglodytes magellanicus* of Gould, 1837, which was probably based on specimens collected by Captain King (*pace* Warren & Harrison 1971). Material: skin/mount, female. Status: missing.

***Troglodytes aedon chilensis* (Lesson, 1830).** HOUSE WREN [CD 1831]. Z: 74: *Troglodytes magellanicus* Gould. Loc.: Chile: Tierra del Fuego: Port Famine. February 1834. Remark: Darwin's specimens are not the types of *Troglodytes magellanicus* of Gould, 1837, which was probably based on specimens collected by Captain King (nec. Warren & Harrison 1971). Material: skin/mount. Status: missing.

MIMIDAE:

***Mimus thenca* (Molina, 1782).** CHILEAN MOCKINGBIRD CD 2169. Z: 61: *Mimus thenca* G. R. Gray. Loc.: Chile: Valparaso. August–September

1834. Ex. coll. ZSL. Material: ex mount, male [on label], female [CD], ad. Status: BMNH 1855.12.19.230. Cat. VI: 345: e/f [see following].

***Mimus thenca* (Molina, 1782).** CHILEAN MOCKINGBIRD [CD 2170]. Z: 61: *Mimus thenca* G. R. Gray. Loc.: Chile: Valparaso. August–September 1834. Ex. coll. ZSL. Material: ex mount,? male [CD], ad. Status: BMNH 1855.12.19.226. Cat. VI: 345: e/f [see previously].

***Mimus saturninus modulator* (Gould, 1836).** CHALK-BROWED MOCKINGBIRD [CD 1620]. Z: 60: *Mimus orpheus* G. R. Gray. Loc.: Uruguay: Montevideo: banks of the Plata. November 1833. Ex. coll. ZSL. Remark: The types of *Orpheus modulator* of Gould, 1836, had probably been collected by King, not by Darwin, nec. Warren & Harrison 1971. Material: ex mount, ad. Status: BMNH 1855.12.19.229. Cat. VI: 348: a.

***Mimus saturninus modulator* (Gould, 1836).** CHALK-BROWED MOCKINGBIRD [CD 1213]. Z: 60: *Mimus orpheus* G. R. Gray. Loc.: Uruguay: Maldonado: banks of the Rio Plata. 1833. Ex. coll. ZSL. The types of *Orpheus modulator* of Gould, 1836, were probably collected by King, not by Darwin, contra Warren & Harrison 1971. Material: ex mount, ad. Status: BMNH 1855.12.19.227. Cat. VI: 348: d.

***Mimus patagonicus* (Lafresnaye & d'Orbigny, 1837).** PATAGONIAN MOCKINGBIRD *MIMUS PATAGONICUS* [? CD 2011, see following]. Z: 60–61: *Mimus patagonicus* G. R. Gray. Loc.: Argentina: Santa Cruz: April 1834. Ex. coll. ZSL. Remark: Cat. erroneously listed this specimen as syntype of *Mimus patagonicus* G. R. Gray. Material: ex mount, ad. Status: BMNH 1855.12.19.221. Cat. VI: 352: a/b [see following].

***Mimus patagonicus* (Lafresnaye & d'Orbigny, 1837).** PATAGONIAN MOCKINGBIRD *MIMUS PATAGONICUS* [? CD 2011, see previously]. Z: 60–61: *Mimus patagonicus* G. R. Gray. Loc.: Argentina: Santa Cruz: April 1834. Ex. coll. ZSL. Remark: Cat. erroneously listed this specimen as syntype of *Mimus patagonicus* G. R. Gray. Material: ex mount, ad. Status: BMNH 1855.12.19.311. Cat. VI: 352: a/b [see previously].

***Mimus patagonicus* (Lafresnaye & d'Orbigny, 1837).** PATAGONIAN MOCKINGBIRD *MIMUS PATAGONICUS* [CD 1461]. Z: 60–61: *Mimus patagonicus* G. R. Gray. Loc.: Argentina: Rio Negro. August 1833. Remark: Cat. erroneously listed this specimen as syntype of *Mimus patagonicus* G. R. Gray. Material: skin/mount. Status: missing.

***Nesomimus trifasciatus parvulus* (Gould, 1837).** GALAPAGOS MOCKINGBIRD *MIMUS PARVULUS* [CD 3349]. Z: 63: *Mimus parvulus* G. R. Gray. Holotype *Orpheus parvulus* Gould, 1837. Loc.: Ecuador: Galápagos Archipelago: Albemarle Island. October 1835. Ex. coll. ZSL. Material: ex mount, female, ad. Status: BMNH 1855.12.19.92. Cat. VI: 350: a.

***Nesomimus trifasciatus personatus* (Ridgway, 1890)**. GALAPAGOS MOCKINGBIRD *MIMUS PARVULUS* [CD 3350]. Z: 62: *Mimus melanotis* G. R. Gray.? Syntype *Orpheus melanotis* Gould, 1837. Loc.: Ecuador: Galápagos Archipelago: [James Island]. October 1835. Ex. coll. ZSL. Material: ex mount, male, ad. Status: BMNH 1855.12.19.223. Cat. VI: 350:?b.

***Nesomimus trifasciatus trifasciatus* (Gould, 1837).** FLOREANA MOCKINGBIRD [CD 3306]. Z: 62: *Mimus trifasciatus* G. R. Gray. Holotype *Orpheus trifasciatus* Gould, 1837. Loc.: Ecuador: Galápagos Archipelago: Charles Island. October 1835. Ex. coll. ZSL. Material: ex mount, male, ad. Status: BMNH 1855.12.19.225. Cat. VI: 346: a.

***Nesomimus trifasciatus melanotis* (Gould, 1837).** SAN CRISTOBEL MOCKINGBIRD *MIMUS MELANOTIS* [CD 3307]. Z: 62: *Mimus melanotis* G. R. Gray. Syntype *Orpheus melanotis* Gould, 1837. Loc.: Ecuador: Galápagos Archipelago: Chatham Island. October 1835. Ex. coll. ZSL. Remark: BMNH 1881.2.18.80 originates from Covington's collection (cf. Sulloway 1982). Material: ex mount, male, ad. Status: BMNH 1855.12.19.228. Cat. VI: 350: a.

TURDIDAE:

***Turdus rufiventris rufiventris* (Vieillot, 1818).** RUFOUS-BELLIED THRUSH [? CD 1460/1470, see following]. Z: 59: *Turdus rufiventer* Licht. Loc.: Argentina: Rio Negro. August 1833. Material: skin/mount. Status: missing.

***Turdus rufiventris rufiventris* (Vieillot, 1818).** RUFOUS-BELLIED THRUSH [? CD 1274/1233, see following]. Z: 59: *Turdus rufiventer* Licht. Loc.: Uruguay: Maldonado. 1833. Material: skin/mount. Status: missing.

***Turdus rufiventris rufiventris* (Vieillot, 1818).** RUFOUS-BELLIED THRUSH [? CD 1233/1274, see previously]. Z: 59: *Turdus rufiventer* Licht. Loc.: Uruguay: Maldonado. 1833. Material: skin/mount. Status: missing.

***Turdus falcklandii falcklandii* (Quoy & Gaimard, 1824).** AUSTRAL THRUSH [CD 1900]. Z: 59: *Turdus falklandicus* [*sic*] Quoy & Guim. [*sic*]. Loc.: Falkland Islands/Islas Malvinas: East Falkland Island. March 1834. Material: skin/mount. Status: missing.

***Turdus falcklandii magellanicus* (King, 1831).** AUSTRAL THRUSH [CD not traced]. Z: 59: *Turdus falklandicus* [*sic*] Quoy & Guim. [*sic*]. Loc.: Chile: Tierra del Fuego. December 1832–February 1833 or January–March 1834. Material: skin/mount. Status: missing.

***Turdus falcklandii magellanicus* (King, 1831).** AUSTRAL THRUSH [CD 2125]. Z: 59: *Turdus falklandicus* [*sic*] Quoy & [*sic*]. Loc.: Chile: Isla de Chiloé. July 1834. Material: skin/mount. Status: missing.

***Turdus falcklandii pembertoni* (Wetmore, 1923).** AUSTRAL THRUSH [? CD 1470/1460, see previously]. Z: 59: *Turdus falklandicus* [*sic*] Quoy &

Guim. [*sic*]. Loc.: Argentina: Rio Negro. August 1833. Material: skin/mount. Status: missing.

PLOCEIDAE:

***Passer hispaniolensis hispaniolensis* (Temminck, 1820).** SPANISH SPARROW CD 189 [original field label]. Z: 95: *Passer hispaniolensis* G. R. Gray. Loc.: Portugal: Cape Verde Islands: So Tiago: Porto Praia. January 1832. Ex. coll. Gould, ex. ZSL. Remark: In Darwin's hand "X [/] loc [/]] S Jago" [in pencil] on label. Material: skin, male, ad. Status: BMNH 1881.5.1.2117. Cat. XII: 319: i.

***Passer iagoensis iagoensis* (Gould, 1838).** IAGO SPARROW [CD 190]. Z: 95: *Passer jagoensis* [*sic*] Gould. Holotype *Pyrgita Iagoensis* Gould, 1838 [the issue containing 77–79 of the Proceedings of the Zoological Society of London for the year 1837 was published in 1838]. Loc.: Portugal: Cape Verde Islands: So Tiago: Porto Praia. January 1832. Ex. coll. Gould, ex. ZSL. Remark: BMNH 1881.5.1.2124 [2133] has no type status. Material: ex mount, male, ad. Status: BMNH 1867.3.16.79. Cat. XII: 324: a.

MOTACILLIDAE:

***Anthus lutescens lutescens* (Pucheran, 1855).** YELLOWISH PIPIT [CD 685]. Z: 85: *Anthus chii* Licht. Loc.: Brazil: Rio de Janeiro. 5 April–5 July, 1832. Ex. coll. ZSL. Material: ex mount, ad. Status: BMNH 1855.12.19.185. Cat. X: 608: h.

***Anthus furcatus furcatus* (Lafresnaye & d'Orbigny, 1837).** SHORT-BILLED PIPIT [CD 1202]. Z: 85: *Anthus furcatus* D'Orb. & Lafr. Loc.: Uruguay: Maldonado: northern bank of the Plata. 1833. Material: skin/mount. Status: missing.

***Anthus furcatus furcatus* (Lafresnaye & d'Orbigny, 1837).** SHORT-BILLED PIPIT [?? CD 1230]. Z: 85: *Anthus furcatus* D'Orb. & Lafr. Loc.: Uruguay: Maldonado: northern bank of the Plata. 1833. Material: skin/ mount. Status: missing.

***Anthus furcatus furcatus* (Lafresnaye & d'Orbigny, 1837).** SHORT-BILLED PIPIT [CD 1592, shared with eggs of *Zonotrichia capensis* & *Molothrus bonariensis*]. Z: 85: *Anthus furcatus* D'Orb. & Lafr. Loc.: Uruguay: Montevideo: northern bank of the Plata. 1833. Material: two eggs. Status: missing.

***Anthus correndera chilensis* (Lesson, 1839).** CORRENDERA PIPIT [CD 2181]. Z: 85: *Anthus correndera* Vieill. Loc.: Chile: Valparaso. August–September 1834. Ex. coll. ZSL. Material: skin, male, ad. Status: BMNH 1841.1.18.22. Cat. X: 610: m.

***Anthus correndera chilensis* (Lesson, 1839).** CORRENDERA PIPIT [CD 2182]. Z: 85: *Anthus correndera* Vieill. Loc.: Chile: Valparaso. August–September 1834. Material: skin/mount, male. Status: missing.

***Anthus correndera grayi* (Bonaparte, 1850).** CORRENDERA PIPIT [CD 1898]. Z: 85: *Anthus correndera* Vieill. Loc.: Falkland Islands/Islas Malvinas: East Falkland Island. March 1834. Material: skin/mount. Status: missing.

***Anthus correndera correndera* (Vieillot, 1818).** [CD 1246]. CORRENDERA PIPIT Z: 85: *Anthus correndera* Vieill. Loc.: Uruguay: Maldonado. 1833. Ex. coll. ZSL. Material: ex mount, ad. Status: BMNH 1855.12.19.131. Cat. X: 610: q.

FRINGILLIDAE:

***Carduelis barbata* (Molina, 1782).** BLACK-CHINNED SISKIN *SPINUS BARBATA* CD 2195. Z: 89: *Chrysometris* [*sic*] *campestris* Gould. Loc.: Chile: Valparaso [Maldonado on label]. September 1834. Ex. coll. ZSL via Darwin. Material: ex mount, male, ad. Status: BMNH 1856.3.15.5. Cat. XXII: 217: f.

***Carduelis barbata* (Molina, 1782).** BLACK-CHINNED SISKIN *SPINUS BARBATA* [CD 1830]. Z: 89: *Chrysometris* [*sic*] *campestris* Gould. Loc.: Chile: Tierra del Fuego: Port Famine: forests. February 1834. Material: skin/mount. Status: missing.

***Carduelis magellanica magellanica* (Vieillot, 1805).** HOODED SISKIN *SPINUS MAGELLANICA* [CD 1465]. Z: 97: *Chrysomitris magellanica* Bona Loc.: Argentina: Rio Negro. May 1833. Material: skin/mount. Status: missing.

***Carduelis magellanica magellanica* (Vieillot, 1805).** HOODED SISKIN *SPINUS MAGELLANICA* [CD 1209]. Z: 97: *Chrysomitris magellanica* Bona Loc.: Uruguay: Maldonado. May 1833. Material: skin/mount. Status: missing.

PARULIDAE:

***Geothlypis aequinoctialis velata* (Vieillot, 1808).** MASKED YELLOWTHROAT [CD 1215]. Z: 87: *Trichas velata* G. R. Gray. Loc.: Uruguay: Maldonado: shot in a garden. June 1833. Material: skin/mount. Status: missing.

***Dendroica petechia aureola* (Gould, 1839).** MANGROVE WARBLER *SETOPHAGA PETECHIA* CD 3347. Z: 86: *Sylvicola aureola* Gould. Holotype *Sylvicola aureola* Gould, 1839 [pl. XXVIII]. Loc.: Ecuador: Galápagos Archipelago. September 1835. Ex. coll. ZSL via Darwin. Material: ex mount, male, ad. Status: BMNH 1856.3.15.14. Cat. X: 283: i.

ICTERIDAE:

***Dolichonyx oryzivorus* (Linnaeus, 1758).** BOBOLINK CD 3374 [original field label]. Z: 106: *Dolichonyx oryzivorus* Swains. Loc.: Ecuador: Galápagos Islands: James Island. October 1835. Ex. coll. Gould, ex. ZSL. Material: skin, imm. Status: BMNH 1881.5.1.2394. Cat. XI: 332: c'.

***Dolichonyx oryzivorus* (Linnaeus, 1758).** BOBOLINK [CD 1309, specimens in spirit]. Z: 106: *Dolichonyx oryzivorus* Swains. Loc.: Ecuador: Galápagos

Islands: James Island. October 1835. Remark: Body of BMNH 1881.5.1.2394. Material: specimen in alcohol. Status: missing.

***Pezites militaris militaris* (Linnaeus, 1771).** RED-BREASTED BLACKBIRD *STURNELLA MILITARIS* [CD 1784]. Z: 110: *Sturnella militaris* Vieill. Loc.: Chile: Straits of Magellan. 1834. Material: skin/mount. Status: missing.

***Pezites militaris falklandicus* (Leverkhn, 1889).** RED-BREASTED BLACKBIRD *STURNELLA MILITARIS* [CD 1146]. Z: 110: *Sturnella militaris* Vieill. Loc.: Falkland Islands/Islas Malvinas: East Falkland Island. March 1833. Material: skin/mount. Status: missing.

***Molothrus bonariensis bonariensis* (Gmelin, 1789).** SHINY COWBIRD [CD 1211]. Z: 107–109: *Molothrus niger* Gould. Loc.: Uruguay: Maldonado. 1833. Material: skin/mount. Status: missing.

***Molothrus bonariensis bonariensis* (Gmelin, 1789).** SHINY COWBIRD [CD 1212]. Z: 107–109: *Molothrus niger* Gould. Loc.: Uruguay: Maldonado. 1833. Material: skin/mount. Status: missing.

***Molothrus bonariensis bonariensis* (Gmelin, 1789).** SHINY COWBIRD [CD 1592, shared with eggs of *Anthus furcatus* & *Zonotrichia capensis*]. Z: 107–109: *Molothrus niger* Gould. Loc.: Uruguay: Montevideo. 1833. Material: single egg. Status: missing.

***Curaeus curaeus curaeus* (Molina, 1782).** AUSTRAL BLACKBIRD [CD 2186]. Z: 107: *Agelaius chopi* Vieill. Loc.: Chile: Valparaso. August–September 1834. Material: skin/mount. Status: very unlikely the unregistered BMNH specimen of coll. Eyton.

***Amblyramphus holosericeus* (Scopoli, 1786).** SCARLET-HEADED BLACKBIRD [CD 1244]. Z: 109–110: *Amblyramphus ruber* G. R. Gray. Loc.: Uruguay: Maldonado. 1833. Material: skin/mount. Status: missing.

***Molothrus badius badius* (Vieillot, 1819).** BAYWING *AGELAIOIDES BADIUS*? CD 1242/1418, see previously & following]. Z: 107: *Agelaius fringillarius* G. R. Gray. Loc.: Uruguay/Argentina: Maldonado or banks of the Parana. 1833. Material: skin/mount. Status: missing.

***Agelaius thilius thilius* (Molina, 1782).** YELLOW-WINGED BLACKBIRD *AGELASTICUS THILIUS* CD 2187 [original field label]. Z: 106: *Xanthornus chrysopterus* G. R. Gray. Loc.: Chili: Valparaso. August–September 1834. Ex. coll. Gould, ex. ZSL. Material: skin, female, ad. Status: BMNH 1858.6.25.27. Cat. XI: 344: j.

***Agelaius thilius petersii* (Laubmann, 1934).** YELLOW-WINGED BLACKBIRD *AGELASTICUS THILIUS* [? CD 1242/1418/1426, see following]. Z: 106: *Xanthornus chrysopterus* G. R. Gray. Loc.: Argentina or Uruguay: La Plata. 1833. Material: skin/mount. Status: missing.

***Xanthopsar flavus* (Gmelin, 1788).** SAFFRON-COWLED BLACKBIRD CD 1217. Z: 107: *Xanthornus flavus* G. R. Gray. Loc.: Uruguay: Maldonado. 1833. Ex. coll. Salvin & Godman, ex. Gould, ex. ZSL. Material: skin, female, ad. Status: BMNH 1885.11.2.301. Cat. XI: 346: c.

***Xanthopsar flavus* (Gmelin, 1788).** SAFFRON-COWLED BLACKBIRD [CD 1218]. Z: 107: *Xanthornus flavus* G. R. Gray. Loc.: Uruguay: Maldonado. 1833. Ex. coll. Salvin & Godman, ex. Gould, ex. ZSL. Material: skin, male, ad. Status: BMNH 1885.11.2.300. Cat. XI: 346: b.

***Pseudoleistes virescens* (Vieillot, 1819).** BROWN-AND-YELLOW MARSHBIRD [CD 1201]. Z: 107: *Leistes anticus* G. R. Gray. Loc.: Uruguay: Maldonado: La Plata. 1833. Material: skin/mount. Status: missing.

Icteridae [one of the previously mentioned species] [?? CD 1242/1418/1426, see previously]. Loc.: Uruguay/Argentina: Maldonado, Buenos Aires or Bajada. 1833. Material: skin/mount. Status: missing.

EMBERIZIDAE:

***Zonotrichia capensis subtorquata* (Swainson, 1837).** RUFOUS-COLLARED SPARROW [CD 1592, is shared with eggs of *Anthus furcatus* & *Molothrus bonariensis*]. Z: 91: *Zonotrichia matutina* G. R. Gray [later referred to as *Zonotrichia ruficollis*, 108]. Loc.: Uruguay: Montevideo. 1833. Material: three eggs. Status: missing.

***Zonotrichia capensis subtorquata* (Swainson, 1837).** RUFOUS-COLLARED SPARROW [CD 1615]. Z: 91: *Zonotrichia matutina* G. R. Gray. Loc.: Uruguay: Montevideo: banks of the Plata. November 1833. Material: skin/mount. Status: missing.

***Zonotrichia capensis subtorquata* (Swainson, 1837).** RUFOUS-COLLARED SPARROW [CD 683]. Z: 91: *Zonotrichia matutina* G. R. Gray. Loc.: Uruguay: Montevideo: banks of the Plata. August 1832. Material: skin/mount. Status: missing.

***Zonotrichia capensis australis* (Latham, 1790).** [CD 750]. Z: 91: *Zonotrichia matutina* G. R. Gray. Loc.: Argentina: Bahia Blanca. September 1832. Ex. coll. Gould, ex.? ZSL. Material: skin, ad. Status: probably BMNH 1857.10.16.51. Cat. XII: 610: a.

***Zonotrichia capensis australis* (Latham, 1790).** RUFOUS-COLLARED SPARROW CD 1826. Z: 91–92: *Zonotrichia canicapilla* Gould. Syntype *Zonotrichia canicapilla* Gould, 1839. Loc.: Chile: Tierra del Fuego: Port Famine. February 1834. Ex. coll. Salvin & Godman, ex. ZSL. Material: skin, juv. Status: BMNH unregistered specimen. Cat. XII: 610: c.

***Zonotrichia capensis australis* (Latham, 1790).** RUFOUS-COLLARED SPARROW [CD 1704]. Z: 91–92: *Zonotrichia canicapilla* Gould. Syntype

Zonotrichia canicapilla Gould, 1839. Loc.: Argentina: Port Desire. January 1834. Material: skin/mount. Status: missing.

***Zonotrichia capensis australis* (Latham, 1790).** RUFOUS-COLLARED SPARROW [? CD 902/904/1001, see following]. Z: 91–92: *Zonotrichia canicapilla* Gould. Syntype *Zonotrichia canicapilla* Gould, 1839. Loc.: Chile: Tierra del Fuego. December 1832–February 1833. Ex. coll. ZSL. Material: ex mount, ad. Status: BMNH 1855.12.19.388. Cat. XII: 610: b.

***Zonotrichia capensis australis* (Latham, 1790).** RUFOUS-COLLARED SPARROW [CD 1771]. Z: 91–92: *Zonotrichia canicapilla* Gould. Syntype *Zonotrichia canicapilla* Gould, 1839. Loc.: Argentina: Port St. Julian. January 1834. Material: skin/mount. Status: missing.

***Zonotrichia capensis australis* (Latham, 1790).** RUFOUS-COLLARED SPARROW [CD 1710, number shared with eggs of *Falco femoralis*, see previously]. Z: 91–92: *Zonotrichia canicapilla* Gould. Paratype *Zonotrichia canicapilla* Gould, 1839. Loc.: Argentina: Port St. Julian. January 1834. Material: egg(s). Status: missing.

***Zonotrichia capensis chilensis* (Meyen, 1834).** RUFOUS-COLLARED SPARROW [CD 2299]. Z: 91: *Zonotrichia matutina* G. R. Gray. Loc.: Chile: Valparaso. August–September 1834. Material: skin/mount. Status: missing.

***Ammodramus humeralis xanthornus* (Gould, 1839).** GRASSLAND SPARROW [CD 1262]. Z: 90: *Ammodramus manimbe* G. R. Gray. Holotype *Ammodramus xanthornus* Gould, 1839 [pl. XXX]. Loc.: Uruguay: Maldonado. June 1833. Ex. coll. Gould, ex. ZSL. Remark: FitzRoy's specimens BMNH 1837.2.21.328 = Cat. XII: 693: z & BMNH 1837.2.21.291 = Cat. XII: 693: a' are not the types. Material: skin. Status: VMM B19633.

***Aimophila strigiceps strigiceps* (Gould, 1839).** STRIPE-CAPPED SPARROW *RHYNCHOSPIZA STRIGICEPS* [? CD 1414/1415/1416, see previously & following]. Z: 92: *Zonotrichia strigiceps* Gould. Holo/Syntype *Zonotrichia strigiceps* Gould, 1839. Loc.: Argentina: Santa Fé. October 1833. Ex. coll. Sclater, ex. Gould, ex. ZSL. Material: skin, ad. Status: BMNH 1885.2.10.447. Cat. XII: 608: a.

***Gubernatrix cristata* (Vieillot, 1817).** YELLOW CARDINAL [? CD 1417]. Z: 89: *Emberiza gubernatrix* Temm. Loc.: Argentina: banks of the Parana near Santa Fé. 1833. Material: skin/mount. Status: missing.

THRAUPIDAE

***Thraupis bonariensis bonariensis* (Gmelin, 1789).** BLUE-AND-YELLOW TANAGER [CD 1229]. Z: 97–98: *Aglaia striata* D'Orb. & Lafr. Loc.: Uruguay: Maldonado. 1833. Remark: BMNH holds a specimen of Covington's collection, which might be in fact this missing Darwin bird (BMNH 1839.6.8.2.

Cat. XI: 164: e), nec. MNHN Paris No. 3068. This is not a type of *Tanagra Darwinii*. Material: skin/mount. Status: probably missing.

***Pipraeidea melanonota melanonota* (Vieillot, 1819).** FAWN-BREASTED TANAGER [?? CD 1245]. Z: p 98: *Aglaia vittata* [Temm.] Loc.: Uruguay: Maldonado. 1833. Material: skin/mount. Status: missing.

***Phrygilus gayi caniceps* (Burmeister, 1860).** GREY-HOODED SIERRA-FINCH [CD 2017]. Z: 93: *Fringilla gayi* Eyd. & Gerv. Loc.: Argentina: Santa CruZ: April 1834. Material: skin/mount, male. Status: missing.

***Phrygilus gayi caniceps* (Burmeister, 1860).** GREY-HOODED SIERRA-FINCH [CD 2018]. Z: 93: *Fringilla gayi* Eyd. & Gerv. Loc.: Argentina: Santa Cruz: April 1834. Ex. coll. ZSL. Material: ex mount, female, ad. Status: 1855.12.19.42. Cat. XII: 784: h.

***Phrygilus patagonicus* (Lowe, 1923).** PATAGONIAN SIERRA FINCH [? CD 1818]. Z: 93–94: *Fringilla formosa* Gould. Syntype *Fringilla formosa* Gould, 1839 & Syntype *Phrygilus gayi patagonicus* Lowe, 1923. Loc.: Chile: Tierra del Fuego: Port Famine. Early February 1834. Material: ex mount, female, ad. Status: BMNH 1856.3.15.12. Cat. XII: 782: i.

***Phrygilus patagonicus* (Lowe, 1923).** PATAGONIAN SIERRA FINCH [? CD 902/904/1001, see previously & following]. Z: 93–94: *Fringilla formosa* Gould. Syntype *Fringilla formosa* Gould, 1839 & Syntype *Phrygilus gayi patagonicus* Lowe, 1923. Loc.: Chile: Tierra del Fuego. December 1832?–February 1833. Ex. coll. ZSL. Material: ex mount, ad. Status: BMNH 1855.12.19.162. Cat. XII: 782: g.

***Phrygilus patagonicus* (Lowe, 1923).** PATAGONIAN SIERRA FINCH [? CD 902/904/1001, see previously]. Z: 93–94: *Fringilla formosa* Gould. Syntype *Fringilla formosa* Gould, 1839 & Syntype *Phrygilus gayi patagonicus* Lowe, 1923. Loc.: Chile: Tierra del Fuego. December 1832 February 1833. Ex. coll. ZSL. Material: ex mount, male, ad. Status: BMNH 1855.12.19.24. Cat. XII: 782: f.

***Phrygilus fruticeti fruticeti* (Kittlitz, 1833).** MOURNING SIERRA FINCH [CD 2829]. Z: 94: *Fringilla fruticeti* Kittl. Loc.: Chile: Coquimbo. 1835. Ex. coll. ZSL. Material: ex mount, female, ad. Status: BMNH 1855.12.19.16. Cat. XII: 791: g.

***Phrygilus fruticeti fruticeti* (Kittlitz, 1833).** MOURNING SIERRA FINCH [CD 2016]. Z: 94: *Fringilla fruticeti* Kittl. Loc.: Argentina: Santa Cruz: April 1834. Material: skin/mount, female. Status: missing.

***Phrygilus fruticeti fruticeti* (Kittlitz, 1833).** MOURNING SIERRA FINCH [CD 2015]. Z: 94: *Fringilla fruticeti* Kittl. Loc.: Argentina: Santa Cruz: April 1834. Ex. coll. ZSL. Material: ex mount, female/imm. Status: BMNH 1855.12.19.45. Cat. XII: 791: l.

***Phrygilus carbonarius* (Lafresnaye & d'Orbigny, 1837).** CARBONATED SIERRA FINCH [CD 1466]. Z: 94: *Fringilla carbonaria* G. R. Gray. Loc.: Argentina: between Rio Negro and Colorado. August 1833. Material: skin/mount. Status: missing.

***Phrygilus alaudinus alaudinus* (Kittlitz, 1833).** BAND-TAILED SIERRA FINCH [CD 2177]. Z: 94: *Fringilla alaudina* Kittl. Loc.: Chile: Valparaso. August–September 1834. Ex. coll. ZSL. Material: ex mount, female, [male in CD], ad. Status: BMNH 1855.12.19.390. Cat. XII: 795: t.

***Phrygilus alaudinus alaudinus* (Kittlitz, 1833).** BAND-TAILED SIERRA FINCH [CD 2178]. Z: 94: *Fringilla alaudina* Kittl. Loc.: Chile: Valparaso. August–September 1834. Ex. coll. ZSL. Material: ex mount, male, ad. Status: BMNH 1855.12.19.41. Cat. XII: 795: s.

***Melanodera melanodera melanodera* (Quoy & Gaimard, 1824).** WHITE-BRIDLED FINCH [? CD 1879/1046, see following]. Z: 95–96: *Chlorospiza? melanodera* G. R. Gray. Loc.: Falkland Islands/Islas Malvinas: East Falkland Island. March 1833/1834. Ex. coll. ZSL. Material: ex mount, imm. Status: BMNH 1855.12.19.50. Cat. XII: 788: o.

***Melanodera melanodera melanodera* (Quoy & Gaimard, 1824).** WHITE-BRIDLED FINCH [?? CD 1701]. Z: 95–96: *Chlorospiza? melanodera* G. R. Gray. Loc.: Argentina: Santa Cruz or Port Desire. January or April 1834. Ex. coll. Salvin & Godman, ex. ZSL. Remark: error of locality in CD or Z: or not entered in CD or specimen not a Darwin bird. Material: skin, female, ad. Status: probably BMNH 1885.12.14.807. Cat. XII: 788: x.

***Melanodera melanodera melanodera* (Quoy & Gaimard, 1824).** WHITE-BRIDLED FINCH [?? CD 1468]. Z: 95–96: *Chlorospiza? melanodera* G. R. Gray. Loc.: Argentina: Santa Cruz [? Rio Negro]. April 1834. Ex. coll. Salvin & Godman, ex. ZSL. Remark: not under this loc. in CD or specimen not a Darwin bird. Material: skin, male, ad. Status: probably BMNH 1885.12.14.806. Cat. XII: 788: w.

***Melanodera xanthogramma xanthogramma* (Gray, 1839).** YELLOW-BRIDLED FINCH [?? CD 1003, see following]. Z: 96–97: *Chlorospiza? xanthogramma* G. R. Gray. Syntype *Chlorospiza xanthogramma* G. R. Gray, 1839. Loc.: Chile: Tierra del Fuego. February 1833. Ex. coll. ZSL via Darwin. Remark: One of these two specimens cannot be traced in CD (see following). The authorship is attributed to Gray on the basis that his name appears after the new name; however, plate XXXIII, which has been published together with the text in the November 1839 issue, was probably sketched by John Gould and executed on stone by Elizabeth Gould. This matter would need further investigations. Material: skin, female, ad. Status: BMNH 1841.11.18.24. Cat. XII: 790: e [see following].

Melanodera xanthogramma xanthogramma **(Gray, 1839).** YELLOW-BRIDLED FINCH [?? CD 1003, see previously]. Z: 96–97: *Chlorospiza? xanthogramma* G. R. Gray. Syntype *Chlorospiza xanthogramma* G. R. Gray, 1839. Loc.: Chile: Tierra del Fuego. February 1833. Ex. coll. ZSL. Remark: One of these two specimens cannot be traced in CD (see previously). The authorship is attributed to Gray on the basis that his name appears after the new name; however, plate XXXIII, which has been published together with the text in the November 1839 issue, was probably sketched by John Gould and executed on stone by Elizabeth Gould. This matter would need further investigations. Material: ex mount, imm. Status: BMNH 1855.12.19.181. Cat. XII: 790: e [see previously].

Melanodera xanthogramma xanthogramma **(Gray, 1839).** YELLOW-BRIDLED FINCH [? CD 1046/1879, see previously]. Z: 96–97: *Chlorospiza? xanthogramma* G. R. Gray. Syntype *Chlorospiza xanthogramma* G. R. Gray, 1839. Loc.: Falkland Islands/Islas Malvinas: East Falkland Island. March 1833/1834. Material: skin/mount. Status: missing.

Melanodera melanodera/xanthogramma **(Quoy & Gaimard, 1824)/ (Gray, 1839).** WHITE-BRIDLED/YELLOW-BRIDLED FINCH [?? CD 1919/ 1920/1923/1047, see following & previously]. Z: 95–97: *Chlorospiza? melanodera* or *xanthogramma* G. R. Gray.? Syntype *Chlorospiza xanthogramma* G. R. Gray, 1839. Loc.: Falkland Islands/Islas Malvinas: East Falkland Island. March 1833/1834. Material: skin/mount. Status: missing.

Melanodera melanodera/xanthogramma **(Quoy & Gaimard, 1824)/ (Gray, 1839).** WHITE-BRIDLED/YELLOW-BRIDLED FINCH [?? CD 1919/ 1920/1923/1047, see following & previously]. Z: 95–97: *Chlorospiza? melanodera* or *xanthogramma* G. R. Gray.? Syntype *Chlorospiza xanthogramma* G. R. Gray, 1839. Loc.: Falkland Islands/Islas Malvinas: East Falkland Island. March 1833/1834. Material: skin/mount. Status: missing.

Melanodera melanodera/xanthogramma **(Quoy & Gaimard, 1824)/ (Gray, 1839).** WHITE-BRIDLED/YELLOW-BRIDLED FINCH [?? CD 1919/ 1920/1923/1047, see following & previously]. Z: 95–97: *Chlorospiza? melanodera* or *xanthogramma* G. R. Gray.? Syntype *Chlorospiza xanthogramma* G. R. Gray, 1839. Loc.: Falkland Islands/Islas Malvinas: East Falkland Island. March 1833/1834. Material: skin/mount. Status: missing.

Melanodera melanodera/xanthogramma **(Quoy & Gaimard, 1824)/ (Gray, 1839).** WHITE-BRIDLED/YELLOW-BRIDLED FINCH [?? CD 1919/ 1920/1923/1047, see previously]. Z: 95–97: *Chlorospiza*? *melanodera* or *xanthogramma* G. R. Gray.? Syntype *Chlorospiza xanthogramma* G. R. Gray, 1839. Loc.: Falkland Islands/Islas Malvinas: East Falkland Island. March 1833/ 1834. Material: skin/mount. Status: missing.

***Melanodera melanodera/xanthogramma* (Quoy & Gaimard, 1824)/ (Gray, 1839).** CANARY-WINGED/YELLOW-BRIDLED FINCH [CD 1922]. 95–97: *Chlorospiza*? *melanodera* or *xanthogramma* G. R. Gray.? Syntype *Chlorospiza xanthogramma* G. R. Gray, 1839. Loc.: Falkland Islands/Islas Malvinas: East Falkland Island. March 1834. Material: skin/mount, female. Status: missing.

***Donacospiza albifrons* (Vieillot, 1817).** LONG-TAILED REED FINCH [? CD 1605/1611/1612/1614/1616/1617, see following]. Z: 90: *Ammodramus longicaudatus* Gould. Syntype *Ammodramus longicaudatus* Gould, 1839 [pl. XXIX]. Loc.: Uruguay: Montevideo. November 1832. Ex. coll. ZSL via Darwin. Material: ex mount, ad. Status: BMNH 1856.3.15.13. Cat. XII: 767: i.

***Donacospiza albifrons* (Vieillot, 1817).** LONG-TAILED REED FINCH [?? CD 1297]. Z: 90: *Ammodramus longicaudatus* Gould. Syntype *Ammodramus longicaudatus* Gould, 1839 [pl. XXIX]. Loc.: Uruguay: Maldonado. June 1833. Ex. coll. ZSL via Darwin. Material: ex mount, ad. Status: BMNH 1856.3.15.9. Cat. XII: 767: h.

***Diuca diuca chiloensis* (Philippi & Pena, 1964).** COMMON DIUCA FINCH [CD 2132]. Z: 93: *Fringilla diuca* Mol. Loc.: Chile: Isla de Chiloé. July 1834. Material: skin/mount, male. Status: missing.

***Diuca diuca chiloensis* (Philippi & Pena, 1964).** COMMON DIUCA FINCH [CD 2131]. Z: 93: *Fringilla diuca* Mol. Loc.: Chile: Isla de Chiloé. July 1834. Material: skin/mount. Status: missing.

***Diuca diuca chiloensis* (Philippi & Pena, 1964).** COMMON DIUCA FINCH [CD 2133]. Z: 93: *Fringilla diuca* Mol. Loc.: Chile: Isla de Chiloé. July 1834. Ex. coll. ZSL. Material: ex mount, female, ad. Status: BMNH 1855.12.19.187. Cat. XII: 801: e.

***Diuca diuca diuca* (Molina, 1782).** COMMON DIUCA FINCH [CD 2320]. Z: 93: *Fringilla diuca* Mol. Chile: Valparaso. September–October 1834. Material: egg clutch & nest. Status: missing.

***Poospiza nigrorufa nigrorufa* (d'Orbigny & Lafresnaye, 1837).** BLACK-AND-RUFOUS WARBLING FINCH [? CD 1241]. Z: 98: *Pipillo* [*sic*] *personata* Swains. Loc.: Uruguay: Maldonado. 1833. Ex. coll. ZSL. Material: skin, male, ad. Status: BMNH 1858.4.3.120. Cat. XII: 641: a.

***Poospiza nigrorufa nigrorufa* (d'Orbigny & Lafresnaye, 1837).** BLACK-AND-RUFOUS WARBLING FINCH CD 1234. Z: 98: *Pipillo* [*sic*] *personata* Swains. Loc.: Uruguay: Maldonado. 1833. Ex. coll. Tristram, ex. ZSL. Material: skin, male. Status: LIVCM 14875.

***Sicalis flaveola pelzelni* (Sclater, 1872).** SAFFRON FINCH [? CD 1247]. Z: 88: *Crithagra*? *brasiliensis* [Spix]. Uruguay: northern bank of the Plata near Maldonado. June 1833. Material: skin/mount. Status: missing.

***Sicalis flaveola pelzelni* (Sclater, 1872).** SAFFRON FINCH [? CD 1605/1611/1612/1614/1616/1617, see following & previously]. Z: 88: *Crithagra? brasiliensis* [Spix]. Loc.: Uruguay: northern bank of the Plata near Montevideo. November 1833. Material: skin/mount. Status: missing.

***Sicalis [?] luteola luteiventris* (Meyen, 1834).** GRASSLAND YELLOW FINCH *SICALIS LUTEOLA* [? CD 1232]. Z: 88–89: *Crithagra*? *brevirostris* Gould. Syntype *Crithagra brevirostris* Gould, 1839. Loc.: Uruguay: Maldonado. May 1833. Remark: The original description would fit to *Sicalis luteola luteiventris* (Meyen, 1834), but as the type is lacking, some doubts remain. Material: skin/mount. Status: missing.

***Sicalis [?] luteola luteiventris* (Meyen, 1834).** GRASSLAND YELLOW FINCH [? CD 2196]. Z: 88–89: *Crithagra? brevirostris* Gould. Syntype *Crithagra brevirostris* Gould, 1839. Chile: Valparaso. September 1834. Remark: The original description would fit to *Sicalis luteola luteiventris* (Meyen, 1834), but as the type is lacking, some doubts remain. Material: skin/mount, [?] male. Status: missing.

***Sicalis luteola luteiventris* (Meyen, 1834).** GRASSLAND YELLOW FINCH [CD 2019]. Z: 89: *Emberiza luteoventris* [*sic*] G. R. Gray. Loc.: Argentina: Santa Cruz: April 1834. Material: skin/mount, male. Status: missing.

***Embernagra platensis platensis* (Gmelin, 1789).** PAMPA FINCH CD 683. Z: 98: *Emberizoides poliocephalus* G. R. Gray. Syntype *Emberizoides poliocephalus* G. R. Gray, 1841. Loc.: Uruguay: Montevideo: northern shore of the Plata. July–August 1832. Ex. coll. Salvin & Godman, ex. Gould, ex. ZSL. Remark: The only type for which Gray's authorship seems to be without any doubts. Material: skin, ad. Status: BMNH 1885.12.14.1325. Cat. XII: 759: d.

***Embernagra platensis platensis* (Gmelin, 1789).** PAMPA FINCH CD 1207 [original field label]. Z: 98: *Emberizoides poliocephalus* G. R. Gray. Syntype *Emberizoides poliocephalus* G. R. Gray, 1841. Loc.: Uruguay: Maldonado: northern shore of the Plata. 1833. Ex. coll. Gould, ex. ZSL. Remark: The only type for which Gray's authorship seems to be without any doubts. Material: ex mount. Status: VMM No. B19600.

***Volatinia jacarina jacarina* (Linnaeus, 1766).** BLUE-BLACK GRASSQUIT [? CD No. between 412–446, number not in Barlow 1963 & Keynes 2000]. Z: 92: *Passerina jacarina* Vieill. Loc.: Brazil: Rio de Janeiro. April–July 1832. Ex. coll. BMNH, ex. Gould,? ex. ZSL. Material: ex mount, male, ad. Status: MANCH B.7528, ex.1857.11.28.251. Cat. XII: 155: h.

***Sporophila caerulescens caerulescens* (Vieillot, 1823).** DOUBLE-COLLARED SEEDEATER [? CD 1605/1611/1612/1614/1616/1617, see following & previously]. Z: 88: *Spermophila nigrogularis* Gould. Syntype *Spermophila nigrogularis* Gould, 1839. Loc.: Uruguay: Montevideo. November 1832/1833. Ex. coll. ZSL via Darwin. Material: skin, female, ad. Status: BMNH 1841.1.18.26. Cat. XII: 127: o.

***Sporophila caerulescens caerulescens* (Vieillot, 1823).** DOUBLE-COLLARED SEEDEATER [? CD 1605/1611/1612/1614/1616/1617, see following & previously]. Z: 88: *Spermophila nigrogularis* Gould. Syntype *Spermophila nigrogularis* Gould, 1839. Loc.: Uruguay: Montevideo. November 1832/1833. Ex. coll. ZSL. Material: ex mount, male, ad. Status: BMNH 1855.12.19.200. Cat. XII: 127: n.

Geospiza Darwin discusses genus in Z 98–100.

***Geospiza magnirostris magnirostris* (Gould, 1837).** LARGE GROUND FINCH [? CD 3312-19/24-29/32-36/38/39/41, see following]. Z: 100: *Geospiza magnirostris* Gould. Syntype *Geospiza magnirostris* Gould, 1837. Ecuador: Galápagos Archipelago. October 1835. Ex. coll. ZSL. Material: ex mount, female, ad. Status: BMNH 1855.12.19.113. Cat. XII: 8: b.

***Geospiza magnirostris magnirostris* (Gould, 1837).** LARGE GROUND FINCH [? CD 3312-19/24-29/32-36/38/39/41, see previously & following]. Z: 100: *Geospiza magnirostris* Gould. Syntype *Geospiza magnirostris* Gould, 1837. Loc.: Ecuador: Galápagos Archipelago. October 1835. Ex. coll. ZSL. Material: ex mount, male, ad. Status: BMNH 1855.12.19.80. Cat. XII: 8: a.

***Geospiza magnirostris magnirostris* (Gould, 1837).** LARGE GROUND FINCH [?? CD 3312-19/24-29/32-36/38/39/41, see following & previously]. Z: 100: *Geospiza magnirostris* Gould. [?] Syntype *Geospiza magnirostris* Gould, 1837. Loc.: Ecuador: Galápagos Archipelago. October 1835. Ex. coll. Salvin & Godman, ex. Eyton, ex.? Gould, ex.? ZSL. Remark: probably from Covington's private collection, cf. Sulloway 1982. Material: skin, male, ad. Status: BMNH 1885.12.14.280. Cat. XII: 8: c/d [see following].

***Geospiza magnirostris magnirostris* (Gould, 1837).** LARGE GROUND FINCH [? CD 3312-19/24-29/32-36/38/39/41, see following & previously]. Z: 100: *Geospiza magnirostris* Gould. [?] Syntype *Geospiza magnirostris* Gould, 1837. Loc.: Ecuador: Galápagos Archipelago. October 1835. Ex. coll. Salvin & Godman, ex. Eyton, ex.? Gould, ex.? ZSL. Remark: Sulloway (1982) believed that this specimen is from Covington's collection; numbers in CD reveal that it is Darwin's. Material: skin, [?] female, ad. Status: BMNH 1885.12.14.281. Cat. XII: 8: c/d [see previously].

***Geospiza magnirostris strenua* (Gould, 1837).** LARGE GROUND FINCH [? CD 3312-19/24-29/32-36/38/39/41, see following & previously]. Z: 100–101: *Geospiza strenua* Gould. Syntype *Geospiza strenua* Gould, 1837. Loc.: Ecuador: Galápagos Archipelago. October 1835. Ex. coll. ZSL. Material: ex mount, female, ad. Status: BMNH 1855.12.19.114. Cat. XII: 9: b/c [see following].

***Geospiza magnirostris strenua* (Gould, 1837).** LARGE GROUND FINCH [? CD 3312-19/24-29/32-36/38/39/41, see previously & following].

Z: 100–101: *Geospiza strenua* Gould. Syntype *Geospiza strenua* Gould, 1837. Loc.: Ecuador: Galápagos Archipelago. October 1835. Ex. coll. ZSL. Material: ex mount, male, ad. Status: BMNH 1855.12.19.81. Cat. XII: 9: a.

***Geospiza fortis* [cf. Sulloway 1982] (Gould, 1837).** MEDIUM GROUND FINCH [? CD 3312-19/24-29/32-36/38/39/41, see previously & following]. Z: 100–101: *Geospiza strenua* Gould. Syntype *Geospiza strenua* Gould, 1837. Ecuador: Galápagos Archipelago. October 1835. Ex. coll. ZSL. Material: ex mount/ skin, female, ad. Status: BMNH 1855.12.19.83. Cat. XII: 9: b/c [see previously].

***Geospiza fortis* (Gould, 1837).** MEDIUM GROUND FINCH [? CD 3312-19/24-29/32-36/38/39/41, see previously & following]. Z: 101: *Geospiza fortis* Gould. Syntype *Geospiza fortis* Gould, 1837. Loc.: Ecuador: Galápagos Archipelago. October 1835. Ex. coll. ZSL. Remark: Sulloway (1982) discussed that this specimen might derive from the private collection of Covington, which is here not followed. Material: ex mount, female, ad. Status: BMNH 1855.12.19.82. Cat. XII: 11: a.

***Geospiza fortis* (Gould, 1837).** MEDIUM GROUND FINCH [? CD 3312-19/ 24-29/32-36/38/39/41, see previously & following]. Z: 102: *Geospiza dentirostris* Gould. Holotype *Geospiza dentirostris* Gould, 1837. Loc.: Ecuador: Galápagos Archipelago. October 1835. Ex. coll. ZSL. Material: ex mount, male [fide Sulloway 1982], juv. Status: BMNH 1855.12.19.176. Cat. XII: 12: a.

Geospiza fortis [fide Swarth 1931] **(Gould, 1837).** MEDIUM GROUND [? CD 3312-19/24-29/32-36/38/39/41, see previously & following]. Z: 103: *Geospiza dubia* Gould. Holotype *Geospiza dubia* Gould, 1837. Loc.: Ecuador: Galápagos Archipelago. October 1835. Ex. coll. ZSL. Material: skin/mount, female. Status: missing.

***Geospiza fuliginosa* (Gould, 1837).** SMALL GROUND FINCH [? CD 3312-19/24-29/32-36/38/39/41, see previously & following]. Z: 101–102: *Geospiza fuliginosa* Gould. Syntype *Geospiza fuliginosa* Gould, 1837. Loc.: Ecuador: Galápagos Archipelago. October 1835. Ex. coll. ZSL. Material: ex mount,? female [fide Sulloway 1982], ad. Status: BMNH 1855.12.19.44. Cat. XII: 13: a.

***Geospiza fuliginosa* (Gould, 1837).** SMALL GROUND FINCH [? CD 3312-19/24-29/32-36/38/39/41, see previously & following]. Z: 101–102: *Geospiza fuliginosa* Gould. Syntype *Geospiza fuliginosa* Gould, 1837. Loc.: Ecuador: Galápagos Archipelago. October 1835. Ex. coll. Gould, ex. ZSL. Material: skin,? female [fide Sulloway 1982], ad. Status: BMNH 1857.11.28.247. Cat. XII: 13: b.

***Geospiza fuliginosa* (Gould, 1837).** SMALL GROUND FINCH [? CD 3312-19/24-29/32-36/38/39/41, see previously & following]. Z: 101–102:

Geospiza fuliginosa Gould. Syntype *Geospiza fuliginosa* Gould, 1837. Loc.: Ecuador: Galápagos Archipelago. October 1835. Ex. coll. Salvin & Godman, ex. Eyton, ex.? Gould, ex.? ZSL. Remark: Sulloway (1982) believed that this specimen is from Covington's collection; numbers in CD reveal that it is Darwin's. Material: skin, female, ad. Status: BMNH 1885.12.14.320. Cat. XII: 13: c.

***Geospiza fuliginosa* (Gould, 1837).** SMALL GROUND FINCH [? CD 3312-19/24-29/32-36/38/39/41, see previously & following]. Z: 101–102: *Geospiza fuliginosa* Gould. Syntype *Geospiza fuliginosa* Gould, 1837. Loc.: Ecuador: Galápagos Archipelago. October 1835. Ex. coll. Frank, ex. Gould, ex. ZSL. Material: skin, male. Status: RMNH Cat. 2 [purchase of 1863].

***Geospiza fuliginosa* (Gould, 1837).** SMALL GROUND FINCH [? CD 3312-19/24-29/32-36/38/39/41, see previously & following]. Z: 101–102: *Geospiza fuliginosa* Gould. Syntype *Geospiza fuliginosa* Gould, 1837. Loc.: Ecuador: Galápagos Archipelago. October 1835. Ex. coll. Frank, ex. Gould, ex. ZSL. Material: skin, female. Status: RMNH Cat. 3 [purchase of 1863].

***Geospiza fuliginosa* (Gould, 1837).** SMALL GROUND FINCH [? CD 3312-19/24-29/32-36/38/39/41, see previously & following]. Z: 101–102: *Geospiza fuliginosa* Gould. Syntype *Geospiza fuliginosa* Gould, 1837. Loc.: Ecuador: Galápagos Archipelago. October 1835. Ex. coll. Frank, ex. Gould, ex. ZSL. Material: skin, male. Status: RMNH Cat. 4 [purchase of 1863].

***Geospiza nebulosa debilirostris* (Ridgway, 1894).** SHARP-BEAKED [FLOREANA] GROUND FINCH *GEOSPIZA DIFFICILIS* [? CD 3312-19/24-29/32-36/38/39/41, see previously & following]. Z: 100–101: *Geospiza strenua* Gould. Syntype *Geospiza strenua* Gould, 1837. Loc.: Ecuador: Galápagos Archipelago. October 1835. Ex. coll. ZSL via Darwin. Material: ex mount, female [fide Sulloway 1982]. Status: BMNH 1856.3.15.4. Cat. XII: 12: a.

Geospiza nebulosa nebulosa [fide Sulloway 1982] **(Gould, 1837).** SHARP-BEAKED [FLOREANA] GROUND FINCH *GEOSPIZA DIFFICILIS* [CD 3323]. Z: 104–105: *Cactornis scandens* Gould. Syntype *Cactornis scandens* Gould, 1837. Loc.: Ecuador: Galápagos Archipelago. October 1835. Ex. coll. ZSL. Material: ex mount, female, ad. Status: BMNH 1855.12.19.20. Cat. XII: 20: b.

***Geospiza? nebulosa nebulosa* (Gould, 1837).** SHARP-BEAKED [FLOREANA] GROUND FINCH *GEOSPIZA DIFFICILIS* [? CD 3312-19/24-29/32-36/38/39/41, see previously & following]. Z: 101: *Geospiza nebulosa* Gould. Syntype *Geospiza nebulosa* Gould, 1837. Loc.: Ecuador: Galápagos Archipelago. October 1835. Ex. coll. ZSL. Remark: Grant et al., 1985, accidentally cited original Darwin finches at Stockholm Museum, which is an error for Leiden Museum. These are not the missing birds listed here. Material: skin/mount. Status: former BMNH 1855.12.19.43, missing.

***Geospiza scandens scandens* (Gould, 1837).** COMMON CACTUS FINCH [CD 3320]. Z: 104–105: *Cactornis scandens* Gould. Syntype *Cactornis scandens* Gould, 1837. Loc.: Ecuador: Galápagos Archipelago. October 1835. Ex. coll. ZSL. Material: ex mount, male, ad. Status: BMNH 1855.12.19.125. Cat. XII: 20: a.

***Geospiza scandens scandens* (Gould, 1837).** COMMON CACTUS FINCH [? CD 3321/3322, see following]. Z: 104–105: *Cactornis scandens* Gould. Syntype *Cactornis scandens* Gould, 1837. Loc.: Ecuador: Galápagos Archipelago. October 1835. Ex. coll. Frank, ex. Gould, ex. ZSL. Material: skin, male [fide Sulloway 1982], juv. Status: RMNH Cat. 1 [purchase of 1863].

***Geospiza scandens? rothschildi* [fide Sulloway 1982] (Heller & Snodgrass, 1901).** COMMON CACTUS FINCH [? CD 3321/3322, see previously]. Z: 105: *Cactornis assimilis* Gould. Syntype *Cactornis assimilis* Gould, 1837. Loc.: Ecuador: Galápagos Archipelago. October 1835. Ex. coll. ZSL. Material: ex mount, male [fide Sulloway 1982], juv. Status: BMNH 1855.12.19.15. Cat. XII: 18: a.

***Geospiza* s** [CD 3337]. Syntype [of one of the species]. Loc.: Ecuador: Galápagos Archipelago. October 1835. Ex. coll.? ZSL. Remark: Grant et al., 1985, cited original Darwin finches at Stockholm Museum, which is an error for Leiden Museum. These are not the missing birds listed here. Material: skin/mount, upper mandible broken. Status: missing.

***Geospiza* s** [CD 3361]. Part of Syntype [of one of the species] Loc.: Ecuador: Galápagos Archipelago. October 1835. Ex. coll.?ZSL. Material: Upper mandible of CD 3337. Status: missing.

***Camarhynchus crassirostris* (Gould, 1837).** VEGETARIAN FINCH *PLATYSPIZA CRASSIROSTRIS* [? CD 3312-19/24-29/32-36/38/39/41, see previously & following]. Z: 103–104: *Camarhynchus crassirostris* Gould. Syntype *Camarhynchus crassirostris* Gould, 1837. Loc.: Ecuador: Galápagos Archipelago. October 1835. Ex. coll. Frank, ex. Gould, ex. ZSL. Material: skin, male [fide Sulloway 1982], juv. Status: RMNH Cat. 2 [purchase of 1863].

***Camarhynchus psittacula psittacula* (Gould, 1837).** LARGE TREE FINCH [CD 3331]. Z: 103: *Camarhynchus psittaculus* [*sic*] Gould. Syntype *Camarhynchus psittacula* Gould, 1837. Loc.: Ecuador: Galápagos Archipelago. October 1835. Ex. coll. ZSL. Material: ex mount, female, ad. Status: BMNH 1855.12.19.22. Cat. XII: 17: a/b [see following].

***Camarhynchus psittacula psittacula* (Gould, 1837).** LARGE TREE FINCH [CD 3330]. Z: 103: *Camarhynchus psittaculus* [*sic*] Gould. Syntype *Camarhynchus psittacula* Gould, 1837. Loc.: Ecuador: Galápagos Archipelago. October 1835. Ex. coll. ZSL. Remark: specimen has last been seen by Swarth 1931: 215. Material: skin/mount, [?] male [Sulloway 1982 female] ad. Status: former BMNH 1855.12.19.12, now missing. Cat. XII: 17: a/b [see previously].

***Camarhynchus parvulus parvulus* (Gould, 1837).** SMALL TREE FINCH [? CD 3312-19/24-29/32-36/38/39/41, see previously & following]. Z: 102: *Geospiza parvula* Gould. Syntype *Geospiza parvula* Gould, 1837. Ecuador: Galápagos Archipelago. October 1835. Ex. coll. ZSL. Material: ex mount, female, ad. Status: BMNH 1855.12.19.167. Cat. XII: 14: b.

***Camarhynchus parvulus parvulus* (Gould, 1837).** SMALL TREE FINCH [? CD 3312-19/24-29/32-36/38/39/41, see previously & following]. Z: 102: *Geospiza parvula* Gould. Syntype *Geospiza parvula* Gould, 1837. Loc.: Ecuador: Galápagos Archipelago. October 1835. Ex. coll. ZSL. Material: ex mount, male, ad. Status: BMNH 1855.12.19.194. Cat. XII: 14: a.

***Certhidea olivacea olivacea* (Gould, 1837).** GREEN WARBLER-FINCH [? CD 3310/3346/3348, see following]. Z: 106: *Certhidea olivacea* Gould. Syntype *Certhidea olivacea* Gould, 1837. Loc.: Ecuador: Galápagos Archipelago. October 1835. Ex. coll. BMNH, ex. Gould, ex. ZSL. Material: skin. Status: MANCH B. 3089, ex. BMNH 1857.11.28.248.

***Certhidea olivacea olivacea* (Gould, 1837).** GREEN WARBLER-FINCH CD 3340. Z: 106: *Certhidea olivacea* Gould. Syntype *Certhidea olivacea* Gould, 1837. Loc.: Ecuador: Galápagos Archipelago. October 1835. Ex. coll. ZSL. Material: ex mount, male, ad. Status: BMNH 1855.12.19.126. Cat. XI: 28: a–c [see following].

***Certhidea olivacea olivacea* (Gould, 1837).** GREEN WARBLER-FINCH [? CD 3310/3346/3348, see following & previously]. Z: 106: *Certhidea olivacea* Gould. Syntype *Certhidea olivacea* Gould, 1837. Loc.: Ecuador: Galápagos Archipelago. October 1835. Ex. coll. ZSL. Material: ex mount, ad. Status: BMNH 1855.12.19.164. Cat. XI: 28: a–c [see following/previously].

***Certhidea olivacea olivacea* (Gould, 1837).** GREEN WARBLER-FINCH [? CD 3310/3346/3348, see previously]. Z: 106: *Certhidea olivacea* Gould. Syntype *Certhidea olivacea* Gould, 1837. Loc.: Ecuador: Galápagos Archipelago. October 1835. Ex. coll. ZSL. Material: ex mount, ad. Status: BMNH 1855.12.19.127. Cat. XI: 28: a–c [see previously].

CARDINALIDAE

***Saltator coerulescens coerulescens* (Vieillot, 1817).** GREYISH SALTATOR [? CD 1414/1415/1416, see previously]. Z: 97: *Pitylus superciliaris* [Spix]. Loc.: Argentina: Santa Fé. 1833. Material: skin/mount. Status: missing.

The following CD numbers were not linked: 779, 1026, 1027, 1404, 1451, 1462, 1468, 1829, 2189, 2300; one of 751 & 828, 1430 & 1431 and 1604 & 1613; three of 1423, 1425, 1434, 1450; and two of 1605, 1611, 1612, 1614, 1616, and 1617.

APPENDIX 4

John Gould as an Anti-Darwinian Visual Propagandist

It has recently been suggested that John Gould designed and published the ornithological composition of his coloured illustrations in a couple of his large books about birds as visual propaganda against the implications of Darwin's theory of the origin of species through natural selection (Smith 2006: 99–114), as mentioned in chapter 8. I here review and address this surprising claim in the detail that is required to address it adequately.

THE BIRDS OF GREAT BRITAIN

When John Gould prepared his drawing compositions for his great book *The Birds of Great Britain*, the species involved therein were by far the most intimately known birds on earth (Gould 1862–1873), given the advanced state of British ornithology. Thus, at that time in history, that specific work, like few if any others (possibly only also one on North American birds), could illustrate so many nests, eggs, and nestlings of nidicolous bird species and the hatchlings of precocial ones—for precious little was then known about the nidification of significant proportions of avifaunas beyond Great Britain. Thus this circumstance, and the fact that Gould would presumably have wanted to include more rather than less information about the biology of each species in his images, could alone account for their bird "family" content. Moreover, such content also provided highly useful components for designing more interesting compositions for each species' illustration. Also, as an astute businessman sensitive to the sensibilities and wants of his predominantly wealthy and perhaps more sentimental clients, Gould might very well have simply painted images appropriate for his market as he perceived it to be—with no philosophical or theological strings attached. As for him often showing a parental pair of birds at the nest, he did no more than reflect the deeply ingrained perception of the role of the sexes of his social era.

That Gould sometimes included a bird hovering over its nest can be seen as no more than using this as a useful and artful means of showing additional parts of the plumage characters of a species, such as the underwing or undertail, not visible in birds depicted as perched or standing—or simply as enhancing his compositions with some life and movement. It is therefore difficult to objectively justify the somewhat bizarre interpretation of Gould's British bird colour plate "happy family" contents as representing a visual repudiation of Darwin's natural selection theory as Jonathan Smith does (cf. J. Smith 2006; see the "John Gould as an Anti-Darwinian" section in chapter 8). I seriously doubt that making such an oh-so-subtle, even if oft-repeated, "visual statement" would ever have crossed Gould's pragmatic and busy mind. I also seriously doubt that it would have stuck the wealthy owners of his book on British birds as representing any such message at any level, conscious or subconscious.

Having mounted his argument for John Gould as a proactive anti-Darwinian visual propagandist, Smith somewhat contradicts this by noting that "Gould's plates do contain violence, but it is domesticated, almost invariably presented in the form of parents feeding their young ; and a parent kills a bird of another species only to feed it to its young" (Smith 2006: 103–4). In this he appears to be bending over backward to press his case—for other than to feed themselves or their offspring—or (rarely) in defence of themselves, their offspring, or food resources—birds but only rarely kill individuals of other species; and Darwin never suggested otherwise.

Smith's argument thus appears quite spurious, as does his attempt to put thoughts and motivation in Gould's mind over his depiction of true shrikes (of the family Lanidae) at their food "larders." These shrikes catch prey animals and impale them onto the thorns of spiny plants to anchor them as they tear them apart, or to store them there for later consumption; and Gould describes them in his text as "tyrannical and cruel" for doing so. This is, however, no more than a reflection of the kind of anthropomorphism so typically applied to the lives of animals by authors at that time. That said, Gould goes on to objectively describe how nestling birds and shrews are also killed and torn apart by shrikes, apparently with no qualms about shocking his readers with these facts—facts that might readily bring to mind the "survival of the fittest" or the notion of natural selection. Moreover, Smith records that Gould illustrates a "gray shrike" (Northern, or Great Grey, Shrike [*Lanius excubitor*]) with a dead shrew in its beak and another together with a dead Blue Tit (*Parus caeruleus*) impaled on its "larder" thorns. Another plate shows a parasitic Common Cuckoo's nestling ejecting a nestling of its foster parents (a Meadow Pipit [*Anthus pratensis*]) with two previously ejected siblings below the nest that have obviously died of starvation or exposure.

A review of the content of but a few of Gould's British bird plates results in the following: A Golden Eagle (*Aquila chrysaetos*) with a bleeding dead rabbit in its talons; Eurasian Sparrow Hawk (*Accipiter nisus*) grabbing a live House Sparrow (*Passer domesticus*) in its talons in flight; Western Osprey (*Pandion haliaetus*) with a Scottish Trout in its talons; Gyrfalcon (*Falco rusticolus*) with a bleeding drake Mallard (*Anas platyrhynchos*) in its talons; Peregrine Falcon (*Falco peregrinus*) having just killed a drake Mallard in flight, which is plummeting vertically to the ground head first leaving a trail of feathers behind it; Eurasian Hobby (*Falco subbuteo*) with a large dragonfly in its talons; Merlin (*Falco columbarius*) with a dead Yellowhammer (*Emberiza citrinella*) in its talons; Common Kestrel (*Falco tinnunculus*) with a dead bleeding rodent in its talons; White Stork (*Ciconia ciconia*) with a frog in its bill tips, about to be fed to a nestling; Eurasian Oystercatcher (*Haematopus ostralegus*) with a starfish in its bill tips; Eurasian Eagle-Owl (*Bubo bubo*) with a dead young rabbit (*Oryctolagus cuniculus*) hanging from its beak; Tawny Owl (*Strix aluco*) with a rodent in its beak; Little Owl (*Athene noctua*) with a rodent in its talons; Common Oystercatcher (*Haematopus ostralegus*) with a starfish in its bill tip; Common Tern (*Sterna hirundo*) feeding a fish to its young; Atlantic Puffin (*Fratercula arctica*) feeding a beak full of fishes to its young; several other water and sea birds holding or swallowing fish; Hooded Crow (*Corvus cornix*) taking an egg from a plover's nest; and a good number of other song birds holding or feeding invertebrate prey to their offspring. The preceding surely paints Gould as an objective ornithologist shining a light on the biology of birds regardless of any perceived or real "violence and competition" and not one going to extraordinary lengths to incorporate a subtle alternative view to Charles Darwin's theory of natural selection, as is perceived by Smith to be visibly apparent in his images. In fact, Gould clearly far from avoids the natural realities of violent life and death among birds in his paintings.

It is also surprising that Smith makes the remarkable observation that "Gould's 'practical experience' with hummingbirds—the exhibition of his collection had displayed him as the expert on them, and his *Monograph* had confirmed it—was far more valuable than Darwin's armchair theorizing" without further comment (Smith 2006: 97). Gould's expertise of hummingbirds was primarily of their taxonomy based on specimens in European museum and private collections, and in this his knowledge was far greater than that of Darwin: but Gould gained it predominantly as an "armchair" ornithologist. Gould did, however, visit America in the 1850s specifically to see living hummingbirds, first seeing a Ruby-throated (*Archilochus colubris*) in Bartram's Gardens, Philadelphia, in 1857 before seeing many in the grounds of the Capitol in Washington. He also took some live hummingbirds back to England, where they soon died (Clemency Fisher personal communication, April 2015, C. Fisher–C.B. Frith). Darwin's knowledge of hummingbirds was, however, from far more extensive "practical experience" of observing living individuals of many species in the wilds of South America.

Smith also states "hummingbirds came to occupy a prominent place in *The Descent of Man*" (Smith 2006: 98), whereas they are in fact mentioned extremely briefly and but in passing within a line or two on seven pages of discussion of general avian topics. They occupy 21, 20, 15, 9, 5, and 4 lines, plus a brief footnote on seven other pages, plus two illustrations of pairs to show sexual dimorphism in two species. Thus, hummingbirds (or Gould's book on them) hardly occupy a "prominent place" in Darwin's 954 page book *The Descent of Man*.

In early 1864, Gould privately expressed his opinion that Darwin's theory of the origin of species by natural selection would be generally adopted before long (see the "John Gould as an Anti-Darwinian" section in chapter 8). In the light of this stated opinion, it is hard to imagine that he would have gone to the great lengths that Smith (2006) attributes to him as going to with the supposedly subtle theoretical message content and composition of his paintings in *The Birds of Great Britain* three years earlier. It is, moreover, difficult to see any merit in Smith's hypothesized case for John Gould using the composition of his illustrations of British birds for anti-Darwinian propaganda, restricted as the publication was to less than 500 very wealthy subscribers (Sauer 1982: 74) who, in all probability, would not have noticed any such, entirely hypothetical, extremely subtle, subliminal visual message.

Had John James Audubon's famous *The Birds of America* (Audubon 1827–1838), with its plates depicting birds living in "Nature red in tooth and claw," appeared shortly after Darwin's (1859) *Origin of Species*, might Jonathan Smith have interpreted Audubon's motive in so depicting bird life (objectively, as it is in nature) as an attempt at visual propaganda in support of Darwin's ideas?

Given Gould's clear efforts to avoid being seen as supporting Darwin's theory of the origin of species by natural selection, it is most ironic that their collaboration in writing up the specimens of finches collected on the Galapagos Islands is seen as being "the final catalyst in his [Darwin's] conversion to the theory of evolution" (Sulloway 1982c: 369).

Bowerbirds

I must, as a student of bowerbirds for the past several decades, take issue with Smith's statement, given in support of what I see as a spurious argument, that Gould's paintings in his *The Birds of Australia* (Gould 1840–1848) "are clearly designed to make the bowers *look* like nests" (Smith's emphasis). This is an entirely erroneous interpretation that insults the scientific integrity of a great ornithologist and objective ornithological artist. Males of the majority of the bowerbird family (Ptilonorhynchidae) build structures of sticks, grasses, orchids, or other vegetation that they then decorate with stones, snail shells, fruits, flowers, and numerous other objects exclusively to visually impress females and court them at. The nests of the bower building species are built in trees by the females only and are shallow, open, cup-shaped structures not remotely like bowers (Frith and Frith 2004, 2008, 2009b).

Smith inexplicably goes on: "The plate of the satin [Satin Bowerbird (*Ptilonorhynchus violaceus*)] . . . is even more nest-like; it contains two young birds as well as the parents, again with the female inside the bower. That this was Gould's intent is evident from an early sketch of a young bird with its parents, one of whom has food in its beak. . . . Like the later plates in *The Birds of Great Britain* these are domestic scenes of married life, not courtship" (J. Smith 2006: 120). This extraordinary brief text is full of scientific error and misinterpretation: the illustrated bower is not made to look nest-like but is a not unreasonable depiction of a bower of the species; it (the bower) is vaguely implied by Gould to contain a known female, but from the image, it can only be said to be a female-plumaged bird of unknowable age and sex. It cannot possibly be known from only the visual content of the colour plate that Smith alludes to that two of the four birds in the image (see colour figure A.1) are the parents of one or both of the other two individuals. All that can be said about them *from the visual content of the plate* (emphasis mine) is that one (the all blue-black individual) is an adult male, one a subadult male (approximately six to seven years of age, in green female-like plumage but with the blue-black adult male plumage much intruding into it), and that the other two individuals are of unknowable age and sex in green plumage typical of immature birds of both sexes but that Gould does indicate as being immature males. Indeed, the specimens that Gould used to paint the plate in question are now held at the Academy of Natural Sciences in Philadelphia, Pennsylvania, and they consist of an adult male, an adult female, and two immature males (Clemency Fisher, personal communication, April 2015, C. Fisher–C. B. Frith). Gould makes no claim, however, about any genetic relationship between the four birds in his illustration (and could never make such claim).

But far more telling than the preceding is that Gould wrote of the bowers of the Satin Bowerbird, under the name *Ptilonorhynchus holosericeus*, "they are certainly not used as a nest, but a place of resort for many individuals of both sexes, which, when they are assembled, run through and around the bower in a sportive and playful manner, and that so frequently that it is seldom entirely deserted" (Gould 1848b: plate 10). Thus, John Gould clearly did not attempt to make bowers look like nests in his paintings as has been claimed (cf. J. Smith 2006: 120).

The earlier sketch of Satin Bowerbirds alluded to and illustrated by Jonathan Smith shows an adult male and two birds of unknowable age and sex (unless Gould recorded such data on the sheet of paper), and the assumption that they are a parental pair with their young is completely unjustified. Male bowerbird parents have nothing whatsoever to do with their offspring, and nothing in the sketch indicates the sex or the age of the other two birds depicted. This is but another of several similar attempts by Smith to convince his readers that Gould used his paintings (primarily for *The Birds of Great Britain*) as showing "happy families" as a visual repudiation of Darwin's theory of natural and sexual selection—one that would involve him being less than honest as an objective ornithologist and one made so terribly subtlety, at considerable effort, to an extremely limited audience as to be quite futile.

The suggestion by Smith that Gould's plate of the Spotted Bowerbird (*Chlamydera maculata*) in his *The Birds of Australia* (volume 4, plate 8) is ambiguous because the text states that the sexes have *similar* plumage (emphasis mine), "but clues in both image and text suggest that the female is in the bower," lacks justification because Gould states in his text that "The plate represents the bower, with two birds, a male and a female, all of the natural size" (colour figure A.2). As the bird shown standing outside the bower is fully crested, and the individual within the bower lacks a crest, there can be no doubt about which is which. There are in fact, however, and contrary to Smith's words, no "clues in both image and text" in Gould's painting or his text as to the sex of the bird shown in the bower because birds in such plumage cannot be sexed except by dissection (cf. Gould 1848b: plate 8; J. Smith 2006: 120).

The preceding strongly indicates that no convincing evidence exists that could suggest to an objective observer that a covert plot was carefully designed and implemented by John Gould to undermine Darwin's theory of the origin of species by natural selection through the medium of the composition of his published bird paintings. It is a colourful conspiracy theory that lacks, I think, any convincing evidence and is far more novel than it is factual.

GLOSSARY

agglutinated: to cause to adhere, as with a glue.
anthropological: of humankind.
anthropomorphic: to attribute human behaviour or values to animals.
archipelago: a group of islands.
Assyria: ancient kingdom of north Mesopotamia.
avifauna: all birds of a particular region, species and individuals.
baleen: strictly whalebone, but herein the plates in the mouth of whales to filter food from water.
bolas: two or more heavy balls on a cord used to hurl at running animals to entangle their legs as so catch them—of South American origin.
brace: two or a pair (of birds shot).
carex: a vast genus of over 2,000 species of grassy plants in the family Cyperaceae, commonly known as sedges.
carnivorous: flesh, or meat, eating—including insect-eating plants.
carrion: dead and rotting flesh.
carunculated skin: a fleshy outgrowth, as on the heads of some birds.
clutch: the complete number of eggs laid in a bird nest for simultaneous incubation.
collagen: a fibrous scleroprotein of connective tissue and bones rich in glycine and proline.
conspecific: belonging to the same species.
convergent evolution: the independent evolution of similar traits in different lineages resulting from adaptations to similar ecology, lifestyles, etc.
cordillera: a series of parallel mountain ranges.
Crustacea: a mainly aquatic class of invertebrate animals with a hardened carapace, known as crustaceans.
dimorphism: existing in two (or more) distinct morphological forms, which typically involves dimorphism between the sexes but also ages.
Diptera: the insect order of flies, mosquitoes, and gnats with a single pair of wings.

DNA: deoxyribonucleic acid—DNA-DNA hybridization being the molecular technique used to estimate the genetic, and thus evolutionary, distance between taxa.

ecological isolation: reproductive separation of populations due to ecological barriers.

endemic: an organism that is peculiar and restricted to a particular area.

fitness: an individual's relative success in contributing its genes to the next generation; commonly taken as the number of offspring of an individual that survive to themselves reproduce.

gallinaceous: belonging to the avian order Galliformes, which was previously different and broader than today.

Gauchos: cattlemen of the South American pampas.

gene flow: the exchange of genetic material between separate populations due to dispersal and interbreeding.

genealogical: with regard to the direct descent of an individual or group from ancestors.

genetic fitness: an individual's relative success in contributing its genes to the next generation; commonly taken as the number of offspring of an individual that survive to themselves reproduce.

genetically monogamous: or genetic monogamy—the mating system in which the sexes of a pair associate to rear offspring for at least one breeding season and remain faithful to their partner; as opposed to social monogamy.

genomic: genomics is a discipline in genetics that applies recombinant DNA-DNA sequencing methods, and bioinformatics to sequence, assemble, and analyse the function and structure of genomes (being the complete set of DNA within a single cell of an organism).

genus: a taxon consisting of one or several similar species considered mutually more closely related than any one of them is to any other species—genera in plural.

graminivorous: grass-eating.

granivorous: seed-eating.

holotype: in taxonomy, the specimen designated to represent a typical example of the species when it was named and described for the first time; also called the type specimen.

Homeric era: that period of Homer of ancient Greece.

intraspecific: within a species.

isolating mechanism: intrinsic or extrinsic barriers that prevent interbreeding between populations.

lamellae: numerous plates or comb-like "teeth" on the inner edge of a bird's beak (e.g., of ducks and flamingos) that function to filter food from water or mud.

lamelli-rostral aquatic birds: also called the lamellirostres, which refers to the waterfowl of the duck family Anatidae.

lek(king): a site to which plural promiscuous males attract and court females.

lithography: a method of printing from a stone surface on which the lines or marks to be printed are not raised but are made ink receptive.

mammae: the milk-secreting organs of female mammals.

millennia: thousands of years—the singular being millennium.

monogamous: in practising monogamy—in which a male and female bond as a pair for one or more breeding seasons to raise offspring.

monograph: a paper or book concerned with a single subject, such as a bird species or group.

monomorphism: in the ornithological context, the state in which the adult sexes of a species are indistinguishable in morphology, although monochromatic is a more accurate word because sexual dimorphism in size is typically the case.

morphology: the external and internal appearance, form, and structure of organisms.

moult: the growth of new, replacement, feathers, usually through regular renewal—the word covering the loss of old and the growing of new plumage.

natural selection: the mechanism of Darwinian evolution by the survival of the genetically fittest individuals over less fit ones.

Neotropics: zoogeographical region comprising the Americas south of subtropical Mexico, including the West Indies.

nest parasitism: typically involves one species of bird laying an egg(s) in the nest of another (foster) species that then incubates its egg(s) and raises the offspring.

nidicolous: nestling birds hatched naked, blind, and helpless, as are those of passerines.

nidification: the nesting morphology (of nest and egg) and biology of any bird.

nuptial plumage: breeding plumage.

oesophagus: part of the upper alimentary canal, between the pharynx and stomach.

oil-gland: actually paired or bilobed, uropygial or preen, glands at the base of the upper tail in the vast majority of birds.

ornithologist: a scientist that applies scientific principles to the study of birds.

osteology: the study of skeletal anatomy and of bones.

pace: with due deference to (e.g., in politely noting that an author has made an error).

palaeontologist: a scientist that studies fossil organisms.

pappus: a ring of fine feathery hairs surrounding the fruit in composite plants.

paratype: in taxonomy, any specimen in the type series except the holotype or lectotype.

passerine(s): a bird belonging to the order Passeriformes—also loosely referred to as perching, or song, birds.

pelage: the coat of a mammal, of hair, fur, wool, etc.

Pentateuchal = Pentecostal: of or relating to Christian groups emphasizing charismatic aspects of their religion and adopting a fundamental adherence to the Bible content.

pollinators: animals that disperse plant pollen, and thus cross-fertilize them.

polygyny, polygynous(ly): a mating system in which males and/or females have more than one sexual partner in any one breeding season.

precocial: of those birds hatching in a more advanced state than do naked and helpless (nidicolous) ones.

progenitor: an ancestor or an ancestral form.

quadrupeds: animals, particularly mammals, with four walking legs.

quarto: a book size resulting from folding sheets of paper, usually crown or demy, into four leaves (eight pages).

runaway sexual selection: female animals that select to mate with males with exaggerated secondary sexual characters produce more offspring and their sons inherit the male traits and their daughters the female preference for them—this leading to accelerating enhancement, via positive feedback, over generations and hence "runaway" selection.

secondary sexual characters: sex-specific external morphological traits such as plumage and bare skin characters and colours.

sexual selection: an evolutionary process in animals in which selection by females of males with certain secondary sexual characters as mates results in the preservation of such characters in the species.

sexually dichromatic: species are those in which the sexes differ in colouration, such species exhibiting sexual dichromatism.

sexually dimorphic: the form of dimorphism in which each sex of a species has a species-specific size, form, or external (typically adult) morphology or appearance.

socially monogamous: or social monogamy—the mating system in which the sexes of a pair associate to rear offspring for at least one breeding season, but may or may not be faithful to their partner; as opposed to true or genetic monogamy.

sympatric: two or more species are sympatric when sharing the same, entire, or part geographical area.

synonymy: the scientific names of an organism considered by an author to have previously been applied to it but which are subsumed by the current name.

syntype: in taxonomy, if the first description and naming of a taxon was based on a series of specimens these are all called syntypes; unless a particular one was designated as the type specimen or holotype.

systematics: the study of evolutionary relationships; the study of the diversity and relationships of organisms.

taxa, taxon: one or a group of organisms, respectively, regardless of taxonomic rank.

taxonomy: the theory and practice of classifying organisms into taxa.

Tipulidae: the insect family consisting of the crane flies and daddy longlegs.

topography: the physical nature, or the study or description, of the surface of a geographical area or region.

vertebrates: chordate animals of the subphylum Vertebrata, with a bony or cartilaginous skeleton and well-developed brain.

vicariance: when closely related taxa or populations are separated by natural barriers, such as oceans, mountain ranges, or rivers.

volant: flying; capable of flight.

zoogeography: the science of studying patterns of geographical distributions of animals, past and present.

REFERENCES

Abzhanov, A., Kuo, W. P., Hartmann, B., Grant, R., Grant, P. R. and Tabin, C. J. (2006). The Calmodulin pathway and evolution of elongated beak morphology in Darwin's Finches. *Nature*, **442**, 563–7.

Ackerman, J. (1998). Dinosaurs Take Wing. *National Geographic*, **194**, 74–99.

Aikman, H. and Miskelly, C. (2004). *Birds of the Chatham Islands*. Department of Conservation, Wellington, New Zealand.

Andersson, M. (1982). Female choice selects for extreme tail length in a widowbird. *Nature*, London, **299**, 818–20.

Andersson, M. (1994). *Sexual Selection*. Princeton University Press, Princeton, NJ.

Andersson, S. (1992). Female preference for long tails in lekking Jackson's widowbirds: Experimental evidence. *Animal Behaviour*, **43**, 379–88.

Andersson, S. and Andersson, M. (1994). Tail ornamentation, size dimorphism and wing length in the genus *Euplectes* (Ploceinae). *Auk*, **111**, 80–86.

Andrewartha, H. G. and Birch, L. C. (1954). *The Distribution and Abundance of Animals*. University of Chicago Press, Chicago.

Arcese, P. (1989). Intrasexual competition and the mating system in primarily monogamous birds: The case of the song sparrow. *Animal Behaviour*, **38**, 96–111.

Armstrong, E. A. (1975). *The Life & Lore of the Bird: In Nature, Art, Myth, and Literature*. Crown Publishers Inc., New York.

Audebert, J. B. and Vieillot, L. P. (1802). *Histoire naturelle et générale des Colibris, oiseaux-mouches, jacamars et promerops*. Desray, Paris.

Audubon, J. J. (1826). An account of the habits of the turkey buzzard (*Vultur aura*), particularly with the view of exploding the opinion generally entertained of its extraordinary power of smelling. *Edinburgh New Philosophical Journal*, **2**, 172–84.

Audubon, J. J. (1827–1838). *The Birds of America*. 4 volumes. R. Havell, London.

Audubon, J. J. and MacGillivray, W. (1831–1839). *Ornithological biography, or an account of the habits of the birds of the United States of America: Accompanied by descriptions of the objects represented in the work entitled The Birds of America, and interspersed with delineations of American scenery and manners*. 5 volumes. A. and C. Black, Edinburgh, Scotland.

Aydon, C. (2002). *Charles Darwin: The Naturalist Who Started a Scientific Revolution*. Carrol & Graf, New York.

Azara, Félix Manuel de (1809). *Voyage dans l'Amerique meridionale depuis 1781 jusqu'en 1801*. 5 volumes. Dentu, Paris.

Baker, A. J., Haddrath, O., McPherson, J. D., and Cloutier, A. (2014). Genomic support for a moa-tinamou clade and adaptive morphological convergence in flightless ratites. *Molecular Biology and Evolution*, **31**, 1686–96.

Baker, E. C. S. (1913). The evolution of adaptation in parasitic cuckoos' eggs. *Ibis*, **55**, 384–98.

Bannerman, D. A. and Bannerman, W. M. (1968). *History of the Birds of the Cape Verde Islands: Birds of the Atlantic Islands*. Volume 3. Oliver & Boyd, Edinburgh, Scotland.

Barash, D. P. and Lipton, J. E. (2001). *The Myth of Monogamy*. Henry Holt, New York.

Barlow, N. (1963). Darwin's Ornithological Notes. *Bulletin of the British Museum (Natural History) Historical Series*, **2**(7), 201–78.

Barlow, N. ed. (1967). *Darwin and Henslow: The Growth of an Idea*. Bentham-Moxon Trust, John Murray, London.

Barrett, P. H. and Freeman, R. B. eds. (1988). *The Works of Charles Darwin*. 29 volumes. New York University Press, New York.

Bennett, A. T. C., Cuthill, I. C., Partridge, J. C. and Maler, E. J. (1996). Ultraviolet vision and mate choice in zebra-finches. *Nature, London*, **380**, 433–5.

Bennett, P. M. and Owens, I. P. F. (2002). *Evolutionary Ecology of Birds*. Oxford University Press, Oxford.

Benton, T. (2013). *Alfred Russel Wallace: Explorer, Evolutionist, Public Intellectual—A thinker for our own times?* Siri Scientific Press, Manchester, UK.

Beolens, B. and Watkins, M. (2003). *Whose Bird? Men and Women Commemorated in the Common Names of Birds*. Christopher Helm, London.

Berger, A. J. (1964). Evolutionary trends in the avian genus *Clamator*. *Smithsonian Miscellaneous Collections*, **146**, 1–127.

Berger, A. J. (1968). The evolutionary history of the avian genus *Chrycococcyx*. *United States National Museum Bulletin*, **265**, 1–137.

Berry, A. ed. (2002). *Infinite Tropics: An Alfred Russel Wallace Anthology*. Verso, London.

Bewick, T. (1797–1804). *A History of British Birds*. Beilby & Bewick, London.

BirdLife International. (2000). *Threatened Birds of the World*. Lynx Edicions and BirdLife International, Barcelona and Cambridge.

Birkhead, T. (1991). *The Magpies*. T. & A. D. Poyser, London.

Birkhead, T. (2012). *Bird Sense: What It's Like to Be a Bird*. Bloomsbury, London.

Birkhead, T. R. (2003). *The Red Canary: The Story of the First Genetically Engineered Animal*. Weidenfeld & Nicolson, London.

Birkhead, T. and Gallivan, P. T. (2012). Alfred Newton's contribution to ornithology: A conservative quest for facts rather than grand theories. *Ibis*, **154**, 887–905.

Birkhead, T., Hemmings, N., Spottiswoode, C. N., Mikulica, O., Moskat, C., Bán, M. and Schulze-Hagen, K. (2011). Internal incubation and early hatching in brood parasitic birds. *Proceedings of the Royal Society B*, **278**, 1019–24.

Birkhead, T., Wimpenny, J. and Montgomerie, B. (2014). *Ten Thousand Birds: Ornithology Since Darwin*. Princeton University Press, Princeton, NJ.

Black, J. M. ed. (1996). *Partnerships in Birds*. Oxford University Press, Oxford.

Blainville, H. M. D. de, Blainville, M. M. de, Brongniart, A. and Savary, C. (1834). Rapport sur les resultants scientifiques du voyage de M. Alcide d'Orbigny [. . . .]. *Nouvelles annales du Muséum National*, **3**, 84–115.

Bock, W. J. (1994). History and nomenclature of avian family-group names. *Bulletin of the American Museum of Natural History*, **222**, 1–281.

Bodio, S. L. (2009). Darwin's other Bird. *Living Bird*, **28**, 26–33.

Bonaparte, C. L. (1850, 1857). *Conspectus generum avium*. 2 volumes, unfinished. E. J. Brill, Leiden, Netherlands.

Bonaparte, C. L. (1855). *Coup d'oeil sur l'ordre des pigeons*. Imprimerie de Mallet Bachelier, Paris.

Bourne, W. R. P. (1992). FitzRoy's foxes and Darwin's finches. *Archives of Natural History*, **19**, 29–37.

Bowman, R. I. (1961). Morphological differentiation and adaptation in the Galápagos finches. *University of California Publications in Zoology*, **58**, 1–302.
Bowman, R. I. (1963). Evolutionary patterns in Darwin's finches. *Occasional Papers of the California Academy of Sciences*, **44**, 107–40.
Brehm, A. E., Schmidt, E. O., and Taschenberg, E. L. (1876–1879). *Brehms Tierleben. Allgemeine Kunde des Tierreichs*. 2nd, expanded, ed. 10 volumes. Bibliographisches Institut, Leipzig, reprinted 1882–1884.
Brent, B. P. (1864). *The Canary, British Finches, and Some Other Birds. Journal of Horticulture & Cottage Gardener*. London.
Brent, P. (1981). *Charles Darwin: "A Man of Enlarged Curiosity."* Heinemann, London.
Brisson, M.-J. (1760). *Ornithologie ou Méthode contenant la division des oiseaux en Ordes, Sections, Genres, Espèces & Leurs Variétés*. 6 volumes. Bauche, Paris.
Brooke, M. (2004). *Albatrosses and Petrels across the World*. Oxford University Press, Oxford.
Brooker, M. G., Rowley, I., Adams, M. and Baverstock, P. R. (1990) Promiscuity: An inbreeding avoidance mechanism in a socially monogamous species? *Behavioural Ecology and Sociobiology*, **26**, 191–9.
Brown, L. and Amadon, D. (1968). *Eagles, Hawks and Falcons of the World*. 2 volumes. Country Life Books, Feltham, England.
Browne, J. (1995). *Charles Darwin: Voyaging*. Jonathan Cape, London.
Browne, J. (2002). *Charles Darwin: The Power of Place*. Jonathan Cape, London.
Bruning, D. F. (1974). Social structure and reproductive behavior in the greater rhea. *Living Bird*, **13**, 251–94.
Buffon, G-L. L. Comte de (1770–1783). *Histore Naturelle des Oiseaux*. 10 volumes. Imprimerie Royale, Paris.
Burchell, W. J. (1822–1824). *Travels in the interior of Southern Africa*. 2 volumes. Longman, Hurst, Rees, Orme and Brown, London.
Burkhardt, F. et al. eds. (1985–2014). *The Correspondence of Charles Darwin*. 21 volumes. Cambridge University Press, Cambridge.
Carlquist, S. (1974). *Island Biology*. Columbia University Press, New York.
Castro, I. and Phillips, A. (1996). *A Guide to the Birds of the Galapagos Islands*. Christopher Helm, London.
Chancellor, G. and Wyhe, J. van (2009). *Charles Darwin's Notebooks from the Voyage of the Beagle*. Cambridge University Press, Cambridge.
Chancellor, G., DiMauro, A., Ingle, R. and King, G. (1988). Charles Darwin's Beagle Collections in the Oxford University Museum. *Archives of Natural History*, **15**(2), 197–231.
Chance, E. (1922). *The Cuckoo's Secret*. Sedgwick & Jackson Ltd., London.
Chance, E. (1940). *The Truth about the Cuckoo*. Country Life, London.
Chantler, P. (1999). Family Apodidae (Swifts). Pp. 388–457 in: del Hoyo, J., Elliot, A. and Sargatal, J., eds. *Handbook of the Birds of the World*. Volume 5: *Barn-owls to Hummingbirds*. Lynx Edicions, Barcelona, Spain.
Christidis, L. and Boles, W. E. (2008). *Systematics and Taxonomy of Australian Birds*. CSIRO Publishing, Melbourne, Australia.
Chubb, C. (1913). Exhibit and description of two new forms of Rhea. *Bulletin of the British Ornithologists' Club*, **33**, 79.
Cinat-Tomson, H. (1926). Die geschlechtliche Zuchtwahl beim Wellensittich (*Melopsittacus undulatus* Shaw). *Biologisches Zentralblatt*, **46**, 543–52.
Clements, J. (2009). *Darwin's Notebook: The Life, Times, and Discoveries of Charles Robert Darwin*. Running Press, Philadelphia.

Clutton-Brock, T. H. (1986). Sex ratio variation in birds. *Ibis*, **128**, 317–29.

Cockburn, A., Brouwer, L., Double, M. C., Margraf, N. and va de Pol, M. (2013). Evolutionary origins and persistence of infidelity in *Malurus*: The least faithful birds. *Emu*, **113**, 208–17.

Cocker, M. (2013). *Birds and People*. Jonathan Cape, London.

Cody, M. L. (2005). Family Mimidae (Mockingbirds and Thrashers). Pp448–95 in: del Hoyo, J., Elliot, A. and Christie, D. A. eds. *Handbook of the Birds of the World*. Volume 10: *Cuckoo-shrikes to Thrushes*. Lynx Edicions, Barcelona, Spain.

Collar, N. J. (2005). Family Turdidae (Thrushes). Pp. 514–807 in: del Hoyo, J., Elliot, A. and Christie, D. A., eds. *Handbook of the Birds of the World*. Volume 10: *Cuckoo-shrikes to Thrushes*. Lynx Edicions, Barcelona, Spain.

Cooper, J. H. (2010). Charles Darwin, the bird curator. *Journal of Afrotropical Zoology*, Special Issue, **6**, 23–9.

Cooper, J. H. (2014). Pigeon post: contributions from India to Charles Darwin's domestic bird research. In: Ray, R., Chattopadhyay, D. and Banerjee, S., Eds. *Darwin and Human Evolution: Origin of Species Revisited*. The Asiatic Society, Kolkata, India.

Couve, E. and Vidal, C. (2003). *Birds of Patagonia, Tierra del Fuego & Antarctic Peninsula, the Falkland Islands & South Georgia*. Editorial Fantástico Sur Birding Ltda., Punta Arenas, Chile.

Cramp, S. ed. (1985). *Handbook of the Birds of Europe the Middle East and North Africa*. Volume 4. Oxford University Press, Oxford.

Cramp, S. ed. (1988). *Handbook of the Birds of Europe the Middle East and North Africa*. Volume 5. Oxford University Press, Oxford.

Crook, J. H. (1961). The fodies of the Seychelles Islands. *Ibis*, **103a**, 517–48.

Cuthill, I. C. (2006). "Colour perception." Pp. 3–40 in: Hill, G. E. and McGraw, K. eds. *Bird Coloration: Mechanisms and Measurements*. Harvard University Press, Cambridge, MA.

Cuvier, G. L. C. F. D. Baron (1817). *Le Régne Animal distriubé d'aprés son organization pour servir de base à l'historie naturelle des Animaux et d'introduction à l'Anatomie Comparée, nouvelle edn, volume 1 (mammiféres et oiseaux)*. Déterville, Paris.

Cuvier, G. L. C. F. D. Baron (1829). *Le Régne Animal distriubé d'aprés son organization pour servir de base à l'historie naturelle des Animaux et d'introduction à l'Anatomie Comparée, nouvelle edn, volume 1 (mammiféres et oiseaux)*. Déterville, Paris.

Cuvier, G. L. C. F. D. Baron et. al. (1816–1845). *Dictionnaire des Sciences Naturelles*. Le Normat, Paris.

Dakin, R. and Montgomerie, R. (2011). Peahens prefer peacocks displaying more eye-spots, but rarely. *Animal Behaviour*, **82**, 21–8.

Darwin, C. (1826). Manuscript Collection, University Library, Cambridge. 129.

Darwin, C. R. (1837). Notes on *Rhea americana* and *Rhea darwinii*. *Proceedings of the Zoological Society of London*, **5**, 35–6.

Darwin, C. R. ed. (1838–1841). *The Zoology of the H. M. S. Beagle, Under the Command of Captain Fitzroy, R. N. During the Years 1832 to 1836. Part III: Birds*. Smith, Elder & Co., Cornhill, London.

Darwin. C. (1839a). In FitzRoy, *Voyages of the Adventure and Beagle Between the Year 1826 and 1836 . . . Volume 3. Journal and Remarks 1832–1836*. Henry Colburn, London.

Darwin. C. (1839b). *Journal of Researches into the Geology and Natural History of the Various Countries Visited by H. M. S. Beagle Under the Command of Captain FitzRoy, R. N. from 1832 to 1836*. Henry Colburn, London.

Darwin, C. R. (1839c). *Questions About the Breeding of Animals.* Stewart & Murray, London.

Darwin, C. R. (1842). *The Structure and Distribution of Coral Reefs. Being the First Part of the Geology of the Voyage of the* Beagle, *under the Command of Capt. Fitzroy, R.N. During the Years 1832 to 1836.* Smith Elder, London.

Darwin, C. R. (1844). *Geological Observations on the Volcanic Islands Visited During the Voyage of H.M.S. Beagle, Together with Some Brief Notices of the Geology of Australia and the Cape of Good Hope. Being the Second Part of the Geology of the Voyage of the Beagle, under the Command of Capt. Fitzroy, R.N. During the Years 1832 to 1836.* Smith Elder, London.

Darwin. C. (1845). *Journal of Researches into the Natural History and Geology of the Countries Visited During the Voyage of H. M. S. Beagle Round the World Under the Command of Captain FitzRoy, R. N.* Second edition. John Murray, London.

Darwin, C. R. (1846). *Geological Observations on South America. Being the Third Part of the Geology of the Voyage of the* Beagle, *under the Command of Capt. Fitzroy, R.N. During the Years 1832 to 1836.* Smith Elder, London.

Darwin, C. R. (1849). Section VI: Geology. In Herschel, J. F. W. ed. *A Manual of Scientific Enquiry: Prepared for the Use of Her Majesty's Navy and Adapted for Travellers in General.* London: John Murray, pp. 156–95.

Darwin, C. R. (1851a). Fossil Cirripedia of Great Britain: A monograph on the fossil Lepadidae, or pedunculated cirripedes of Great Britain. London: *Palaeontographical Society.* Volume 1.

Darwin, C. R. (1851b). Living Cirripedia, A monograph on the sub-class Cirripedia, with figures of all the species. The Lepadidae; or, pedunculated cirripedes. London: *The Ray Society.* Volume 1.

Darwin, C. R. (1854a). Living Cirripedia, The Balanidae (or sessile cirripedes); the Verrucidae. London: *The Ray Society.* Volume 2.

Darwin, C. R. (1854b). A monograph on the fossil Balanidae and Verrucidae of Great Britain. London: *Palaeontographical Society.* Volume 2.

Darwin, C. R. (1859). *On the Origin of Species by Means of Natural Selection, or the Preservation of Favoured Races in the Struggle for Life.* John Murray, London. Facsimile edition with an introduction by Ernst Mayr. Harvard University Press, Cambridge, MA, 1964 facsimile.

Darwin. C. (1860). *Journal of Researches into the Natural History and Geology of the Countries Visited During the Voyage of H. M. S. Beagle Round the World Under the Command of Captain FitzRoy, R. N.* A new edition. [Tenth Thousand.] John Murray, London.

Darwin, C. R. (1862). Penguin ducks. *Journal of Horticulture and Cottage Gardener,* **3**(30 December): 797.

Darwin, C. R. (1868a). *The Variation of Animals and Plants Under Domestication.* Volume 1. John Murray, London.

Darwin, C. R. (1868b). *The Variation of Animals and Plants Under Domestication.* Volume 2. John Murray, London.

Darwin, C. R. (1870). Notes on the habits of the pampas woodpecker (*Colaptes campestris*). *Proceedings of the Zoological Society of London,* **47**, 705–6.

Darwin, C. R. (1871). *The Descent of Man and Selection in Relation to Sex.* John Murray, London.

Darwin, C. R. (1876). *The Effects of Cross and Self Fertilisation in the Vegetable Kingdom.* London: John Murray.

Darwin, C. R. (1877). *The Different Forms of Flowers on Plants of the Same Species.* John Murray, London.

Darwin, C. (1881a). *The Formation of Mould, through the Action of Worms, with Observations on Their Habits*. John Murray, London.

Darwin, C. R. (1881b). The parasitic habits of Molothrus. *Nature. A Weekly Illustrated Journal of Science*, **25**, 51–2.

Darwin, C. R. (1889). *A Naturalist's Voyage. Journal of Researches into the Natural History and Geology of the Countries Visited During the Voyage of H. M. S.* "Beagle" *Round the World. Under the Command of Capt. FitzRoy, R. N.* John Murray, London. New edition [first published 1839].

Darwin, C. R. (1901). *The Descent of Man and Selection in Relation to Sex*. John Murray, London 1901. Second edition [first published 1871].

Darwin, C. R. (1902). *The Origin of Species by Means of Natural Selection or the Preservation of Favoured Races in the Struggle for Life*. December 1902 reprint with "Additions and Corrections" and "An Historical Sketch." John Murray, London. Sixth edition [first published 1859].

Darwin, C. R. (1904). *The Expression of the Emotions in Man and Animals*. Edited by Francis Darwin. John Murray, London. Popular Edition [first published 1872].

Darwin, C. R. (2002). *Voyage of the Beagle*. Dover Publications, New York. A facsimile reprint of the 1909 volume 29 of the Harvard Classics series published by P. F. Collier & Son Co., New York, based on the 1845 second edition published by John Murray, London.

Darwin, C. R. and Wallace A. R. (1858). On the tendency of species to form varieties; and on the perpetuation of varieties and species by natural means of selection. *Journal of the Proceedings of the Linnean Society of London; Zoology*, **3**, 45–62.

Darwin, E. (1794–1796). *Zoonomia; Or the Laws of Organic Life*. 2 volumes. London.

Darwin, E. (1803). *The Temple of Nature: or, the Origin of Society: A Poem, with Philosophical Notes*. Canta 1, Production of Life. J. Johnson, London.

Darwin, F. ed. (1887). *Life and Letters of Charles Darwin, including an Autobiographical Chapter*, 3 volumes. John Murray, London.

Darwin, F. ed. (1950). *Charles Darwin's Autobiography*. Henry Schuman, New York.

Darwin, R. W. (1787). *Principia Botanica; Or, a Concise and Easy Introduction to the Sexual Botany of Linnaeus*. Newark.

Datta, A. (1997). *John Gould in Australia: Letters and Drawings*. Miegunyah Press, Melbourne.

D'Aubenton, E.-L. (1771–1786). *Planches Enluminées*. 10 volumes. Paris.

Davies, N. B. (2000). *Cuckoos, Cowbirds and Other Cheats*. T. & A. D. Poyser, London.

Davies, N. B. (2015). *Cuckoo—Cheating by Nature*. Bloomsbury, London.

Davies, S. J. J. F. (2002). *Ratites and Tinamous: Tinamidae, Rheidae, Dromaiidae, Casuariidae, Apterygidae, Struthionidae*. Oxford University Press, Oxford.

Davison, G. W. H. (1985). Avian spurs. *Journal of Zoology*, **206**, 353–66.

Dawkins, R. (1986). *The Blind Watchmaker*. Longman, London.

De Beer, G. (1968). *Charles Darwin: Evolution by Natural Selection*. Nelson, Melbourne, Australia.

Delacour, J. (1954). *The Waterfowl of the World*. Volume 1. Country Life Limited, London.

del Hoyo, J. (1992). Family Phoenicopteridae (Flamingos). Pp. 508–26 in: del Hoyo, J., Elliot, A., Sargatal, J. and Christie, D. A. eds. *Handbook of the Birds of the World*. Volume 1: *Ostrich to Ducks*. Lynx Edicions, Barcelona, Spain.

del Hoyo, J. and Collar, N. J. (2014). *HBW and BirdLife International Illustrated Checklist of the Birds of the World*. Volume 1: Non-passerines. Lynx Edicions, Barcelona, Spain.

De Queiroz, A. (2014). *The Monkey's Voyage*. Basic Books, New York.

De Roy, T., Jones, M. and Cornthwaite, J. (2013). *Penguins: Their World, Their Ways*. Christopher Helm, London.

Desmond, A. and Moore, J. (1991). *Darwin*. Michael Joseph, London.

Desmond, A. and Moore, J. (2009). *Darwin's Sacred Cause: Race, Slavery and the Quest for Human Origins*. Allen Lane, London.

Diamond, J. (1978). Niche shifts and the rediscovery of interspecific competition. *American Scientist*, **66**, 322–31.

Dobrizhoffer, M. (1749). *Historia de Abiponibus, equestri bellicosaque Paraquariae natione, locupletata copiosis barbarorum gentium, urbium, fluminum, ferarum, amphibiorum, insectorum, serpentium praecipuorum, piscium, avium, arborum, plantarum aliarumque ejusdem provinciae proprietatum observationibus*. Hand written in Latin, found in Vienna.

Donald, P. F. (2007). Adult sex ratios in wild bird populations. *Ibis*, **149**, 671–92.

Donohue, K. ed. (2011). *Darwin's Finches; Readings in the Evolution of a Scientific Paradigm*. University of Chicago Press, Chicago.

d'Orbigny, A. C. V. M. D. (1834). *La Relation du Voyage dans l'Amérique Méridionale*. Travels, volume 2. Pitois-Levrault, Paris.

Dutson, G. (2011). *Birds of Melanesia*. Christopher Helm, London.

Erritzøe, J. and Erritzøe, H. B. (1998). *Pittas of the World: A Monograph on the Pitta Family*. The Lutterworth Press, Cambridge, England.

Erritzøe, J., Kampp, K., Winker, K. and Frith, C. B. (2007). *The Ornithologist's Dictionary*. Edicions, Barcelona, Spain.

Erritzøe, J., Mann, C. F., Brammer, F. P. and Fuller, R. A. (2012). *Cuckoos of the World*. Christopher Helm, London.

Eyton, T. C. (1859). On the different methods of preparing natural skeletons of birds. *Ibis*, **1**, series 1, 55–7.

Farber, P. L. (1982). *The Emergence of Ornithology as a Scientific Discipline: 1760-1850*. D. Reidel, Dordrecht, Netherlands.

Feduccia, A. (1996). *The Origin and Evolution of Birds*. Yale University Press, New Haven & London.

Fergusen-Lees, J. and Christie, D. A. (2001). *Raptors of the World*. Christopher Helm, London.

Fernández, G. J. and Reboreda, J. C. (1998). Effects of clutch size and timing of breeding on reproductive success of greater rheas. *Auk*, **115**, 340–8.

Fisher, R. A. (1930). *The Genetical Theory of Natural Selection*. Clarendon, Oxford.

Fisher, R. A. (1958). *The Genetical Theory of Natural Selection*. Second edition. Dover, New York.

FitzRoy, R. (1839). *Narrative of the Surveying Voyages of H.M.S.* Adventure *and* Beagle, *between the Years 1826 and 1836*. Vol 1. *Proceedings of the First Expedition, 1826–1830, under the Command of Captain P. Parker King*. Edited by Robert FitzRoy. Vol. 2. *Proceedings of the Second Expedition, 1831–1836, under the command of Captain R. FitzRoy*. Vol. 3. *Journal and remarks, 1832–1836*. By Charles Darwin. 3 volumes. London.

Fjeldså, J. (2013). Avian classification in flux. Pp. 77–146 in: del Hoyo, Elliott, A., Sargatal, J. and Christie, D. A. eds. *Handbook of Birds of the World. Special Volume: New Species and Global Index*. Lynx Edicions, Barcelona, Spain.

Forster, J. R. (1778). *Observations Made During a Voyage Round the World (in H. M. S. Resolution) on Physical Geography, Natural History and Ethic Philosophy* [. . .]. Robinson, London.

Freeman, A. R. and Hare, J. F. (2015). Infrasound in mating displays: a peacock's tale. *Animal Behaviour*,**102**, 241–50.

Frith C. B. (1976). A twelve-month field study of the Aldabran Fody *Foudia eminentissima aldabrana*. *Ibis*, **118**, 155–78.

Frith, C.B. (1994). Adaptive significance of tracheal elongation in manucodes (Paradiseaidae). *Condor*, **96**, 552–5.

Frith, C. B. (1997). Huia (*Heteralocha acutirostris*, Callaeidae) -like sexual bill dimorphism in some birds of paradise (Paradisaeidae) and its significance. *Notornis*, **44**, 177–84.

Frith, C. B. (2013). *The Woodhen: A Flightless Island Bird Defying Extinction*. CSIRO Publishing, Melbourne, Australia.

Frith, C. B. and Beehler, B. M. (1998). *The Birds of Paradise: Paradisaeidae*. Oxford University Press, Oxford.

Frith, C. B. and Frith, D. W. (1997). Biometrics of the birds of paradise (Aves: Paradisaeidae): With observations on variation and sexual dimorphism. *Memoirs of the Queensland Museum*, **42**, 159–212.

Frith, C. B. and Frith, D. W. (2004). *The Bowerbirds: Ptilonorhynchidae*. Oxford University Press, Oxford.

Frith, C. B. and Frith, D. W. (2008). *Bowerbirds: Nature, Art & History*. Frith & Frith, Malanda.

Frith, C. B. and Frith, D. W. (2009a). Family Paradisaeidae (Birds-of-paradise). Pp. 404–93 in: del Hoyo, J., Elliot, A. and Christie, D. A. eds. *Handbook of the Birds of the World*. Volume 14: *Bush-shrikes to Old World Sparrows*. Lynx Edicions, Barcelona, Spain.

Frith, C. B. and Frith, D. W. (2009b). Family Ptilonorhynchidae (Bowerbirds). Pp. 344–403 in: del Hoyo, J., Elliot, A. and Christie, D. A. eds. *Handbook of the Birds of the World*. Volume 14: *Bush-shrikes to Old World Sparrows*. Lynx Edicions, Barcelona, Spain.

Frith, C. B. and Frith, D. W. (2010). *Birds of Paradise: Nature, Art & History*. Frith & Frith, Malanda.

Gause, G. F. (1934). *The Struggle for Existence*. Williams and Wilkins, Baltimore.

Gifford, E. W. (1919). Field notes on the land birds of the Galapagos Islands and of Cocos Island, Costa Rica. *Proceedings of the Californian Academy of Science*, **2**, 189–258.

Gill, F. and Donsker, D. eds. (2015). International Ornithologists' Union IOC World Bird List (v. 5.2). http//dx.doi.org/10.14344/IOC.ML.5.2.

Gould, J. (1830–1833). *A Century of Birds from the Himalaya Mountains*. 5 volumes. Author, London.

Gould, J. (1832–1837). *The Birds of Europe*. Author, London.

Gould, J. (1833–1835). *A Monograph of the Ramphasidae, or Family of Toucans*. Author, London.

Gould, J. (1836). [Untitled: but observations on some bird specimens] *Proceedings of the Zoological Society of London*, **4**, 6.

Gould, J. (1836–1838). *A Monograph of the Trogonidae, or Family of Trogons*. Author, London.

Gould, J. (1837a). On a new Rhea (R. Darwinii). *Proceedings of the Zoological Society of London*, **5**, 35–6.

Gould, J. (1837b). Remarks on a group of ground finches from Mr. Darwin's collection. *Proceedings of the Zoological Society of London*, **5**, 4–7.

Gould, J. (1837c). [Exhibition of] Three species of the genus *Orpheus*, from the Galapagos, in the collection of Mr. Darwin. *Proceedings of the Zoological Society of London*, **5**, 27.
Gould, J. (1837–1838a). *The Birds of Australia and the Adjacent Islands*. Author: London.
Gould, J. (1837–1838b). *Icones Avium, or Figures and Descriptions of New and Interesting Species of Birds from Various parts of the World*. Author: London.
Gould, J. (1837–1838c). *A Synopsis of the Birds of Australia and the Adjacent Islands*. Author: London.
Gould, J. (1840–1848). *The Birds of Australia*. 7 volumes. Author, London.
Gould, J. (1843). On nine new birds collected during the Voyage of the H. M. S. Sulphur. *Proceedings of the Zoological Society of London*, **11**, 103–6.
Gould, J. (1844–1850). *A Monograph of the Odontophorinae, or Partridges of America*. Author, London.
Gould, J. (1848a). *The Birds of Australia*. Volume 4. Author, London.
Gould, J. (1848b). *An Introduction to the Birds of Australia*. Author, London.
Gould, J. (1849–1861). *A Monograph of the Trochilidae, or Family of Humming-birds*, 5 volumes. Author, London.
Gould, J. (1850–1883). *The Birds of Asia*. 7 volumes. Author, London.
Gould, J. (1851–1869). *Supplement to the Birds of Australia*. Author, London.
Gould, J. (1852–1854). *A Monograph of the Ramphastidae, or Family of Toucans*. Second edition. Author, London.
Gould, J. (1855). *Supplement to the First Edition of A Monograph of the Ramphastidae, or Family of Toucans*. Author, London.
Gould, J. (1858–1875). *A Monograph of the Trogonidae or Family of Trogons*. Second edition. Author, London.
Gould, J. (1861). *An Introduction to the Trochilidae, or family of Humming-birds*. Author, London.
Gould, J. (1862–1873). *The Birds of Great Britain*. 5 volumes. Author, London.
Gould, J. (1865). *Handbook to the Birds of Australia*. 2 volumes. Author, London.
Gould, J. and Darwin, C. eds. (1838–1841). *The Zoology of the Voyage of H. M. S. Beagle, under the Command of Captain Fitzroy, R. N., During the Years 1832 to 1836. Part III: Birds*. Pages 1–164. Smith, Elder & Co., London.
Gould, S. J. (1980). *The Panda's Thumb*. W. W. Norton & Co., New York.
Gould, S. J. (1985). *The Flamingo's Smile*. Penguin Books, Harmondsworth, England.
Gould, S. J. (1991). *Ever Since Darwin*. Penguin Books, Harmondsworth, England.
Grant, K. T. and Estes, G. B. (2009). *Darwin in the Galápagos: Footsteps to a New World*. Princeton University Press, Princeton, NJ.
Grant, P. R. (1986). *Ecology and Evolution of Darwin's Finches*. Princeton University Press, Princeton, NJ.
Grant, P. R. and Grant, B. R. (2008). *How and Why Species Multiply: The Radiation of Darwin's Finches*. Princeton University Press, Princeton, NJ.
Grant, P. R. and Grant, B. R. (2010). Natural selection, speciation and Darwin's finches. *Proceedings of the California Academy of Sciences*, **61**(Supplement 2), 245–60.
Gribbin, J. and Gribbin, M. (2003). *FitzRoy*. Review, London.
Griffith, E., Pidgeon, E. and Gray, J. E. (1828). The Class Aves Arranged by the Baron Cuvier, with Specific Descriptions. In Griffith, E. (Editor), *The animal kingdom arranged in conformity with its organization, by the baron Cuvier, Member of the Institute of France, &c. [.] .with additional descriptions of all the species hitherto*

named, and of many not before noticed [.]. Volume 1 [*The Class Aves*]–Volume 6 [*The Animal Kingdom*]. Whittaker, Treacher, London.

Grouw, H. van and Steinheimer, F. D. (2008). Charles Darwin's lost Cinereous Harrier found in the collection of the National Museum of Natural History Leiden. *Zoologische Mededelingen Leiden*, **82**(48), 31, xii.

Hackett, S. J., Kimball, R. T., Reddy, S., Bowie, R. C. K., Braun, E. L., Braun, M. J., Chojnowski, J. L., Cox, W. A., et al. (2008). A phylogenomic study of birds reveals their evolutionary history. *Science*, **320**, 1763–8.

Haeckel, E. (1866). *Generelle Morphologie der Organismen*. G. Reimer, Berlin.

Hamilton, W. D. and Zuk, M. (1982). Heritable true fitness and bright birds: A role for parasites. *Science*, **218**, 384–7.

Hancock, J. A., Kushlan, J. A. and Kahl, M. P. (1992). *Storks, Ibises and Spoonbills of the World*. Academic Press, London.

Hanson, T. (2011). *Feathers: The Evolution of a Natural Miracle*. Basic Books, New York.

Harrison, P. (1985). *Seabirds: An Identification Guide*. Revised edition. Croom Helm, London.

Harrison, P. W., Wright, A. E., Zimmer, F., Dean, R., Montgomery, S. H., Pointer, M. A. and Mank, J. E. (2015). Sexual selection drives evolution and rapid turnover of male gene expression. *Proceedings of the Natural Academy of Science*, **112** (14), 4393–8.

Harshman, J. et al. (2008). Phylogenomic evidence for multiple losses of flight in ratite birds. *Proceedings of the National Academy of Sciences*, **105**, 13462–7.

Hastings Belshaw, R. H. (1985). *Guinea Fowl of the World*. Nimrod Book Services, Liss, Hampshire, England.

Haupt, L. L. (2006). *Pilgrim on the Great Bird Continent*. Little, Brown and Company, New York.

Healey, E. (2001). *Emma Darwin: The Inspirational Wife of a Genius*. Headline, London.

Heinroth, O. (1910). Beiträge zur Biologie, namentlich Ethologie und Psychologie der Anatiden. *Verhandlungen des V. Internationalen Ornithologen Kongresses*, Berlin, 1910: 589–702.

Heinroth, O. (1924). Lautäusserungen der Vögel. *Journal für Ornithologie*, **72**, 223–44.

Heinroth, O. and Heinroth, M. (1924–1928). *Die Vögel Mitteleuropas*. Lichtenfelde, Berlin.

Heinzel, H. and Hall, B. (2000). *Galápagos Diary: A Complete Guide to the Archipelago's Birdlife*. University of California Press, Berkley.

Hellmayr, C. E. (1938). Catalogue of birds of the Americas and the adjacent islands in Field Museum of Natural History. Part XI: Ploceidae-Catamblyhynchidae-Fringillidae. *Zoological Series, Field Museum of Natural History*, **13**, 130–46.

Hervieux, de Chanteloup (1719). *Nouveau Traité des serins de canarie*. Paris.

Hesse, R., et. al. (1937). *Ecological Animal Geography*. John Wiley, New York.

Hill, G. E., Montgomerie, R., Roeder, C. and Boag, P. (1994). Sexual selection and cuckoldry in a monogamous songbird: Implications for sexual selection theory. *Behavioral Ecology and Sociobiology*, **35**, 193–9.

Houlihan, P. (1986). *Birds of Ancient Egypt*. Aris and Phillips, Warminster, UK.

Houston, D. C. (1986). Scavenging efficiency of Turkey Vultures in tropical forest. *Condor*, **88**, 318–23.

Houston, D. C. (1994). Family Cathartidae (New World Vultures). Pp. 24–41 in: del Hoyo, J., Elliot, A. and Sargatal, J. eds. *Handbook of the Birds of the World. Volume 2: New World Vultures to Guineafowl*. Lynx Edicions, Barcelona.

Howard, H. E. (1920). *Territory in Bird Life*. John Murray, London.

Howard, H. E. (1907–1914). *The British Warblers*. R. H. Porter, London.

Hudson, W. H. (1870). A third letter on the ornithology of Buenos Ayres. *Proceedings of the Zoological Society of London*, **38**, 158–60.

Humboldt, F. H. A. von (1814–1829). *Personal narrative of travels to the equinoctial regions of the New Continent, during the years 1799–1804, by Alexander de Humboldt, and Aimé Bonpland; with maps, plans, etc. written in French by Alexander de Humboldt, and translated into English by Helen Maria Williams*, 7 volumes. Longman, Hurst, Rees and Brown, London.

Hume, J. P. and Walters, M. (2012). *Extinct Birds*. T & AD Poyser, London.

Huxley, J. and Kettlewell, H. B. D. (1974). *Charles Darwin and his World*. Book Club Associates, London.

Huxley, T. H. (1867). On the Classification of Birds. *Proceedings of the Zoological Society of London*, **1867**, 415–71.

Ings, S. (2007). *The Eye: A Natural History*. Bloomsbury, London.

Irwin, R. (1994). The evolution of plumage dichromatism in the new world blackbirds: Social selection on female brightness? *American Naturalist*, **144**, 890–907.

Jackson, C. E. (1999). *Dictionary of Bird Artists of the World*. Antique Collectors' Club, Woodbridge, England.

Jackson, C. E. and Lambourne, M. (1990). Bayfield—John Gould's unknown colourer. *Archives of Natural History*, **17**, 189–200.

Jaramillo, A. (2011). Family Emberizidae (Buntings and New World Sparrows). Pp. 675–83 in: del Hoyo, J., Elliot, A. and Christie, D. A. eds. *Handbook of the Birds of the World*. Volume 16: *Tanagers to New World Blackbirds*. Lynx Edicions, Barcelona, Spain.

Jaramillo, A. and Burke, P. (1999). *New World Blackbirds: The Icterids*. Christopher Helm, London.

Jarvis, E. D., Mirarah, S., Aherer, A. J., Li, B. Houde, P., Ho, S. Y. W., Faircloth, B. C., Nabholz, B., et al. (2014). Whole-genome analyses resolve early branches in the tree of life of modern birds. *Science*, **346**, 1320–31.

Jerdon, T. C. (1862–1864). *The Birds of India*. Author, Calcutta.

Jobling, J. A. (2010). *The Helm Dictionary of Scientific Bird Names*. Christopher Helm, London.

Johnsgard, P. A. (1994). *Arena Birds: Sexual Selection and Behavior*. Smithsonian Institution Press, Washington.

Johnston, R. F. (1992). Evolution in the Rock Dove: Skeletal morphology. *Auk*, **109**, 530–42.

Jones, S. (2008). *Darwin's Island: The Galapagos in the Garden of England*. Little, Brown, London.

Jordania, J. (2011). *Peacock's Tail: Tale of Beauty and Intimidation*. Pp. 192–6 in: *Why Do People Sing? Music in Human Evolution*. Logos, South Australia, Australia.

Joseph, L. and Buchanan, K. L. (2015). A quantum leap in avian biology. *Emu*, **115**, 1–5.

Jukema, J. and Piersma, T. (2006). Permanent female mimics in a lekking shorebird. *Biology Letters*, **2**, 161–4.

Kear, J. ed. (2005). *Ducks, Geese and Swans*. Oxford University Press, Oxford.

Keynes, R. ed. (2000). *Charles Darwin's Zoology Notes & Specimen Lists from H. M. S.* Beagle. Cambridge University Press, Cambridge.

Keynes, R. (2002). *Darwin, His Daughter and Human Evolution*. Riverhead Books, NY.

Keynes, R. D. (1979). *The Beagle Record*. Cambridge University Press, Cambridge, NY.

Keynes, R. D. (2003). *Fossils, Finches and Fuegians: Darwin's Adventures and Discoveries on the Beagle*. Oxford University Press, Oxford.

Keynes, R. D. (1997). Steps on the path to the origin of species. *Journal of Theoretical Biology*, **187**, 461–71.

King, P. P. (1827). *Narrative of a Survey of the Intertropical and Western Coasts of Australia—Performed between the Years 1818 and 1822. With an Appendix, Containing Various Subjects Relating to Hydrography and Natural History.* Murray, London.

Kirwan, G. M. (1996). Family Rostratulidae (Painted-snipes). Pp. 292–301 in: del Hoyo, J., Elliot, A. and Sargatal, J. eds. *Handbook of the Birds of the World.* Volume 3: *Hoatzin to Auks*. Lynx Edicions, Barcelona, Spain.

Kirwan, G. M. and Green, G. (2011). *Cotingas and Manakins.* Christopher Helm, London.

König, C., Weick, F. and Becking, J.-H. (1999). *Owls: A Guide to the Owls of the World.* Pica Press, Mountfield, England.

Krištin, A. (2001). Family Upupidae (Hoopoe). Pp. 396–411 in: del Hoyo, J., Elliot, A. and Sargatal, J. eds. *Handbook of the Birds of the World.* Volume 6: *Mousebirds to Hornbills*. Lynx Edicions, Barcelona, Spain.

Kruuk, H. (2003). *Niko's Nature: A Life of Niko Tinbergen and His Science of Animal Behaviour.* Oxford University Press, Oxford.

Ksepka, D. T. (2014). Flight performance of the largest volant bird. *Proceedings of the National Academy of Science of the United States of America*, **111**, 10624–29.

Labillardière, J. J. H. de (1799–1800). *Relation du voyage à la recherche de La Pérouse: fait par order de l'Assemblé constituante, pendant les années 1791, 1792, et pendant la lère.et la 2de. année de la République française.* 2 volumes and atlas, Jansen Paris.

Lack, D. (1945). The Galápagos finches (Geospizinae): A study in variation. *Occasional Papers of the California Academy of Sciences*, **21**, 1–159.

Lack, D. (1947). *Darwin's Finches.* Cambridge University Press, Cambridge, NY.

Lack, D. (1964). Darwin's Finches. in: Thomson, A. L. ed. *A New Dictionary of Birds.* Nelson, London.

Lack, D. (1969). Subspecies and sympatry in Darwin's Finches. *Evolution*, **23**, 252–63.

Lack, D. (1971). *Ecological Isolation in Birds.* Blackwell, Oxford.

Lack, D. (1976). *The Life of the Robin.* Fifth edition. H. F. & G. Witherby, London.

Laman, T. and Scholes, E. (2012). *Birds of Paradise: Revealing the World's Most Extraordinary Birds.* National Geographic Society, Washington, DC.

Lambert, F. and Woodcock, M. (1996). *Pittas, Broadbills and Asities.* Pica Press, Sussex, England.

Lambourne, M. (1987). *John Gould—Bird Man.* Osberton Productions Ltd., Milton Keynes, England.

Lane, B. A. and Rogers, D. I. (2000). The Australian Painted Snipe *Rostratula* (*benghalensis*) *australis*: An endangered species. *Stilt*, **36**, 26–34.

Latham, J. (1781–1785). *A General Synopsis of Birds.* 3 volumes. J. van Voorst, London.

Latham, J. (1787–1801). *Supplement I and II to A General Synopsis of Birds.* Leigh, Sotheby and Son, London.

Latham, J. (1821–1828). *General History of Birds.* 10 volumes. Author, Winchester, England.

Lee, M. and Worthy, T. (2014). Flight of the kiwi. *Australasian Science*, **35**, 22–4.

Lesson, R. P. (1828). *Manuel d.ornithologie, ou, Description des genres et des principales espèces d' oiseaux.* 2 volumes. Roret, Paris.

Ligon, J. D. (1999). *The Evolution of Avian Breeding Systems.* Oxford University Press, Oxford.

Lill, A. (2004). Family Menuridae (Lyrebirds). Pp. 484–95 in: del Hoyo, J., Elliot, A. and Christie, D. A. eds. *Handbook of the Birds of the World*. Volume 9: *Cotingas to Pipits and Wagtails*. Lynx Edicions, Barcelona, Spain.

Linnaeus, C. (1758). *Systema Naturae per Regna Tria Naturae*. 10th edition. Laurentii Salvii, Holmiae (Stockholm).

Linnaeus, C. (1766). *Systema Naturae*. 12th edition. Holmiae (Stockholm), Sweden.

Low, T. (2014). *Where Song Began*. Viking, Melbourne, Australia.

Lowe, P. R. (1936). The finches of the Galápagos in relation to Darwin's conception of species. *Ibis*, ser. 13, **6**, 310–21.

Loyau, A., Petrie, M., Saint Jalme, M. and Sorci, G. (2008). Do peahens not prefer peacocks with more elaborate trains? *Animal Behaviour*, **76**, e5–e9.

Lyell, C. (1830–1833). *Principles of geology, being an attempt to explain the former changes of the earth's surface, by reference to causes now in operation*. 3 volumes. London.

MacGillivray, W. (1837–1852). *A History of British Birds*. 5 volumes. Author, London.

Malthus, T. R. (1826). *An Essay on the Principle of Populations: Or, a View of Its Past and Present Effects on Human Happiness, with Inquiry into Our Prospects Respecting the Future Removal or Mitigation of the Evils which it Occasions*. Sixth Edition, 2 volumes. Murray, London.

Manning, J. T. (1989). Age-advertisement and the evolution of the peacock's train. *Journal of Evolutionary Biology*, **2**, 379–84.

Manning, J. T. and Hartley, M. A. (1991). Symmetry and ornamentation are correlated in the peacock's train. *Animal Behaviour*, **42**, 1020–1.

Marchant, J. ed. (1916). *Alfred Russel Wallace: Letters and Reminiscences*. 2 volumes. Cassell, London.

Marchant, S. and Higgins, P. J. eds. (1993). *Handbook of Australian, New Zealand and Antarctic Birds*. Volume 2: *Raptors to Lapwings*. Oxford University Press, Melbourne, Australia.

Marks, J. S., Cannings, R. J. and Mikkola, H. (1999). Family Strigidae (Typical Owls). Pp. 76–242 in: del Hoyo, J., Elliot, A. and Sargatal, J. eds. *Handbook of the Birds of the World*. Volume 5: *Barn-owls to Hummingbirds*. Lynx Edicions, Barcelona, Spain.

Mayr, E. (1935). How many birds are known? *Proceedings of the Linnaean Society of New York*, **45–46**, 19–24.

Mayr, E. (1963). *Animal Species and Evolution*. Harvard University Press, Cambridge, MA.

Mayr, E. and Diamond, J. (2001). *The Birds of Northern Melanesia: Speciation, Ecology, & Biogeography*. Oxford University Press, New York.

Mayr, E. and Greenway, J. C. Jr. (1960). *Check-list of Birds of the World*. Volume 9. Museum of Comparative Zoology, Cambridge, MA.

Mayr, G. (1999). *Pumiliornis tessellatus* n. gen. n. sp., a new enigmatic bird from the Middle Eocene of Grube Messel (Hessen, Germany). *Courier Forschungsinstitut Senckenberg*, **216**, 75–83.

Mayr, G. and Wilde, V. (2014). Eocene fossil is earliest evidence of flower-visiting by birds. *Biology Letters*, **10**, 5.

McCalman, I. (2009). *Darwin's Armada: Four Voyagers to the Southern Oceans and their Battle for the Theory of Evolution*. Simon and Schuster, London.

McCarthy, D. (2009). *Here Be Dragons*. Oxford University Press, Oxford.

McCarthy, E. M. (2006). *Handbook of Avian Hybrids of the World*. Oxford University Press, New York.

McDonald, R. (1998). *Mr Darwin's Shooter*. Random House, Sydney.

McKie, R. (2004). Evolution of radar points to HMS Beagle's resting place. *The Guardian*, London. http://www.theguardian.com/science/2004/feb/15/sciencenews.science

McKinney, H. L. (1972). *Wallace and Natural Selection*. Yale University Press, New Haven.

McNeillie, A. (1976). *Guide to the Pigeons of the World*. Elsevier, Oxford.

Mearns, B. and Mearns, R. (1998). *The Bird Collectors*. Academic Press, San Diego.

Mendel, G. L. (1866). *Versuche über Pflanzen-Hybriden. Verhandlungen des naturforschenden Vereines in Brünn IV* (1865), 3–47.

Mitchell, K. J., Llamas, B., Soubrier, J., Rawlence, N. J., Worthy, T. H., Wood, J., Lee, M. S. Y. and Cooper, A. (2014). Ancient DNA reveals elephant birds and kiwis are sister taxa and clarifies ratite bird evolution. *Science*, **344**, 898–900.

Molina, G. I. (1809). *The Geographical, Natural, and Civil History of Chili*. Translated from the original Italian: To which are added notes from the Spanish and French versions, and two appendixes, by the English editor; the first an account of Chiloé, from the Description historical of P. F. Pedro Gonzalcz de Agucros; the second, an account of the native tribes who inhabit the southern extremity of South America, extracted chiefly from Faulkner's description of Patagonia, 2 volumes. Longman, Hurst, Rees, & Orme, London.

Møller, A. P. (1991). Sexual selection in the monogamous barn swallow (*Hirunda rustica*). I. Determinants of tail ornament size. *Evolution*, **45**, 1823–36.

Møller, A. P. (1992). Female swallow preference for symmetrical male sexual ornaments. *Nature*, **357**, 238–40.

Møller, A. P. and Birkhead, T. M. (1994). The evolution of plumage brightness in birds is related to extra-pair paternity. *Evolution*, **48**, 1089–100.

Montagu, G. (1802). *Ornithological Dictionary; or, Alphabetical Synopsis of British Birds*. J. White, London.

Montagu, G. (1813). *Supplement to the Ornithological Dictionary; or Synopsis of British Birds*. J. White, London.

Moore, J. (1735). *Columbarium: or, the pigeon-house: Being an introduction to a natural history of tame pigeons. Giving an account of the several species known in England, with the method of breeding them, their distempers and cures*. J. Wilford, London.

Moore, J. R. (1982). Charles Darwin lies in Westminster Abbey. *Biological Journal of the Linnean Society*, **17**, 97–113.

Moorehead, A. (1969). *Darwin and the Beagle*. Book Club Associates, London.

Mulder, R. A., Dunn, P. O., Cockburn, A., Lazenby-Cohen, K. A. and Howell, M. J. (1994). Helpers liberate female fairy-wrens from constraints on extra-pair mate choice. *Proceedings of the Royal Society of London, Series B*, **255**, 223–9.

Müller, S. (1846). *Bijdragen tot de kennis van Sumatra: Bijzonder in geschiedkundig en ethnographisch opzigt*. Luchtmans, Leiden.

Navarro, J. L. and Martella, M. B. (1998). Fertility of greater rhea orphan eggs: Conservation and management implications. *Journal of Field Ornithology*, **69**, 117–20.

Newton, A. (1893–1896). *A Dictionary of Birds*. Adam and Charles Black, London.

Newton, I. (1979). *Population Ecology of Raptors*. T. & A. D. Poyser, Berkhamsted, England.

Newton, I. (1986). *The Sparrowhawk*. T. & A. D. Poyser, Waterhouses, UK.

Nicholas, F. W. and Nicholas, J. M. (2002). *Charles Darwin in Australia*. Cambridge University Press, Cambridge, NY.

Nichols, P. (2003). *Evolution's Captain: The Dark Fate of the Man Who Sailed Charles Darwin Around the World*. Harper Collins, New York.

Nilsson, D. E. and Pelger, S. (1994). A pessimistic estimate of the time required for an eye to evolve. *Proceedings of the Royal Society of London B*, **256**, 53–8.
Nisbett, A. (1976). *Konrad Lorenz*. Harcourt Brace Jovanovich, New York.
Olsen, S. L. and Feduccia, A. (1980). Relationships and evolution of flamingos (Aves: Phoenicopteridae). *Smithsonian Contributions to Zoology*, **316**, 1–73.
Orta, J. (1992). Families Phaethontidae, Sulidae and Fregatidae. Pp. 280–5, 312–15, 362–74 in: del Hoyo, J., Elliot, A. and Sargatal, J. eds. *Handbook of the Birds of the World*. Volume 1: *Ostrich to Ducks*. Lynx Edicions, Barcelona, Spain.
Owen, R. (1837). On the habits of *Vultur aura*. *Proceedings of the Zoological Society of London*, **5**, 33–5.
Owen, R. (1838). On the anatomy of the Southern Apteryx (*Apteryx australis*, Shaw). *Transactions of the Zoological Society of London*, **2**, 257–301.
Paley, W. (1828). *Natural Theology*. Second edition. J. Vincent, Oxford.
Pallas, P. S. (1811). *Zoographica Rosso-asiatica, sistens omnium animalium in extenso Imperio Rossico et adjacentibus Maribus* 3 volumes. Academiae scientiarum. Petropoli in officina Caes.
Pauly, D. (2004). *Darwin's Fishes: An Encyclopedia of Ichthyology, Ecology, and Evolution*. Cambridge University Press, Cambridge, NY.
Payne, R. B. (2005). *The Cuckoos*. Oxford University Press, Oxford.
Paynter, R. A. Jr. (1970). Subfamily Emberizinae. In: R. A. Paynter Jr., ed. *Checklist of Birds of the World*. Volume 13. Museum of Comparative Zoology, Cambridge, Massachusetts.
Peters, J. L. et al. (1934–1987). *Check-list of Birds of the World*. Volumes 1–16. Harvard University Press, Cambridge, MA.
Petren, K., Grant, B. R. and Grant, P. R. (1999). A phylogeny of Darwin's finches based on microsatellite DNA length variation. *Proceedings of the Royal Society of London B*, **266**, 321–30.
Petren, K., Grant, P. R., Grant, B. R. and Keller, L. F. (2005). Comparative landscape genetics and the adaptive radiation of Darwin's finches: the role of peripheral isolation. *Molecular Ecology*, **14**, 2943–57.
Petrie, M. (1992). Peacocks with low mating success are more likely to suffer predation. *Animal Behavior*, **44**, 585–6.
Petrie, M. (1993). Do peacock's trains advertise age? *Journal of Evolutionary Biology*, **6**, 443–8.
Petrie, M. (1994). Improved growth and survival of offspring of peacocks with more elaborate trains. *Nature*, London, **371**, 598–9.
Petrie, M. and Halliday, T. (1994). Experimental and natural changes in the peacock's (*Pavo cristatus*) train can affect mating success. *Behavioral Ecology and Sociobiology*, **35**, 213–7.
Petrie, M., Halliday, T. and Sanders, C. (1991). Peahens prefer peacocks with elaborate trains. *Animal Behaviour*, **41**, 323–31.
Petrie, M. and Williams, A. (1993). Peahens lay more eggs for peacocks with larger trains. *Proceedings of the Royal Society of London B*, **251**, 127–31.
Phillips, M. J., Gibb, G. C., Crimp, E. A. and Penny, D. (2010). Tinamous and moa flock together: Mitochondral genome sequence analysis reveals independent losses of flight among ratites. *Systematic Biology*, **59**, 90–107.
Podulka, S., Rohrbaugh, R. W. Jr. and Bonney, R. eds. (2004). *Handbook of Bird Biology*. Cornell Lab of Ornithology, Ithaca, NY.
Porter, D. M. (1985). The Beagle collector and his collections. Pp. 973–1019 in: Kohn, D. ed. *The Darwinian Heritage*. Princeton University Press, Princeton, NJ.

Quammen, D. (1996). *The Song of the Dodo: Island Biogeography in an Age of Extinctions*. Hutchinson, London.

Quoy, J. R. C. and Gaimard, J. P. (1824). Zoologie. In Freycinet, M. L. de (ed.), *Voyage autour du monde, entrepris par ordre du roi [. .].cxćcueć sur les corvettes de S. M. l'Uranic et la Physicicnne, pendant les années 1817, 1818, 1819 et 1820*. Volume 6. Pillet, Paris.

Remsen, J. V. (2003). Family Furnariidae (Ovenbirds). Pp.162–357 in: del Hoyo, J., Elliot, A., and Christie, D. A. eds. *Handbook of the Birds of the World*. Volume 8: *Broadbills to Tapaculos*. Lynx Edicions, Barcelona, Spain.

Rice, T. (2000). *Voyages of Discovery: Three Centuries of Natural History Exploration*. Scriptum Editions, London.

Ridgway, R. (1897). Birds of the Galapagos Archipelago. *Proceedings of the U. S. National Museum*, **19**, 459–670.

Robson, J. and Lewer, S. H. (1911). *Canaries, hybrids, and British birds in cage and aviary*. Cassell, London.

Rothschild, W. and Hartert, E. (1899). A review of the ornithology of the Galápagos Islands. With notes on the Webster-Harris Expedition. *Novitates Zoologicae*, **6**, 85–205.

Rothschild, W. and Hartert, E. (1902). Further notes on the Fauna of the Galapagos Islands. *Novitates Zoologicae*, **9**, 373–418.

Rowley, I. (1991). Petal-carrying by Fairy-wrens of the genus *Malurus*. *Australian Bird Watcher*, **14**, 75–81.

Rowley, I. and Russell, E. (1997). *Fairy-wrens and Grasswrens: Maluridae*. Oxford University Press, Oxford.

Saint Vincent, B. de, Audouin, J. V., Bourdon, J. B. I. and Marcellin, J. B. G. eds. (1822–1831). *Dictionnaire classique d'histoire naturelle*. 17 volumes. Rey et Gravier, Paris.

Salvin, O. (1876). On the avifauna of the Galapagos Archipelago. *Transactions of the Zoological Society of London*, **9**, 447–510.

Sauer, G. C. (1982). *John Gould the Bird Man: A Chronology and Bibliography*. Lansdowne Editions, Melbourne, Australia.

Schorger, A. W. (1966). *The Wild Turkey: Its History and Domestication*. University of Oklahoma Press, Norman.

Schuchmann, K. L. (1999). Family Trochilidae (Hummingbirds). Pp. 468–680 in: del Hoyo, J., Elliot, A. and Sargatal, J. eds. *Handbook of the Birds of the World*. Volume 5: *Barn-owls to Hummingbirds*. Lynx Edicions, Barcelona, Spain.

Sclater, P. L. (1858). On the general geographical distribution of the members of the class Aves. *Proceedings of the Linnean Society: Zoology*, **2**, 130–45.

Sclater, P. L. and Salvin, O. (1870). Characters of new species of birds collected by Dr. Habel in the Galapagos. *Proceedings of the Zoological Society of London*, **38**, 322–3.

Secord, J. A. (1981). Nature's fancy: Charles Darwin and the breeding of pigeons. *Isis*, **72**(262), 163–86.

Secord, J. A. ed. (2010). *Charles Darwin Evolutionary Writings*. Oxford University Press, Oxford.

Sharpe, R. B. et al. (1874–1895). *Catalogue of the Birds in the British Museum*. Volumes 1–27. Trustees of the British Museum (Natural History), London.

Sherborn, C. D. (1897). Notes on the dates of "The Zoology of the Beagle." *The Annals and Magazine of Natural History*, 7th series, **20**, 483.

Shirihai, H. (2008). *The Complete Guide to Antarctic Wildlife*. Second Edition. Princeton University Press, Princeton, NJ.

Short, L. L. (1982). *Woodpeckers of the World*. Delaware Museum of Natural History, Greenville.
Sibley, C. G. and Ahlquist, J. E. (1990). *Phylogeny and Classification of the Birds of the World*. Yale University Press, New Haven, CT.
Sibley, C. G., Ahlquist, J. E. and Monroe, B. L. (1988). A classification of the living birds of the world based on DNA-DNA hybridisation studies. *Auk*, **105**, 409–23.
Sibley, C. G. and Monroe, B. L. (1990). *Distribution and Taxonomy of Birds of the World*. Yale University Press, New Haven, CT.
Sick, H. (1993). *Birds in Brazil*. Princeton University press, Princeton, NJ.
Simms, E. (1979). *The Public Life of the Street Pigeon*. Hutchinson, London.
Skutch, A. F. (1991). *Life of Pigeons*. Cornell University Press, Ithaca, NY.
Smith, C. H. and Beccaloni, G. (2008). *Natural Selection & Beyond: The Intellectual Legacy of Alfred Russel Wallace*. Oxford University Press, Oxford.
Smith, J. (2006). *Charles Darwin and Victorian Visual Culture*. Cambridge University Press, Cambridge.
Smith, J. V., Braun, E. L. and Kimball, R. T. (2013). Ratite nonmonophyly: Independent evidence from 40 novel loci. *Systematic Biology*, **62**, 35–49.
Smith, L. H. (1988). *The Life of the Lyrebird*. William Heinemann, Melbourne, Australia.
Snodgrass, R. E. (1903). Notes on the anatomy of *Geospiza, Cocornis* and *Certhidea*. *Auk*, **20**, 402–17.
Snodgrass, R. E. and Heller, E. (1904). Papers from the Hopkins-Stanford Galapagos Expedition, 1898–1899. *Proceedings of the Washington Academy of Sciences*, **5**, 231–372.
Snow, B. and Snow, D. (1988). *Birds and Berries*. T. & A. D. Poyser, Calton, England.
Snow, D. W. (1976). *The Web of Adaptation*. Collins, London.
Snow, D. W. (1982). *The Cotingas*. British Museum (Natural History), London.
Snow, D. W. (2004). Family Pipridae (Manakins), Pp. 110–69 and Cotingidae (Cotingas), Pp. 32–108 in: del Hoyo, J., Elliot, A. and Christie, D. A. eds. *Handbook of the Birds of the World*. Volume 9: *Cotingas to Pipits and Wagtails*. Lynx Edicions, Barcelona, Spain.
Spix, J. B. von. (1824–1825). *Avium Species Novae*. 2 volumes. Monachii.
Spix, J. B. von and Martius, C. F. P. von (1824). *Travels in Brazil, in the years 1817–1820: Undertaken by command of His Majesty the King of Bavaria*. Longman, Hurst, Rees, Orme, Brown and Green, London.
Stauffer, R. C. ed. (1975). *Charles Darwin's Natural Selection: Being the Second Part of His Big Species Book Written from 1856 to 1858*. Cambridge University Press, Cambridge, NY.
Stearn, W. T. (1998). *The Natural History Museum at South Kensington: A History of the Museum 1753–1980*. The Natural History Museum, London.
Steinheimer, F. (2003). Darwin, Rüppell, Landbeck & Co.—Important Historical Collections at the Natural History Museum, Tring. *Bonner Zoologische Beiträge*, **51**(2–3), 175–188.
Steinheimer, F. D. (2004). Charles Darwin's bird collection and ornithological knowledge during the voyage of H. M. S. "Beagle", 1831–1836. *Journal fur Ornithologie*, **145**, 300–20.
Steinheimer, F. D., Dickinson, E. C. and Walters, M. (2006). The zoology of the voyage of HMS Beagle. Part III: Birds: New avian names, their authorship and their dates. *Bulletin of the British Ornithologists' Club*, **126**, 171–93.
Stott, R. (2003). *Darwin and the Barnacle*. Faber and Faber, London.

Stresemann, E. (1951). *Die Entwicklung der Ornithologie von Aristotcles bis zur Gegenwart*. Peters, Berlin.

Suh, A., Paus, M., Kiefmann, M., Churakov, G., Franke, F. A., Brosius, J., Kriegs, J. O. and Schmitz, J. (2011). Mesozoic retroposons reveal parrots as the closest living relatives of passerine birds. *Nature Communications*, **2**, 443. DOI: 10.1038/ncomms1448 www.nature.com/naturecommunications.

Sulloway, F. J. (1979). Geographic Isolation in Darwin's Thinking: The Vicissitudes of a Crucial Idea. *Studies in History of Biology*, **3**, 23–65.

Sulloway, F. J. (1982a). The *Beagle* collections of Darwin's finches (Geospizinae). *Bulletin of the British Museum (Natural History), Zoology*, **43**, 49–94.

Sulloway, F. J. (1982b). Darwin and his finches: The evolution of a legend. *Journal of the History of Biology*, **15**, 1–53.

Sulloway, F. J. (1982c). Darwin's conversion: The *Beagle* voyage and its aftermath. *Journal of the History of Biology*, **15**, 325–88.

Sundevall, C. J. (1871). On birds from the Galapagos Islands. *Proceedings of the Zoological Society of London*, **1871**, 126, 129.

Sushkin, P. P. (1925). The evening grosbeak (*Hesperiphona*), the only American genus of a Palaearctic group. *Auk*, **42**, 256–61.

Sushkin, P. P. (1929). On some peculiarities of adaptive radiation presented by insular faunae. *Verhandlungen des VI. Internationalen Ornithologen Kongresses*, Copenhagen, **1926**, 375–8.

Swarth, H. S. (1931). The avifauna of the Galápagos Islands. *Occasional Papers of the California Academy of Sciences*, **18**, 1–299

Takahashi, M., Arita, H., Hiraiwa-Hasegawa, M. and Hasegawa, T. (2008). Peahens do not prefer peacocks with more elaborate trains. *Animal Behaviour*, **75**, 1209–19.

Taylor, B. and van Perlo, B. (1998). *Rails: A Guide to the Rails, Crakes, Gallinules and Coots of the World*. Pica Press, Mountfield, England.

Tegetmeier, W. B. (1857). Exhibition of a collection of skins of new varieties of domestic fowls, the property of Charles Darwin Esq. *Proceedings of the Zoological Society of London*, **25**, 46.

Tegetmeier, W. B. (1867). *The Poultry Book: Comprising the Breeding and Management of Profitable and Ornamental Poultry, Their Qualities and Characteristics*. George Routledge and Sons, London.

Temminck, C. J. and Laugier de Chartrouse, M. (1820–1839). *Nouveau recueil de planches coloriées d'oiseaux pour servir de suite et de complément aux planches enluminées de Buffon, d'après dessins de Huet et Prête*. Paris.

Thomson, K. S. (1995). *HMS Beagle: The story of Darwin's ship*. W. W. Norton, London.

Thomson, K. S. and Rachootin, S. P. (1982). Turning points in Darwin's life. *Biological Journal of the Linnean Society*, **17**, 23–37.

Thompson, H. (2005). *This Thing of Darkness*. Headline Review, London.

Travers, H. H. (1869). On the Chatham Islands. *Transactions and Proceedings of the New Zealand Institution*, **1**, 173–80.

Tree, I. (1991). *The Ruling Passion of John Gould: A Biography of the Bird Man*. Barrie & Jenkins, London.

Tristram, H. B. (1859). On the ornithology of Northern Africa. Part **3**. *Ibis*, 429–33.

van Oosterzee, P. (1997). *Where Worlds Collide*. Cornell University Press, Ithaca, NY.

Van Rhijn, J. (1991). *The Ruff*. T & A. D. Poyser, London.

Wallace, A. R. (1855). On the law which has regulated the introduction of new species. *Annals and Magazine of Natural History* (ser. 2), **16**, 184–96.

Wallace, A. R. (1860). On the zoological geography of the Malay Archipelago. *Proceedings of the Linnean Society: Zoology*, **4**, 172–84.

Wallace, A. R. (1868). On birds' nests and their plumage; or the relation between sexual differences of colour and the mode of nidification in birds. *Report of the British Association for the Advancement of Science*, **37**(2), 97.
Wallace, A. R. (1869). *The Malay Archipelago: The Land of the Orang-Utan and the Bird of Paradise*. Macmillan, London.
Wallace, A. R. (1876). *The Geographical Distribution of Animals*. 2 volumes. Macmillan, London.
Wallace, A. R. (1880). *Island Life*. Macmillan, London.
Wallace, A. R. (1883). *The Malay Archipelago: The Land of the Orang-Utan and the Bird of Paradise*. Macmillan, London [first published 1869].
Wallace, A. R. (1889). *Darwinism: An Exposition of the Theory of Natural Selection with some of its Applications*. Macmillan, London.
Walters. M. (2003). *A Concise History of Ornithology*. Helm, London.
Weiner, J. (1994). *The Beak of the Finch*. Jonathan Cape, London.
Weld, C. R. (1862). "Humming-Birds." *Fraser's*, **65**, 457–68.
Werness, H. ed. (2004). *The Continuum Encyclopedia of Animal Symbolism in Art*. Continuum, New York.
White, G. (1789). *Natural History and Antiquities of Selborne*. London.
Whittell, H. M. (1954). *The Literature of Australian Birds*. Paterson Brokensha, Perth, Australia.
Wied-Neuwied, Prince A. P. and Maximilian zu. (1820–1821). *Reise nach Brasilien*. Frankfurt.
Williams, A. R. (2015). Earliest Bird Pollinator. *National Geographic*, **226**, 22.
Williams, L. E. (1981). *The Book of the Wild Turkey*. Winchester Press, Tulsa, OK.
Williams, T. D. (1995). *The Penguins: Spheniscidae*. Oxford University Press, Oxford.
Willughby, F. (1676). *Ornithologiae*. London.
Winkler, H. and Christie, D. A. (2002). Family Picidae (Woodpeckers). Pp. 296–555 in: del Hoyo, J., Elliot, A. and Sargatal, J. eds. *Handbook of the Birds of the World*. Volume 7: *Jacamars to Woodpeckers*. Lynx Edicions, Barcelona, Spain.
Winkler, H., Christie, D. A. and Nurney, D. (1995). *Woodpeckers: A Guide to the Woodpeckers, Piculets and Wrynecks of the World*. Pica Press, Mountfield, England.
Woods, R. W. (1988). *Guide to Birds of the Falkland Islands*. Nelson, Oswestry, England.
Wyhe, J. van. (2012). Where do Darwin's Finches come from? *Evolutionary Review*, **3**, 185–95.
Wyhe, J. van. (2013). "My Appointment Received the Sanction of the Admiralty": Why Charles Darwin really was the naturalist on HMS Beagle. *Studies in History and Philosophy of Biological and Biomedical Sciences*, **44**, 316–26.
Wyhe, J. van. (2014). *Charles Darwin in Cambridge*. World Scientific Publishing Company, Singapore.
Wyhe, J. van. and Rookmaaker, K. eds. (2013). *Alfred Russel Wallace: Letters from the Malay Archipelago*. Oxford University Press, Oxford.
Wyllie, I. (1981). *The Cuckoo*. B. T. Batsford Ltd., London.
Yarrell, W. (1837–1843). *History of British Birds*. 3 volumes. John van Voorst, London.
Zahavi, Amotz (1977). The cost of honesty (further remarks on the handicap principle). *Journal of Theoretical Biology*, **67**, 603–5.
Zahavi, A. and A. (1997). *The Handicap Principle*. Oxford University Press, Oxford.
Zhang, G., Li, C., Li, Q., Li, B., Larkin, D. M., Lee, C, Storz, J. F., Antunes, A., et al. (2014). Comparative genomes reveals insights into avian genome evolution and adaptation. *Science*, **346**, 1311–20.
Zink, R. M. (2002). A new perspective on the evolutionary history of Darwin's finches. *Auk*, **119**, 864–71.

INDEX

Most entries relate to birds: Thus "colour perception" means colour perception *in birds*, unless entries are otherwise qualified. The location of black and white figures within text pages are indicated by boldface page numbers. The numbers of colour figures, which appear in a discrete section of glossy pages, also appear in boldface type.